INTERSTELLAR GAS DYNAMICS

INTERNATIONAL ASTRONOMICAL UNION
UNION ASTRONOMIQUE INTERNATIONALE

SYMPOSIUM No. 39

PROCEEDINGS OF THE SIXTH SYMPOSIUM
ON COSMICAL GAS DYNAMICS ORGANIZED JOINTLY BY THE
INTERNATIONAL ASTRONOMICAL UNION AND THE
INTERNATIONAL UNION OF THEORETICAL AND APPLIED MECHANICS
YALTA, THE CRIMEA, U.S.S.R., 8–18 SEPTEMBER 1969

INTERSTELLAR GAS DYNAMICS

EDITED BY

H. J. HABING

*Astronomy Department, University of California, Berkeley, Calif., U.S.A., and
Sterrenkundig Laboratorium Kapteyn, Groningen, The Netherlands*

D. REIDEL PUBLISHING COMPANY

DORDRECHT-HOLLAND

1970

Published on behalf of
the International Astronomical Union
by
D. Reidel Publishing Company, Dordrecht, Holland

Library of Congress Catalog Card Number 78–124849

ISBN-13:978-94-010-3331-2   e-ISBN-13:978-94-010-3329-9
DOI: 10.1007/978-94-010-3329-9

Dedicated to

JAN HENDRIK OORT

on the occasion of his seventieth birthday
in grateful appreciation for his work with J. M. Burgers
in starting and maintaining this series of symposia

let me ask you,
from where came our universe?

Ya. B. Zel'dovich (during the discussions)

1. Lumley
2. Weymann
3. Shikin
4. Bisnovatyi-Kogan
5. Karpman
6. Gough
7. Lüst
8. Zimmermann
9. Rozhkovskii
10. Greenberg
11. Toomre
12. Woltjer
13. Gordon
14. Busemann
15. Mustel'
16. Lebovitz
17. Kaplan
18. Spiegel
19. Tsytovich
20. Pottasch
21. Ray
22. Pikel'ner
23. Thomas
24. Kadomtsev
25. Sunyaev
26. Verschuur
27. Pacheco
28. Shklovskii
29. Mrs. Csada
30. Csada
31. Konyukov
32. Goldsworthy
33. Kardashev
34. Belyakina
35. Spandaryan
36. Miss Csada
37. Zahn
38. Galkin
39. Shakhovskoi
40. Shcherbakov
41. Kopylov
42. Dobronravin
43. Mrs. Boyarchuk
44. V. Pronik
45. Spivak
46. Krat
47. Kuznetzov
48. Vasil'eva
49. Dibai
50. Zasov
51. Mrs. Weaver
52. Zel'dovich
53. Dyson
54. Dolginov
55. I. I. Pronik
56. van Leer
57. Brodskaya
58. Varshalovich
59. Dobronravina
60. Silk
61. Chuvaev
62. Dubov
63. Field
64. Nikolaiev
65. Kurt
66. Nikonov
67. van de Hulst
68. Habing
69. Gopasyuk
70. Gorbatskii
71. Khromov
72. Blinova
73. Weliachew
74. G. I. Petrov
75. Stecher
76. Malov
77. Unknown
78. Ozernoi
79. Sorochenko
80. Blinov
81. P. P. Petrov
82. Lotova
83. Mestel
84. Grebinskii
85. Pilyugin
86. Gershberg
87. Lyubimkov
88. Zhuzavlev
89. Meyer
90. Sigmatullin
91. Boyarchuk
92. Davis
93. Radler
94. Unknown
95. Shabanskii
96. Kumaigorodskaya
97. Kuvshinov
98. Shulman
99. Lynds
100. Parker
101. Metik
102. Buggisch
103. van Woerden
104. Podgornii
105. Kranosbaev
106. Kotov
107. Syrovat-skii
108. Willis
109. Mogilevskii

TABLE OF CONTENTS

EDITOR'S PREFACE

The following text forms the proceedings of a conference. It is supposed to contain what was actually reported and discussed, though it does this, one hopes, in a polished and organized way. A sense of actuality, a reporting quality, makes this book different from a collection of review papers as, for example, a book in the series on *Stars and Stellar Systems*. All Invited Reports have been included as the Reporters wrote them. The Editor's task has been restricted to improving the presentation, a process which in most cases involved only minor revisions. In a few Reports the Editor did some heavy rewriting; in those cases he checked with the Reporters. Obviously a different course had to be taken with respect to the Discussions. They were recorded on tape, transcribed verbatim and then passed back to the discussants. After the discussants returned their versions, the Editor rearranged and condensed the texts and made a considerable effort to provide references. (Unfortunately he was not able to locate all relevant Russian papers from 1968 and 1969.) The Editor takes the responsibility for mistakes made in this process, which may have produced occasionally his own 'mix-master Universe'. Actually only a few discussion remarks were rejected, more often because of incomprehensibility, rather than because the remark was far from the subject of the Symposium, or was too long, or was too trivial. A few very long remarks have been condensed and put at the end of a Discussion. The transposition of the Russian names has been done in the same way as is followed in the translation of *Astronomicheskii Zhurnal*. Dutch names are in the Dutch spelling.

The Editor acknowledges the work of Mrs. Robert J. Low, formerly at J.I.L.A., Boulder, Colorado, who organized and supervised the transcription and the corrections of the Discussions. Even though she was assisted by such hardworking helpers as Dr. Katharine Gebbie, Dr. Beverly Lynds, and Miss Bess Alta Thomas, it is owing to her organizational talents that the final editing could be done in a reasonably short time (neither sadly, nor slowly). It is a pleasure to thank Dr. Harold Weaver for providing the facilities of the Radio Astronomy Laboratory in Berkeley, which was a condition *sine qua non* for a completion of the Editor's task. Mrs. Karin Schaar supervised very carefully and very efficiently the typing and final editing; gratitude is owed also to Mrs. Nancy Schorn, Mrs. Mary Neighbor, and Miss Gayle Peabody. Dr. Lynds and Mrs. Low read almost all the Reports and the Discussions in order to make a few next-order approximations toward perfect English. The Editor thanks Drs. Dieter, Colgate, Field, Kuhi, Minkowski, Parker, Spinrad, Wallerstein, and Weaver for advice concerning scientific matters.

The Editor's stay in Berkeley has been financed by several sources: The National Science Foundation, the Fullbright Foundation, and the Netherlands Organization for the Advancement of Pure Research (Z.W.O.).

H. J. HABING

Berkeley, January 21, 1970

INTRODUCTION

1. Arrangements

The Sixth Symposium on Cosmical Gas Dynamics, the fourth in the series to deal with gas dynamics of the interstellar medium but the first on that subject since 1957, was held near Yalta, the Crimea, U.S.S.R., at the Sanatorium PARUS, from 8–18 September, 1969, under the joint sponsorship of the IAU and the IUTAM. The host was the Crimean Astrophysical Observatory, whose Director is Professor A. B. Severnyi. It was most fitting that the first meeting of the Cosmical Gas Dynamics series in the U.S.S.R. was held very near this observatory, which, under the directorship of the late Professor Shajn, became an outstanding center in the field of interstellar gas dynamics.

Approximately 140 scientists attended all or part of the sessions. Of these, 23 were from the U.S.A. and Canada, 26 from Europe outside the U.S.S.R., 1 from Australia, and 96 (among them 15 young auditors) from the U.S.S.R. Participants were housed in local sanatoria; thus the living aspects were those of a vacation resort, while the participants could attend the formal and informal sessions without loss of time.

In addition to the IAU-IUTAM support, financial and organizational aid came from the U.S.S.R. Academy of Sciences, the Crimean Astrophysical Observatory, the Yalta Town Soviet, the Joint Institute for Laboratory Astrophysics in Boulder, Colorado, and the Office of Scientific Research of the U.S. Air Force. In each case, support was in the form of a grant or organizational help, given without restrictions for the planning and mechanics of the Symposium.

The chairmen of the scientific organizing committee were Dr. A. B. Severnyi (for the IAU) and Dr. L. I. Sedov (for the IUTAM); members were Drs. S. B. Pikel'ner, F. D. Kahn, J.-P. Zahn, and R. N. Thomas (for the IAU) and Drs. H. F. Clauser, J. M. Giraud, and M. J. Lighthill (for the IUTAM). Drs. A. A. Boyarchuk and R. E. Gershberg were in charge of the local scientific arrangements. Dr. P. P. Dobronravin and Mr. E. Nikolaev, Assistant Directors of the Crimean Astrophysical Observatory, had overall direction of the Symposium arrangements. They were assisted by Mr. O. Blinov, Mr. A. Pilyugin, and others of the technical staff of the Observatory. Overall, the scientific and general program planning were handled efficiently and graciously by Drs. A. B. Severnyi and S. B. Pikel'ner. All participants will long remember their hospitality and scientific organizational ability. English was the only official language during the Symposium. We, the writers of this introduction, know that we speak on behalf of all participants from outside the U.S.S.R. when we express our gratitude to all our hosts for sponsoring so enthusiastically a Symposium in a language different from their own.

As at the previous two Symposia in the series, each problem chosen was introduced by a summary speaker with a long discussion following. All discussions were recorded, the tapes transcribed verbatim, and the transcriptions given to the participants for on-the-spot editing. We were successful in transcribing and distributing records of all sessions during the meeting, save those of the last day. This success is due to the organizational talents of Mrs. Robert J. Low, to the help of Dr. Katharine Gebbie, Dr. Beverly Lynds, and Miss Bess Alta Thomas, and to the efforts of Dr. Severnyi's staff.

The participants voted to dedicate the volume of the proceedings to Professor J. H. Oort on the occasion of his 70th birthday and his retirement as Director of the Leiden Observatory, in grateful appreciation of his work with Professor J. M. Burgers in starting and maintaining this series of Symposia.

2. Scientific Program

The impetus to this series of Symposia came from the proposal by Burgers and Oort that the interstellar medium be treated as an aerodynamical continuum, in which the interstellar clouds would appear as velocity fluctuations in a generalized turbulence. Earlier the clouds had been pictured as essentially discrete individual entities. The first three Symposia in the series considered detailed questions arising from this suggestion; for example: what instability – or instabilities – might produce such clouds?; what kind of a turbulence description might be suitable?; how must we incorporate the large density fluctuations associated with the velocity fluctuations into an aerodynamical turbulence picture? The earlier Symposia are largely characterized by the fact that answers suggested at one Symposium were negated in succeeding ones. Two examples: first, the suggestion, and apparent observational verification, of a Kolmogoroff-type turbulence was later rejected because of better observations and the difficulty of incorporating the associated large density fluctuations. Second, there was the suggestion that differential galactic rotation would drive the clouds. However, later suggestions implied acceleration of interstellar material by hot stars, an entirely different kind of mechanism in which the motions of the clouds seem to be only random, because of the apparent (semi-) random distribution of the hot stars, not because of turbulence in the more common aerodynamical sense. On the other hand, Zanstra's suggestion during the 2nd Symposium of a two-phase equation of state for those astronomical phenomena where the radiation field plays a significant role, the early arguments for the importance of plasma-hydromagnetic effects, and the suggestions of the existence and importance of collisionless shocks seem to have persisted. Apparently we were in the situation where certain physical effects looked important, but where we did not know how to use these effects to explain or systematize the array of astronomical observations. In planning the present Symposium it was hoped that the first part would bring into focus these considerations from the earlier Symposia, assess their present status, and place in perspective possible future work. The wealth of problems to be solved presents a great challenge.

The second part of the Symposium concentrated on mass interchange between stars and interstellar medium. This subject gives some continuity with the last two Symposia on aerodynamical phenomena in stellar atmospheres. In the present Symposium, as in the first ones in the series, interest lay in the amount, and the consequences, of such mass flow on the state of the interstellar medium. A third aspect of the Symposium concentrated on the formation of stars, on interstellar grains, and on interstellar molecules. The table of contents gives the planned program.

3. Some Remarks in Evaluation of the Symposium

1. PARTICIPATION

The original aim of Burgers and Oort had been to start a discussion between dynamicists and astronomers so that the astronomical problems might be clarified in an aerodynamical context. The emphasis in these Symposia lies, then, not on solving any problems during the Symposia, but on defining what the problems are and making some suggestions for future investigations. Already in 1948 it became clear that plasma physicists, as much as aerodynamicists, would be valuable participants.

In a narrow sense, the Symposium was not very successful in attracting a balanced representation of aerodynamicists. Most of the aerodynamicists were new; we lacked people who had attended earlier Symposia, and were reasonably familiar with the astronomical 'language'. As a result there was not as strong an interchange between aerodynamicists and astronomers as at previous Symposia. On the other hand, Severnyi, Pikel'ner, and Sedov had an extraordinary success in attracting a large number of young astronomers and plasma physicists who were very eager to apply their experience to a variety of astronomical problems. In a broader sense the Symposium was very successful in bringing into fruitful contact those groups with a common interest in interstellar gas dynamics, but not always very extensive contact possibilities. These groups are indicated by the contrasts East-West and young-old. Most participants appeared to be enthusiastic about the discussions, formal and informal, and many participants felt stimulated to continue working in the area of cosmic gas dynamics. The fruitfulness of these contacts was made possible, in our view, by the schedule of the Conference, which called for (and achieved) some three hours of free discussion (no reading of papers) per invited summary Report.

This brings us to the following conclusions. (i) Further discussion between aerodynamicists (or even magnetohydrodynamicists), plasma physicists, and astrophysicists remains a very desirable goal (for convincing evidence see the remarks by Goldsworthy, p. 94). However, the discussion needs stimulation. At one point at this Symposium it was noted that aerodynamicists need information on individual objects – but that astronomers usually provide only general averages, which tell very little about the aerodynamical model to be adopted. Since the observational techniques are rapidly improving in quality and increasing in variety, Goldsworthy's suggestion is to be kept in mind. It might be worthwhile to submit for publication summaries of high quality observations and tentative interpretations on individual objects in such journals as *Physics of Fluids* and the like. In the second place the IUTAM member of the organizing committee of a future Symposium in this series should make an effort to bring in aerodynamicists who already have some background in the subject and can help trigger the discussion. They should be invited to give an introductory report. (ii). The schedule of this Symposium seems very satisfactory to many participants. It appears to us that the policy of suppressing the reading of short communications is a very healthy one. From the present Symposium it is evident that fruitful discussions can take place, even in a group of 100 to 150 persons. (iii) Younger astronomers certainly have a need to attend meetings like this where it is possible to meet senior people. It seems very desirable to organize in various countries similar working symposia, composed of perhaps 20 to 25 percent active workers from abroad, the rest, people from within the country who, for one reason or another, have not had much chance for participation in such symposia outside their country. Emphasis would be on cross-discipline subjects; naturally, we are prejudiced towards astronomy as the central theme, with its boundaries defining the border fields. Informal East-West contacts should be stimulated, improved, and extended.

2. SCIENTIFIC CONTENT

Of course, any evaluation and impression is highly personal. To us the first and main part of the Symposium is characterized by noting that both theory and observation presented new and very fundamental material, but that we do not yet know how to interpret, understand, and use it. In the observational approach we are rapidly coming to the point where the whole sky will have been observed in the 21-cm line with a beamwidth of $0°.6$. This set of data may be compared with the Palomar Sky Atlas; although there are only 100 independent data points per field of the dimension of one sky survey plate, a third dimension (radial velocity) has been added. The material reminds one of the phenomenological investigations in the early studies of laboratory turbulence and may well provide us with the same solid basis for a discussion, although the interpretation is considerably more complicated than in the laboratory. One has to unscramble excitation, composition, and radiative transfer effects to get a geometrical picture. Obviously, 21-cm line observations with larger angular resolution will show new phenomena, as Verschuur emphasized several times; but there is reason to believe that much phenomenological insight on the larger scale structures (20 to 100 pc) can be obtained from the data already available. On the theoretical side at least three fundamental effects have been discussed in recent years. Two of them are the thermal instability leading to a two-phase system (the concept has a history back to Spitzer at the first Cosmical Gas Dynamic Symposium and to Zanstra at the second), and a variant of the Rayleigh-Taylor instability, discussed by Parker. Both effects are thought to be extremely fundamental, but both have been studied only in linear approaches and one has sought for static solutions, whereas certainly the theory has to be nonlinear and one should look for dynamical equilibria. Therefore a clear assessment of the value of these two concepts is not yet obtained. The third theoretical achievement of recent years is the spiral-wave theory. It was definitely under-discussed at this meeting, due to the fact that a preceding Symposium (IAU No. 38) was dedicated to spiral structure. Also, in this case there is room for much clarification of the basic concepts of the theory. During one of the informal, unrecorded meetings, Pikel'ner attempted to give a synthesis of the way these three mechanisms might interact (see Field's summary at p. 102). Although the picture given is extremely tentative it implies many suggestions for future research – both theoretical and observational (see the Final Discussion, p. 362). If the picture is correct it is remarkably similar to that discussed at the first Symposium: The interstellar motions represent a sort of 'turbulence' triggered ultimately by the differential rotation of the Galaxy. In this sense the present Symposium is a true child of the previous ones.

The Introductory Reports on mass-ejection, in either the quasistatic situations or the not-so-static situations, emphasized the rudimentary state of our knowledge. The observational material has increased many-fold; the big problem is how we interpret these observations. Even among the theoreticians, there was a tendency to confuse mass-loss possibly associated with a stellar wind with a mass-loss visualized in the earlier models, not based on such an underlying physical picture. The problem is tied to the general aerodynamical phenomena in stellar atmospheres so that the two kinds of Symposia held thus far in the series overlap.

At the present Symposium the discussions of physical background revolved around plasma turbulence and hydromagnetic problems; the Reports were excellent and most stimulating. However, one also had the feeling that attempts to make hydromagnetic models were stretched quite a bit – that one had a technique looking for an application rather than a great deal of insight into the astronomical problem and the boundary conditions to be satisfied. But overall, one was impressed by the number of people working in various aspects of plasma physics and trying very hard to apply their knowledge to astrophysics, to really clarify the problems there. Again, this effort promises interesting future developments.

In summary, it is our impression that this Sixth Cosmical Gas Dynamics Symposium was both timely and useful. Recent developments were discussed, both theoretical and observational, which may prove to be fundamental breakthroughs. A new group was formed willing to continue to explore these breakthroughs. All in all, cosmical gas dynamics, in the words of Busemann, remains a domain where one is "not hindered by too much information, but has to rely on best guesses, and where one can return to a state of innocence, proper for a scientist".

Berkeley, H. J. HABING

January 22, 1970   *Boulder*, R. N. THOMAS

LIST OF PARTICIPANTS

Australia:
B. Y. Mills

Brazil:
J. F. Pacheco

Bulgaria:
I. M. Jankulova

Canada:
A. K. Ray

France:
G. Monnet
L. Weliachew
J. P. Zahn

Germany (Federal Republic):
H. Buggish
R. Lüst
F. Meyer
P. Mezger

Germany (Democratic Republic):
G. Helmis
H. Radler
H. Zimmermann

Hungary:
I. K. Csada

Italy:
A. P. Blagkikh

The Netherlands:
H. J. Habing
H. C. van de Hulst
A. van Leer

The Netherlands (Cont.):
S. R. Pottasch
H. van Woerden

Poland:
J. Stodolkiewita

United Kingdom:
J. Dyson
P. A. Feldman
F. A. Goldsworthy
D. Gough
L. Mestel

U.S.A.:
B. Burke
A. Busemann
S. A. Colgate
L. Davis
G. B. Field
K. B. Gebbie
J. M. Greenberg
N. R. Lebovitz
J. Lumley
B. T. Lynds
T. K. Menon
E. N. Parker
J. Silk
E. A. Spiegel
T. P. Stecher
R. N. Thomas
J. Toomre
G. L. Verschuur
H. F. Weaver
R. J. Weymann
D. R. Willis
L. Woltjer

U.S.S.R.:

B. P. Artamonov
V. I. Ariskin
V. B. Baranov
A. A. Barmin
T. S. Belyakina
G. S. Bisnovatyi-Kogan
A. A. Boyarchuk
M. E. Boyarchuk
E. S. Brodskaya
V. I. Burnashov
Ya. N. Chekvadze
A. D. Chernin
K. K. Chuvaev
E. A. Dibai
P. P. Dobronravin
A. Z. Dolginov
E. M. Drobishevskii
E. E. Dubov
A. S. Dvoryashin
Yu. S. Efimov
N. N. Eryushev
L. S. Galkin
R. E. Gershberg
Yu. I. Glushkov
O. P. Gollandskii
A. N. Golubyatnikov
S. I. Gopasyuk
V. G. Gorbatskii
I. M. Gordon
A. S. Grebinskii
Gribkov
R. E. Guseinov
V. S. Imshennik
B. B. Kadomtsev
S. A. Kaplan
N. S. Kardashev
V. I. Karpman
Yu. K. Khodarev
G. S. Khromov
M. V. Konyukov
I. M. Kopylov

U.S.S.R. (Cont.):

V. A. Kotov
Krasnobaev
V. A. Krat
A. P. Kropotkin
A. G. Kulikovskii
R. N. Kumaigorodskaya
V. G. Kurt
V. M. Kuvshinov
V. A. Kuznetzov
L. S. Levitskii
V. A. Liperovskii
N. A. Lotova
L. S. Lyubimkov
I. F. Malov
L. S. Marochnik
N. S. Melnikova
L. P. Metik
V. A. Minin
E. I. Mogilevskii
I. G. Moiseev
É. R. Mustel'
Yu. M. Nikolaiev
V. B. Nikonov
L. M. Ozernoi
G. I. Petrov
P. P. Petrov
S. B. Pikel'ner
A. G. Pilyugin
I. M. Podgornii
V. I. Pronik
I. I. Pronik
D. A. Rozhkovskii
L. I. Sedov
A. B. Severnyi
V. P. Shabanskii
N. M. Shakhovskoi
A. G. Shcherbakov
I. S. Shikin
I. S. Shklovskii
L. M. Shulman
N. P. Sigmatullin

U.S.S.R. (Cont.):

R. L. Sorochenko
A. A. Spandaryan
R. A. Sunyaev
S. I. Syrovat-skii
V. N. Tsytovich
D. A. Varshalovich
G. Ya. Vasil'eva

U.S.S.R. (Cont.):

V. V. Vitkevich
I. M. Yavorskaya
A. V. Zasov
Ya. B. Zel'dovich
V. P. Zommer
B. N. Zhuzavlev

PART I

DESCRIPTION AND GENERAL THEORY OF
THE INTERSTELLAR MEDIUM

1. REVIEW OF COSMICAL GAS DYNAMICS

Introductory Report

(Monday, September 8, 1969)

H. C. VAN DE HULST

Sterrewacht, Leiden, The Netherlands

1. To Pick up the Trail

The subject allotted to me in the program is the 'Summary of Symposia I–III: problems considered and present status of them'. Perhaps I should have been more cautious when I agreed to this title about two years ago. For it means that, if I do my job well, you will know the present status of all problems and the Symposium is finished today. However, I am sure that, no matter how I try, there will remain enough for other speakers and for other days of discussion. For this reason, in surveying the field, I shall not consciously try to avoid the topics of other speakers. Overlap will do no harm.

I should like to transmit to the new generation – and indeed 20 years *is* a generation – the spirit in which this series was undertaken. The three earlier joint IAU-IUTAM Symposia referred to in the title were (see literature list for full references):

A   1948, Paris, *Cosmical Aerodynamics*
Publication, 1951 (Editors J. M. Burgers and H. C. van de Hulst)
35 published papers and discussion sessions, 237 pages.
B   1953, Cambridge, England, *Gas Dynamics of Cosmic Clouds*
Publication, 1955 (Editors J. M. Burgers and H. C. van de Hulst)
44 published papers and discussion sessions, 247 pages.
C   1957, Cambridge, Massachusetts, *Cosmical Gas Dynamics*
Publication, 1958 (Editors J. M. Burgers and R. N. Thomas)
47 published papers and discussion sessions, 204 pages.

I shall refer to these publications below as A, B, C. For updating references I shall occasionally refer to

D   1965, *Galactic Structure* (Editors A. Blaauw and M. Schmidt).
E   1968, *Nebulae and Interstellar Matter* (Editors B. Middlehurst and L. H. Aller).
F   1967, *Radio Astronomy and the Galactic System* (Editor H. van Woerden).

In addition to these Symposium reports and edited volumes, mention should be made of monographs by Pikel'ner (1964), Kaplan (1966), and Spitzer (1968).

The aim of the first Symposium and, likewise, of the two following ones, was to make aerodynamicists acquainted with some of the intriguing astronomical objects to which their theories might be applied and to give astronomers a feel for these theories. This was a fascinating but not entirely easy task. The final discussion of the second Symposium, after the mutual teaching had had time to sink in, starts off with

Habing (ed.), Interstellar Gas Dynamics, 3–17. All Rights Reserved. Copyright © 1970 by IAU

two grave warnings to the astronomers: do not misuse shockwaves (Liepmann, B, p. 241); do not misuse turbulence (Frenkiel, B, p. 241). Conversely, the complexity of the astronomical objects was difficult to grasp for some aerodynamicists. Von Kármán makes a reasonable suggestion and immediately gets an answer explaining why this is astronomically impossible. To which answer he deftly replies: "My imagination is not handicapped by any knowledge of the facts" (B, p. 180). I have sometimes wondered if this might not be a correct description of all theoretical astrophysicists. But no, on second thoughts I don't think that in the absence of any observations even their imagination would have been bold enough to visualize such things as stars, galaxies, spiral arms, the solar corona, comets, supernovae, quasars, and pulsars. Sometimes the strange facts dazzle us for decades before they find a proper explanation, as for example the solar corona or the white light from the Crab Nebula.

Our main task is to give intelligent comments on the observed facts, assuring full consistency with whatever we know about the laws of physics. New facts seldom reduce the number of problems; they add new ones and shed new light on old ones.

2. Growth of Data Pile

I think we are all aware of the enormous progress, in the quality and quantity, of the observational data over the past decades. Entirely new avenues relevant to our discussion are:

(a) the enormous progress of X-ray astronomy and a clear start of gamma-ray astronomy;

(b) successful surveys both in the near and the far infrared;

(c) first direct measurements of Ly-α and of the ultraviolet extinction curve;

(d) detection and extensive surveys of OH and detection of interstellar H_2O and NH_3;

(e) detection and extensive surveys of high recombination lines yielding densities and radial velocities;

(f) direct space observations of cosmic rays below the GeV energy range permitting conclusions on cosmic-ray heating;

(g) three more ways to measure the magnetic field, viz. Zeeman effect, Faraday effect, and radio-continuum polarization.

I shall return to some of these points below but wish to recall first that the progress in the now 'classical' fields has also been enormous since this Symposium series was started.

The first Symposium took place several years before the 21-cm line was observed. Nothing was known at that time about the spiral arms in our Galaxy and a possible gaseous halo became a topic of discussion only in 1955 (Baldwin, 1957). The volume of the second Symposium contains as novelty several full-page reproductions from sheets of the Palomar Sky Atlas, which had not yet become publicly available (B, pp. 8, 10, and 12); to meet Minkowski's demands I personally supervised the running off of these plates at the printer.

The situation with magnetic fields was amusing. A month before the Symposium, Oort said to me as a young staff member: "Alfvén, who has worked so much with those magnetohydrodynamic waves, is coming and I am afraid that nobody will take up this subject. Why don't you study it a little?" I did. But the Symposium happened to be a few months after the discovery of interstellar polarization* and after Fermi proposed his mechanism for the acceleration of cosmic rays (Fermi, 1949; see also Fermi, 1954). The theorists were well aware of the possible influence of magnetic field on turbulence and shock waves. So, for several days, before it was my turn to speak, everybody had discussed magnetic fields!

Before I leave the memoirs let me try to give you a quantitative idea of observational progress. The accumulated output of the 21-cm line observations, exclusively from the Netherlands, is given below:

1953: 50 'clean' profiles with 7-m dish; first sketch of spiral arms (Van de Hulst, 1953);

1957: 700 'clean' profiles with 7-m dish; 3-dimensional map of spiral structure that has gone into many textbooks (Muller and Westerhout, 1957);

1969: some 40 000 'clean' profiles with 25-m dish (P. Katgert, private communication).

Here a 'clean' profile is the curve showing intensity versus radial velocity (frequency) at one point of the sky, obtained by averaging two to three profiles taken on different nights. The Maryland survey (Westerhout, 1969) of the strip between $b = -1°$ and $b = +1°$ alone contains the equivalent of over 10 000 profiles. With similarly impressive output from other places, the quantitative gain over 1953 is certainly larger than 1000.

In the field of optical observation, really very little was known about the distribution of faint nebulosity over the sky. I should like to mention specifically the work of the late Professor Shajn, Director of the Crimean Observatory, in photographing these nebulae and in pointing out their possible elongation in a galactic magnetic field (B, p. 37). Now, comparing the number of bits of information on the Palomar Sky Atlas, permanently available at many observatories, to the corresponding number on earlier atlases, I would estimate a gain factor 50 (image quality) × 2 (two colors) × 4 (better dynamical range) = 400.

The world number of available coudé spectra, essential in the optical study of interstellar cloud motions, has from the classical collection of Adams (300 stars in 1949) in recent years been gaining at a pace of perhaps 100 a year, again a substantial total factor, although far less than in the preceding examples owing to the time-consuming observing method.

One cannot help wondering, in each of these examples, if the progress made in the interpretation is at all commensurate with the progress in available data.

---

* The announcement (Hiltner, 1949; Hall, 1949) appeared in February 1949 but at the time of the Paris Symposium in August, 1948 it was already taken for granted that the observations were not spurious.

3. Definition of Topics

So much for memoirs. We have to come to the topics and the first problem is how to distinguish one topic from another. In all sciences this problem occurs sooner or later. One can study a lot of biology before encountering the problem of how to separate animals and plants, or one can study a lot of astronomy before the difference between a star and a planet becomes an urgent problem. In interstellar gas dynamics we are faced with such problems from the outset: what is a cloud, what is the normal interstellar medium, what is a spiral arm, what is the halo? Some of these concepts have well-defined meanings in the context of a particular interpretation of a particular set of observations. Others, like 'clouds', are used with a variety of meanings.

Let us first exclude some topics. One of these is the gasdynamics of stellar atmospheres and interiors. It was found useful to discuss those at separate Symposia in Varenna, 1960 (Thomas, 1961) and Nice, 1965 (Thomas, 1967), which were also joint meetings of IAU and IUTAM.

I shall leave aside also the outermost part of the Sun, namely the solar wind or, what comes to the same thing, the dynamics of the interplanetary gas. Yet this is the only part of cosmic space now accessible to observation by direct measurements and even to experimentation. The barium cloud released on 18 March 1969 from the ESRO satellite HEOS I (see, p. 241) was at the altitude of 'only' 75000 km above the earth, i.e., barely outside the magnetosphere, but there is no technical reason why such ion diffusion experiments cannot be performed anywhere in interplanetary space. Rossi has several times pointed out that one significant reason for performing measurements and experiments in interplanetary space is that they permit us to learn more about large-scale physical phenomena, which in turn will permit us to understand better the phenomena occurring on an even larger scale in the Galaxy and in the Universe.

This relevance of interplanetary space was realized even in the pre-history of space flight. For as early as 1953 there was a clear discussion between Gold and Liepmann about the collisionless shock in interplanetary space (but not under this name, B, pp. 103–104). In the same discussion Menzel remarked: "The magnetic field of the Earth will act as a kind of 'bumper' with which the wave coming from the Sun would collide, at a distance of roughly five times the radius of the Earth" (B, p. 104). At that time this was still a completely theoretical concept, dating back to the work of Chapman and Ferraro in 1931. Now the bow shock of the Earth is a known object about which numerous experimental and theoretical studies have been made.

A third subject which I shall leave aside is intergalactic matter. There are a number of ways to estimate its density now, but no direct means for measuring its present state of motion and density fluctuations. The spatial arrangement of galaxies contains information about the state of motion at some time in the past, but in this field even speculative papers are scarce. [See, however, Ozernoi's contribution in this volume, p. 216, Ed.]

A further topic to be excluded is spiral structure in our and other galaxies. This at least was the division of subjects the IAU planned between the present Symposium and the immediately preceding one in Basel (IAU Symposium No. 38). This, of course, is a fully impossible division. For where shall we find the typical interstellar matter that does not have some relation with spiral arms? That which for convenience used to be called 'normal' interstellar space, does not exist any more. In any discussion of observations we must now be careful to specify at least whether space inside or outside a spiral arm is meant, and finer distinctions probably are required.

Furthermore the character of the motion changes (gradually or abruptly) if we approach the galatic center. Near the Sun rotation along almost circular orbits in the galactic plane predominates. Near the center, complicated expansion motions are seen in addition. These by themselves form an important topic certainly worthy of the attention of this audience, but I hesitate to include it in this review.

Finally, even in the solar neighborhood, the vertical structure across the galactic disk is not simple. Matter clearly associated with spiral arms is seen even at $z = 1$ to 3 kpc (Oort, IAU Symposium No. 38), although the general estimate of 200 pc for the 'thickness of the disk' is still a good figure.

Having covered the subjects that I exclude, I now come to the subjects that must be included. Please do not be angry with me if this review seems disorganized. It is difficult to give a regular account of an irregular subject.

Before the 1953 Symposium we exchanged among the participants a dozen preparatory studies. One of these was a list of 'Problems and Suggested Solutions', which I made up from the experience gained with the first Symposium. This list was printed in the volume (B, p. 42). I started out to take this list as a guideline for the present review but it turned out to be too rigid a frame.

The same difficulty of sorting out the basic questions occurred during the third Symposium. After the confusion had been steadily rising for some days we decided to insert an unscheduled 'mid-symposium summary plus general discussion' (C, p. 994) which again reviews the toughest unsolved questions.

I have tried to sort the basic questions of this Symposium and shall briefly go over the most important problem areas in the following sections.

4. Mass Balance

This question was summarized in the form of a double one: There is interstellar gas. Why? There are young hot stars. Why?

The answer, then tentative and now generally accepted is that there is a balance, gas being used by condensation into stars and gas being replenished by various processes from stars. Without details, the numbers describing this 'mass balance' for the entire Galaxy were estimated at (Biermann, B, p. 212):

(a) loss from stars in shells or explosions 0.02 $M_\odot$ yr^{-1} (this now seems an underestimate by at least a factor 3);

(b) continuous loss from stars 1 to 10 $M_\odot$ yr^{-1};

(c) gas condensation into stars $1 \, M_\odot \, yr^{-1}$.

Taking $3 \, M_\odot \, yr^{-1}$ as the correct turnover rate and $0.3 \times 10^{10} \, M_\odot$ as the total mass of interstellar gas in our Galaxy (Kerr and Westerhout, D, p. 199), the average cycle during which an atom passes from the interstellar gas into a star, and conversely, is $10^9 \, yr$.

It is perhaps telling that neither C nor D nor E contains an updating of these estimates. Yet the subject has greatly evolved in a somewhat different context (Salpeter, this volume, p. 221). Astronomers probably are more reluctant now to assume that estimates made for the solar neighborhood hold everywhere in the Galaxy. In addition we know of other factors influencing the mass balance: An inflow from outside the Galaxy, which I estimate again at the order of $1 \, M_\odot \, yr^{-1}$ (Oort, F, p. 279) and a strong suggestion of expulsion of mass from the galactic nucleus, of about $10^7 \, M_\odot$ some 10^7 years ago (Oort, IAU Symposium No. 38), i.e., again of the same magnitude.

5. Dark Matter

Either I am too conservative, or we had pretty well grasped the problem at that time, for I can almost copy the 12 lines from the 1953 Symposium without change.

"Dark matter (dust) is mainly a nuisance for galactic research and its direct dynamical effects are almost nil, although earlier researches have made an important point of the radiation pressure on the grains. Yet the dust is possibly important as a cooling agent for the gas and as an absorber of Ly-α quanta and thus indirectly for dynamics. Further the dust is important as a tracer of

(a) structural details of gas clouds, for dust and gas go mostly together;

(b) magnetic fields, if a magnetic theory of interstellar polarization holds;

(c) relative motions of gas and dust, if a wind theory of interstellar polarization holds."

I should add now that point (a) has been discussed at great length (B, Ch. 40 and 41) until Bondi closed the 'vacuum-cleaner discussion'. Dark clouds and lanes are conspicuous both in a direct look at the Milky Way and in inspecting the finest details on the Palomar sky map. A lot of structure is made visible by the dust just as structure in microscopic images is made visible by certain dyes. Yet a judicious interpretation remains necessary. The common assumption that a dust region is automatically an HI region is not always correct, for dust has also been found in HII regions (see Mathews and O'Dell, 1969). And the earlier suspicion that very dense dust regions do not contain a proportional amount of atomic hydrogen is fully confirmed by recent measurements. How much hydrogen is still present in molecular form remains an open question.

The number of data on interstellar polarization has greatly increased and we shall definitely hear about these in the discussion about the magnetic field. Although the third theory (c) has recently been revived (Salpeter and Wickramasinghe, 1969) it seems as unlikely as ever.

6. Energy Balance

Taking 10^{-12} erg cm$^{-3} = 0.6$ eV cm^{-3} as the unit, we find for the solar neighborhood the well-known approximate equality of energy density in four reservoirs:

star light	0.8
random gas flow or cloud motions	6 (for $\langle V^2 \rangle^{1/2} = 14$ km sec^{-1}, 1 atom cm^{-3})
cosmic rays	1.5
magnetic field	0.4 (for $B = 3$ μG).

The mechanisms of exchange and possible reasons for an approximate equipartition have often been discussed and will again come up during this Symposium. The question put forth by the organizers of the first few Symposia forms part of this problem but was posed in a more vital form: How is it at all possible that the kinetic energy in the gas is maintained? It may directly lose energy to cosmic rays by Fermi acceleration and to magnetic fields by dynamo effects and energy may be lost by radiation in cloud collisions or after thermalization by viscosity. If the turnover time is 3×10^7 yr $= 10^{15}$ sec, the supply needed to keep these motions up is 10^{-26} erg cm^{-3} sec^{-1}.

Initial hopes were expressed to get this energy from the much richer reservoir contained in galactic rotation. Differential rotation would cause turbulence and compressible turbulence would create clouds and cloud motions. This explanation has now been dropped (Parker, C, p. 959). In astronomical terms: normal (low-velocity) objects cannot do much on their small epicycles; high-velocity objects coming from kiloparsecs nearer or farther from the center could play havoc but are just too few in number.

Later preference shifted distinctly to nuclear power as the main supply. It could be released in two forms: (1) Gently, in young OB stars causing a giant expansion and subsequent break-up of the cloud complex in which these stars were born; and (2) violently, in the explosion of nova and supernova shells. Both subjects will again be discussed here. In the earlier Symposia no agreement was reached on which contribution was most important; their sum hardly seemed sufficient to supply the demand.

Some numbers from Parker's 1957 summary (C, pp. 958–59) are:

from supernovae	10^{-30} erg cm^{-3} sec^{-1},
from novae	10^{-27} erg cm^{-3} sec^{-1} (but parceled out in quantities with too small momentum),
from OB stars	10^{-28} erg cm^{-3} sec^{-1}, whereas there is required 10^{-27} erg cm^{-3} sec^{-1} (but my estimate above is a factor 10 higher).

The situation depicted in this table, which is in essence the situation at the end of our last Symposium, is far from satisfactory. Jumping from there to the present time I hesitate to say that the present situation is now clear. I shall try to sketch some important developments.

First, many authors have tried to understand the detailed processes which occur when stellar energy is fed into kinetic energy of interstellar gas flow via an expanding H II region. The list is long but I should certainly mention Oort and Spitzer, Vandervoort, Mathews, and Lasker. In a general review of the evolution of diffuse nebulae Mathews and O'Dell (1969) conclude that the general features are well understood.

Second, attempts have been made at a careful updating of the various estimates of energy supply. The general review by Kahn and Dyson (1965) and the review of the energy supply from supernovae by Kahn and Woltjer (F, p. 117) may give sufficient clues to the recent literature.

Third, there is the highly interesting development by Parker and co-workers. They started out from a somewhat different question: What are the dynamical properties of the cosmic-ray gas in the Galaxy and how are these linked to the dynamics of the interstellar gas and magnetic fields? For those who lack the courage or time to read the original papers (I count 22 references over the three years 1966–68 by Lerche and/or Parker), there are two excellent reviews by Parker himself (1969a, b). The picture developed is that the cosmic rays form a fluid flowing with a speed $\leqslant 60$ km sec^{-1}. The combined magnetic field and cosmic rays form an unstable system in which large clouds are formed. This instability resembles gravitational instability but is much stronger. The clouds thus formed derive their kinetic energy from the gravitational potential perpendicular to the galactic plane.

About a year ago Parker (1968) discussed how the older ideas of expanding H II regions should be matched to this new picture. I must confess that I still have trouble understanding the complete picture. In this paper the 'disruptive forces' of the H II regions are described as counteracting the attractive forces which form the clouds; in the Oort-Spitzer picture, on the contrary, the H II regions are the agents which lead to compression forming the clouds. Also, I do not see how the energy balance works. Traditionally, the gas clouds were thought to oscillate like the stars back and forth across the galactic plane, constantly exchanging kinetic and gravitational energy. Parker modifies this picture, taking the energy for the motions out of the gravitational energy, but I have not yet seen from where he resupplies the energy. The rates, of course, are similar to those mentioned above, or even higher. Cosmic rays alone need a supply of 5×10^{-26} erg cm^{-3} sec^{-1}.

Perhaps I should go home to read; but being here I hope to have the benefit of a direct explanation.

7. Temperature and Density

For gas dynamics we need values of temperature and density and we also need something equivalent to an equation of state. We shall not dwell on this last point but simply recall what Burgers said in his summary (B, p. 228): "The interstellar gas, considered energetically, is not 'self-contained', but finds itself between a powerful source of energy, formed by high-temperature stars, and a sink, represented by the almost empty intergalactic space. ... The majority of interesting cases are influenced by energy exchange." Further interesting cases will be forthcoming during this Symposium.

We should, however, say something about the uncertainty in the temperature and density estimates. Those in the H$\textsc{ii}$ regions can be relatively well determined from a variety of optical line ratios, radiocontinuum, high radio lines. $T = 8000$ K and $n = 10$ atom cm^{-3} are typical values. The variations are large and can often be individually determined. (See the Report by Mezger, p. 336.) A typical turnover time is 10^4 yr, in which the energy content of 30×10^{-12} erg cm^{-3} is turned over at a gain and loss rate of 10^{-22} erg cm^{-3} sec^{-1}. An H$\textsc{ii}$ region is like a rich man having and gaining much but spending it rapidly.

In contrast, an H$\textsc{i}$ region is like a poor man for whom every bit of earning counts, who never has much and who has to spend it slowly. Typical values based on the classical work of Spitzer are: energy content 10^{-14} erg cm^{-3} turned over in 3×10^7 yr at a gain and loss rate of 10^{-29} erg cm^{-3} sec^{-1}.

I do not think the densities can be questioned much. The claim that the 21-cm line can be explained by much lower densities because of maser effect does not seem to work (Van Bueren and Oort, 1968) and the discrepancy with the Ly-α strength in some six stars may have a different explanation. On the contrary, unnoticed saturation effects caused by temperature and density fluctuations could, according to Schmidt, well cause the actual mean density to be two times the traditional average of 0.5 atom cm^{-3}.

The temperature is expected to vary because part of the heating is supposed to occur at occasional incidents called 'cloud collisions'. But it is hard to lay hands on good evidence regarding the temperature variation. The (harmonic?) average is 125 K. But in places we know T must be down to about 30 K; the best clue to this is in the width of the narrowest 21-cm absorption and emission peaks.

The fact that H$\textsc{i}$ regions are susceptible to small gains also brings cosmic rays into the picture as a heating agent. The important energy range is around 10 MeV and the particles in this range have been referred to as sub-cosmic rays or as suprathermal particles. The main problem (see review by Meyer, 1969) is that they cannot be observed close to the earth because of geomagnetic cut-off and the measurements which have been made from space probes still require a correction factor between 10 and 10 000 to allow for solar wind modulation! Anyhow, it seems that this energy source is important for heating the H$\textsc{i}$ regions.

If this is true, another intriguing possibility enters. Following earlier work, e.g., by Pikel'ner (1967), detailed computations have been made (Field *et al.*, 1969; Goldsmith *et al.*, 1969) of the heating and cooling in a wide range of temperatures and densities. These authors find that in a certain range the pressure may drop if the density rises. Thereby they have revived a type of condensation theory which (in a different context) received much attention in our earlier Symposia (Zanstra, B, p. 70). I quote from Burgers' summary (B, p. 231): "Zanstra has shown that there may correspond *three* values of the density to a single value of the pressure (depending on certain factors). The intermediate value of the density is unstable, but the maximum and the minimum values can co-exist, in which case a state is obtained with part of the gas condensed relative to the rest. Evidently, this possibility will be of great influence on the behavior of the gas; it can be compared with the condensation which under laboratory conditions may occur in water vapor."

In an earlier summary (Van de Hulst, 1969) I arrived at typical figures, which I here compare with numbers cited from Radakrishnan and Murray (1969) (Table I).

TABLE I

	n_e/n_H	n_H	T	nT
		cm^{-3}	K	cm^{-3} K
H I standard cloud	0.02	10	100	1000
Same, Radakrishnan and Murray		10	50	500
H II standard cloud	1	10	10^4	2 × 10^5
H II hypothetical intercloud gas (Spitzer)	1	0.05	10^4	1000
H I hot-phase intercloud gas (Field *et al.*)	0.2	0.05	10^4	600
Same, Radakrishnan and Murray		0.5	10^3	500

I bet there will be further modifications before this conference is over. I should like to point out that the new hot H I gas of Field *et al.* is rather similar to the hypothetical H II intercloud gas of Spitzer.

8. Close-ups

An astronomical photographer is not unlike a press photographer. He takes a shot from a fair distance and chooses carefully which part to blow up and publish. The alternative method, to get a close-up, permits less of a selection but shows better details.

The interstellar close-ups refer to objects several 100 pc away, or even closer. Let me review a few of them without observing any special order.

The Orion Nebula (distance 420 pc) is the nearest very young H II region containing the famous trapezium as exciting stars. It has been a choice object both for observers and theorists. One of the early observations is a drawing made by Huygens at Leiden in 1694. There are several experts on the more recent work at this Symposium. The only warning I should like to give is that the Orion Nebula is not typical for H II regions in general.

The giant planetary nebula in Aquarius NGC 7293 (distance 145 pc). This is not strictly interstellar matter, but it is one of the few cases where a planetary nebula is close enough to show its details well. The finest details seen are some comet-like condensations, observed by Baade, with typical diameters of 0.001 pc.

Interstellar matter in the vicinity of Nova Persei 1901 (distance 500 pc). The strong light pulse emitted by this nova has successively illuminated the wisps of interstellar matter in its vicinity up to a distance of about 50 pc. The most prominent feature is a thin sheet in front of the star. This is the only known case where the distribution in three space dimensions has been observed (Oort, 1946). The resolution is determined by the duration of the light pulse, some 20 days, or 0.02 pc.

Nearby atomic hydrogen. It is important to repeat that other so-called three-dimensional pictures are always obtained indirectly. The three coordinates of six-dimensional phase space which can be measured are the two positional components across and the one velocity component along the line-of-sight. In some cases (expanding shells, differential galactic rotation) the velocity component can be converted by a plausible model into a distance and a three-dimensional space picture emerges. This conversion, however, introduces a certain smoothing. In the 21-cm line this smoothing is caused by the random cloud velocities and leads to a resolution of the order of 500 pc in the line of sight, independent of distance. Hence, to get a detailed space distribution of the nearby hydrogen atoms by this method is completely impossible.

Nearest dark clouds. Some dozen striking dark nebulae are seen at distances of 100 to 300 pc (Lynds, E, p. 119). In the analysis of these nebulae by the time-honored method of star counts a distance resolution of the order of even 100 pc is considered quite good. Evidently this is of no use for our purpose of discussing structure.

One could think of a scheme whereby a three-dimensional map of interstellar extinction within 100 pc is made, starting from individual distances and extinction values for many stars. The uncertainty of intrinsic colors would hamper this scheme. However, with very accurate polarization measurements it can be made to work, and I still have the impression that much more can be done along the line that Behr started many years ago. In his review Verschuur will mention Mathewson's recent work in this field (see p. 150).

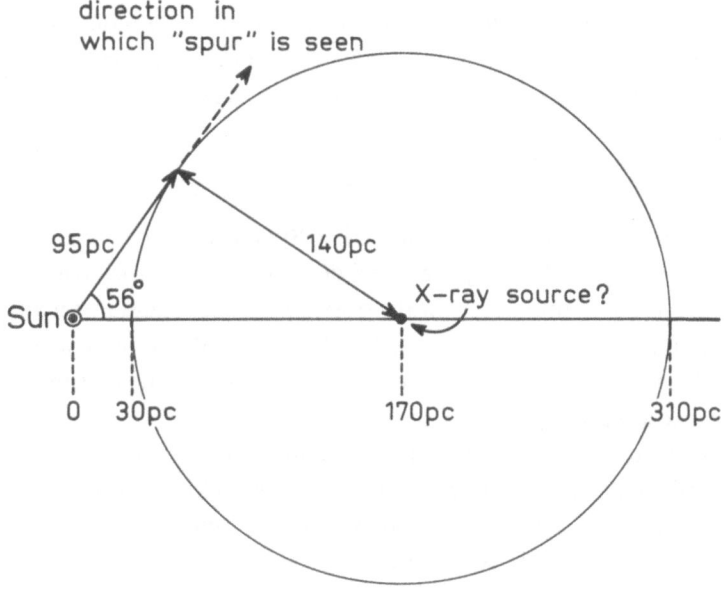

Fig. 1. Model to explain the north galactic spur and associated phenomena. Based on work by Berkhuijsen, Haslam and Salter.

The north galactic spur. This object, visible as an odd spike on the earliest radio-maps of the Galaxy by Reber, had led to many speculations about its origin and distance. There are some indications that this object may be correctly placed among the objects seen in 'close-up'. The geometry in cross-section may be about as shown in Figure 1.

The most striking feature, a roughly ring-shaped object following a small circle on the sky with radius 56° would be the locus of directions where we look tangentially to the thin shell. Enhancements in the 21-cm line at the outer edges of the ring suggest that neutral hydrogen to the amount of some $10^4 M_\odot$ is pushed ahead of it (Berkhuijsen *et al.*, 1970). Inside the shell concentric 'ridges' of continuum radiation are seen (Large *et al.*, 1966; Merkelijn and Davis, 1967); their width is of the order of $1° = 2$ pc or smaller. The short distance is mostly based on the fact that Mathewson finds a clear association between the spur and the optical polarization of stars between 50 and 200 pc. He interprets this association in terms of a helix field but it could be reinterpreted in terms of the shell model. Recently, an X-ray source has been discovered near the center of the ring ($l^{II} = 332°$, $b^{II} = +23°$).

All of this requires confirmation and further study but it is too attractive to be withheld in this review. Several similar but smaller arcs are known from the radio studies.

The Cygnus Loop. Other 'loops' have been observed optically. Among these, the Cygnus loop or Veil nebula (distance 500 to 800 pc) has been known from the earliest days in which photographs of the Milky Way were made. Two curved arcs very strongly suggest parts of a shell of radius $1°2 \approx 10$ pc. Some individual wisps in this veil have a thickness $2'' = 0.005$ pc. This object deserves special mention, because discussions between Oort and Burgers about the way to explain the compression into thin sheets as a collision effect led to the 1948 Symposium. A picture of this nebula was chosen as the frontispiece of the first Symposium (A). Minkowski rediscussed it (C, p. 1048).

9. What is a Cloud?

I return to a very central topic of the earlier Symposia. What is a cloud? The obvious answer is: a region of locally increased density. But this does not tell us much of the physics. From the point of view of the observer, 'local' could mean anything from 1000 to $\frac{1}{1000}$ pc (Van de Hulst, C, p. 922). For a gas-dynamicist *local* means any distance which can be travelled by the speed of sound within the lifetime of the object. This would make anything < 100 pc local.

The values assumed for 'standard' clouds come mostly from spectra, i.e., from a resolution in velocity space, not in configuration space. This subject has been reviewed so often that I shall not spell it out again. Spitzer's chapter in E is a very good reference. Van Woerden (F, p. 3) compares and analyzes the results obtained by some 20 authors. I summarize from his summary the values given in Table II (omitting in each case the extreme low and the extreme high estimate!). The variation may be disappointing to those who had hopes that nature would comply with their model

TABLE II

Cloud diameter	5 to 40 pc
Number on line of sight	0.7 to 11 kpc^{-1}
Mass	20 to 15 000 $M_\odot$
n_H	2 to 70 cm^{-3}
N_H	$(0.5–30) \times 10^{20}$ cm^{-2}
Dispersion of cloud velocities in one component	7 km sec^{-1}
Dispersion of internal velocities	0.8 to 6 km sec^{-1}
Corresponding T if internal velocities were thermal	80 to 4000 K

assumptions. The actual range may be even larger, for there is some truth in the statement that the size of the smallest cloud is always equal to the resolution of the instrument. The data cited are based partly on Ca$^+$ but mostly on 21-cm H lines.

Aerodynamicists may wonder what deep thought there is behind the distinction of velocities of clouds and velocity in clouds. There is none. In the analysis of the line profiles certain convenient entities are called clouds and to each cloud is assigned a speed. Whatever is then left of the velocity field is called an internal cloud motion. Most of it must have the character of gas flow because the temperatures to which this motion would correspond are too high. But if we continue to ask if the external and internal flow could be part of a continuous spectrum of turbulence, I don't think we could give a conclusive answer from these observations.

Turbulence was indeed a central topic in the earlier Symposia. It was discussed with and without compressibility, with and without magnetic fields. Gradually the views developed to consider three causes for the origin of the clouds:

(a) *Gravitational instability* is of no use for standard clouds because it starts to work effectively only at higher densities (formation of protostars) and does not lead to sharp edges.

(b) *Turbulence.* Incompressible turbulence was well explained but is not applicable to the interstellar situation. Highly compressible turbulence may be applicable. The theory is not very complete but certainly leads both to strong density fluctuations and to sharp edges. To some extent it resembles an assembly of shock waves (B, Ch. 22, 42).

(c) *Shockwaves* may form sheets, not clouds, but there is a fair chance that most of the things we call clouds are sheets. They certainly can explain sharp transitions. The sharpness of the order of the mean free path is 0.1 pc at 10 atom cm^{-3}, which is barely sufficient. However, collision-free shocks (not then known by this name) may be a great deal sharper, because the gyro-radius of an electron at 3 μG is only 10 km at 10 000 K.

We conclude that both (b) and (c) are eligible explanations but examination of the kind of energy supply (see earlier section) favors (c). But, although this answers our initial questions, we can hardly be happy with this conclusion.

Upon looking at the photographs, the sharpness of the transition between the cloud and interspace is to me as startling as the density contrast. The shockwave hypothesis may in some examples offer the explanation but in other examples the explanation may be different. We know pretty well, for instance, that the sharp transition between a dark cloud and its bright rim, which is often seen in a large H II region, is simply an ionization front. It only marks the boundary where the ionizing quanta run out and the sharpness is determined by the mean-free path of these quanta, which is 0.005 pc for a density of 10 atom cm^{-3}.

A further competitor may be spontaneous separation between 'two phases' of H I. If the theory of thermal instability, which I briefly mentioned above (see Field, this volume, p. 51), is correct, we still have to find what determines the shapes and sizes of the condensed regions and the sharpness of the transition.

Finally, I wish to point out that in any case a number of different explanations may be required because the objects are so different. Smooth 'globules' may be simple to conceive but the similarly small pitch-dark cloudlets with ragged edges (Bok, B, p. 34) are to me still quite puzzling.

There is a lot to do.

References

Volumes referenced by letter in the text

A Burgers, J. M. and Van de Hulst, H. C. (eds.): 1951, *Cosmical Aerodynamics*, Air Force Documents Office, Dayton, Ohio.
B Burgers, J. M. and Van de Hulst, H. C. (eds.): 1955, *Gas Dynamics of Cosmic Clouds*, North-Holland Publishing Company, Amsterdam.
C Burgers, J. M. and Thomas, R. N. (eds.): 1958, *Cosmical Gas Dynamics*; *Rev. Mod. Phys.* **30**, 905.
D Blaauw, A. and Schmidt, M. (eds.): 1965, *Galactic Structure*, University of Chicago Press, Chicago.
E Middlehurst, B. and Aller, L. H. (eds.): 1968, *Nebulae and Interstellar Matter*, University of Chicago Press, Chicago.
F Van Woerden, H. (ed.): 1967, *Radio Astronomy and the Galactic System*, Academic Press, New York.

Papers referenced by author

Baldwin, J. E.: 1957, IAU Symposium No. 4, *Radio Astronomy*, Ed. H. C. van de Hulst, University Press, Cambridge, p. 233.
Berkhuijsen, E. M., Haslam, G., and Salter, C.: 1970, *Nature* **225**, 364.
Fermi, E.: 1949, *Phys. Rev.* **75**, 1169.
Fermi, E.: 1954, *Astrophys. J.* **119**, 1.
Field, G. B., Goldsmith, D. W., and Habing, H. J.: 1969, *Astrophys. J. Lett.* **155**, L149.
Goldsmith, D. W., Habing, H. J., and Field, G. B.: 1969, *Astrophys. J.* **158**, 173.
Hall, J. S.: 1949, *Science* **109**, 167.
Hiltner, W. A.: 1949, *Science* **109**, 165.
Kahn, F. D. and Dyson, J. E.: 1965, *Ann. Rev. Astron. Astrophys.* **3**, 47.
Kaplan, S. A.: 1966, *Interstellar Gas Dynamics*, Pergamon Press, London.
Large, M. I., Quigley, M. J. S., and Haslam, C. G. T.: 1966, *Monthly Notices Roy. Astron. Soc.* **131**, 335.
Mathews, W. G. and O'Dell, C. R.: 1969, *Ann. Rev. Astron. Astrophys.* **7**, 67.
Merkelijn, J. and Davis, M.: 1967, *Bull. Astron. Inst. Netherl.* **19**, 246.
Meyer, P.: 1969, *Ann. Rev. Astron. Astrophys.* **7**, 1.
Muller, C. A. and Westerhout, G.: 1957, *Bull. Astron. Inst. Netherl.* **13**, 151.
Oort, J. H.: 1946, *Monthly Notices Roy. Astron. Soc.* **106**, 159.
Parker, E. N.: 1968, *Astrophys. J.* **154**, 875.
Parker, E. N.: 1969a, in *Cosmic Ray Studies in Relation to Recent Developments in Astronomy and Astrophysics* (ed. by R. R. Daniel, P. J. Kavakare, and S. Ramadurai), Tata Institute, Bombay.
Parker, E. N.: 1969b, *Space Sci. Rev.* **9**, 651.

Pikel'ner, S. B.: 1964, Fundamentals of Cosmic Electrodynamics, NASA Technical Translation TTF-175 (original text 1961, Moscow).

Pikel'ner, S. B.: 1967, *Astrophys. Lett.* **1**, 43.

Radakrishnan, V. and Murray, J. D.: 1969, *Proc. Astron. Soc. Australia* **1**, 215.

Salpeter, E. E. and Wickramasinghe, N. C.: 1969, *Nature* **222**, 442.

Spitzer, L.: 1968, *Diffuse Matter in Space*, Interscience Publishers, New York.

Thomas, R. N. (ed.): 1961, *Aerodynamic Phenomena in Stellar Atmospheres, IAU Symposium No. 12*; *Nuovo Cim. Suppl.* **22**, 1.

Thomas, R. N. (ed.): 1967, *Aerodynamic Phenomena in Stellar Atmospheres, IAU Symposium No. 28*, Academic Press, London.

Van Bueren, H. G. and Oort, J. H.: 1968, *Bull. Astron. Inst. Netherl.* **19**, 414.

Van de Hulst, H. C.: 1953, *Observatory* **73**, 129.

Van de Hulst, H. C.: 1969, in *Cosmic Ray Studies in Relation to Recent Developments in Astronomy and Astrophysics* (ed. by R. R. Daniel, P. J. Lavakare, and S. Ramadurai), Tata Institute, Bombay.

Westerhout, G.: 1969, *Maryland-Greenbank Galactic 21-cm Line Survey*, 2nd ed., University of Maryland, College Park, Md.

2. DISCUSSION FOLLOWING VAN DE HULST'S REPORT

(Monday September 8, 1969)

Chairman: R. N. THOMAS

Editor's remarks: This discussion on the opening day of the Symposium started in the afternoon and was consequently rather short. The version presented here is even shorter since some extensive remarks by Zel'dovich and Sunyaev have been transferred to a more proper location, the discussion following the Reports by Weaver and by Field (p. 77).

Pikel'ner: First, van de Hulst mentioned in passing the hypothetical concept of an H_{II} intercloud medium. I think this possibility can now be ruled out. Observations of the 21-cm line show the existence of a hot, neutral gas between the clouds. Second, as for the heating of the gas, Faraday rotation and pulsar dispersion measurements show that the interstellar electron density is much higher now than was assumed 10 years ago. Therefore the cooling is much faster and cloud collisions cannot explain the high gas temperatures.

Van Woerden: Van de Hulst mentioned a kinetic energy density W for the interstellar gas equal to 6×10^{-12} erg cm^{-3}. This number seems too large. In a volume of, say, 300 pc diameter around the Sun, the average hydrogen density is $\langle n_H \rangle = 0.5$ cm^{-3} and $\langle \varrho \rangle = 1 \times 10^{-24}$ g cm^{-3} (helium included, no molecular hydrogen). With a velocity dispersion in one coordinate of $u = 6$ km sec^{-1}, I obtain $W = 0.5 \times 10^{-12}$ erg cm^{-3}. Inside clouds, hydrogen densities are much larger ($n_H \approx 10$ cm^{-3}), but internal velocities are much smaller, say, $u = 2$ km sec^{-1}; inside clouds, we therefore have $W = 1 \times 10^{-12}$ erg cm^{-3}.

Pikel'ner: Much of space is filled with the intercloud medium, which has a low density. The value $W = 6 \times 10^{-12}$ erg cm^{-3} refers to the cloud volume and cannot be compared with, for instance, the magnetic energy density, which fills all of space. In addition I think that a proper density for standard clouds is 2 to 3 atom cm^{-3}, instead of 10 atom cm^{-3}.

Syrovat-skii: It seems to me that van de Hulst underestimated the energy supplied by supernovae. He gave 10^{-30} erg cm^{-3} sec^{-1}. But, if per supernova the energy release is about 10^{49} erg or more and if there is one supernova each 30 years, then the energy supply for the galactic volume of $\lesssim 10^{67}$ cm^3 is about 10^{-27} erg cm^{-3} sec^{-1} or more.

Van de Hulst: My figure contains only that part of the energy that is indeed going into the cloud motion. Therefore my calculation contains an efficiency factor.

Woltjer: Kahn and I estimated this efficiency, using a rather extreme model (Kahn and Woltjer, 1967). My numbers indicate an energy per supernova outburst of a few times 10^{49} erg and a conversion efficiency into interstellar cloud motion of about five

Habing (ed.), Interstellar Gas Dynamics, 18–21. All Rights Reserved. Copyright © 1970 by IAU

per cent. [Kahn, F. D. and Woltjer, L.: 1967, *IAU Symposium No. 31, Radio Astronomy and the Galactic System* (ed. by H. van Woerden), Academic Press, New York, p. 117.]

Syrovat-skii: Then there is a real discrepancy. For one needs an efficiency factor of one per cent or less, and a supernova frequency of one per 300 yr (ten times larger than is assumed at present) in order to obtain van de Hulst's estimate. In addition we have to neglect gas heating and expansion by supernova radiation and to neglect the action of supernova cosmic rays.

Shklovskii: I think van de Hulst's estimate is closer to reality than Syrovat-skii's. Shock waves will convert most of the energy of the explosion into thermal energy of the gas and transform it into UV and X-ray radiation. Most of the mechanical energy goes into X-ray emission.

Woltjer: The five per cent figure I gave is the fraction of the original mechanical energy that goes into interstellar motion. If originally there were more energy coming out as X-ray or UV radiation, then the total energy would go up, the efficiency would go down, but the input of kinetic energy would remain the same.

Colgate: I disagree with Shklovskii. Most of the energy ends up as kinetic energy as a result of adiabatic expansion. X-rays, gamma-rays, and optical light form only a small fraction of the total energy.

Menon: Dr. van de Hulst, you gave parameters for clouds of low temperature and density, but you did not discuss the possible range of their masses and whether or not there is any correlation between the densities and the masses. In particular, do the high density clouds have lower masses, and the low density clouds higher masses?

Van de Hulst: I cited mass values from van Woerden's review paper. Perhaps he can say more on this question. [Van Woerden, H.: 1967, *IAU Symposium No. 31, Radio Astronomy and the Galactic System* (ed. by H. van Woerden), Academic Press, New York, p. 3.]

Van Woerden: The data I assembled do not indicate a correlation such as Menon suggests. If anything, there is a slight positive correlation between density and mass.

Menon: The few cases where a high density has been observed are H II regions. We know, from various mass estimates, that these objects have a fairly low mass. Therefore, the question is, do the low density clouds have large masses?

Van Woerden: The large masses I listed came from large clouds or cloud complexes where the density is not low (i.e., 10 to 50 atom cm^{-3}). The total mass of a cloud depends to a large extent on how one defines the cloud. My summary did not include masses and densities of H II regions, and it contains no objects with densities larger than 100 atom cm^{-3}.

Menon: To make an extreme statement: The only clouds with reliably known dimensions and densities are dense H II regions. For all other clouds we lack a suitable, accepted definition and the published values of the cloud parameters have been obtained by widely differing methods.

Van Woerden: I disagree with Menon in two ways. Among the high-density H II regions, many are ionization bounded, not density bounded, so that the total mass of the cloud cannot be determined, but only the mass of the ionized part. On the other hand, as far as HI regions are concerned, I think there are obvious cases of well-bounded clouds.

Verschuur: I wish to raise the question of the term *standard cloud*. It seems to me a useless concept. For example, knowing what the average person in this room is like may tell us nothing about any individual member. Should we any longer use the term standard cloud? Does the term cloud still have some significance? I think we should observe specific regions only and discuss these, not the average ones.

Van de Hulst: I agree that the standard cloud is just a concept for limited use, for instance, in the context of components in a spectrum. We should always be very careful, first to define how we will use a model, and then to see whether or not the picture fits the observations.

Thomas: For the sake of mutual understanding I hope that at this Symposium we will give attention to questions of semantics as well as to the usual questions of physics. As an example the word 'cloud' was adopted historically to indicate that the interstellar medium was not homogeneous. The early question about this non-homogeneous medium concerned how one obtained density fluctuations. One answer was that perhaps turbulence variations could produce them. But then one has to define what is meant by turbulence, which in turn means that one must define velocity fluctuations. I should like to ask: How do I describe the interstellar medium? Can I talk about an aerodynamic continuum? Can I talk about a density fluctuation and a velocity fluctuation? Does it make any sense to talk about clouds at all? Should we rather talk about density and velocity fields?

Weaver: In my Report I will show pictures of 21-cm observations demonstrating the existence of interstellar clouds. But they are not what in the past has been called a standard cloud.

Spiegel: Could van de Hulst clarify his remarks on the distinction between velocities inside and velocities outside the clouds?

Van de Hulst: The question was: What is actually the difference between (i) a continuous velocity field and (ii) moving clouds with internal velocities? My point was that the observations themselves were really not quite sufficient to distinguish between these two possibilities. Historically it starts with the surprising results obtained by Adams that the interstellar lines could be separated into neat components. If you call each component a cloud, then you get a set of numbers representing 'internal' and 'external velocities'. But the more subtle question I now ask is whether or not such a simple interpretation is justified.

Field: I would suggest that a cloud is a mass of gas moving supersonically with respect to its own internal sound speed, but only subsonically with respect to the external sound speed. The external sound speed is much larger (ten times) than the internal sound speed because of a phase transition between a cold and a hot gas. This transition is not a shock wave, but rather a contact discontinuity across which there is pressure equilibrium. It is a critical point whether or not there can be lasting supersonic motions within this well-defined cloud. The observations suggest that the internal motions are in fact supersonic, because the observed widths of the lines exceed the sound speed by a factor of two or three. I wish to suggest a model in which cloud collisions maintain supersonic turbulence within individual clouds. This supersonic turbulence is caused by cloud collisions, which set the gas into motion on the largest scale within the cloud with the Mach number between 1 and 10. Zimmermann has shown how this process happens for central collisions. In the case of collisions off-center, multiple shocks will form and will continue to cause supersonic motion for long periods. For example, rotation induced by the angular momentum of relative motion will result in a field of supersonic motion, with both rotational and compressive components. The compressive components will dissipate first, but the rotational ones may last longer and give the observed, mildly supersonic line widths. (Zimmermann, H.: 1968, *Astron. Nachr.* **290**, 193, 211.)

Spiegel: It is very difficult to maintain supersonic turbulence inside a cloud and to keep the gas H I, since it will heat up. Hence it seems unlikely that the observed velocity dispersions can be internal cloud turbulence.

Field: I agree with Spiegel that there is a problem. Perhaps it will take much longer to dissipate the rotational part than the compressive part of such turbulence. As the time between cloud collisions is only ten times the time of dissipation of compressive motions, perhaps the longer time to dissipate rotational motions will explain their presence in many clouds. Also, in the presence of a magnetic field cloud, collisions could generate large-amplitude Alfvén waves. These would add significantly to the observed widths.

3. SOME CHARACTERISTICS OF INTERSTELLAR GAS IN THE GALAXY

Introductory Report

(Tuesday, September 9, 1969)

HAROLD F. WEAVER

Radio Astronomy Laboratory, University of California, Berkeley, Calif., U.S.A.

1. Task of the Review

The task assigned for this review is to discuss structural features of the Galaxy as they may involve or relate to gas dynamics. Since a topic of such great breadth would permit discussion of essentially any aspect of the Galaxy, we shall narrow the view and direct attention primarily towards the larger-scale features of the system. In particular, effort will be made to place emphasis on those areas of the subject in which problems of interpretation exist and in which new theoretical models are needed.

2. The Galaxy Among Others

Examination of photographs of nearby galaxies similar to our own (see, for example, *The Hubble Atlas of Galaxies*, Sandage, 1961) reveals a number of general characteristics of such systems that should guide investigations of the large-scale properties of our own Galaxy.

(i) A galaxy normally exhibits a grand design, a two-armed (in some instances multi-armed) spiral winding outward from a spheroidal nucleus.

(ii) A spiral arm in this grand design is not a continuous structure; it appears to be composed of segments. On the largest scale, arms of some galaxies appear to be made of a series of almost straight-line sections; universally, on a smaller scale, the arms appear to be mottled and irregular, composed of clumps. Arms frequently are split and bifurcated.

(iii) In the majority of galaxies the arms are not more than 2π or 3π in angular length.

(iv) Not every feature in a galaxy fits into the grand design. There are many non-conforming structures, very often in the form of spurs (also called 'branches' or 'twigs' by some investigators) or interarm features. Spurs normally appear to originate on the outside of an arm and show a larger pitch angle (a factor of two is not unusual) than is exhibited by an arm in the grand design.

(v) Dark material is preferentially concentrated, often in a clumpy manner, at the inside edges of spiral arms. The outside edges of the arms are often less well-defined than the inner edges. The outside edges appear composed of material that is drawn out, coarsely 'brushed', from the inner edges. The 'brushed-out' clumps of material show larger pitch angles than the arm as a whole. In the inner parts of a galaxy the spiral arms may be delineated primarily by dark material.

Habing (ed.), Interstellar Gas Dynamics, 22–50. All Rights Reserved. Copyright © 1970 by IAU

(vi) Arm and interarm regions appear to differ in character. Specifically, an arm does not appear to be just a region of slightly greater-than-average density.

(vii) Spiral arms and other features in a galaxy are generally narrow. Such generally compact delineation of the spiral arms must indicate that conditions suitable for stellar birth (at least for very massive stars) exist over very limited regions of the Galaxy – those same regions in which the massive stars are seen. During their main-sequence lifetimes most massive blue stars move less than 100 pc (much less than the thickness of a spiral arm) from their places of birth, and even 'runaway' stars move only a kpc.

Clearly there must be much physics in progress in the arms of a galaxy. There the interstellar material is most plentiful. In the arms condensation takes place, stars are formed; we see the arms outlined by young stars, which are the principal spiral tracers. The interstellar medium may be expected to be quite different in character inside and outside the arms. To investigate such differences and to examine the gas structures that form the arms, we must turn to the Galaxy.

3. The Grand Design in the Galaxy

There are now two basic methods of establishing the nature of the grand design in our Galaxy.

(a) From optical observations we outline the arms by means of spiral tracers – young stars, H II regions, and the like – for which photometric distances can be derived.

(b) In the radio range we observe neutral hydrogen (which, by empirical evidence, is concentrated in the arms) and employ the galactic rotation curve to infer distances to the gas concentration from their observed Doppler velocities. (As a variant, in place of observations of H I, one may observe radio recombination lines which originate in H II regions.)

Neither of these methods is without shortcomings. Because of interstellar obscuration we can observe spiral tracers in the optical range over a very limited region of the Galaxy, within a distance of about 3 kpc from the Sun, hence no truly large-scale pattern of arms can be delineated by optical techniques. The 1420 MHz line of hydrogen can be observed throughout the entire extent of the Galaxy, but the gas in the system is not without radial motions of an apparently regional systematic character. Such non-circular motions are directly reflected in erroneous kinematic distances, hence the gas-derived picture of spiral structure will be regionally distorted by unknown amounts. In spite of this difficulty in distance determination, we must use H I as our probe if we are to determine the overall galactic spiral pattern. We have, in this case, no alternative to a kinematic distance scale.

If the pattern of spiral arms in the Galaxy is reasonably regular, we can determine one characteristic of the system, the pitch of the spiral arms, without recourse to any velocity measurement. From that pitch angle, we can provide an overall smooth representation of the Galaxy.

In order to find the pitch angle we have only to determine the longitudes at which the line of sight is tangent to the next inner (Sagittarius) arm. We observe the longitudes of tangency to be $50°.5$ and $284°.0$; the derived pitch angle is $12°.5$ (Weaver, 1970). We take the distance, R_0, from the galactic center to the Sun to be the standard value, 10 kpc, and employ the longitudes of tangency to find the constants of the equiangular spiral $R = 2.00 + 2.20\ (\theta + \frac{0}{\pi})$ by which we represent the galactic spiral. Here θ

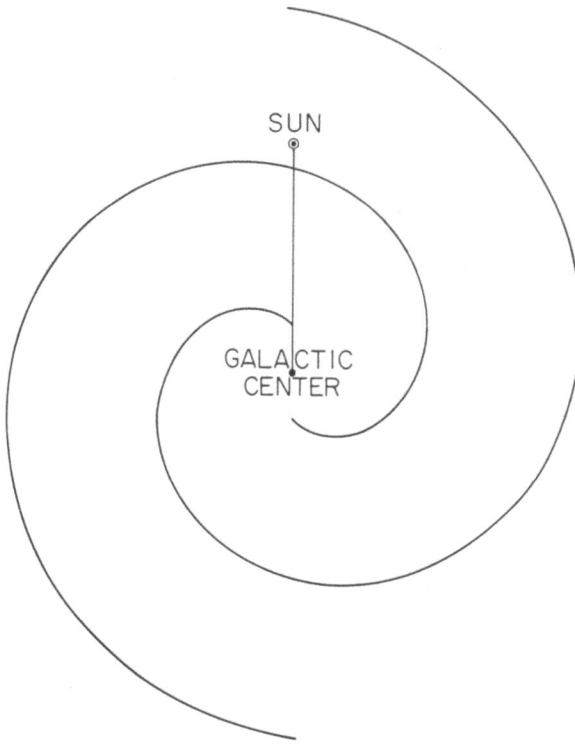

Fig. 1.   Smooth, two-arm spiral with pitch angle $12°.5$ representing the overall picture of our Galaxy.

represents galactocentric longitude. The smooth spiral galaxy thus derived is shown in Figure 1.

The real Galaxy is far more complex than the idealized system shown in Figure 1. We employ the extensive observations of H I now available and utilize the rotation curve established by Schmidt (1965) to derive the galactic model shown in Figure 2a (Weaver, 1970). Shown in Figure 2b for comparison is the observed distribution of young stars and H II regions (Schmidt-Kaler, 1964).

In Figure 2 we note that the left and right sides of the spiral are quite different; more detail is shown on the right than on the left. This is observational only; it is not real. At present, hydrogen observations from the northern hemisphere are far more extensive and detailed than those from the southern hemisphere. The left-hand side

Fig. 2a. Diagram representing the structure of the Galaxy as derived from H<small>I</small> observations. The left-hand side of the diagram is schematic. See the text for explanation of the difference between the two sides of the picture, which is not a real effect in the Galaxy.

of the figure is derived from southern-hemisphere observations limited in coverage. There is in reality much more detail present on the left side of the figure than is shown in the diagram, but that detail cannot now be delineated with the certainty desired because of the lack of a sufficiently close net of data points. Details were therefore omitted from the left side of the picture. Details on the right side can be drawn in with some assurance; the left side is a very rough sketch only.

In Figure 2 the arms are shown broken into bits and pieces. This characteristic may, in part, arise from regional non-circular motions of the type mentioned earlier. Clearly, however, there are many spurs and interarm links. The gas is concentrated in large complexes spirally distributed, but not all belonging to a smooth two-arm spiral.

There has long been a disagreement in arm pitch angle derived from stars and from

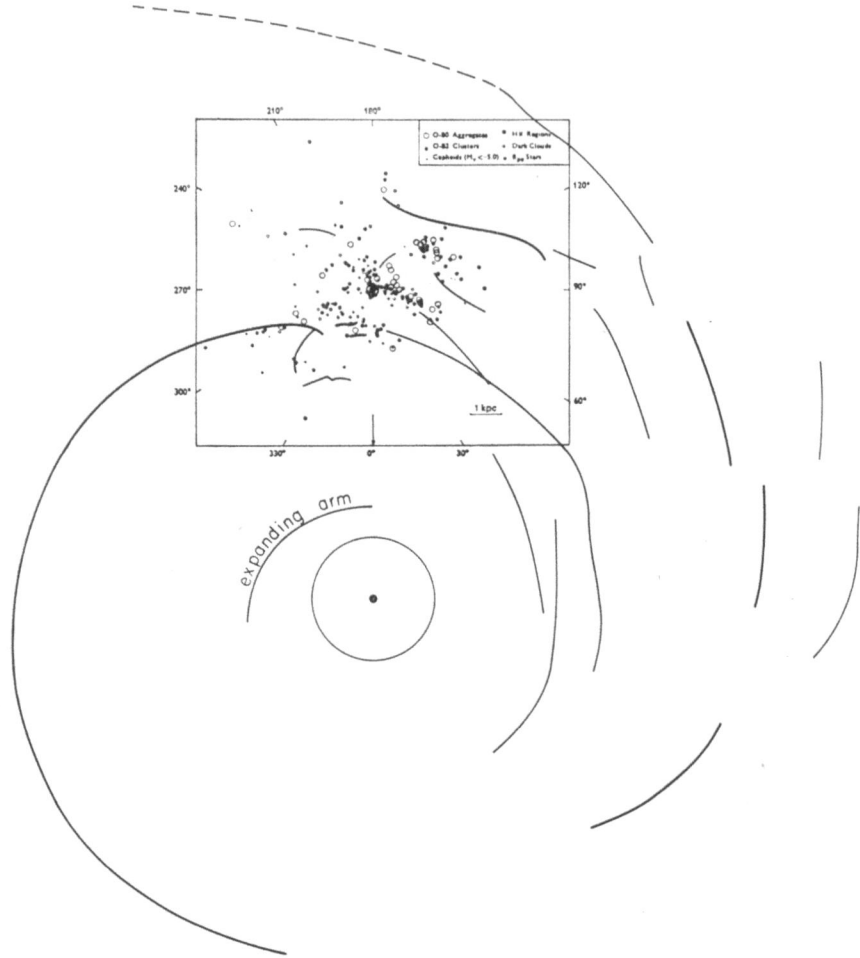

Fig. 2b. Picture identical with Figure 2a except that the spiral arms as outlined by stars is shown superimposed on the H I arms. See text for more complete explanation.

gas. The extensive observations now available make it clear that the Sun is not located in a major spiral arm, but in a spur or offshoot of the Sagittarius arm. If the pitch of the spur is derived from stellar observations and from radio observations, there is no disagreement. In the past, optical observations of the local (spur) structure have been compared with galactic-scale hydrogen observations. In such a comparison, disagreement is inevitable. A more complete discussion of this situation has been given by Weaver (1970).

Not given in Figure 2 is any indication of the cross-sectional thickness of the arm structures or the well-known warp of the galactic plane. Kerr (1969) treats the latter phenomenon.

4. Some Characteristics of the Gas in the System

A. THE NATURE OF AN ARM

For the present we largely ignore the non-gaseous components of the Galaxy and use the term spiral arm to refer to a large-scale elongated gas structure of spiral form which fits into the grand design. Spurs and interarm features are numerous; presumably, these are structures similar to the arms in physical properties.

In the definition we use here we describe an arm in space in accord with the usual photographs we are accustomed to see. Operationally, we use a different coordinate system in studying spiral structures. We observe these gas structures in position on the sky (angular coordinates) with velocity replacing radial distance. This results in many troublesome ambiguities. Observations provide intensity or antenna temperature, T_A, of the radiation at each radial velocity, v_r, over some bandwidth, Δv_r, arising

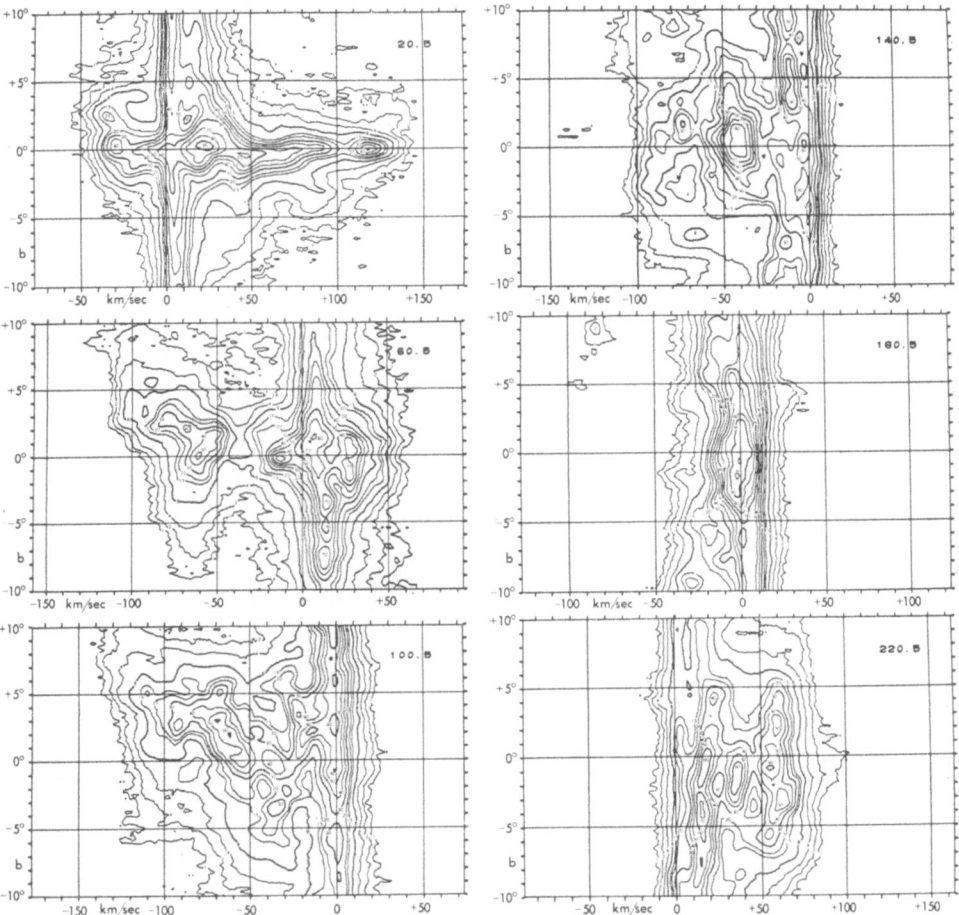

Fig. 3.   Examples of contour maps $T_A(v_r, b \mid l)$ for a variety of l-values.

from a point on the sky at longitude, l, and latitude, b. We display the data in the form of contour maps of antenna temperature as a function of radial velocity and one angle, keeping the second angle fixed. Alternatively, we may display the data as contour maps of antenna temperature as a function of two angles for a fixed velocity centered in bandwidth Δv_r. This latter display is a picture of gas intensity seen on the sky in a specific velocity range.

In Figure 3 we show, for several l-values, contour maps in the form T_A as a function of v_r and b for fixed l. We designate such a map symbolically as the function $T_A(v_r, b \mid l)$. Such maps show clearly velocity cross cuts of spiral arms and spurs at a given l. From these particular maps the general nature of the gaseous arm structures is well seen; arm structures are separate entities (at least in velocity) within which there are subconcentrations of gas. From a series of such maps closely spaced in l, it is clear that the major structure we would term an arm persists over a long range in l, while any individual subconcentration within the arm persists over a very limited range in l and b. The latter phenomenon is readily demonstrated by means of a $T_A(l, b \mid v_r)$ map, Figure 4, which shows the l, b extent of the individual members of a group of sub-concentrations.

Contour maps of these two kinds suggest several basic, gas-dynamically related questions that we now consider:

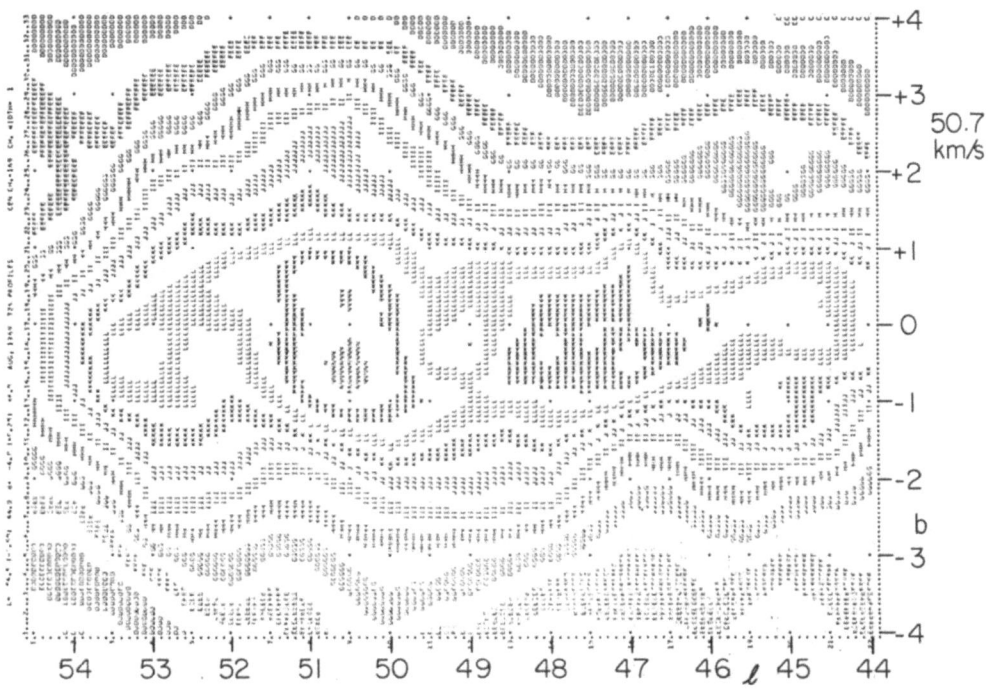

Fig. 4. $T_A(l, b \mid v_r)$ map for longitude range 44° to 54°.5 illustrating clumpiness within a section of a spiral arm which shows continuity in longitude.

(1) What is the arm/interarm contrast in number density of H I atoms?

(2) What are the statistical properties – linear scales, number densities, masses – of the subconcentrations within the arm and other large-scale structures in the Galaxy?

(3) What are the velocity characteristics of the gas and the subunits of the gas forming an arm?

B. ARM/INTERARM CONTRAST

We can determine the arm/interarm contrast directly by measuring absorption by the H I gas in the arm and interarm regions along the line of sight to a strong radio source.

A concentration of cool H I gas in the line of sight will attenuate the radiation of the radio source lying beyond it. (For simplicity in this discussion we assume the source and the H I cloud to be centered one over the other and of the same angular size and shape. We thus eliminate geometrical factors not necessary to the physical argument.) From the measured attenuation we compute the effective optical depth, $\tau_{eff}(\nu_0)$, of the H I concentration at ν_0, the frequency of maximum absorption. The optical depth $\tau_{eff}(\nu_0)$ is related to the physical parameters of the H I gas by the equation

$$\tau_{eff}(\nu_0) = qL \left\langle \frac{n_H}{T_{ex}} \right\rangle. \tag{1}$$

Here q is a numerical factor composed of atomic and numerical constants, L represents the path length in the absorbing hydrogen, n_H is the number density of the hydrogen, and T_{ex} is the excitation temperature, which describes the relative populations of two energy levels involved in the transition. For H I T_{ex} is equal to T_{kin}. We shall take T_{ex} to be uniform throughout the H I concentration producing the absorption. Under this circumstance $\langle n_H/T_{ex} \rangle = \langle n_H \rangle / T_{ex}$.

The value of T_{ex} (as well as $\tau_{eff}(\nu_0)$) can be determined from the absorption measurements. Specifically,

$$T_{ex} = [T_A(\nu_0)/\Delta T_A(\nu_0)] T_{B,s}(\nu_0), \tag{2}$$

where $T_A(\nu_0)$ represents the antenna temperature one would observe for the H I if there were no absorption, $\Delta T_A(\nu_0)$ the depth of the absorption, and $T_{B,s}(\nu_0)$ the brightness temperature of the source, all at frequency ν_0. From the derived values of $\tau_{eff}(\nu_0)$ and T_{ex}, one can determine $\langle n_H \rangle$ provided an estimate of L is available. If one makes observations of the absorption arising from the gas in arm and interarm regions along the line of sight, one can then determine the arm-interarm contrast, $\langle n_H \rangle_{arm}/\langle n_H \rangle_{interarm}$, which is what we require.

Alternatively, if pressure equilibrium exists between the arm and the interarm gas, as seems likely, one can estimate the arm-interarm contrast from the relation

$$\langle n_H \rangle_{arm}/\langle n_H \rangle_{interarm} = T_{ex, interarm}/T_{ex, arm}. \tag{3}$$

We shall use both methods of estimating the contrast.

Figure 5 shows a contour map, $T_A(v_r, b \,|\, l)$ for the longitude of Cas A observed at Hat Creek (Weaver and Williams, 1970). We note immediately the strong absorp-

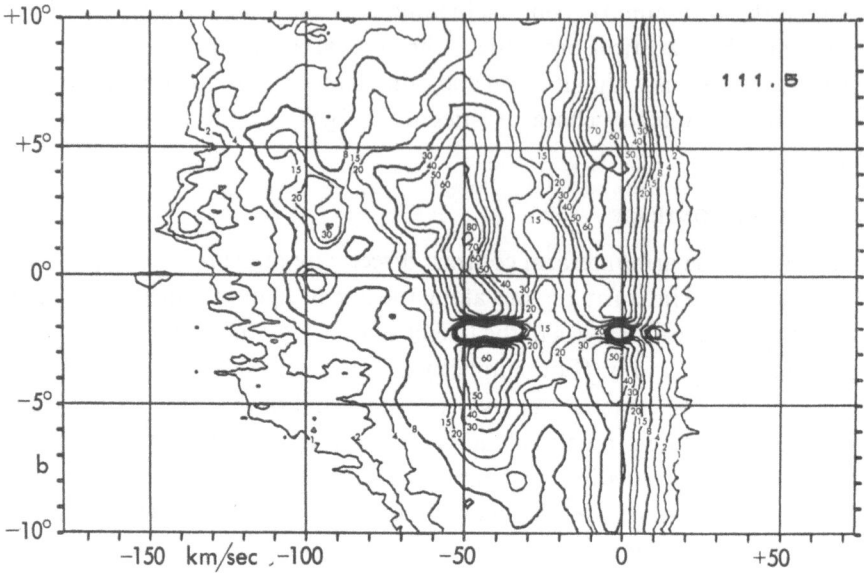

Fig. 5.   Contour map showing H I absorption resulting from the source Cas A.

tion in the arms, the weak absorption in the interarm region. Williams (1969) has analyzed the observations in the vicinity of Cas A and has found the T_{ex}, $\tau_{eff}(\nu_0)$ values along the line of sight to Cas A. He finds (see Figure 6) that in the arm T_{ex} is low, the T_{ex} values derived from the negative velocity lines being 70 K and 85 K. The zero-velocity line shows a temperature of 93 K, while the $+9$ km sec^{-1} line corresponds to a temperature of 160 K. By comparing widths of H and OH lines from the same source, Barrett *et al.* (1964) found for the components of the zero-velocity line (the line shows as a close-spaced doublet on OH) kinetic temperatures of 90 K and 120 K. Some lack of confidence has been expressed for temperatures derived from H/OH line widths because of maser action frequently shown by the OH molecule. It is not likely that such trouble has sensibly affected the 1667 MHz line involved in this comparison, though the satellite lines show peculiar intensities. The temperatures derived from absorption measurements and from H/OH line width comparison are in excellent agreement. In the interarm region Williams finds that T_{ex} is high, ≈ 1000–3000 K.

With the data derived by Williams we find for the arm/interarm regions crossed by the line of sight to Cas A

$$\langle n_H \rangle_a / \langle n_H \rangle_{ia} \approx T_{ia}/T_a \approx 1500/90 \approx 17 \quad \text{(by Equation (3))}$$

$$\langle n_H \rangle_a / \langle n_H \rangle_{ia} \approx \left(\frac{\tau_a}{\tau_{ia}} \right) \left(\frac{L_{ia}}{L_a} \right) \left(\frac{T_{ex,\,a}}{T_{ex,\,ia}} \right) \approx \frac{0.6}{0.01} \times \frac{5}{1} \times \frac{80}{1500} \approx 16$$

$$\text{(by Equations (1, 2))}.$$

The arm-interarm contrast is in the range 15 to 20.

From similar observational data by Williams (1969) for the Tau A absorption, we estimate that the arm/interarm contrast along the line of sight to Tau A is > 20.

Thus the conclusion from these two representative regions is that the arm-interarm contrast is generally high, ≈ 20, and possibly even higher. However, this conclusion is open to the criticism that the values derived represent extreme upper limits since (a) the arms are represented by the lowest temperature objects within them and (b) the interarm regions may not have been adequately separated from influences of arm/interarm interface regions; the temperature taken for the interarm region may be too low. Higher temperatures are recorded in the interarm regions; we employed an average that may not be representative. What is wanted for the determination of contrast, of course, is $\langle n_H \rangle$ over a representative volume of the arm which includes concentrations and ambient gas, and a similarly representative volume of an interarm region. Adequately accurate values of these quantities are not now available; observers must make a strong effort to obtain such values. Very preliminary values estimated by methods of differing reliabilities suggest values for arm/interarm contrast which, though relatively high, are possibly half the value quoted. In no instance is there suggested a contrast ratio as low as, say, 3.

C. REMARKS ON THE OBSERVED TEMPERATURES OF THE GAS IN THE ARM AND INTERARM REGIONS

Though we have some knowledge of the temperatures of H I concentrations in the arms, much remains to be learned. Observational knowledge is almost completely lacking about the temperature of (a) the gas in the arms in which the concentrations appear to be embedded, and (b) the interarm gas, which appears to be more generally free of concentrations than is the gas in the arms.

Under the gross simplification that within an arm or arm-like structure n_H and T_{ex} were space invariant, van de Hulst et al. (1954) and later Schmidt (1957) employed observed maximum antenna temperatures in directions little influenced by galactic rotation (Cygnus, the galactic center, and the galactic anti-center) to determine T_{ex}. They estimated path lengths, L, within the relevant gas in each of the three named directions; Schmidt derived a universal T_{ex} of 125 K.

However, at approximately the same time, Heeschen (1955) investigated hydrogen absorption in an extended cloud in the general direction of the galactic center and found $T_{ex} \approx 40$ K. Davies (1956, 1958) derived similar temperatures for a cloud in Auriga. Later, Riegel and Jennings (1969) found an upper limit of 42 K and a more probable temperature of 20 K for an extension of the same object investigated by Heeschen. Clearly, objects of low temperature exist; the excitation temperature is not universal. From Figure 6 it is clear that regions of quite different temperatures exist within an arm or arm-like structure.

Shuter and Verschuur (1964), from a study of absorption lines, determined that H I concentrations in the so-called 'Orion Arm' (the local large-scale spur that contains the Sun) had temperatures ranging from 25 K to 120 K, with a harmonic mean temperature of 56 K, a value quite different from the then frequently quoted and generally

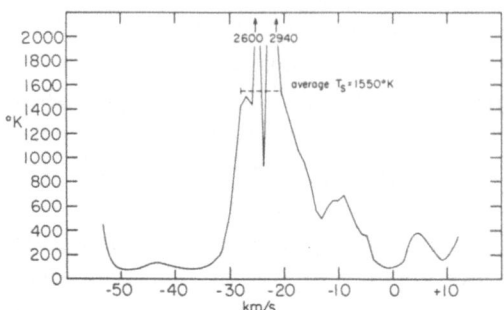

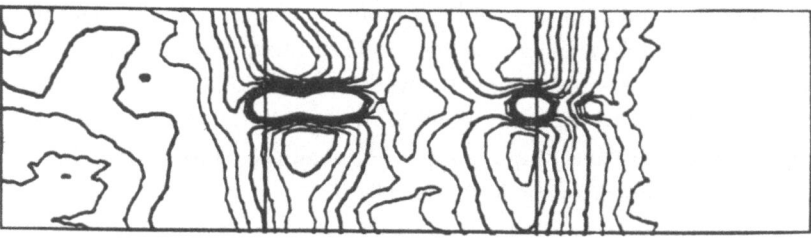

Fig. 6.   Temperatures of concentrations of H$_I$ producing absorption in Cas A as determined by Williams. Lower figure: enlarged portion of Cas A absorption. (Taken from Figure 5.) Upper figure: plot of observed T_{ex} values of features shown in the lower map.

assumed universal value of 125 K. In their discussion Shuter and Verschuur showed that in absorption, cold clouds are far more visible than hot clouds. The distribution of cloud temperatures they found was therefore biased. To reproduce the then-accepted value of 125 K, Shuter and Verschuur had to add to their observed distribution of cloud temperatures about 20 percent more clouds with temperatures > 120 K. Presumably, these hotter clouds had been missed in the absorption studies.

Clark (1965), in discussing his interferometer measurements of H$_I$ absorption lines, proposed an alternative to the isolated cloud model in terms of which earlier investigators had tended to interpret their results. Clark wrote: "...it is worthwhile to at least examine another alternative to this cloud explanation. This is to assume that the hydrogen seen in emission is not entirely the same hydrogen as that seen in absorption. This can come about when part of the hydrogen is at very high temperature, and thus has a very low absorption coefficient. In order to make it invisible in the absorption spectrum, it must have a temperature in excess of 1000 K. Let us now assume that this hot hydrogen is a continuous, hot medium surrounding clouds somewhat cooler than 100 K."

Clark supported his model by a variety of observational evidence involving both velocity and temperature measurements. The Clark model does indeed appear to be of the correct character. High temperatures have recently been measured in the ambient interstellar gas by several observers; for example, by Mebold (1969), by Radhakrishnan and Murray (1969), and by Williams (1969). There can be little doubt that the interstellar medium is a mixture of hot gas and embedded cold regions. Radhakrishnan and Murray speak of such a mixture as a 'raisin pudding' model of the interstellar medium.

Firm knowledge of the properties of the 'raisin pudding' is seriously lacking; observational clarification of the characteristics of the interstellar medium is crucial if proper theoretical insights are to be gained. Questions that require answers are, for example:

(1) Is the enveloping gas (the 'pudding') distributed on the large scale more or less smoothly in R and z, with concentrations of cold hydrogen (the 'raisins') being everywhere present except that they are far more numerous in the spiral arms than elsewhere, or

(2) Are the arm and arm-like structures composed of generally cool gas in which still cooler concentrations of hydrogen exist, with the cool, large-scale arm structures being surrounded by hot gas which, certainly in the interarm regions, and possibly elsewhere, reaches temperatures greater than 1000 K?

(3) What is the nature of the boundary of a concentration of cool gas in the arms? Is the boundary between the hot surrounding medium and the cool concentration sharply defined, a narrow transition zone between two phases of the gas, or

(4) Are there, in relatively cool arms of the Galaxy, temperature and density fluctuations associated with a turbulence spectrum?

(The question of the sharpness of the edge of concentrations of cool gas is crucial. A strong answer to this question might well establish the class of physical mechanisms responsible for formation of the concentrations.)

These questions which arise immediately in connection with a description of the interstellar gas, lead, of course, to more basic and fundamental questions of what causes the cool gas to be concentrated in spiral structures – spiral structures which appear to differ in overall character and not just in density of material from the surrounding interarm gas. Specifically, such questions relate to the principal problems involving current theory and observations of the interstellar gas. An important part of our discussion should determine how current observations fit into the two-phase model of the interstellar gas (Goldsmith et al., 1969), the density-wave model of spiral structure (Lin, 1967), and Roberts' (1969, 1970) discussion of density-wave induced shock phenomena in the gas on the inside edge of spiral arms.

D. SOME CHARACTERISTICS OF THE GAS CONCENTRATIONS WITHIN SPIRAL ARMS

The patchy, sharp-edged appearance of obscuring clouds of dark material seen so frequently on photographs of the Milky Way combined with the sharpness of the multiple interstellar lines seen in many stellar spectra no doubt gave rise to the hypothesis that the interstellar medium is composed quite generally of discrete spherical 'clouds' to which, as a first approximation, were assigned standard size, mass, density, and so forth. The numerous and still rapidly accumulating observations of neutral hydrogen do not lend support to such a hypothesis of isolated clouds of standard properties. The HI gas, as described earlier, appears, on the galactic scale, to be concentrated in very large structures – spiral arms, spurs, interarm links. Within these major, large-scale structures there appears to be a continuum of gas in which there are smaller-scale concentrations having a variety of characteristics.

The picture of a wide variety of concentrations in a continuum of neutral hydrogen which formed a major structural unit in the Galaxy was first clearly brought to view by Heiles (1967a), who used the 300-foot antenna at Green Bank to investigate the region $l = 100°$ to $140°$, $b = +13°$ to $+17°$. The hydrogen he observed constitutes a portion of the local spur in which the Sun is located. Broadly, Heiles found that in the area he investigated about three-fourths of the total 21-cm emission came from a generally smooth background, the ubiquitous gas surrounding the Sun. Within that background Heiles found what he describes as two sheets of gas, that is, concentrations of hydrogen with different mean velocities and small velocity dispersions, and with continuity of density and mean velocity over large angular extent, particularly in longitude. Such sheet-like structure may or may not be common or frequent; that aspect of the picture given by Heiles is not of primary concern in this discussion. Our interest centers on structures of smaller scale than the two sheets of gas. Within the sheets of gas, Heiles found smaller-scale concentrations of several types:

(1) a few concentrations of gas generally resembling 'standard' clouds having characteristic radii of 10 pc, number densities of 6 H atom $^{-3}$, and masses $\approx 10^3 \, M_\odot$;

(2) a small number of concentrations of larger dimensions, characteristically 20 pc radii, number densities ≈ 2 H atom cm^{-3}, and masses of the order of a few $\times 10^3$ $M_\odot$; and

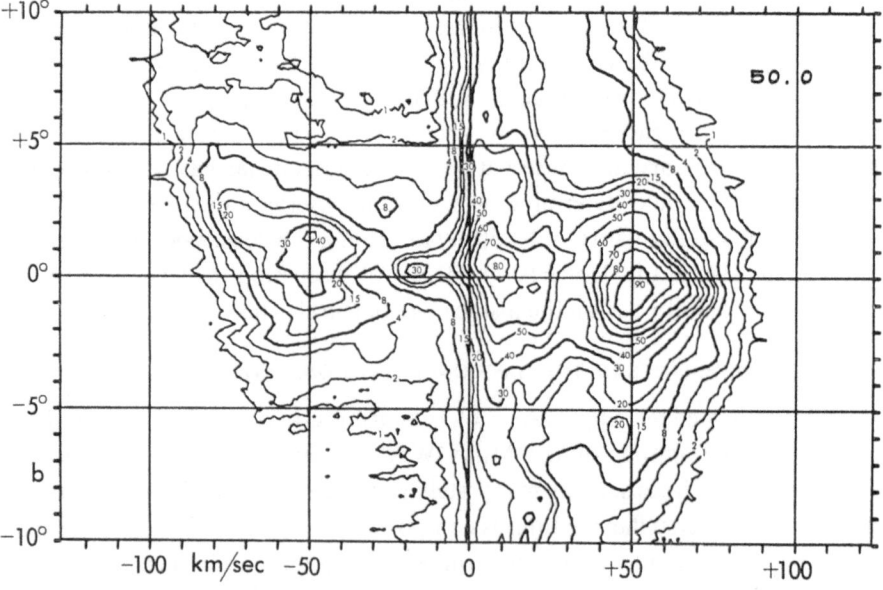

Fig. 7.   Contour map $T_A(v_r, b \mid l = 50°)$ from the Hat Creek Survey of HI.

(3) very many 'cloudlets' having, at the mode, radii of 2 pc, number densities of 2 H atom cm^{-3}, and masses $\approx 4\ M_\odot$.

The Hat Creek observations (Weaver and Williams, 1970) can be used to test Heiles' hypothesis of many concentrations or density fluctuations in a continuous medium. The test area arbitrarily selected for this test extends from $l=44°$ to 54°, $b=-4°$ to $-10°$. An observed contour map, $T_A(v_r, b \mid l=50°)$, Figure 7, will serve to characterize the general area; we look through the local gas to more distant spiral features. A set of closely spaced contour maps, $T_A(l, b \mid v_r)$, Figure 8 shows, on the sky, the hydrogen intensity in a series of narrow velocity ranges. These $T_A(l, b \mid v_r)$ maps are, essentially, pictures of the hydrogen brightness in a narrow distance range at fixed distance from the Sun. Examination of a velocity sequence of such maps shows clearly (a) the existence of numerous concentrations or

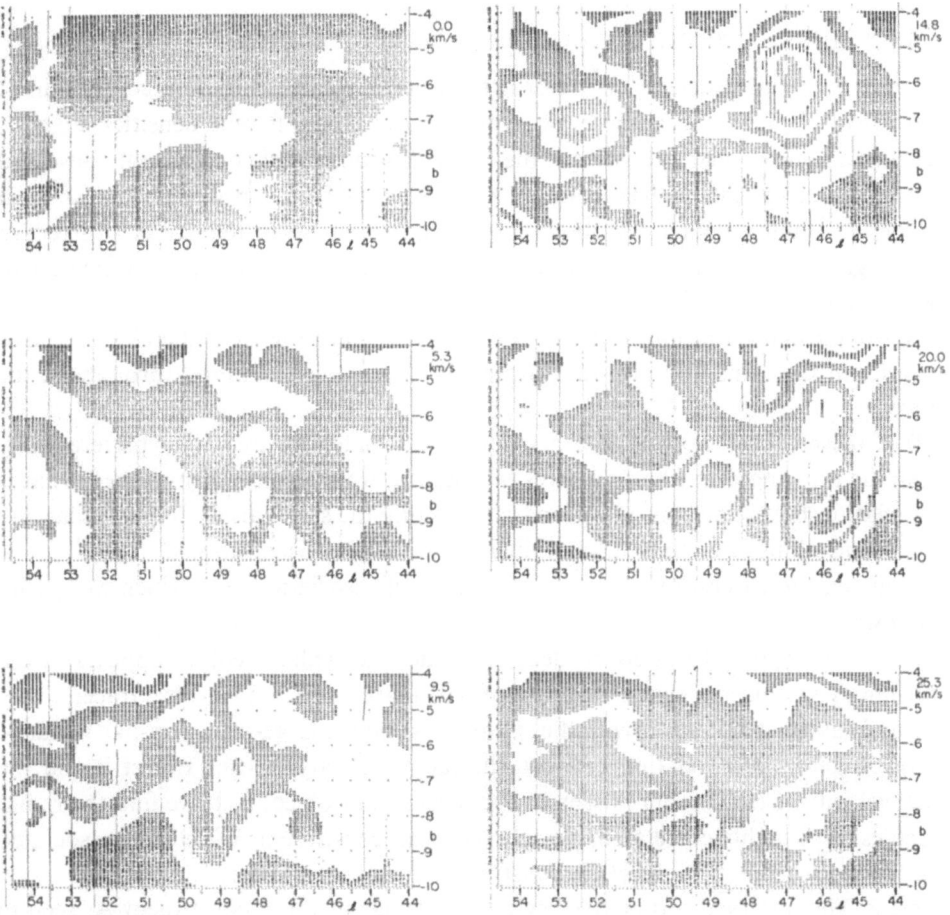

Fig. 8.   Set of $T_A(l, b \mid v_r)$ maps for ranges $l = 44°$ to 54°.5, $b = -4°$ to $-10°$ at the velocity values shown in the diagram. The velocity range covered in each diagram is 2.11 km sec^{-1}.

'cloudlets' of small angular extent and narrow velocity dispersion, and (b) a variety of gas concentrations of larger angular size but small velocity dispersion.

The numerical data for these various concentrations or density fluctuations are in excellent agreement with the data tabulated by Heiles. For examples, we find (Figure 8) for the largest concentration centered at $l=47°$, $b=-6°$, $v=14.8$ km sec^{-1}, assumed distance = 750 pc:

$$\text{radius} \approx 22 \text{ pc}$$
$$\langle n_H \rangle \approx 5 \text{ atom cm}^{-3}$$
$$\text{mass} \approx 4500 \, M_\odot$$
$$\sigma_v \approx 3.2 \text{ km sec}^{-1}.$$

The general background in which the concentration occurs is 0.5 to 1.0 H atom cm^{-3}. The concentration of gas at $l=47°$, $b=-6°$, $v=14.8$ km sec^{-1} is not gravitationally bound. It, like the other gas concentrations pictured, must form and 'evaporate' with time.

E. OTHER CHEMICAL CONSTITUENTS OF THE INTERSTELLAR MEDIUM

The observational evidence for major arm structures consisting of a continuum of neutral hydrogen in which there are concentrations of different densities suggests questions about the characteristics of the concentrations of other commonly observed constituents of the interstellar medium: Ca II, Na I, OH, ... (In this review we shall consider only interstellar lines observable with ground based telescopes.) In regard to these gaseous constituents of the Galaxy other than hydrogen we investigate two closely related questions:

(1) Are the concentrations of these other gaseous constituents of the Galaxy found primarily in the arms and arm-like structures or are they scattered throughout the Galaxy?

(2) If these concentrations of other interstellar materials are found preferentially within the arm and arm-like structures (as, indeed, we shall find them to be), what is their relationship to the concentrations of H I within the spiral arms? In particular, are the concentrations of Ca II, Na I, ..., coincident with H I concentrations?

To answer these questions, we employ a series of association tests which we specify as follows.

Test 1. Absorption from constituent A of the interstellar medium appears in the spectrum of an object of type B. Objects of type B are known to be spiral tracers; they are closely associated in space with the spiral arms and the large-scale arm-like features. We investigate the correlation between the radial velocity of each object B and the radial velocity of constituent A seen in the spectrum of object B. If we find the radial velocities to be highly correlated, we conclude that constituent A is closely associated with spiral structure; like objects B, constituent A is a spiral tracer.

Test 2. The observed radial velocity of interstellar medium constituent A, seen in direction α, is compared with the expected velocity of objects of class B in direction α. Class B objects are known spiral tracers. If we find a high positive correlation between the values compared, taking into account the known dispersion in the expected velocity of objects B and the error of measurement of the velocity of A, we conclude that constituent A is a spiral tracer also.

Test 3. Absorption lines from chemical constituents A and B of the interstellar medium appear in the spectrum of object C, which serves merely as a source of continuum radiation. If the observed radial velocities of constituents A and B are the same within measuring error, we assume that the concentrations of constituents A and B are coincident in space. As a secondary criterion of coincidence, we should find similar profiles for the lines arising from constituents A and B if the materials are coexistent and well mixed in a single concentration in space.

Test 4. We observe that spectral lines from constituents A and B of the interstellar medium are observed in one and the same precisely defined area of the sky and are not found outside of that area. The velocities of A and B are found to be the same within measuring error. Again, but of secondary character, we find the line shapes to be the same. We conclude that the concentrations of constituents A and B from which the spectral lines arise are coincident in space.

Test 5. A spectral line arising from constituent A of the interstellar medium is observed over an area of the sky identical with that occupied by some visible constituent B of the interstellar medium (a dark obscuring cloud, for example). We do not observe the spectrum of A outside the area covered by B. We conclude that the concentration constituents A and B responsible for the features observed are coexistent in space.

Tests 1 and 2 relate to the establishment of certain observed gaseous constituents of the Galaxy as spiral tracers; Tests 3, 4, and 5 relate to the establishment of coincidence of different gaseous constituents within a single concentration in space.

We consider first the question: are the concentrations of interstellar gas that give rise to the optical interstellar lines spiral tracers?

The close association between spiral arms as defined by $H\text{I}$ and the gas concentrations giving rise to the strongest and most frequently seen interstellar lines (those of $Ca\text{II}$ and $Na\text{I}$) is clearly demonstrated by Münch's (1965) diagram showing the observed relation between radial velocity of interstellar $Ca\text{II}$ and/or $Na\text{I}$ and galactic longitude. To a very high degree the observed velocity-longitude for the interstellar lines agrees with the analogous velocity-longitude diagram found for $H\text{I}$ by Weaver (1970). On the basis of Test 2, we conclude that the concentrations of $Ca\text{II}$ and $Na\text{I}$ from which originate the observed interstellar lines are closely associated with the concentrations of $H\text{I}$ which define the gaseous spiral arms. Concentrations of interstellar gas which produce lines of $Ca\text{II}$ and $Na\text{I}$ are spiral tracers.

We consider Test 1, and, in Figure 9, illustrate the correlation between radial velocities derived from the interstellar lines of $Ca\text{II}$ and $Na\text{I}$ and the velocities of the OB stars in which the interstellar lines are observed. The correlation is positive and high. Howard *et al.* (1963) found a correlation coefficient of $+0.97$, while Takakubo (1967) found a correlation coefficient of $+0.85$ from studies of this kind. Münch (1965) pointed out that in some areas of the sky interstellar lines show a systematically nega-

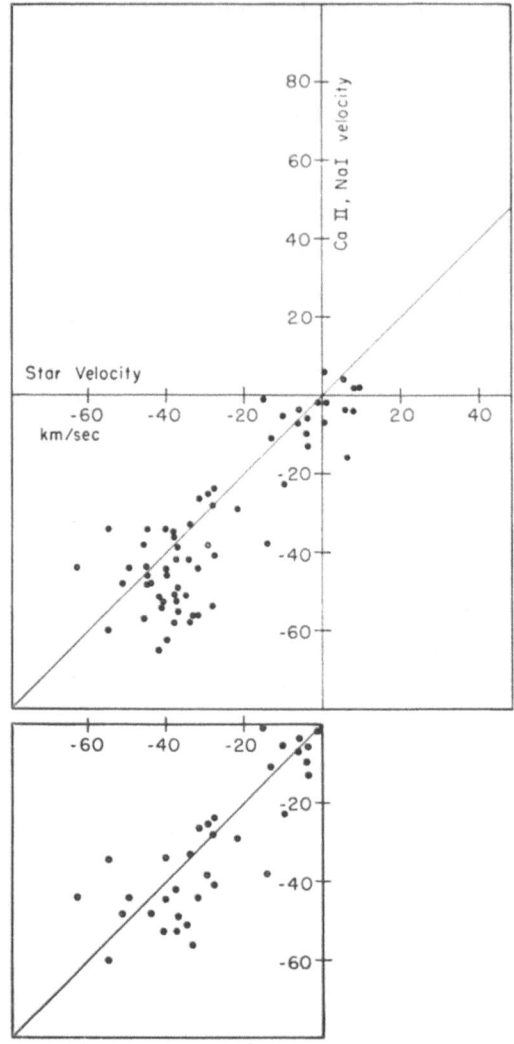

Fig. 9.   The relationship between velocities of interstellar Ca II and the B stars in whose spectra the interstellar lines appear. The lower diagram is the same as a section of the upper one except for the omission of a group of stars from a region in the sky known to show systematically negative velocities. The larger-than-normal negative velocities of the Ca II lines shown by the omitted stars is readily apparent.

tive radial velocity with respect to the associated stellar radial velocities. The explanation proposed by Münch is that the interstellar gas, which lies between the stars and the observer is, in the part of the Galaxy affected, systematically expanding away from the group of stars involved. Such an effect, while interesting and, for some purposes, of dominant importance, does not affect our conclusion, namely that the high positive value of the correlation coefficient represented by Figure 9 (and found numerically by

Howard *et al.* and by Takakubo) clearly indicates the close association between the OB stars and the concentrations of gas giving rise to the interstellar Ca II and Na I lines.

Interstellar absorption lines in the optical range observable from the ground have been identified from Na I, K I, Ca I, Ca II, Ti II, Fe I, CH, CH$^+$, and CN. Concentrations of interstellar gas contain a variety of materials; visibility of particular spectral lines depends upon the atomic or molecular constants of the materials present and upon the physical conditions within each gas concentration. We test the hypotheses that the various interstellar materials identified by their spectral lines in a single star arise from a single concentration of gas. We employ Test 3 and compare radial velocities derived from the interstellar lines. Invariably, we find that the various lines may be grouped together on the basis of agreement of radial velocity. All lines originating in one gas concentration show the same radial velocity within the limits of measuring accuracy. We take as an example Herbig's (1968) observations of ζ Oph, which contains in its spectrum interstellar lines from more than one gas concentration. In one concentration of gas, the star exhibits all identified optical interstellar molecules and atoms except Ti II; from a second gas concentration only a few lines are seen. For each concentration the lines agree in radial velocity within measuring errors. Test 3 leads us to conclude that all lines of one velocity arise in one concentration of gas as, indeed, Herbig assumed in his analysis.

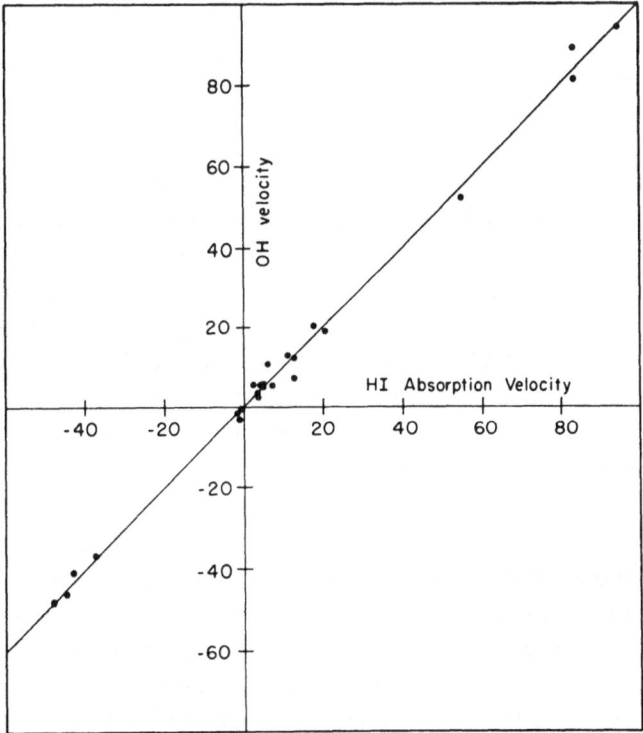

Fig. 10.   Correlation diagram of the velocity of OH absorption and H I absorption observed in one and the same source.

Interstellar lines arising from constituents other than Ca II and Na I can always be associated in a gas concentration of Ca II and Na I on the basis of velocity. Our tests have shown that gas concentrations from which Ca II and Na I lines originate are spiral tracers; we therefore conclude that all interstellar optical lines originate from gas concentrations that are spiral tracers.

The OH molecule is seen in many radio sources distributed around the galactic plane In Figure 10 we apply Test 3 to the radial velocities derived from the *absorption* lines of OH and H I seen in the same source. The data for the comparison are from Goss (1968). Within measuring accuracy the velocity relationship is one to one; the correlation coefficient approaches unity. We conclude that the concentrations of absorption of OH and H I are spatially coexistent; the OH is as much a spiral tracer as the H I.

OH *emission* lines originate in sources of several types.

(1) On the basis of Test 5 Heiles (1968) concluded that there is 'normal' OH emission (presumably OH in thermodynamic equilibrium) associated with dark obscuring clouds. The dark clouds are known spiral tracers; the OH producing the emission is thus closely connected with spiral arms.

(2) Anomalous, highly non-thermal OH emission, generally strong in the principal lines, arises from concentrations of gas of very small physical size which a modified Test 5 shows to be associated with some H II regions. (See, for example, Weaver *et al.*, 1968.) Since H II regions are prime spiral tracers, the OH gas again is found to be associated with the spiral arms.

(3) Anomalous non-thermal OH emission, often mainly visible in the satellite lines, frequently appears to be associated with non-thermal radio sources. (See Turner, 1969.) Test 2 leads to the hypothesis that the OH from which the lines originate is involved in the spiral arms.

(4) Anomalous OH absorption, mainly in the 1612 MHz satellite line appears to be associated with some infrared stars (Wilson and Barrett, 1968). Since this is a stellar phenomenon rather than an interstellar one, we do not consider it further.

The radio lines of H_2O, like those of one of the classes of OH, appear to be closely associated with H II regions. (See, for example, Cheung *et al.*, 1969b; Knowles *et al.*, 1969; and Meeks *et al.*, 1969). Further, again like OH, H_2O is strongly non-thermal, shows maser action, and exhibits pronounced variations in intensity with time. The close association between the H_2O and H II regions indicates location of the gas concentrations showing H_2O in spiral arms.

H_2CO is rather widely observed along the galactic plane. (See Snyder *et al.*, 1969; Palmer *et al.*, 1969; Rydbeck *et al.*, 1969.) H I, OH, and H_2CO are often seen together in absorption (though they are not necessarily, of course, all seen in each source); H_2CO is observed in some dark clouds in which 'normal' OH emission is present. Application of Test 3 leads to the conclusion that the associated lines arise from materials located in single concentrations of gas. Test 4 indicates that in some instances the various constituents are differently localized within the gas concentrations. From each test, however, we reach the strong conclusion that the H_2CO is found in concentrations of gas closely associated with spiral arms.

To date, NH_3 has not been found widely distributed. (See Cheung *et al.*, 1968; Cheung *et al.*, 1969a, c.) This may be instrumental rather than physical; very low noise receivers may be required for extensive discovery of NH_3 sources. Tests 3 and 4 lead to the conclusion that in the places in which NH_3 has been observed, it is closely related to OH absorption which we associate with gas concentrations in spiral arms.

We sum up these various tests by noting simply that the concentrations of gas which give rise to the interstellar lines, both optical and radio, appear to be constituents of the spiral arms of the Galaxy. More specifically, the various constituents producing interstellar lines in the optical range appear to exist together in single condensations as do those producing radio lines. In the next section we examine further relations between the condensations of gas.

F. SOME INTERRELATIONS BETWEEN OBSERVED ATOMIC AND MOLECULAR CONSTITUENTS AND PHYSICAL PROPERTIES OF INTERSTELLAR CONCENTRATIONS OF GAS

An arm of the Galaxy appears to be composed of a continuum of gas in which concentrations of cooler gas are present. The concentrations of H I seen in emission exhibit a variety of sizes and masses. The study by Heiles (1967) of an area of 160 square degrees is currently the only one which directly provides fairly extensive quantitative data on size and mass of the concentrations of interstellar gas. Heiles found only a few large concentrations that generally resemble the 'standard' clouds which have been widely discussed in the literature; he observed very many small concentrations of gas which he termed 'cloudlets'. In Figures 11 and 12 we reproduce the distributions of radii and masses found by Heiles for the 'cloudlets' in the area he investigated. Heiles states that the finite angular resolution of the telescope used for the observations does not distort the distribution of radii found; the instrumental angular resolution corresponds to $R = 0.75$ pc at the average distance of the concentration observed. This value lies on the far left side of the published distribution of R-values.

The concentrations of interstellar gas in which *absorption* lines originate exhibit the same spatial distribution as the concentrations of H I that are seen in emission, and that define the gas arms of the Galaxy. The concentrations of gas in which absorption lines originate are mixtures of atomic and molecular materials; they, like the emitting H I concentrations appear to exhibit a variety of different sizes, densities, and temperatures, though few quantitative data are available. Which particular absorption lines are seen in any concentration depends upon the spectroscopic properties of the atomic and molecular species present and upon the physical conditions within the gas concentration.

In the optical range of the spectrum mainly the strong lines of Ca II and Na I are observed in the interstellar gas concentrations, but, at the same time, it is not uncommon to find, originating from a concentration of gas which shows Ca II and Na I, lines from a variety of other atoms and molecules as well. Indeed, some single concentrations exhibit all identified optical interstellar lines.

In the radio range, absorption lines of various materials likewise originate in a single concentration of gas. H I is often seen together with the molecules OH and

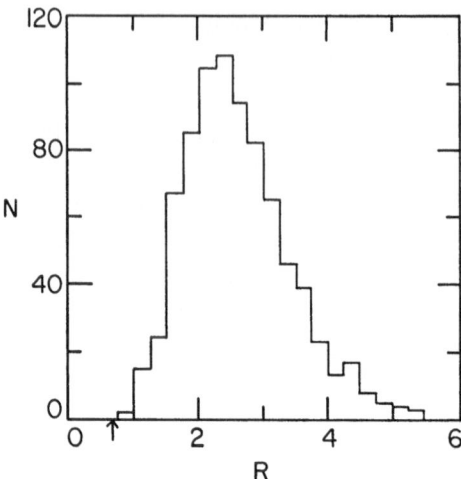

Fig. 11.   Distribution of radii of cloudlets. (From Heiles, 1967a.) Note the position of the arrow on the abscissa scale. The *R*-value indicated by the arrow represents the resolving power of the telescope used for the observations.

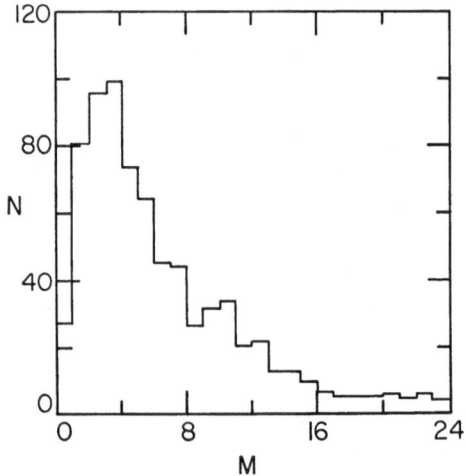

Fig. 12.   Distribution of masses of cloudlets. (From Heiles, 1967a.)

H_2CO, all having origin in a single gas concentration. They are not, of course, seen together in every source; various combinations occur, presumably because of physical conditions in the gas concentration. Anomalous emission from OH and H_2O appear to come from sources of very small size associated with H II regions in special situations. Absorption lines of OH and NH_3 appear together in a few sources; NH_3 is not widely observed. In some dark obscuring clouds of particulate material that strongly attenuates optical radiation, 'normal' OH emission is observed. It is possible that all such clouds are emitting and that the emission would be found if very low noise receivers were available for the observations. Some dark clouds show H_2CO in absorption. Several investigators have found the dark obscuring clouds to be highly deficient in H I 21-cm radiation; see Heiles, 1967a, b; Kerr and Garzoli, 1968; Varsavsky, 1968; and Mészáros, 1968. The most complete investigation is the recent one by Heiles (1969), who surveyed 48 dark clouds in H I emission. He found only one cloud that showed excess H I, and its 35 K emission he assumed to be saturated. Two clouds showed deficiencies in H I, while another showed a complex structure not immediately interpretable. All the rest (44 objects) did not show H I radiation.

A directly related observation of a broader character is the ratio of H I emission to dust, which has been studied by a number of investigators. Bok *et al.* (1955) early demonstrated that the H I/dust ratio decreased in the direction of dark clouds. (See also Garzoli and Varsavsky, 1966.) Pronik (1963) showed that for the Omega Nebula the H I/dust ratio decreased from normal by an order of magnitude.

It is reasonable to assume that in the dark clouds that have been observed, the abundances of the elements is reasonably normal (at least in general order of magnitude) and that physical conditions in the gas are primarily responsible for what spectral lines are observed. One may suppose, with Heiles and other investigators, that in the very dark clouds cooling has gone on to such an extent that, while some *molecules* are present and are obscured, the gas that would normally be seen as H I has been frozen down to H_2, hydrogen mantles on grains, or to some other form that eliminates it as a candidate for observation as H I.

The concentrations of gas in the arms of the Galaxy thus show a wide range of temperatures. At the low end of the temperature range are the dark clouds of particulate material in which molecules are observed but in which atomic hydrogen is not seen, and is presumably frozen down. At the high end of the temperature range are concentrations of gas showing H I excitation temperatures of a few hundred degrees. Preliminary observations indicate that these 'hot' concentrations do not show molecular lines. It is not known with any precision how the temperatures of the concentrations are distributed within this range of temperatures, though gas concentrations are observed throughout the whole temperature range. Different observing techniques emphasize different portions of the range. It is not known where the high-temperature cutoff for concentrations of gas occurs; it does appear that there is gas in the interarm region for which the excitation temperature is > 1000 K. There is, however, no indication that the interarm region contains concentrations of gas of the character observed to occur in large numbers in the arms of the Galaxy. Direct observational evidence relating to the temperature of the intercloud gas in the arms, that is, the continuum of gas in which the concentrations are located, is lacking.

H II regions show both optical and radio lines in emission. Can we find an H I region – that is, a single concentration of cool gas in an arm of the Galaxy – which shows both optical and radio lines? The optical lines would be seen as interstellar absorption lines. Hobbs (1969a, b) has shown that the Na I lines he measured are formed in H I regions; we expect such a situation to be true in general for optical interstellar absorption lines.) The radio lines from the H I region could be seen either in emission (H I, for example) or in absorption. Various investigators have stated that they have found equivalence between optical and radio lines (e.g., Herbig, 1968;

Riegel and Jennings, 1969), but the velocities of optical and radio lines discussed differ by one or two (or more) km sec^{-1}. Such a difference is greater than can be accepted for the association tests applied earlier, and does not constitute what would here be looked upon as proof of spatial coexistence of the radio and optical sources. We do not consider in this context agreement between an optical line velocity and the velocity from a 'gaussian component' of H I as indication of coexistence.

The most extensive test for spatial coexistence of radio and optical interstellar line sources was made by Habing (1969a, b). He investigated 48 cases, conclusively proving that the concentrations from which the optical interstellar lines originate are spiral tracers. (Essentially he used Test 2.) However, he found no normal case in which there is agreement (within observational errors) of optical interstellar velocity and the velocity of a specific H I concentration. (We here omit consideration of two high-velocity cloud cases which might be abnormal.) The lack of precise agreement between velocities derived from optical and radio interstellar lines in H I regions is surprising; its significance requires clarification.

G. VELOCITY DISPERSION AMONG SUB-UNITS OF AN ARM AND IN THE GAS
BETWEEN CONCENTRATIONS

In discussing velocity dispersion, one must be specific as to the type of unit or sub-unit of the Galaxy that is under discussion. In the past, on the basis of the standard cloud model of the Galaxy, reference has been made to 'internal dispersion' and 'external dispersion', that is, dispersion of velocities within a cloud or among clouds. The gaseous medium between the clouds was not considered, nor was the medium between arms. One might therefore wish to add other dispersions in a compilation as, for example, a dispersion estimate relating to the intercloud gas – in the terms we have used in this review, the gas between concentrations – in an arm, or a dispersion estimate relating to the gas in the interarm region.

Van Woerden (1967) has provided useful tabulations of many published values of internal dispersions and external dispersions which, in this discussion will, for precision, be termed 'dispersion within concentrations' and 'dispersion among concentrations'. We shall find, unfortunately, that observational data for any category of structure are not numerous and that much observational work remains to be done.

(1) It is probable that the dispersion within H I concentrations is related to the size of the concentration. Thus, in the H I in our local spur, Heiles (1967) finds:

for 'cloudlets' $\sigma = 0.6$ to 1.2 km sec^{-1},
                             mean $\sigma \approx 0.9$ km sec^{-1},
for 'larger concentrations' $\sigma = 1.2$ to 4.5 km sec^{-1},
                             mean $\sigma \approx 4.0$ km sec^{-1}.

These numerical values overlap older tabulated numerical values in which there has been mixing of a variety of types and sizes of concentrations located in many regions of the Galaxy. Quite frequently estimates of dispersion have been derived from analysis of profiles into gaussian components, little or no attention being given to the size or nature of gas structure. Caution is required if older values of dispersions are used.

Hobbs (1969a, b) investigated many interstellar Na I lines with an interferometer having a bandpass at half-intensity of 0.51 km sec^{-1}. Many lines were found to be multiple. For 15 well-observed single lines Hobbs found values of σ that ranged from 0.64 to 1.48 km sec^{-1}. If the kinetic temperature of the concentration of Na is $\leqslant 100$ K, thermal broadening of the lines is negligible and the measured velocity spread must represent turbulence almost completely. If, on the other hand, there is essentially no turbulence in the concentrations of Na I, or Ca II, the kinetic temperature of the concentrations must range from 1100 to 6100 K. (If one observation is omitted, the upper limit becomes 4400 K.)

The Na I lines observed by Hobbs and the H I 'cloudlets' observed by Heiles have similar dispersions. If the spectroscopic features reported originate in equivalent concentrations of gas, the gas temperature must be quite low. The observed dispersion of velocities would be representative largely of turbulence. However, the precise coincidence in space of, say, Na I and H I condensations remains an important question in need of further investigation.

(2) Very little numerical information on the dispersion among gas concentrations in major arms is available. Kerr (1964), observing H I with the 210-foot antenna in Australia, found, for the outer arm of the Galaxy, mean arm velocity changes of as much as 20 km sec^{-1} in angular distances of 1°. This could indicate rather high dispersion values, at least in some volume elements in an arm.

Hat Creek observations of the velocities among gas concentrations in the major arms indicate that the velocity distribution may not be completely random. Rather, there appears a tendency for parts of arms to split in velocity by 10 km sec^{-1} or more over some angular range, then to coalesce. Much material relating to this question is available; investigations are in progress.

Most data relevant to the dispersion of velocity among gas concentrations refer to the situation within the local spur. Blaauw (1952) investigated the K-line velocities of Ca II measured by Adams. Blaauw showed that an exponential law fits the observed distribution of velocities more closely than a gaussian distribution. From K-line velocities he derived a value of $\eta = 5$ km sec^{-1} ($\sigma = 7$ km sec^{-1}) for the exponential distribution. The value found by Blaauw may represent an upper limit to the dispersion among the gas concentrations that give rise to optical lines; there is some question as to the value of the zero velocity with respect to which individual peculiar velocities should be measured. Takakubo (1967), using gaussian component analysis of the local H I as the basis of his discussion, found $\sigma = 5.6$ km sec^{-1} and noted that this value changes with position on the sky and other properties of the population investigated. The physical significance of these variations is unclear.

(3) The dispersion of velocities frequently quoted for 'the motions of interstellar clouds' is 6 km sec^{-1} (Westerhout, 1957). This value probably refers more nearly to the dispersion of the intercloud gas in the local spur than it does to the motions of concentrations of gas within the local spur. This frequently quoted value was deter-

mined by Pottasch (see Westerhout, 1957), who fitted gaussian curves to the 'forbidden velocities' (that is, the positive wings of the velocity distribution in the l^I-range 60° to 100°, latitudes $b^I = 0°$, $\pm 2.5°$) to find $\sigma = 5.8$ to 9.7 km sec^{-1}, with $\langle\sigma\rangle \approx 6.0$ km sec^{-1}. A similar mean value and range was found in other longitude zones.

In view of the differences in the characters of the gas structures and the natures of the motions investigated, there should be no surprise that Blaauw and Pottasch found different mathematical forms for the distributions they observed. Blaauw found an exponential law most satisfactory for representing the distribution of Ca II velocities,

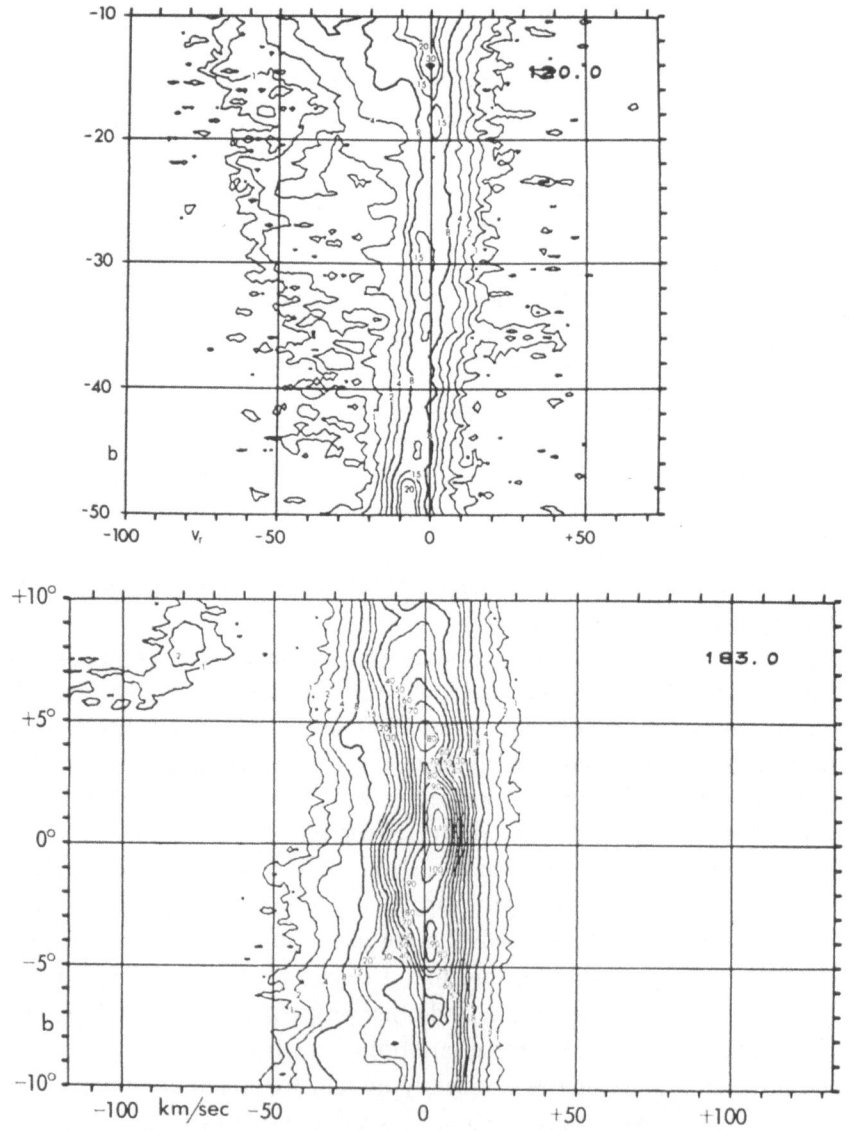

Fig. 13.   Sample contour diagrams of $T_A(v_r, b \mid l)$ showing the preponderance of negative velocities among the low-intensity gas.

while Pottasch determined that a gaussian function provided a better fit to the distribution of neutral hydrogen forbidden velocities.

(4) Systematic information relating to the velocity dispersion of the interarm gas is essentially lacking. From the scanty observational evidence a few trends may be visible.

(a) The dispersion velocity of the interarm gas is higher than that in the spiral arms, and may change from one part of the Galaxy to another in a random way.

(b) In the interarm region (probably more accurately in the region of an arm/

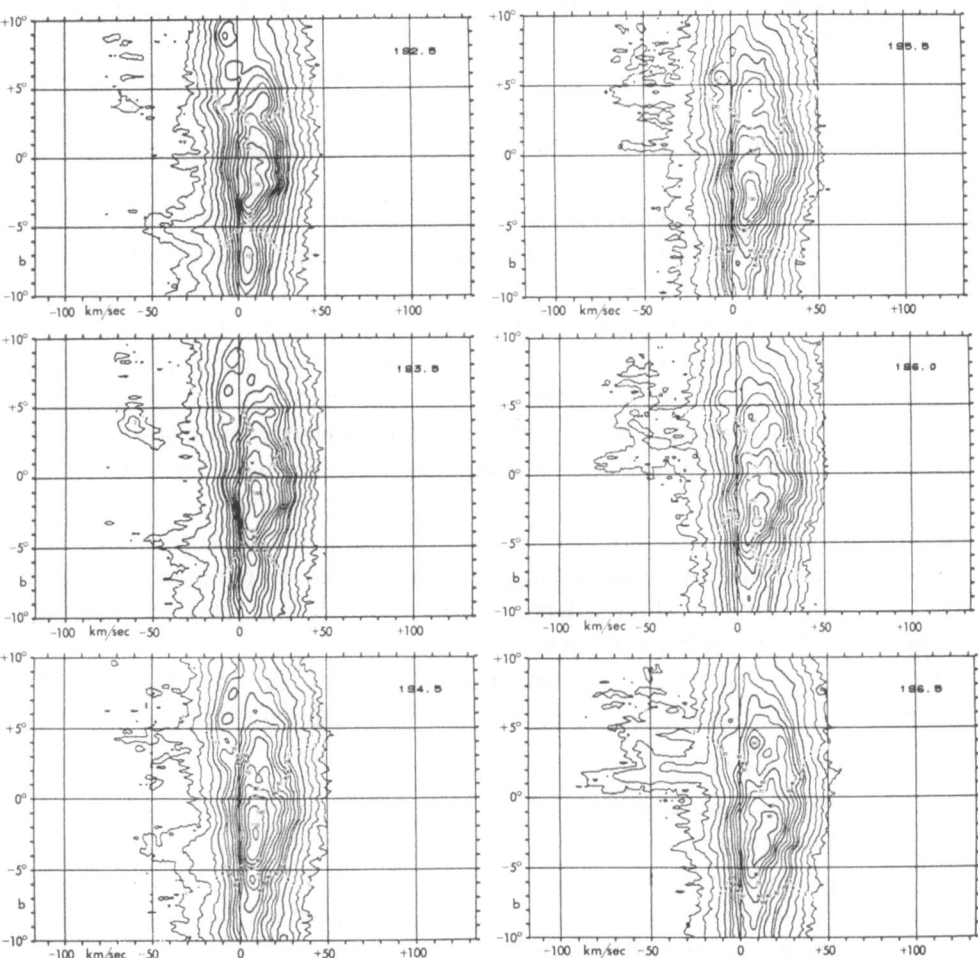

Fig. 14.   Sample of contour diagrams illustrating the existence of 'active regions' in the Galaxy.
Note the jet-like structures at longitude 196.5.

interarm interface) the distribution of velocities may not be isotropic; there is normally a negative velocity tail.

The anisotropy of the observed distribution becomes a dominant factor as we go to
higher velocities and lower intensities. Negative velocities are preponderant. The effect
is not subtle and it must carry important information about the interstellar medium
that we do not yet fully appreciate. Two contour maps (Figure 13) chosen almost at
random will illustrate the point. The negative tails in the velocity distributions are
very pronounced.

Discussion of negative velocities leads logically to the question of high-velocity gas
and the problems related to its nature: is it galactic or extragalactic? Mrs. Dieter
(1969) has recently reviewed the observational aspects of the high-velocity gas; her
review is strongly recommended for further information on this subject. No attempt
will be made in this review to present an abstract of her discussion. Rather, we conclude these remarks on the motion properties of the gas in the Galaxy with a few
details of phenomena that should be of particular interest to members of this conference.

Most studies of high-velocity gas have been in latitudes $>15°$. The extensive
observations in the latitude range $\pm$ 10° recently completed at Hat Creek provide
much information on high-velocity gas in these lower latitudes. We find:

(1) There is much intermediate-velocity and high-velocity gas (preponderantly of
negative sign) in the galactic plane. It is not in the form of isolated clouds or concentrations; it is connected to the lower-velocity gas in the Galaxy.

(2) There is much greater tendency for intermediate-velocity and high-velocity gas
of high intensity to be present in some longitudes than in others. There are 'active'
regions in the Galaxy.

(3) High-velocity features are often jet-like in nature. Jet-like here means that a
mass of gas occupying a very limited angular extent in the sky (an area one to a few
degrees in diameter) will show a large range in velocity extending, perhaps from
-100 km sec^{-1} to the velocity of the normal galactic gas.

An example of a jet-like gas structure mentioned in (3) is shown in Figure 14. The
strong jet visible particularly at $l^{II} = 196°.5$ has the following characteristics:

Effective diameter	$\approx 2°$
Estimated distance	250 pc
Total mass	185 $M_\odot$
Total kinetic energy	5×10^{48} erg
Column density of H I	4×10^{20} H atom cm^{-2}
Average number density of H	$\leqslant 10$ H atom cm^{-3}.

For comparison, we may note that the energy expended by an average nova is 10^{45}
to 10^{46} erg, while that expended by an average supernova is perhaps 10^{50} to 10^{51} erg.

(In the presentation before the Symposium, a computer-produced movie, made
from the Hat Creek observations, was used to illustrate galactic active regions and
jet-like gas structures.)

References

Barrett, A. H., Meeks, M. L., and Weinreb, S.: 1964, *Nature* **202**, 475.

Blaauw, A.: 1952, *Bull. Astron. Inst. Netherl.* **11**, 459.

Bok, B. J., Lawrence, R. S., and Menon, T. K.: 1955, *Publ. Astron. Soc. Pacific* **67**, 108.

Cheung, A. C., Rank, D. M., Townes, C. H., Thornton, D. D., and Welch, W. J.: 1968, *Phys. Rev. Lett.* **21**, 1701.

Cheung, A. C., Rank, D. M., Townes, C. H., Knowles, S. H., and Sullivan, W. T. III: 1969a, *Astrophys. J. Lett.*, **157**, L13.

Cheung, A. C., Rank, D. M., Townes, C. H., Thornton, D. D., and Welch, W. J.: 1969b, *Nature* **221**, 626.

Cheung, A. C., Rank, D. M., Townes, C. H., and Welch, W. J.: 1969c, *Nature* **221**, 917.

Clark, B. G.: 1965, *Astrophys. J.* **142**, 1398.

Davies, R. D.: 1956, *Monthly Notices Roy. Astron. Soc.* **116**, 443.

Davies, R. D.: 1958, *Rev. Mod. Phys.* **30**, 931.

Dieter, N. H.: 1969, *Publ. Astron. Soc. Pacific* **81**, 186.

Field, G. B., Goldsmith, D. W., and Habing, H. J.: 1969, *Astrophys. J. Lett.* **155**, L149. See also Garzoli, S. L. and Varsavaky, C. M.: 1966, *Astrophys. J.* **145**, 79.

Goldsmith, D. W., Habing, H. J., and Field, G. B.: 1969, *Astrophys. J.* **158**, 173.

Goss, W. M.: 1968, *Astrophys. J. Suppl. Ser.* **15**, 131.

Habing, H. J.: 1969a, *Bull. Astron. Inst. Netherl.* **20**, 120.

Habing, H. J.: 1969b, *Bull. Astron. Inst. Netherl.* **20**, 177.

Heeschen, D. S.: 1955, *Astrophys. J.* **121**, 569.

Heiles, C.: 1967a, *Astrophys. J. Suppl. Ser.* **15**, 97.

Heiles. C.: 1967b, *Astrophys. J.* **148**, 299.

Heiles, C.: 1968, *Astrophys. J.* **151**, 919.

Heiles, C.: 1969, *Astrophys. J.* **156**, 493.

Herbig, G.: 1968, *Z. Astrophys.* **68**, 243.

Hobbs, L. M.: 1969a, *Astrophys. J.* **157**, 135.

Hobbs, L. M.: 1969b, *Astrophys. J.* **157**, 165.

Howard, W. E., Wentzel, D. G., and McGee, R. X.: 1963, *Astrophys. J.* **138**, 988.

Kerr, F. J.: 1964, in *IAU Symposium No. 20, The Galaxy and the Magellanic Clouds* (ed. by F. J. Kerr and A. W. Rodgers), Australian Academy of Sciences, Canberra, Australia, p. 81.

Kerr, F. J.: 1969, *Rev. Astron. Astrophys.* **7**, 39.

Kerr, F. J. and Garzoli, S.: 1968, *Astrophys. J.* **152**, 51.

Knowles, S. H., Mayer, C. H., Cheung, A. C., Rank, D. M., and Townes, C. H.: 1969, *Science* **163**, 1055.

Lin, C. C.: 1967, *Ann. Rev. Astron. Astrophys.* **5**, 453 (other earlier references will be found in this publication).

Mebold, U.: 1969, *Beitr. Radioastron.* **1**, 97 (Max-Planck-Institut für Radioastronomie Bonn).

Meeks, M. L., Carter, J. C., Barrett, A. H., Schwartz, P. R., Waters, J. W., and Brown, W. E. III.: 1969 (in press).

Mészáros, P.: 1968, *Astrophys. Space Sci.* **2**, 510.

Münch, G.: 1965, in *Galactic Structure* (ed. by A. Blaauw and M. Schmidt), University of Chicago Press, Chicago, p. 203.

Palmer, P., Zuckerman, B., Buhl, D., and Snyder, L. E.: 1969, *Astrophys. J. Lett.* **156**, L147.

Pronik, I. I.: 1963, *Izv. Krymsk. Astrofiz. Observ.* **30**, 118.

Radhakrishnan, V. and Murray, J. D.: 1969, *Proc. Astr. Soc. Australia* **1**, 215.

Riegel, K. W. and Jennings, M. C.: 1969, *Astrophys. J.* **157**, 563.

Roberts, Jr., W. W.: 1969, *Astrophys. J.* **158**, 123.

Roberts, Jr., W. W.: 1970, in *IAU Symposium No. 38, The Spiral Structure of Our Galaxy* (ed. by W. Becker and G. Contopoulos), Reidel, Dordrecht, The Netherlands, p. 415.

Rydbeck, O. E. H., Ellder, J., and Kollberg, E.: 1969 (in press).

Sandage, A. R.: 1961, *The Hubble Atlas of Galaxies*, Carnegie Institution of Washington.

Schmidt, M.: 1957, *Bull. Astron. Inst. Netherl.* **13**, 247.

Schmidt, M.: 1965, in *Galactic Structure* (ed. by A. Blaauw and M. Schmidt), University of Chicago Press, Chicago, p. 513.

Schmidt-Kaler, T.: 1964, *Trans. IAU* **12B**, 416.

Shuter, W. L. H. and Verschuur, G. L.: 1964, *Monthly Notices Roy. Astron. Soc.* **127**, 387.

Snyder, L. E., Buhl, D., Zuckerman, B., and Palmer, P.: 1969, *Phys. Rev. Lett.* **22**, 679.

Takakubo, K.: 1967, *Bull. Astron. Inst. Netherl.* **19**, 125.

Turner, B. E.: 1969, *Astrophys. J.* **157**, 103. See also Ball, J. A. and Staelin, D. H.: 1968, *Astrophys. J. Lett.* **153**, L41.

Van de Hulst, H. C., Muller, C. A , and Oort, J. H.: 1954, *Bull. Astron. Inst. Netherl.* **12**, 117.

Van Woerden, H.: 1967, in *IAU Symposium No. 31, Radio Astronomy and the Galactic System* (ed. by H. van Woerden), Academic Press, London, p. 3.

Varsavsky, C. M.: 1968, *Astrophys. J.* **153**, 627.

Weaver, H.: 1970, in *IAU Symposium No. 38, The Spiral Structure of Our Galaxy*, Basel, Switzerland (ed. by W. Becker and G. Contopoulos), Reidel, Dordrecht, The Netherlands, p. 126.

Weaver, H. and Williams, D. R. W.: 1970 (in preparation).

Weaver, H., Dieter, N. H., and Williams, D. R. W.: 1968, *Astrophys. J. Suppl. Ser.* **16**, 219.

Westerhout, G.: 1957, *Bull. Astron. Inst. Netherl.* **13**, 201.

Williams, D. R. W.: 1969, private communication.

Wilson, W. J. and Barrett, A. H.: 1968, *Science* **161**, 778, 23.

4. THEORETICAL DESCRIPTION OF THE INTERSTELLAR MEDIUM

Introductory Report

(Tuesday, September 9, 1969)

GEORGE B. FIELD*

Dept. of Applied Mathematics and Theoretical Physics, Cambridge University, Cambridge, U.K.

1. Introduction

As excellent reviews have appeared recently (Spitzer, 1968a; Pikel'ner, 1968a) we shall consider here only some special topics where progress is being made – thermal properties of H I and H II clouds, and shock waves. The components of the medium which are dynamically important are neutral atoms, charged ions and electrons, grains, cosmic rays, magnetic field, and light. In Table I we indicate the degree of

TABLE I

Momentum coupling between components
(S = Strong, M = Medium, W = Weak)

Component	Atoms	Ions	Grains	Cosmic rays	Light	Magnetic field
Atoms	S	S	M	W	W	W
Ions	S	S	M	W	W	S
Grains	M	M	W	W	M	M
Cosmic rays	W	W	W	W	W	S
Light	W	W	M	W	W	W
Magnetic field	W	S	M	S	W	–

momentum coupling among these components. The main mass of the medium, in ions and atoms, is well coupled through collisions, with a mean free path $\lambda \approx 10^{-3}$ pc, and may usually be treated hydrodynamically. Parker (1965) has shown that for many purposes the cosmic-rays can be considered as a separate gas, strongly coupled to the magnetic field. The coupling of the atom-ion gas to the magnetic field is through the magnetic forces on the ions, which transmit the forces to the neutral atoms by collisions unless the differential stresses on the two components are unusually large. Grains are not coupled strongly to any other component, as their stopping distances are of the order of parsecs and their gyro-periods of the order of 10^4 yr. The coupling of grains to radiation pressure can lead to interesting differential effects. Light is important in carrying away thermal energy but does not critically affect the momentum balance of the gas. A possible exception is the Ly-α radiation field discovered by Kurt and Sunyaev (1967) which has a pressure of 10^{-13} dyne cm^{-2}. If this phenomenon is common in interstellar space, it could have effects.

* Usual address: Dept. of Astronomy, University of California, Berkeley, Calif., U.S.A.

Habing (ed.), Interstellar Gas Dynamics, 51–76. All Rights Reserved. Copyright © 1970 by IAU

Most work has been on the large-scale motion of the atom-ion gas, includes magnetic stresses, and assumes that the magnetic field is frozen in. Such studies include equilibrium configuration in the galactic magnetic field, cloud collisions, expansions of HII regions and supernova shells, and perturbation calculations aimed at discovering instabilities (gravitational, thermal, hydromagnetic). Many of these calculations have paid insufficient attention to the effects of cosmic-ray pressure. Similarly, insufficient work has been done on the dynamical effects of radiation pressure on grains.

2. Thermal Properties of HI Regions and Formation of Clouds

Spitzer calculated thermal equilibrium conditions in the 1940's. He concluded that the main heating is due to photo-electrons ejected from carbon atoms, and that the main cooling is by infrared emission following collisional excitation of the fine-structure levels of various ions. Spitzer's equilibrium temperatures (≈ 20 K) are too low to explain the 21-cm result that $\langle T^{-1} \rangle^{-1} \approx 100$ K, therefore Kahn (1955) suggested that the temperature may be maintained by shocks associated with supersonic cloud collisions. We shall see below that this mechanism appears inadequate to maintain the temperature. Hayakawa *et al.* (1961) showed that if low-energy (1 to 100 MeV) cosmic rays are present in sufficient numbers, the heating by the secondary electrons they produce would explain the temperatures of clouds. Field (1962) showed that the observed cosmic rays with $E > 1$ GeV would heat any low-density gas to 10^4 K or above, so that there could be pressure equilibrium between two phases: a rarefied, hot intercloud medium and dense, cool clouds. This explanation of the stability of clouds had been suggested by Spitzer (1951). Pikel'ner (1967) revived the discussion, showing that the heating by cosmic rays and the partial ionization of otherwise neutral HI explains various observations, including the existence of hot HI (21-cm) regions, Faraday rotation, and long-wavelength radio absorption. Pikel'ner included the heating due to Coulomb collisions of cosmic rays with ambient thermal electrons and discussed extensively the possibility and the consequences of the existence of three thermal phases in the interstellar gas: a stable, hot phase of low density ($n \lesssim 10^{-1}$ cm^{-3}, $T \gtrsim 6000$ K), a stable, cool phase of high density ($n \gtrsim 1$ cm^{-3}, $T \lesssim 300$ K), and an intermediate, thermally unstable phase in which $dp/d\varrho < 0$.

Spitzer and Tomasko (1968) investigated the efficiency of cosmic-ray heating by calculating the energy put into the secondary electrons and subtracting the energy lost by the secondaries through excitation of optical transitions. Later, Spitzer and Scott (1969) (to be called SS) extended these calculations to include elastic collisions between the secondary electrons and the ambient thermal electrons, and Coulomb collisions between cosmic rays and ambient electrons.

Field *et al.* (1969) (FGH) and Goldsmith *et al.* (1969) using a different method, also included the efficiency of the secondaries in their calculations and calculated equilibrium temperatures (heating = cooling) using cooling processes and cross-sections somewhat different from those of Pikel'ner and of Spitzer and Tomasko.

They found results qualitatively similar to Pikel'ner's and showed that Pikel'ner's model also explains the electron density observed by pulsar dispersion measurements. They concluded that if agreement with cloud temperatures is to be obtained, depletion of cooling elements on grains is to be postulated. The latter effect might then explain the low abundance of sodium and calcium as deduced from interstellar absorption-line observations. Hjellming *et al.*, (1969) (HGG), using somewhat different cooling processes, applied the efficiency calculations of Spitzer and Scott and obtained thermal equilibria in agreement with FGH. Silk and Werner (1969) have suggested that ionization by X-rays <1 keV could supplement or replace low-energy cosmic rays as a heating and ionizing agent.

Figure 1 shows the amount of energy delivered to the gas and the total number of electrons produced per primary ionization as a function of n_e/n. The values of SS are the most reliable. Figure 2 shows the equilibrium temperatures derived. The results have all been normalized to a particular value of the primary ionization rate ζ of hydrogen found by FGH to be in good agreement with observation. (FGH showed that all processes are proportional to this function of the cosmic-ray spectrum, so that the detailed form of the latter, which is unknown, is irrelevant). Three of the calculations agree fairly well in view of the somewhat different assumptions about ζ_{eff} and

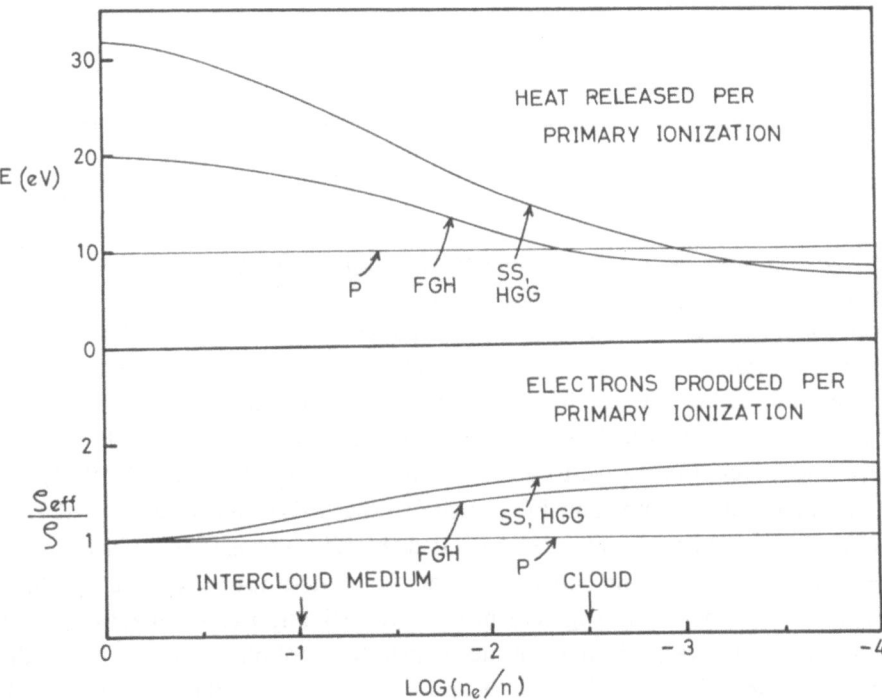

Fig 1. Heat released and number of electrons produced per primary ionization of a neutral hydrogen atom. P = Pikel'ner (1967), FGH = Field, Goldsmith, and Habing (1969), SS = Spitzer and Scott (1969), HGG = Hjellming, Gordon, and Gordon (1969). ζ is the rate of primary ionization per neutral hydrogen atom; ζ_{eff} includes in addition ionization by energetic secondary electrons.

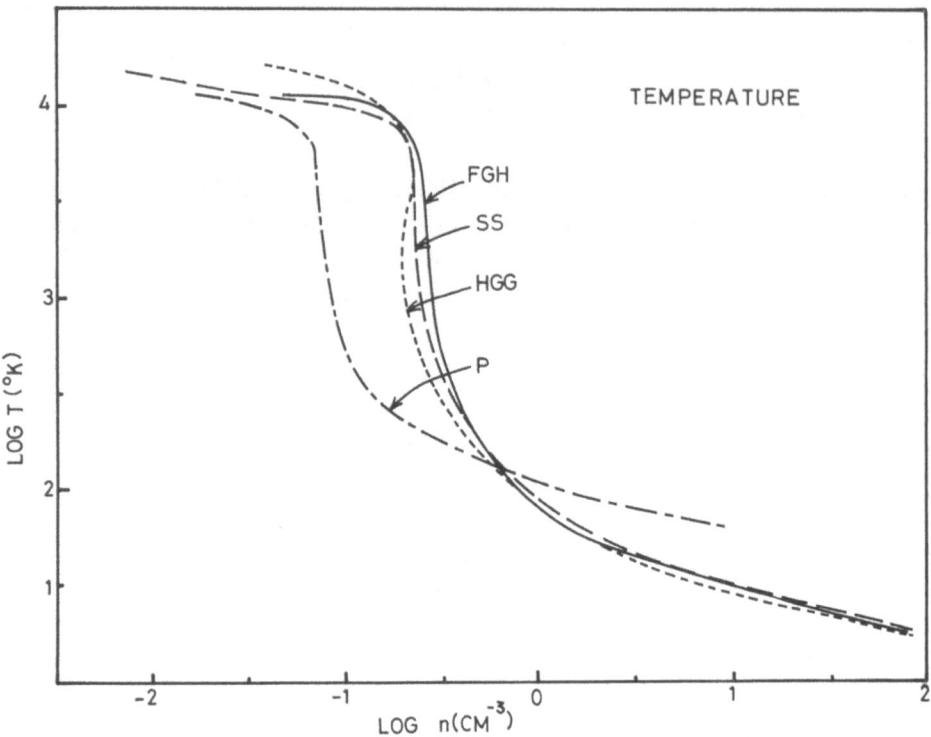

Fig. 2.   Temperature for which the cosmic ray heating equals the cooling by inelastic, thermal collisions as a function of density n. Results of various authors are normalized to one value of $\zeta (= 4 \times 10^{-16} \text{ sec}^{-1})$. See Figure 1 for explanation on the names of the authors.

ΔE (Figure 1), and about the cooling mechanisms. The results of Pikel'ner agree qualitatively, and the quantitative differences may probably be explained by the fact that certain of his cross sections were only estimated.

The curves can be understood qualitatively as follow. As n drops, the cooling $(\propto n^2)$ decreases faster than the heating $(\propto n)$ and T, which influences only the cooling, must rise to compensate for the difference. In the range 10^3 to 5×10^3 K, fine-structure cooling is relatively insensitive to T, and T must rise rapidly to about 10^4 K, where Ly-α cooling becomes important. FGH included the ^3P-^1D transition of OI at 1.96 eV $(E/k = 22\,800\,\text{K})$, while SS and HGG included the ^3P-^1D transition of NII at 1.90 eV $(E/k = 22\,100\,\text{K})$ as well. The conclusion that NII should not contribute significantly (FGH) is strengthened by the agreement of HGG, SS, and FGH. All authors omitted what could possibly be an important mechanism for cooling the intercloud medium, i.e., electron excitation of a $^4F_{9/2}$-level of Fe^+ with $E/k = 2700\,\text{K}$. Pottasch (1968) estimates $\Omega = 15$ for this transition, and if Fe/O $= 0.1$ (Pottasch, 1963; Swings, 1965; Gary *et al.*, 1969), rather than the 0.01 assumed by FGH, we find that this transition should be quite important in the range of several thousand degrees and would thus depress the temperature of the intercloud medium (see below).

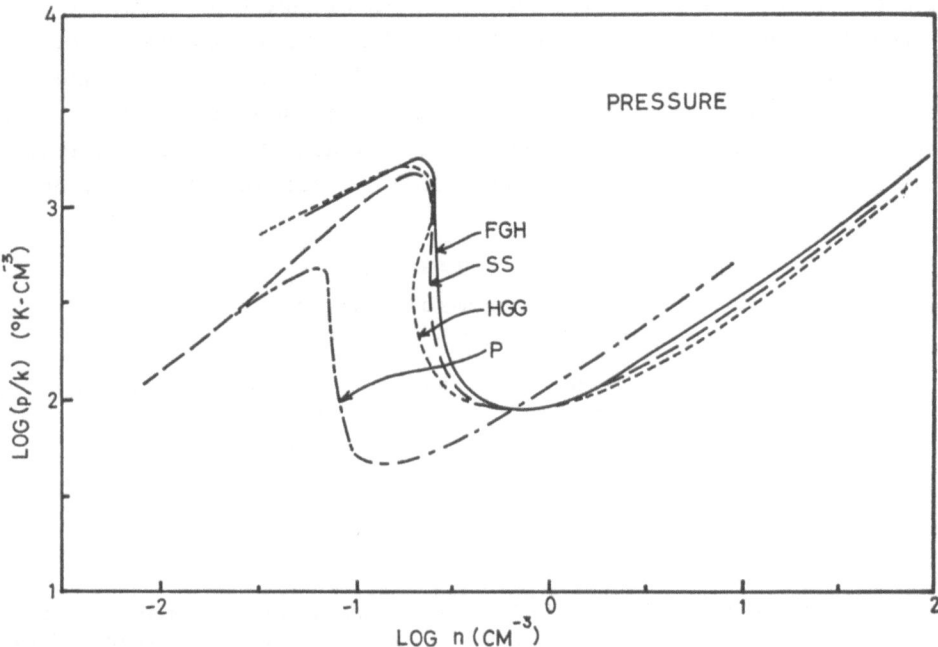

Fig. 3.   The pressure of the interstellar medium as a function of total nuclear density. See Figures 1 and 2 for explanation.

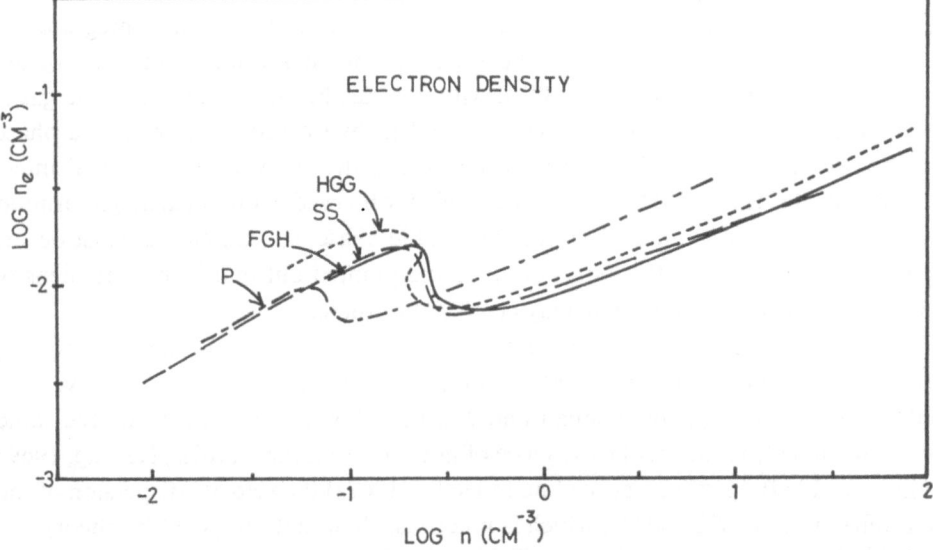

Fig. 4.   Electron density in the interstellar medium as a function of total nuclear density. See Figures 1 and 2 for explanation.

Figure 3 shows the gas pressure $p/k = (n + n_e)T$; and Figure 4, the electron density n_e. The latter is about 10^{-2} cm^{-3}, in agreement with observations of pulsar dispersion and low-frequency radio absorption. In every case there is a pressure maximum; if gas in the high-temperature phase at the left of the maximum were to be slowly compressed, no equilibrium could be found beyond the critical point whose parameters (p_c, n_c, T_c) are listed in Table II. FGH interpret this to mean that in our local neigh-

TABLE II

Parameters of critical point
($\zeta = 4 \times 10^{-16}$ sec^{-1})

Author	n_c(cm^{-3})	T_c(K)	p_c(cm^{-3}K)	n_{ec}(cm^{-3})
P	0.062	7000	430	0.009
FGH	0.21	7800	1800	0.016
SS	0.21	7000	1500	0.015
HGG	0.19	9000	1700	0.016

References: P = Pikel'ner (1967); FGH = Field, Goldsmith, and Habing (1969); SS = Spitzer and Scott (1969); HGG = Hjellming, Gordon, and Gordon (1969).

borhood gas must occur in the low-temperature phase. This gas forms the cool clouds that produce at 21 cm all of the absorption lines and part of the emission lines. The argument is as follows. Consider a particular tube of force, which on the Parker (1968a) model of the magnetic field (see below) loops out of the plane. Along the tube we must have hydrostatic equilibrium in the galactic gravitational field, K_z. At large z, the density is low and the gas is in the hot stable phase. If one can integrate the hydrostatic equilibrium equation to the lowest point on the flux tube without exceeding p_c, the entire tube could be filled with hot gas, but if the weight of the gas is too great, one will reach p_c first. Since $dp/dz < 0$ in hydrostatic equilibrium, a phase transition to the stable cool phase must occur and the gas will form normal interstellar clouds. FGH showed that the amount of gas observed in our local neighborhood is too great to be accommodated in the hot stable phase, so that clouds must occur. Pikel'ner (1967) also took this point of view, and pointed out that if the total density between arms is much lower, there may be no clouds there.

This model predicts the existence of a rarefied, hot H I intercloud medium. Recent observations of the 21-cm line in absorption confirm this prediction and give $T > 1000$ K (Clark 1965; Radhakrishnan and Murray, 1969). Observations of the same line in emission also indicate the existence of hot H I gas (Heiles, 1967a; Habing, 1969; Grahl et al., 1969). However, according to Heiles (1967a) the velocity dispersion of the line profile indicates $T \leqslant 4000$ K, which is much less than that predicted by theory for the stable hot phase ($T > 8000$ K, see Table II). Mechanisms, such as iron cooling, which might reduce T_c (the temperature at the critical point), are therefore of particular interest. HGG suggest that a way out of this problem is to take the intercloud medium to be the intermediate unstable phase. We have two objections to this procedure. Away from the plane, where K_z is significant, hydrostatic equilibrium demands that $dp/dz < 0$, and since $(dp/d\varrho) < 0$ in the intermediate phase, a region containing this phase would have $(d\varrho/dz) > 0$, which is Rayleigh-Taylor unstable (Layzer, 1967) on a time scale $(dK_z/dz)^{-1/2} = 1.3 \times 10^7$ yr. Even if z is small, and $K_z \propto z$, gas in the intermediate phase is thermally unstable (Field, 1965). According to Pikel'ner (1967), intermediate-phase gas may be present while the long wavelengths required to make interstellar clouds are developing, since the time scale for development of long wavelengths is about $\lambda/c \gg$ thermal time scale (c is the sound speed). However, this overlooks the fact that shorter wavelengths will develop faster, namely on the thermal time scale, and it has been shown (Goldsmith, 1969) that this leads to a transition to stable phases in about 10^6 yr.

We now consider an alternative model proposed by Parker (1966). Parker (1968b, 1969) gives references to the many papers by him and Lerche on this model. In 1960 Hoyle and Ireland pointed out that differential rotation of the Galaxy will double the disk magnetic field in about 4×10^8 yr. Since $B^2/(8\pi\varrho z)$ is comparable with K_z, the support of the gas against gravitation is partly due to magnetic forces, and the increase of B will lead to an increase in z. They argued that this process will not be uniform but will proceed fastest where the density is lowest. Further, the process is unstable (Hoyle and Harwit, 1962); since the gas will run from the elevated regions into the depressed regions to form clouds, the elevated regions will be able to expand even faster.

Parker also shows mathematically that, even if one starts with a relatively smooth field, constant in time, and parallel to the disk, the system is unstable toward perturbations in which the field acquires a sinusoidal z-component. The kinetic energy of the unstable gas derives from three sources: the gravitational energy of the gas sliding down the line of force, the magnetic energy released by the upwelling field lines, and the cosmic-ray energy released by the expansion of the field. The effect of gas pressure depends on the sign of $(\gamma - 1)$, where $\gamma = d(\log p)/d(\log \varrho)$ characterizes the thermal behavior of the gas. $\gamma - 1 < 0$ implies destabilization.

Parker's results for the case $K_z = $ constant are easiest to relate, and he showed that they are approximately correct for the more realistic case in which $K_z \propto z$. He finds the following instability criterion:

$$[\tfrac{1}{2}\alpha + \beta + (\alpha + \beta)^2] - [(\gamma - 1)(1 + \tfrac{3}{2}\alpha + \beta)] - 2\alpha\gamma k^2 L^2 > 0, \tag{1}$$

where $\alpha = $ magnetic pressure/gas pressure, $\beta = $ cosmic-ray pressure/gas pressure, $\gamma = d(\log p)/d(\log \varrho)$, $L = $ scale height $= (dp/d\varrho)(1 + \alpha + \beta)/|K_z|$, and $k = $ horizontal wave number of the disturbance along the magnetic field. The first term shows that magnetic and cosmic-ray pressures are always destabilizing. The second shows that gas pressure is also destabilizing if $\gamma < 1$; otherwise it is stabilizing, because a hot gas will resist collapse down the field line more than a cool one. The last term shows that if the gas is thermally stable ($\gamma > 0$) short wavelengths tend to be more stable because of the restoring forces associated with strong gas-pressure gradients in short-wavelength disturbances. If, on the other hand, the gas is thermally unstable the effect of gas pressure is destabilizing, whatever the wavelength.

Parker has discussed a model in which $\alpha = \beta$ (equipartition between cosmic rays and magnetic field, $\beta \approx 3.5\ \mu G$). This model may be close to reality because Anand *et al.* (1968) have demonstrated that proper accounting for solar modulation of energetic electrons lowers the field estimated from the synchrotron emission of the Galaxy to $6\ \mu G$ (from the old value, $20\ \mu G$), in harmony with the values 3 to $6\ \mu G$ from Zeeman and Faraday rotation measurements. Equation (1) then reveals that the medium is unstable for the conditions shown in Table III. Here k_c is a critical wave number

TABLE III

Conditions for Parker instability
($K_z = $ const., cosmic-ray $=$ magnetic pressure)

$\gamma = \mathrm{d}\log p/\mathrm{d}\log\varrho$	α	k
$\gamma \leqslant 0$	All	All
$0 < \gamma < 1$	All	$k < k_c$
$1 < \gamma$	$\alpha > \frac{2}{3}(\gamma - 1)$	$k < k_c$

obtained by replacing in Equation (1) the sign $>$ by $=$. We may apply to Parker's model the calculation of pressure given previously, using the results of FGH, SS, and HGG as an example. These curves are consistent with various observations to a factor of 2, provided one takes $B \approx 3\ \mu G$ in the intercloud medium. From Figure 3 we see that $\gamma > 0$ above $n = 0.8$ nuclei cm^{-3} ($n_H = 0.7$ cm^{-3} when 10 percent helium is taken into account). Since this is close to the observed mean density of the gas, one might consider a model in which all the gas is in the stable cool phase, close to the pressure minimum, with magnetic pressure (5×10^{-13} dyne cm^{-2}) much greater than gas pressure (1.3×10^{-14} dyne cm^{-2}). This would correspond to $\alpha = 40$. Such a disk can be supported in equilibrium by the pressure of magnetic field and cosmic rays, as Parker showed, but Table III shows that the disk is unstable for all wavelengths $\lambda > \lambda_c \approx 2\pi L\sqrt{(2\alpha\gamma)/(1+2\alpha)}$. The value of γ is quite sensitive to the exact value of n, but if we take $\gamma = 0.1$ as an example, $\lambda_c = 33$ pc.

Parker analyzed the situation in which γ is constant. However, as the density changes the pressure will change in a much more complicated way. In fact, since the thermal relaxation time scale is short (10^5 to 10^6 yr) compared to the dynamical time scale (10^7 yr), the gas will either remain close to a stable phase, or, if led into an unstable regime, make a rapid phase transition out of it. While no detailed calculations have been done, one can imagine the following sequence of events as the dynamical instability develops.

Suppose we start with a horizontal flux tube containing a uniform gas in conditions close to that of the pressure minimum in Figure 3. If the flux tube begins to develop toward a configuration like that in Figure 5, its volume will expand. Since the gas cannot leave the tube, the mean density has to drop. This drop will bring the gas in the

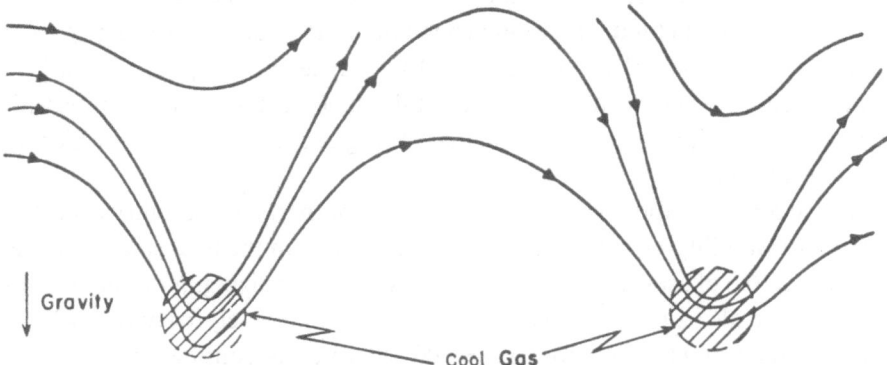

Fig. 5. Sketch of the local state of the lines of force of the interstellar magnetic field and interstellar gas-cloud configuration resulting from the intrinsic instability of a large-scale field along the galactic disk or arm when confined by the weight of the gas (Parker, 1966).

unstable phase. One may suppose that some part of the gas will expand and some part will contract. The contracting gas loses buoyancy and runs down the flux tube. In the expanding gas thermal instability develops rapidly (10^6 yr). Regions less dense than average (presumably those near the top of the flux tube) are below the equilibrium line of Figure 3, and gain heat from the cosmic rays faster than they can radiate it. This increases the pressure and the rarefied gas expands. If one assumes that the gas, which started contracting at the beginning of the expansion occupied about half the volume of the flux tube the density in the hot gas can only fall by a factor of about 2. The hot gas will therefore continue to increase in pressure, and stabilize only when it has reached the hot stable phase at $n \approx 0.2$ cm^{-3}, $p \approx p_c = 2.5 \times 10^{-13}$ dyne cm^{-2}. As the flux tube continues to rise upward, hydrostatic equilibrium gives $p = p_0 \exp(-z/L)$ in the approximation $K_z = $ constant. Here $L[=(\mathrm{d}p/\mathrm{d}\varrho)/K_z \simeq 100$ pc] is the scale height in the hot stable phase. It seems likely that p_0, the pressure at the bottom of the field line, remains close to p_c. This pressure, according to Figure 3, requires $n = 100$ cm^{-3}, $T = 18$ K. While some clouds do have such characteristics, notably those that stand out in 21-cm absorption observations, most clouds are probably hotter and less dense, say 120 K and 15 cm^{-3}. FGH showed that depletion of cooling elements by accretion on grains can account for the increased temperature. A theory of depletion by preferential sticking of charged species with high binding energy to a dielectric grain shows that this takes about 10^8 yr in a cloud having densities of the order of 10 cm^{-3} (Field and Zachariades, 1969).

The final appearance of the system is as shown in Figure 5 (Parker, 1966). The cloud is supported against galactic gravitation by magnetic stresses much as are solar filaments in the theory of Kippenhahn and Schlüter (1957). The detailed magnetic configuration within the cloud is of interest. Pikel'ner (1967) argued that the curvature of the field lines is small, the radius of curvature being some 150 pc. However, if his argument is modified for smaller B and larger density, we obtain a radius of curvature of only a few pc, and this seems too small for the following reason. The scale height of the cool stable phase in the galactic gravitational field at $z=100$ pc is only 1 pc, and in the 10 pc radius of a cloud, one would encounter a Δz of about 3 scale heights. The intercloud pressure would then be only 1.2×10^{-14} dyne cm^{-2}, $n_H < 10^{-2}$ cm^{-3}, and $n_e < 5 \times 10^{-3}$ cm^{-3}. There would be significant disagreement with pulsar observations $(n_e = 2 \times 10^{-2}$ $cm^{-3})$ and observations of hot H_I regions $(n_H \approx 0.2$ $cm^{-3})$ which agree with a model in which $p \approx p_c$.

Finally, we note that the presence of regimes where thermal instability occurs adds to the Parker instability by making $\gamma < 0$ in Equation (1). In the final state, however, if one has only stable phases, the effect of gas pressure is stabilizing. Not only is $\gamma \approx 1$ in both phases but also the gas pressure is no longer completely trivial: $\alpha = 2$ rather than 40. In Equation (1) this leads to $\lambda_c = 300$ pc, which is considerably larger than the mean length of a field line between clouds (≈ 120 pc); so the intercloud medium is stable against further activity.

Instreaming of gas has been observed at both poles of the Galaxy (Dieter, 1969). Dieter (1964) has suggested that this could be cool material condensing from hot gas at high altitude, perhaps by thermal instability. Within the framework of Parker's theory, such gas would stream down along the lines of force looping out of the disk, much like 'coronal rain' in the solar atmosphere. If the gas is much heavier than its surroundings, it will acquire free-fall velocity equal to 10, 20, 30, and 50 km sec^{-1} for origin at 100, 200, 400, and 1000 pc, respectively. As these velocities are of the order of those observed at the pole, the mechanism deserves further attention (Heiles, 1967b). The relative gas density $n(z)/n(0)$ is 0.6, 0.15, and 0.004 at $z = 100$, 200, and 400 pc, respectively, and is negligible at larger z even if T is as high as $T_c = 9000$ K. Two mechanisms could initiate the infall. Parker (1965) has argued that the slow build-up of cosmic-ray pressure due to constant injection can initiate a local rapid expansion of the field. This could cause the relative density at the top of the field line to drop so low (≈ 0.05) that a stable thermal equilibrium is impossible, and thermal instability commences. Alternatively, a finite amplitude disturbance such as a shock wave could propagate along the field line from a supernova, for example. In this case Goldsmith (1969) has shown that a phase transition can be induced (essentially because $p > p_c$), with the result that clumps of cool gas fall down the field line as required. Such mechanisms might account for activity originating up to 400 pc altitude ($v = 30$ km sec^{-1}), but it seems unlikely that activity at 1000 pc (50 km sec^{-1}) would be observable because of the negligible density there on the present model. Of course a hot corona (Spitzer, 1956) would provide the density, but at present there are doubts about the stability of a phase at 10^6 K. In the present picture, gas is not easily visible as it rises because it does so very slowly; it becomes visible only when descending at measurable velocities.

3. Shock Waves and Cloud Collisions

Kahn (1955) proposed that clouds are heated by mutual supersonic collisions. Various authors have treated the cooling of clouds subsequent to such heating, with the simplifying assumption that the gas density is constant. The results depend on certain parameters appropriate for the precise cooling mechanisms used, notably on the density of H_2; values of the cooling time vary from 10^5 to 10^7 yr. If the cooling time is of the same order as the estimated time between successive cloud collisions (10^7 yr), such collisions would be helpful in maintaining the high temperatures (> 100 K) observed for some clouds.

Zimmermann (1967, 1968a, b) described the dynamics of cloud collisions in more detail, neglecting the effects of magnetic field and cosmic rays. First, two clouds come together at velocities ≈ 10 km sec^{-1}. Detailed study was made of the central collision of two spherical clouds with a relative velocity of 18 km sec^{-1}. The initial parameters were $n_H = 10$ cm^{-3} and $T = 100$ K. Second, upon first contact, a shock wave forms and proceeds into each cloud. The conditions immediately behind the shock are adiabatic, and the temperature rises to 3300 K. Third, after a certain time has elapsed, the gas sufficiently far behind the shock has cooled down to 20 K. The density in the cool regions rises in order to keep the pressure constant behind the shock. Fourth, the shock reaches the outer boundary of the cloud. At this point, all the gas in the two clouds has been compressed into a thin cloud of high density. The compression occurs along the line of collision, motion in the transverse dimension being neglected. The gas at the edge of the cloud begins to expand, and an expansion wave proceeds into the cloud at about the speed of sound (1 km sec^{-1}). This expan-

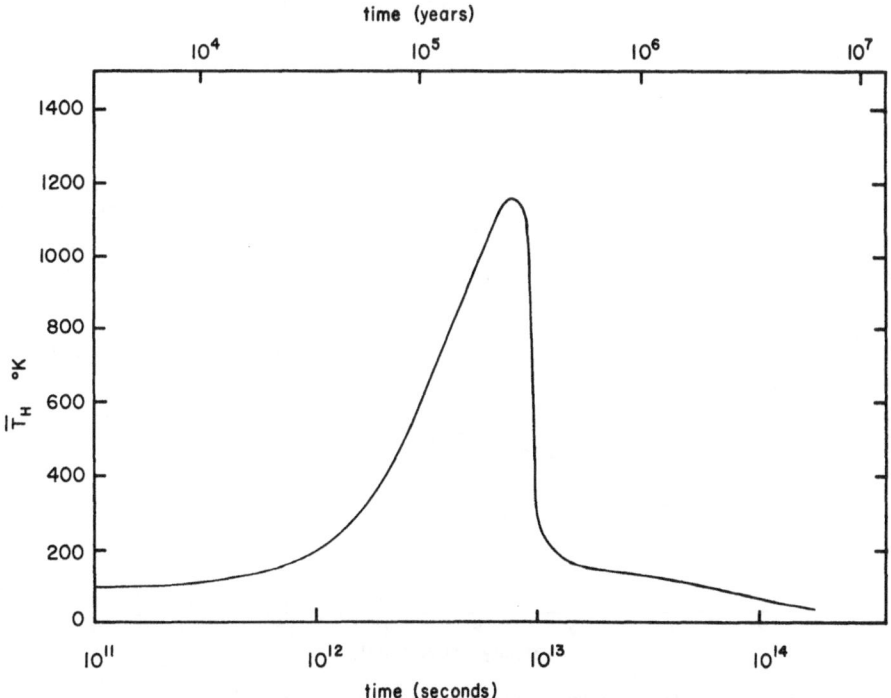

Fig. 6.   The arithmetic mean temperature as a function of time of an H I cloud after it starts colliding with another cloud (Zimmermann, 1968b).

sion proceeds until, fifth, the pressure in the cloud has dropped to the pressure in the intercloud medium. This requires an expansion of about a factor 15 if the expansion is adiabatic (a doubtful assumption); correspondingly, the gas temperature falls to 3 K. Cosmic-ray heating, which would increase the temperature obtained during the expansion phase, was neglected.

Figure 6 shows the mean temperature of the whole cloud (solid curve) as a function of time. One sees first the increase in temperature as more of the cloud has gone through the shock, and then the rapid drop as the shock reaches the opposite side and cooling is completed. Since a large fraction of the cloud is at a very low temperature in the postshock region, the results would presumably be quite different if the harmonic mean temperature were shown in Figure 6, instead of the arithmetic mean temperature. Such a curve would be useful for comparing with 21-cm measurements.

Steady shocks in H_I clouds were studied by Wentzel (1967) and by Field *et al.* (1968), including the effect of a magnetic field parallel to the shock front. Neither paper considered the expansion phase described by Zimmermann. Field *et al.* included a variety of cooling mechanisms; in particular, fine-structure excitation of O_I by hydrogen atoms, and excitation of the fine-structure levels of C^+, Si^+, and Fe^+ by collisions with hydrogen atoms, in addition to those considered by Zimmermann. Figure 7 summarizes the results of a number of calculations of cooling time (defined as the time for the gas to drop back to 100 K from the postshock temperatures) for various shock velocities and magnetic field strengths. This figure is based on $H_2/H = 10^{-3}$ but the results for $H_2/H = 10^{-5}$ are very similar; the latter value is probably more

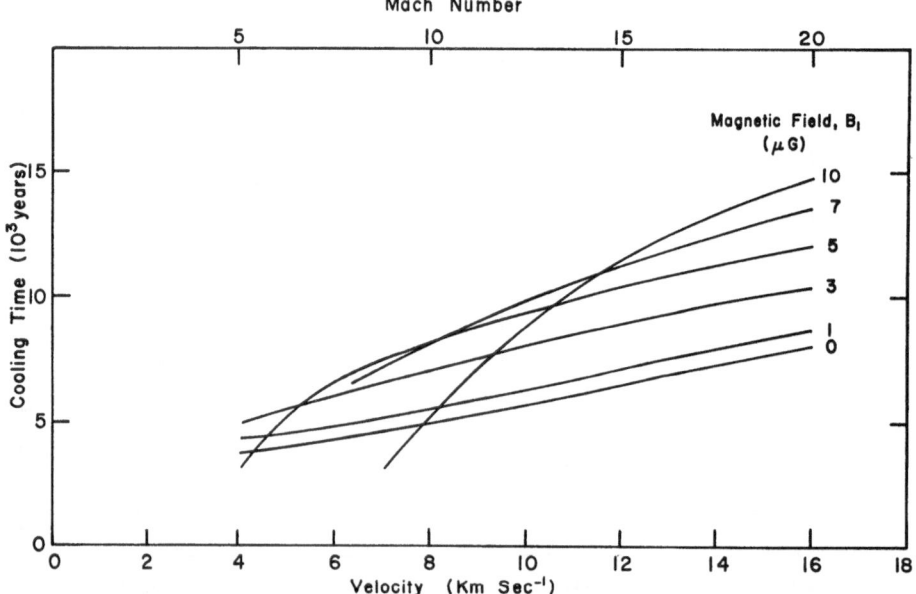

Fig. 7.   Cooling time for H_I gas after passage through a shock front. The magnetic field is parallel to the front. $H_2/H = 10^{-3}$, but a much lower value would give approximately the same results (Field *et al.*, 1968).

realistic (Stecher and Williams, 1967). The magnetic field has two compensating effects on the cooling time: first a decrease in time by reduction of the original post-shock temperature, and second, an increase in time by prevention of a density rise in the cool regions. As a result, the cooling times are relatively insensitive to the value of B over the plausible range 0 to 10 μG. It appears that for shock velocities to be encountered in cloud collisions (half the relative velocity, about 10 km sec^{-1}) the cooling time should be less than 10^4 yr – a significant reduction from the 10^6 yr found by Zimmermann for the case $H_2/H = 0$. The difference is attributed to his neglect of OI cooling, which dominates in every case.

If we could be confident of the present results, we could state definitely that cloud collisions cannot maintain cloud temperatures at the observed values, since 10^4 yr is such a small fraction of the time it takes (10^6 yr) for a 10 km sec^{-1} shock to cross a 10 pc cloud. After the first 10^4 yr has elapsed, a cloud collision actually decreases the cloud temperature by compressing a large fraction of the cloud to such a high density that cosmic rays cannot keep it hot.

Unfortunately, serious revisions are required before we can be certain of this:

(i) The cross section for excitation of O by H should be reduced a factor 3 (Bahcall and Wolf, 1968).

(ii) The free electrons from cosmic-ray ionization should be included in electron-ion cooling.

(iii) Account should be taken of depletion on grains.

(iv) The excitation of the a^4F term of Fe$^+$ should be included (Pottasch, private communication).

Of these, (i) and (iii) will increase T and (ii) and (iv) will decrease it, so the outcome is hard to predict. In addition to these revisions, study should be directed to the effect of an oblique magnetic field, the coupling of the magnetic field to the neutrals through the ions (Wentzel, private communication) and the dynamical effects of cosmic-ray pressure (Hudson, 1965, 1968).

A more general problem is whether cloud collisions actually occur as described. As Pikel'ner (1967) points out, the simple concept of a cloud collision will be modified by the presence of a magnetic field. He estimates that a typical cloud will oscillate perpendicular to the field about its mean position with an amplitude of 10 to 15 pc. Since this distance is considerably less than the mean free path for collisions, the number of collisions will be reduced. Furthermore, collisions perpendicular to the field will tend to be elastic as shown by Pikel'ner (1957) and Kogure (1965); we have mentioned evidence of this above, in the reduced heating by shock waves normal to the field. In summary, collisions across the field will be relatively ineffective in heating the gas.

Pikel'ner has argued that motions along the field are unimpeded, because the depth of the wells in which they settle is small – only a few pc – so that little displacement of the lines of force is necessary for motion to occur. This agrees with the observational fact that random cloud motions seem to occur equally parallel and perpendicular to the field. At first sight it might appear that deep wells (such as might occur in the

Parker model) would keep gas from moving along the field. It is true that energy must be put into the expanded magnetic field in front of the cloud to compress it to the value within the cloud. However, as much of this energy is regained again upon emerging from the rear of the cloud, the net loss of kinetic energy is small. We note that the speed of the cloud is of the same order (≈ 10 km sec^{-1}) as the sound speed in the intercloud medium, so there is probably little dissipation by a shock wave there. We tentatively conclude that collisions along the field can occur in a one-dimensional way, although much analysis remains to be done here.

If the collision event takes place only between clouds (or parts of clouds) on the same lines of force, transverse motion is inhibited, and the description of Zimmermann is quite accurate. While the speed of cooling behind the shocks is still uncertain, we can be confident that it is somewhat faster than the collision itself, so that collisions along the lines of force are quite inelastic. Each cloud therefore grows in mass, and the trend toward larger clouds will continue until some other process intervenes. According to Spitzer (1968a, p. 221), gravitational collapse for masses larger than the Ebert-Bonnor critical mass is one such process. For a pressure in the intercloud medium $p = 2.5 \times 10^{-13}$ dyne cm^{-2} ($nT = 1800$), this critical mass is $\approx 10^3 M_\odot$; Spitzer argues, on the basis of the virial theorem (and more detailed arguments by Field, 1969, confirm this), that a field of 3 μG will increase this critical mass only by a factor of 2 or 3. Presumably fragmentation and star formation is the normal outcome of this process. According to Oort (1954), after formation, the more massive stars will create H II regions in the remaining gas of the cloud and accelerate it outward, accounting for the kinetic energy retained by the clouds in spite of their inelastic collisions. We shall return to the problem of H II regions below, but here we note that in the Parker model, all this would occur in a massive cloud at the bottom of a magnetic well, and kinetic energy would preferentially be transmitted by clouds thrown out along the field.

Field and Saslaw (1965), Field and Hutchins (1968), and Penston *et al.* (1969) proposed and developed a statistical model of the cloud mass spectrum, assuming that it is fed by small units from expanding H II regions and is drained by collapse of massive clouds. They found a quasi-equilibrium spectrum $M^{-3/2}$, which is in fair agreement with observation. The latter paper shows that under certain reasonable assumptions the velocity distribution of clouds should obey equipartition and be Maxwellian. All of this work should be revised to take account of additional phenomena, such as the dynamical interactions with intercloud gas, the generation of new clouds by instabilities, and the constraint of one-dimensional motion imposed by the magnetic field.

All of the work on cloud collisions suffers from inadequate contact with observation. While radio astronomers have given us much data on mean values of temperatures, density, and magnetic field, I am not aware of observations of an individual cloud collision or even an individual shock wave (in the 21-cm line). It is dangerous to proceed with theoretical models without such observational checks. Field *et al.* (1968) estimated the fluxes of infrared lines between 4 and 156 μ, which would be emitted by the post-shock gas, and concluded that these are marginally observable with advanced equipment. Pottasch (1968) and Osterbrock (1969) have performed similar calculations for H I gas under other conditions. If such lines can be observed, information about temperature, density, composition, and spatial configuration can be achieved. Since the emission should be particularly intense from shock waves and other regions where dissipation is occurring, infrared measurements should tell us about the small-scale dynamics of the medium.

4. H II Regions; Expansion and Inhomogeneities

It has long been realized that the high temperature of H II regions ($\lesssim 10\,000$ K) and a similar density with respect to neighboring H I regions provide a pressure gradient which may be dynamically important. In the years since the last Symposium, considerable calculation has clarified this situation.

Spitzer and Savedoff (1950) calculated equilibrium temperatures of H II regions, balancing the gains by photoionization of hydrogen against collisional losses due to excitation of both fine-structure and optical levels of impurity atoms. They obtained values in the range (5 to 13) $\times\,10^3$ K and pointed out that the observed strength in many nebulae of $O^+\lambda 3727$ ($E/k = 39\,000$ K) requires temperatures of this order. Burbidge $et\ al.$ (1963) revised these calculations, emphasizing the importance of fine-structure transitions such as $^2P_{3/2}$-$^2P_{1/2}$ of Ne^+ at 12.8 μ. Taking $Ne/H = 5 \times 10^{-4}$ and $\Omega = 1$ (est.), they found $T = 3500$ K if the central star has a T of 30 000 K. Osterbrock (1965) showed that Ω was only 0.3, so that T increased to 6000 K for the same type of star. He also called attention to the fact that collisional de-excitation of O^{++}, N^+, and C^+ for $n_e > 10^4$ cm^{-3} reduces the cooling rate. Hjellming (1966, 1968) did a complete calculation, taking into account the detailed ionization equilibrium of the elements responsible for cooling, the transfer of Lyman continuum radiation, the hardening of the radiation by absorption as it proceeds through the nebula, the effects of collisional de-excitation, and the detailed flux distribution of the stellar atmosphere. He used modern Ω's (Saraph $et\ al.$, 1969). This work was extended to non-uniform densities by Rubin (1968). For a main-sequence star of 30 000 K Hjellming obtained 5800 K for a low-density region near the star. This temperature drops by about 800 K in the outer parts of the nebulae because of the Lyman-continuum diffuse radiation field. If one raises n_e to $\sim 10^4$ cm^{-3} then T will go up by about 2000 K, whereas if one increases the optical depth τ at the Lyman limit to 2, T goes up by about 1000 K. Hjellming concludes that 6000 K is a typical temperature, with values up to 9000 K being possible in dense or optically thick regions, and values down to 4000 K in low-density outer regions. Inspection of his results indicates that fine-structure transitions often play only a minor role. This would be especially true if the abundance of neon is only 6×10^{-5} (Flower, 1969; Pagel, 1969) rather than 4×10^{-4} as he assumed. We note that in addition to the temperature equilibrium calculations described here, Hjellming computed the structure of various types of ionization fronts.

There is some controversy as to whether theory and observations yield different temperatures. Peimbert (1967) gives optical results for three nebulae, based on the ratios of auroral-to-nebular lines of heavy ions and line-to-continuum intensities of hydrogen. He obtains values between 6000 and 14 000K in the center of Orion, depending on the method. Kaler (1968) reviews other optical data with similar results. Measurements based on the ratio of radio recombination lines to free-free continuum seem to give systematically lower values (Dieter, 1967; Mezger, 1968) but are unreliable because of non-LTE effects. Radio measurements based on fitting the shape of the free-free continuum give very low values in two cases (Terzian *et al.*, 1968). These authors obtained 3000(+2000; −1000)K for Orion. Since the method is sensitive primarily to the temperature of the outer parts, they suggest that the discrepancy between optical and radio measurements reflects a strong inward temperature gradient. This may be in qualitative agreement with the calculations of Hjellming (1968), because the diffuse radiation field tends to decrease T in the outer parts, while collisional de-excitation tends to increase it in the dense inner parts.

One of the aspects of H II regions of interest here is the dynamical expansion caused by high thermal pressure. The above discussion indicates that the theoretical situation

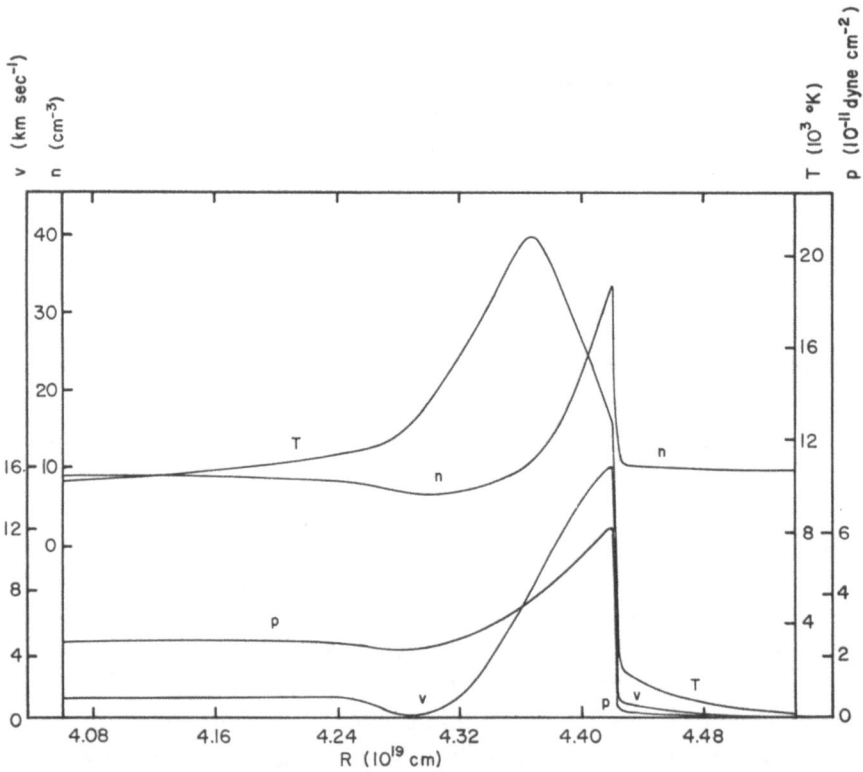

Fig. 8.   Variation of velocity v, number density n, temperature T, and pressure p as a function of radius R from the star in the vicinity of an ionization front moving away from an O star with $T = 42\,000$K after 3×10^4 yr (Mathews, 1965).

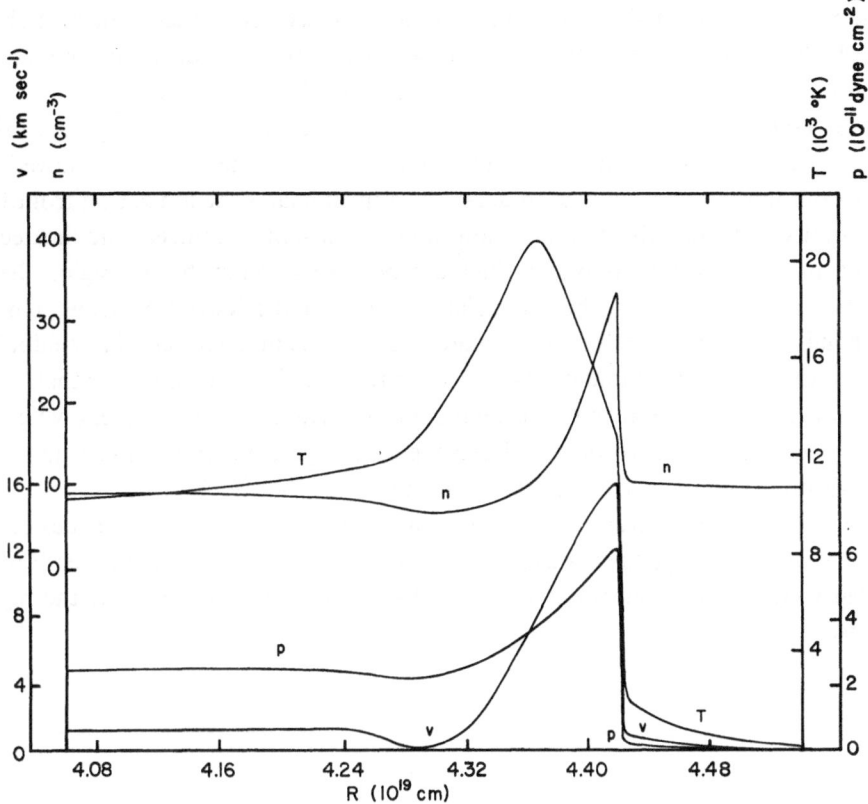

Fig. 9. The same as in Figure 8, but after 6×10^4 yr.

regarding the temperature is well in hand and that there are no glaring disagreements with observation. We now turn to two recent discussions of the expansion problem, by Mathews (1965) and Lasker (1966a, b). (A review of these and early studies of the problem has been given by Mathews and O'Dell, 1969). Earlier work on this problem had been restricted to similarity solutions but the papers of Mathews and of Lasker are based on numerical solution of the partial differential equations involved for specified initial values and use finite-difference methods.

Mathews (1965) treats the evolution of an H I region with $n_H = 10$ cm^{-3} and $T = 100$ K as it is ionized by a 30 $M_\odot$ star, whose main-sequence temperature is 42 000 K. Mathews treats in detail the luminosity of the star during its approach to the main sequence, since the time of approach (6000 yr) is comparable to $1.3/\alpha n = 10^4$ yr, the classical time of formation of the H II region. As predicted from earlier work, an R-type ionization front moves very rapidly into the gas, averaging 1000 km sec^{-1} in the first 10^4 yr. Velocities of the ionized gas remain less than 0.2 km sec^{-1} during this phase, in accordance with prediction for steady R-type fronts. At 6×10^4 yr, the buildup of the pressure wave behind the ionization front has developed into a shock front which leads the ionization front by 3 per cent of the radius. Although the computations were terminated at this point, the much-discussed detachment of a shock front which permits the ionization front to be D-type by virtue of its compression of the neutral hydrogen was demonstrated by Mathews (Figures 8 and 9).

Lasker (1966a) assumed that his star starts to shine instantaneously; his initial conditions are similar to those of Mathews. Unlike Mathews, who followed the Lyman-continuum radiation field in detail, he approximates the ionization front by a discontinuity through which atoms flow and are instantly ionized and heated to 10 000 K by the ultraviolet photons which escape absorption in the H II region; hence his results are unreliable in the immediate vicinity of the ionization front. On the other hand, in certain models, approximate allowance is made for cooling behind the shock (in the H I). His computations extend up to 2×10^6 yr, much longer than those of Mathews, and one can therefore examine the later stages of development which are probably more pertinent to observed H II regions. Figure 10 shows his results for a case with no H I cooling. One sees the detachment of the shock at about 8×10^4 yr, and the progressive acceleration of ever-increasing amounts of H I as time goes on. At about 9×10^5 yr an expansion wave caused by the acceleration of the outer H II has travelled to the center; thereafter the central density of H II decreases as the region expands.

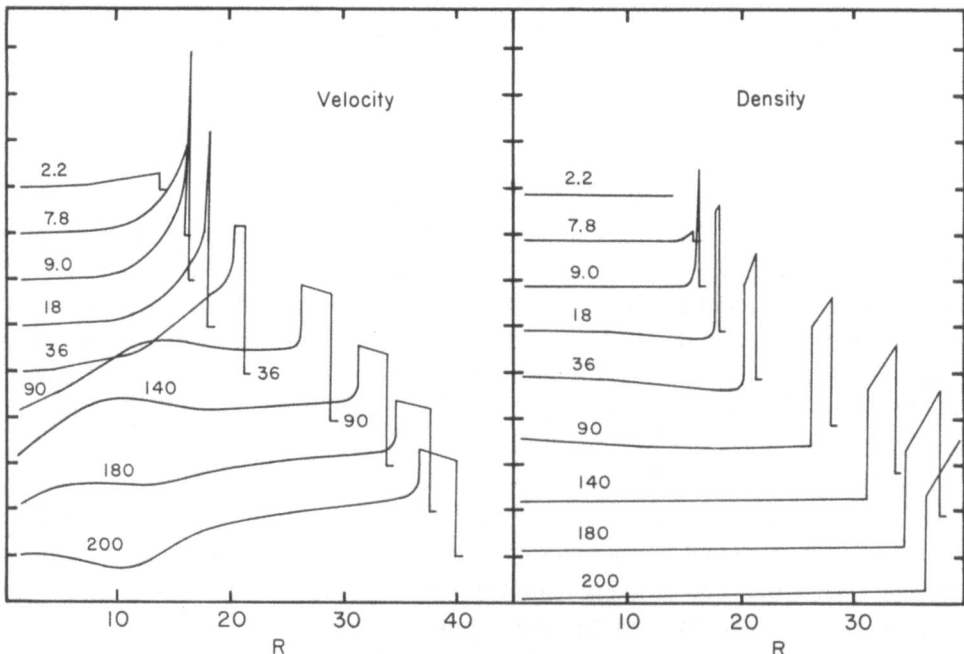

Fig. 10.   Variation of velocity V and density n as a function of radius R from the star for one of the cases of expanding H II regions considered by Lasker (1966a). To each curve is added the time after beginning of the expansion expressed in units of 10^4 yr. The velocity unit is 3.2 km sec^{-1}, the density unit 6.4 cm^{-3}.

The thicknesses Lasker obtained for the H I shock seem unrealistically small. The cooling time of H I (10^4 yr) is small compared to the evolution time (2×10^6 yr). Lasker therefore considered an additional model with an exponential cooling $e^{-t/\tau}$, with $\tau = 5 \times 10^4$ yr, and found that the evolution of the H II region is changed hardly at all. On the other hand, Δr, the thickness of the compressed H I region, decreases from about 10 per cent of the radius of the shock, to only 1 per cent, with a corresponding increase in density. Apparently the I-front acts like a piston which forces a shock at about 10 km sec^{-1} to compress the H I up to about the same pressure as the H II. It does not matter much whether this pressure is due to a high temperature (no cooling) or a high density (cooling). In the absence of magnetic fields, however, one would expect cooling to be effective in bringing about compression, and consequently, that a very thin but dense H I shell should be found around old H II regions. The available

TABLE IV

H II regions with observed H I shells

Nebula	Radius of I-front (pc)	Outer radius (pc)	$\Delta R/R$	Expansion velocity (km sec^{-1})	References
Rosette	20	50	0.6	15	Raimond, 1966
NGC 6910	1.5	5.5	0.7	10	Davies and Tovmassian, 1963
λ Ori	13	30	0.6	8	Wade, 1958
I Mon	40	90	0.7	23	Girnstein and Rohlfs, 1964
NGC 5146	1.5	3.3	0.5	< 6	Riegel, 1967
NGC 281	7.5	15	0.5	< 8	Riegel, 1967
DR 21	0.4	–	–	25	Thompson, Colvin and Hughes, 1969

observations seem at first glance to disagree with this. In Table IV we note some 21-cm observations of expanding H I shells around H II regions. In every case $\Delta R/R$ is much larger, not only than the value of 10^{-2} predicted for H I cooling, but also than the value of 10^{-1} predicted for purely adiabatic shocks. It is interesting that both λ Ori and NGC 5146 have dust shells around them of the order of a few per cent of the radius (Wade, 1958; Riegel, 1967). Possibly this dust (which in places has several magnitudes of extinction) marks the actual compressed region. From this conclusion it would follow that the 21-cm antenna beams are too large to resolve the expanding shells (Lasker, 1966a; Riegel, 1967). We note that infrared observations at the wavelengths of cooling transitions should be able to locate the H I shock unambiguously with high resolution (Field *et al.*, 1968).

Among the mechanisms suggested for increasing the width of the compressed region are turbulent or magnetic pressures in the H I region, a decrease in driving pressure as the star evolves, or expansion of the outer part of the cloud into the intercloud medium. Lasker (1966b) investigated the effects of magnetic fields on I-fronts and concluded that if $Q=B^2/8\pi p\lesssim 1$ ahead of the front, there is little effect (for either R- or D-type fronts). In fact, $Q=1.5$ if $B=3\ \mu G$ in a cloud at the critical pressure found by Field et al. (1969), 2.5×10^{-13} dyne cm^{-2}. After adiabatic compression by a 10 km sec^{-1} shock, Q falls to 0.33, so that an I-front moving into such a region is relatively unaffected. However, cooling behind the shock increases Q considerably (by reducing p). Lasker (1966a) discusses this situation for the two extreme cases: (i) the angle between magnetic field and radius from the star is 0° and (ii) the angle is 90°. Along the magnetic field (zero angle) there is no effect so the compressed H I region remains thin, while at an angle of 90° simple equations shows that the compressed region is about 10 times thicker. As Lasker notes, however, at intermediate angles the thicknesses may not be intermediate, since there can be discontinuous transitions from one branch of solutions to another (corresponding to the fast and slow hydromagnetic waves) as the angle between **B** and the radius vector varies from 0° to 90°. More work on this problem is recommended, both observationally (by seeking thin compressed regions) and theoretically (by verifying the correct branches of solutions for the shock and I-fronts in the context of the actual initial-value problem).

Lasker (1967) summarized numerical results on H II-region evolution in a simple way. The results indicate that (i) the radius r_i of the ionized region very quickly (about $10^4/n$ yr) approaches the value $s_0=(3S/4\ \alpha n_0^2)^{1/3}$ predicted by static Strömgren theory (α is the recombination coefficient, S the rate of emission of ionizing photons from the exciting star), (ii) thereafter r_i grows slowly as the I-front moves into the shocked H I, always satisfying the Strömgren relation for the actual density, n, in the H II region, (iii) n remains rather uniform and the velocities within the H II region remain $<\frac{1}{4}c$ (the isothermal sound speed in the H II region), (iv) the pressure is constant through the I-front after the early formation period, (v) the shock front never moves far ahead of the I-front. The last fact means that $dr_i/dt\approx v_s$ where the shock velocity v_s for strong shocks is given by $(4p'/3p_0)^{1/2}\ c_0$, where p_0 and c_0 refer to the unshocked H I and p' to the shocked H I. Since $p'=p_{II}=\varrho_{II}c^2$ from (iv), where c is the sound speed in the H II region, and $p_0=\varrho_0c_0^2$, it follows that $v_s=c_1(n/n_0)^{1/2}$, where $c_1=2c/\sqrt{3}$. From (ii), $n/n_0=(r_i/s_0)^{-3/2}$, so that

$$dr_i/dt \approx v_s = c_1\left(r_i/s_0\right)^{-3/4}, \tag{2}$$

which has the solution

$$r_i = s_0\left(1 + 7c_1t/4s_0\right)^{4/7}, \tag{3}$$

where t is the time elapsed since the Strömgren region formed ($r_i=s_0$). Hence

$$n = n_0\left(1 + 7c_1t/4s_0\right)^{-6/7}. \tag{4}$$

These expressions are found to represent the numerical results quite well. The I-front can be considered to be a piston exerting a pressure ϱc^2. As it moves, it does work on the H I ahead. Of the energy imparted by the strong shock, half goes into heat and half into kinetic energy. Hence the kinetic energy of the accelerated HI is

$$E_k = \int p \, dV = 2\pi \varrho_0 c^2 s_0^3 \int_{s_0}^{r_i} \left(\frac{n}{n_0}\right)\left(\frac{r_i}{s_0}\right)^2 \frac{dr_i}{s_0}, \tag{5}$$

which from Equations (3) and (4) is equal to

$$E_K = (4\pi/3) \, \varrho_0 c^2 s_0^3 \left[(r_i/s_0)^{3/2} - 1\right]. \tag{6}$$

This is readily shown to be $\frac{2}{3}[1 - (r_i/s_0)^{-3/2}]$ times the thermal energy in the HII region. It is important to compare E_K to the energy available from ultraviolet photons, $E_L = St \, \Delta E$, where ΔE is the average energy going into heat per ionization ($\Delta E \approx kT_*$). By using the Strömgren relation one can show that

$$\varepsilon = \text{efficiency} = \frac{E_K}{E_L} = \frac{2kT}{\Delta E} \frac{1}{\alpha nt} \left(1 - \frac{n}{n_0}\right). \tag{7}$$

For late phases of evolution ($n \ll n_0$) this has the simple interpretation that the total number of recombinations per proton is αnt, and since each is balanced by an ionization involving the expenditure of ΔE, the total energy put into the HII is $\alpha nt \, \Delta E$ erg per proton. Almost all of this energy from the star goes into keeping the gas hot, only the fraction $2kT$ being the pressure (which can do work) available per proton. Lasker shows that when $t =$ main sequence lifetime of the star and $n =$ the corresponding final density starting with $n_0 = 10$ cm^{-3}, ε varies from 0.002 for spectral type O5 to 0.07 for B1. When weighted by the numbers and fluxes of stars of various spectral types, the average efficiency is $\varepsilon = 0.0024$ (as O-stars supply most of the energy). A higher efficiency ($\varepsilon' = 0.014$) is required (Spitzer, 1968a) to account for the motions of interstellar clouds by the kinetic energy of HI shells accelerated by HII regions (Oort-Spitzer mechanism). As the discrepancy is only a factor 6, our tentative conclusion is that a significant fraction of the energy dissipated in cloud collisions, but probably not all, originates in HII regions. The alternatives include inflation of the magnetic field with resulting downward motion of gas clouds (Hoyle and Ireland, 1960; Parker, 1966), infall of intergalactic matter (Oort, 1969), electromagnetic radiation from neutron stars (Hoyle et al., 1964), and expansion of supernova shells (Shklovskii, 1962; Kahn and Dyson, 1965; Kahn and Woltjer, 1967; Spitzer, 1968b).

Kahn and Woltjer (1967) make quantitative estimates of the efficiency of cloud acceleration by supernova shells. They adopt a snowplow model appropriate for rapid cooling of shocked gas, and consider two cases of shell speed. If the speed is relativistic, the efficiency is $\varepsilon_R = \frac{1}{2}(v/c)^{1/2}$, while for non-relativistic shells it is $\varepsilon_{NR} = v/V$, where $v = 10$ km sec^{-1} is the speed of interstellar clouds and $V = 7000$ km sec^{-1} is an estimated shell velocity. Hence $\varepsilon_R = 0.003$ and $\varepsilon_{NR} = 0.0015$, of the same order as found by Lasker (1967) for HII regions. Using the data for dissipation in cloud collisions used above, adopting the non-relativistic model, and taking the supernova rate as 1 per 100 yr, we find that the required kinetic energy of the shell is 3×10^{51}

erg, compared to the observational estimate of 10^{51} erg (Minkowski, 1968). H II regions contribute $\frac{1}{6}$ and supernovae contribute $\frac{1}{3}$ of what is required, so together they may well account for the motions of clouds.

When the acceleration is complete, the radius of the supernova shell is about 50 pc according to Kahn and Woltjer (1967). Shane and Katgert (1967) have tentatively identified such a feature observationally. The large radius means that many clouds are actually members of expanding shells. From the model of Kahn and Woltjer one can estimate that the fraction is roughly 20 per cent. It is interesting that according to Spitzer (1968b), about 10 per cent of the calcium clouds observed are of relatively high velocity and show an excess of negative velocities as if they had been recently accelerated away from the vicinity of the early-type stars against which they are seen.

In closing our discussion of the expansion of H II regions, reference is made to the effects of stellar winds. Mathews (1966) has shown that mass loss of the order $10^{-5}\ M_\odot$ yr^{-1} can, by the pressure it exerts on the interior of the gas, cause the observed hole at the center of the Rosette Nebula, while Pikel'ner (1968b) has constructed a model based on stellar winds to explain the high internal velocities observed in some H II regions (Shcheglov, 1968). Mathews (1967) has also considered the effect of radiation pressure on grains within the H II region. He finds that the electrostatic drag on the charged grains can transmit the pressure effectively to the gas, with the result that the inner part of the Rosette Nebula can be evacuated.

We now turn to our final topic, inhomogeneities in H II regions. There is ample evidence that such inhomogeneities exist: determination of $\langle n_e \rangle$ by O II $\lambda 3727$ yields lower values than determination of $\langle n_e \rangle^{1/2}$ from hydrogen emission; the forbidden lines yield higher temperatures than the hydrogen lines and the radio continuum; the emission lines are broadened by motions comparable to the thermal velocity and, in some cases, are split in local regions. According to Kahn (1967), one can possibly interpret the discrepancy between the ages of the stars in Orion and the calculated expansion age of the Nebula, if the Nebula is actually made up of many dense globules of neutral matter which are being slowly ionized at present. This model can also account for the observed line splitting, by streams of ionized gas emerging from the globules and colliding to form shock waves. No doubt it would also account qualitatively for the variations in n_e and T observed, and possibly for the presence of O I $\lambda 6300$, which has always been hard to understand in a completely ionized region.

Some divergence exists on the theoretical treatment of globules. T Tau stars are often seen imbedded in elephant-trunk or comet-tail structures at the edges of H II regions and Dibai (1963) suggested that this could be explained as the result of a converging shock wave, initiated by the high pressure in the surrounding H II. Later Dibai and Kaplan (1964) studied such a shock wave by the method of similarity solutions and obtained velocities up to two times the speed of sound in H II regions and density increases (after passage of first a converging and then a reflected shock) up to 10^5.

Dyson (1968), on the other hand, analyses the situation by assuming that the globule is in hydrostatic equilibrium under self-gravitation, internal pressure, and a boundary pressure supplied by the H II streaming away from its surface in a D-critical I-front. No shocks are included, and a relation is derived between the mass and radius of the globule for a given external ionizing radiation field. This corresponds to the fact that the I-front (assumed to be D-critical) has a definite value of $p + \varrho v^2$ which must be balanced by the internal gas pressure. The theory of truncated isothermal spheres then leads to the mass-radius relation. As the globule evaporates, gravitational binding usually becomes less and less important; and confinement by external pressure, more important. Gravitational instability does not occur, and the whole globule is evaporated (on a time scale of 2 to 8×10^4 yr per solar mass). On the other hand, globules rather close to instability initially can be set into collapse as time goes on.

The question of whether the front is D-critical was analyzed by Kahn (1969). He finds rather that the front is strong D-type, progressing subsonically into the globule and ejecting ionized gas supersonically into the H II region. He mentions the possibility that the detailed heating and cooling processes within the I-front may require a pressure wave to propagate ahead of the I-front, but otherwise omits deviations from hydrostatic equilibrium within the globule. Mendis (1968, 1969) analyzed the situation further, taking into account the fact that the globule is probably composed of H_2 rather than H at the high densities contemplated (10^6 cm^{-3}). He concludes that the front is strong D-type for all stars earlier than O9.

Considerable effort has been expended on discovering whether elephant trunks (which may later become globules) can arise as a result of instability. The present consensus appears to be that instabilities (Rayleigh-Taylor; ionization front) are present but either yield parameters in disagreement with observation or are too weak (Mathews and O'Dell, 1969). Rather, it seems more natural to suppose that large density fluctuations are already present in the H I region before it is overrun by the I-front. Two possibilities suggest themselves. The first is that dense concentrations (10^4 cm^{-3}) are present in the cloud before the hot stars begin to shine. As their internal pressure must greatly exceed that of the normal intercloud medium, they must be gravitationally bound. In this case, it is hard to see why they do not collapse in the free-fall time (10^6 yr), since isothermal spheres of this kind are unstable. Possibly they are actually collapsing, and represent unstable fragments of the original cloud which would have formed stars if the ultraviolet radiation field had not intervened.

Alternatively, one might identify them with the small cloudlets (1 to 30 $M_\odot$) discovered by Heiles (1967a) in the interstellar medium. While the masses roughly agree, the densities do not. To account for this, we note that by the time the I-front makes the dense clouds visible, they have already been overrun by a shock front which will certainly compress them. The degree of compression can be estimated as $2T_2/T_1 \approx 20\,000/T_1$, where T_2 is the H II temperature and T_1 is the H I temperature. Ordinarily T_1 is taken to be 100 K, so the compression is a factor of 200. In regions which are already denser than average, however, cooling processes are particularly effective, and T_1 might drop as low as 10 K as a result of cooling by C$^\circ$. The resulting compression of a factor of 2000, would result in a final density of 2×10^4 cm^{-3}.

References

Anand, K. C., Daniel, R. R., and Stephens, S. A.: 1968, *Proc. Ind. Acad. Sci.* **67**, 267.
Bahcall, J. N. and Wolf, R. A.: 1968, *Astrophys. J.* **152**, 701.
Burbidge, G. R., Gould, R. J., and Pottasch, S. R.: 1963, *Astrophys. J.* **138**, 945.
Clark, B. G.: 1965, *Astrophys. J.* **142**, 1398.
Davies, R. D. and Tovmassian, H. M.: 1963, *Monthly Notices Roy. Astron. Soc.* **127**, 45.
Dibai, E. A.: 1963, *Astron. Zh.* **40**, 606 (translation: 1964, *Soviet Astron.* **7**, 606).
Dibai, E. A. and Kaplan, S. A.: 1964, *Astron. Zh.* **41**, 652 (translation: 1965, *Soviet Astron.* **8**, 520).
Dieter, N. H.: 1964, *Astron. J.* **69**, 288.
Dieter, N. H.: 1967, *Astrophys. J.* **150**, 435.
Dieter, N. H.: 1969, *Publ. Astron. Soc. Pacific* **81**, 186.
Dyson, J. E.: 1968, *Astrophys. Space Sci.* **1**, 388.
Field, G. B.: 1962, *Interstellar Matter in Galaxies* (ed. by L. Woltjer), Benjamin, New York, p. 183.
Field, G. B.: 1965, *Astrophys. J.* **142**, 531.
Field, G. B.: 1969, Sixteenth Liège Symposium (June 1969), Institut d'Astrophysique, Liège.
Field, G. B. and Hutchins, J.: 1968, *Astrophys. J.* **153**, 737.
Field, G. B. and Saslaw, W. C.: 1965, *Astrophys. J.* **142**, 568.
Field, G. B. and Zachariades, C.: 1969, in preparation.
Field, G. B., Rather, J. D. G., Aannestad, P. A., and Orszag, S. A.: 1968, *Astrophys. J.* **151**, 953.
Field, G. B., Goldsmith, D. W., and Habing, H. J.: 1969, *Astrophys. J. Lett.* **155**, L149.
Flower, D. R.: 1969, Conference on Infra-Red and Microwave Radiation from Nebulae and Galaxies (July 1969), Institute of Theoretical Astronomy, Cambridge, England.
Gary, T., Holweger, H., Koch, M., and Richter, J.: 1969, *Astron. Astrophys.* **2**, 446.
Girnstein, H. G. and Rohlfs, K.: 1964, *Z. Astrophys.* **59**, 83.
Goldsmith, D. W.: 1969, The Formation and Equilibrium of Interstellar Clouds, Ph.D Dissertation, University of California, Berkeley, February, 1969 (also *Astrophys. J.*, in press).
Goldsmith, D. W., Habing, H. J., and Field, G. B.: 1969, *Astrophys. J.* **158**, 173.
Grahl, B. H., Hachenberg, O., and Mebold, U.: 1969, *Beitr. Radioastron.* **1**, 3.
Habing, H. J.: 1969, *Bull. Astron. Inst. Netherl.* **20**, 177.
Hayakawa, S., Nishimura, S., and Takayanagi, K.: 1961, *Publ. Astron. Soc. Japan* **13**, 184.
Heiles, C.: 1967a, *Astrophys. J. Suppl.* **15**, 97.
Heiles, C.: 1967b, *Astron. J.* **72**, 1040.
Hjellming, R. M.: 1966, *Astrophys. J.* **143**, 420.
Hjellming, R. M.: 1968, *Interstellar Ionized Hydrogen* (ed. by Y. Terzian), Benjamin, New York, p. 435.
Hjellming, R. M., Gordon, C. P., and Gordon, K. J.: 1969, *Astron. Astrophys.* **2**, 202.
Hoyle, F. and Harwit, M.: 1962, *Publ. Astron. Soc. Pacific* **74**, 359.
Hoyle, F. and Ireland, J. G.: 1960, *Monthly Notices Roy. Astron. Soc.* **120**, 173.
Hoyle, F., Narlikar, J. V., and Wheeler, T. A.: 1964, *Nature* **203**, 914.
Hudson, P. D.: 1965, *Monthly Notices Roy. Astron. Soc.* **131**, 23.
Hudson, P. D.: 1968, *Monthly Notices Roy. Astron. Soc.* **140**, 255.
Kahn, F. D.: 1955, *Gas Dynamics of Cosmic Clouds* (ed. by J. M. Burgers and H. C. van de Hulst), North-Holland Publishing Co., Amsterdam, p. 60.
Kahn, F. D.: 1967, *Radio Astronomy and the Galactic System* (ed. by H. van Woerden), Academic Press, London, p. 95.
Kahn, F. D.: 1969, *Physica* **41**, 172.
Kahn, F. D. and Dyson, J. E.: 1965, *Ann. Rev. Astron. Astrophys.* **3**, 47.
Kahn, F. D. and Woltjer, L.: 1967, *Radio Astronomy and the Galactic System* (ed. by H. van Woerden), Academic Press, London, p. 117.
Kaler, J. B.: 1968, *Interstellar Ionized Hydrogen* (ed. by Y. Terzian), Benjamin, New York, p. 459.
Kippenhahn, R. and Schlüter, A.: 1957, *Z. Astrophys.* **49**, 73.
Kogure, T.: 1965, *Publ. Astron. Soc. Japan* **17**, 385.
Kurt, V. G. and Sunyaev, R. A.: 1967, *Astron. Zh.* **44**, 1157 (translation: 1968, *Soviet Astron.* **11**, 928).
Lasker, B. M.: 1966a, *Astrophys. J.* **143**, 700.
Lasker, B. M.: 1966b, *Astrophys. J.* **146**, 571.

Lasker, B. M.: 1967, *Astrophys. J.* **149**, 23.
Layzer, D.: 1967, Fourteenth Liège Symposium (June 1966), Institut d'Astrophysique, Liège, p. 315.
Mathews, W. G.: 1965, *Astrophys. J.* **142**, 1120.
Mathews, W. G.: 1966, *Astrophys. J.* **144**, 206.
Mathews, W. G.: 1967, *Astrophys. J.* **147**, 965.
Mathews, W. G. and O'Dell, C. R.: 1969, *Ann. Rev. Astron. Astrophys.* **7**, 67.
Mendis, D. A.: 1968, *Monthly Notices Roy. Astron. Soc.* **141**, 409.
Mendis, D. A.: 1969, *Monthly Notices Roy. Astron. Soc.* **142**, 441.
Mezger, P. G.: 1968, *Interstellar Ionized Hydrogen* (ed. by Y. Terzian), Benjamin, New York, p. 477.
Minkowski, R.: 1968, *Stars and Stellar Systems*, Volume 7: *Nebulae and Interstellar Matter* (ed. by
  B. M. Middlehurst and L. H. Aller), University of Chicago Press, Chicago, p. 623.
Oort, J. H.: 1954, *Bull. Astron. Inst. Netherl.* **12**, 177.
Oort, J. H.: 1969, First Annual Nuffield Lecture (July 1969), Institute of Theoretical Astronomy,
  Cambridge, England.
Osterbrock, D. E.: 1965, *Astrophys. J.* **142**, 1423.
Osterbrock, D. E.: 1969, *Phil. Trans. Roy. Soc. London* **A264**, 241.
Pagel, B.: 1969, Conference on Infra-Red and Microwave Radiation from Nebulae and Galaxies
  (July, 1969), Institute of Theoretical Astronomy, Cambridge, England.
Parker, E. N.: 1965, *Astrophys. J.* **142**, 584.
Parker, E. N.: 1966, *Astrophys. J.* **145**, 811.
Parker, E. N.: 1968a, *Stars and Stellar Systems*, Volume 7: *Nebulae and Interstellar Matter* (ed. by
  B. M. Middlehurst and L. H. Aller), University of Chicago Press, Chicago, p. 707.
Parker, E. N.: 1968b, in *Proc. Intern. Conf. Cosmic Rays* (November 1968), Tata Institute, Bombay.
Parker, E. N.: 1969, this volume, p. 168.
Peimbert, M.: 1967, *Astrophys. J.* **150**, 825.
Penston, M. V., Munday, V. A., Stickland, D. J., and Penston, M. J.: 1969, *Monthly Notices Roy.
  Astron. Soc.* **142**, 355.
Pikel'ner, S.: 1957, *Astron. Zh.* **34**, 314 (translation: 1959, *Soviet Astron.* **1**, 310).
Pikel'ner, S.: 1967, *Astron. Zh.* **44**, 1915 (translation: 1968, *Soviet Astron.* **11**, 737).
Pikel'ner, S.: 1968a, *Ann. Rev. Astron. Astrophys.* **6**, 165.
Pikel'ner, S.: 1968b, *Astrophys. Letters* **2**, 97.
Pottasch, S. R.: 1963, *Astrophys. J.* **437**, 945.
Pottasch, S. R.: 1968, *Bull. Astron. Inst. Netherl.* **19**, 469.
Radhakrishnan, V. and Murray, J. D.: 1969, *Proc. Astron. Soc. Australia* **1**, 215.
Raimond, E.: 1966, *Bull. Astron. Inst. Netherl.* **18**, 191.
Riegel, K.: 1967, *Astrophys. J.* **148**, 87.
Rubin, R. H.: 1968, *Astrophys. J.* **153**, 761.
Saraph, H. E., Seaton, M. J., and Shemming, J.: 1969, *Phil. Trans. Roy. Soc. London* **A264**, 77.
Shane, W. W. and Katgert, P.: 1967, *Radio Astronomy and the Galactic System* (ed. by H. van Woer-
  den), Academic Press, London, p. 127.
Shcheglov, P. V.: 1968, *Astrophys. Lett.* **1**, 145.
Shklovskii, I. S.: 1962, *Astron. Zh.* **39**, 209 (translation: 1962, *Soviet Astron.* **6**, 162).
Silk, J. I. and Werner, M. W.: 1969, *Astrophys. J.* **158**, 185.
Spitzer, L., Jr.: 1951, Problems of Cosmical Aerodynamics, Central Air Documents Office, Dayton,
  Ohio, p. 31.
Spitzer, L., Jr.: 1956, *Astrophys. J.* **124**, 20.
Spitzer, L., Jr.: 1968a, *Diffuse Matter in Space*, Interscience, New York.
Spitzer, L., Jr.: 1968b, *Stars and Stellar Systems*, Volume 7: *Nebulae and Interstellar Matter* (ed. by
  B. M. Middlehurst and L. H. Aller), University of Chicago Press, Chicago, p. 1.
Spitzer, L., Jr. and Savedoff, M. P.: 1950, *Astrophys. J.* **111**, 593.
Spitzer, L., Jr. and Scott, E. H.: 1969, *Astrophys. J.* **157**, 161.
Spitzer, L., Jr. and Tomasko, M. G.: 1968, *Astrophys. J.* **152**, 971.
Stecher, T. P. and Williams, D. A.: 1967, *Astrophys. J. Lett.* **149**, L29.
Swings, P.: 1965, *Ann. Astrophys.* **28**, 703.
Terzian, Y., Mezger, P. G., and Schraml, J.: 1968, *Astrophys. Lett.* **1**, 153.
Thompson, A. R., Colvin, R. S., and Hughes, M. P.: 1969, *Astrophys. J.* **158**, 939.
Wade, C.: 1958, *Rev. Mod. Phys.* **30**, 946.

Wentzel, D. G.: 1967, *Astrophys. J.* **150**, 453.

Zimmermann, H.: 1967, Fourteenth Liège Symposium (June 1966), Institut d'Astrophysique, Liège, p. 285.

Zimmermann, H.: 1968a, *Astron. Nachr.* **290**, 193.

Zimmermann, H.: 1968b, *Astron. Nachr.* **290**, 211.

5. DISCUSSION FOLLOWING THE REPORTS
BY WEAVER AND FIELD
(Tuesday September 9, 1969)

Chairman: B. F. BURKE

Editor's remark: This long, very intensive, and partially very confused discussion, has been rearranged in six sections: (1) Direct and Indirect Evidence of X-rays and Low-Energy Cosmic Rays; (2) The Boundary Layer between the Stable Gas Phases; (3) Theoretical Aspects of Interstellar Gas Dynamics and the Formation of Clouds; (4) Observational Aspects of Interstellar Gas Dynamics and the Formation of Clouds; (5) Observations of the Rarefied, Neutral Intercloud Medium and of the Interstellar Electron Density; (6) The Dynamical Theory of H II Regions. Section 2 has been transferred from the Discussion on Monday, September 8 (Chapter 2). To Section 6 have been added remarks made during various discussions. A couple of remarks have been transferred to other Discussions. Part of the Discussion (in Section 5) was very confused; an attempt has been made to condense and to make as much sense as possible out of what was said. For the convenience of the reader I recapitulate a few concepts, the (mis)-use of which lead partially to the confusion:

(i) The hydrogen surface density or column density $N_H = \int n_H \, dl$ (N_H is sometimes called the hydrogen measure HM).

(ii) The dispersion measure $DM = \int n_e \, dl$ (DM is often called the electron surface density N_e).

(iii) The rotation measure $RM = c_1 \int n_e B_{\parallel} \, dl$.

(iv) The emission measure $EM = \int n_e^2 \, dl$.

In these definitions, n_H represents the hydrogen density, n_e the electron density, $B_{\parallel}$ the component of the magnetic field strength along the line of sight and l the distance along the line of sight. Conventionally n_H and n_e are expressed in cm^{-3}, B in μG and l in pc. In addition there are two combinations of these quantities (and of the electron temperature T) involved in the Discussion:

(v) The free-free absorption coefficient $k(v) = c_2(v) n_e^2 T^{-3/2}$ (v is the frequency).

(vi) The free-free emissivity $\varepsilon(v) = c_3(v) n_e^2 T^{-1/2}$ (only at radio wavelengths).

c_2 and c_3 depend also on T, but only rather weakly.

1. Direct and Indirect Evidence of X-rays and Low-Energy Cosmic Rays

Sunyaev: I want to discuss an observational upper limit on the number of low-energy cosmic rays. Cosmic-ray protons may become neutral atoms without loss of kinetic energy through charge-exchange reactions. Some of these fast atoms will be excited, either because the charge-exchange process leaves the atom in an excited state, or because the atom collides inelastically with thermal neutral atoms. The excited, fast atoms emit Doppler-shifted Ly-α quanta, for which the interstellar optical depth is low. Observation of such a Doppler-shifted flux then yields an upper limit on the cosmic-ray flux. Unfortunately the cross-sections get very small for cosmic-ray energies in excess of 100 keV. For particles in the energy range 25 to 100 keV, Kurt and I found an upper limit $W(E < 100 \text{ keV}) < 5 \times 10^{-3} \text{ eV cm}^{-3}$ from measurements by the 'Venus' space probe. To obtain a limit on the density of cosmic rays around 1 MeV we have to interpolate between low and high energies. Suppose that all cosmic-ray protons are injected with energies $E > E_0 (\gg 100 \text{ keV})$. The spectrum for $E < E_0$ is then determined by ionization losses and can be calculated. From $W (E < 100 \text{ keV}) < 5 \times 10^{-3} \text{ eV cm}^{-3}$, it follows that $W (E < E_0) = (1 \text{ to } 3) E_0^{3/2} \text{ eV cm}^{-3}$, where E_0 is in MeV.

Field: If the cosmic-ray heating is due to 2 MeV particles, the required energy density is only 0.03 eV cm^{-3}, 30 to 100 times less than Sunyaev's upper limit.

Sunyaev: The upper limit, which I mentioned, can be brought down considerably by narrowing the filter of the photometer. 100 keV corresponds to a 17 Å wide band, ten times smaller than the band in the existing experiments.

Silk: Let me make a few more comments on the low-energy cosmic-ray density. Direct observations of these particles have been made, but the numbers found have to be demodulated to correct for the effects of the solar wind. I compare two different demodulations: that by Gloeckler and Jokipii (1967), and the more conservative demodulation by Webber (1968). In order to obtain the ionization rate ζ of 4×10^{-16} ionizations per H-atom per sec, mentioned by Field in his report, we need to extend the Gloeckler-Jokipii demodulation down to 10 MeV, and we obtain a low-energy cosmic-ray energy density of 1 eV cm^{-3}. For the Webber demodulation these numbers are 2 MeV and 3×10^{-3} eV cm^{-3}. In either case one is multiplying the observed cosmic-ray flux at solar minimum by a very large factor (10^4 and 10^3, respectively).

I would now mention recent work on indirect limits on the flux of low-energy cosmic rays. The results are expressed in terms of ζ. The first limit that I mention is due to Fowler, Reeves, and myself. We have considered the spallation of low-energy cosmic rays in the interstellar medium, producing boron and lithium. The threshold for boron production is 5 MeV, and we find that the value of ζ due to cosmic rays above 5 MeV per nucleon is less than 2×10^{-16} sec^{-1}. A similar limit by Solomon and Werner concerns the dissociation of molecular hydrogen by low-energy cosmic rays. The presence of molecular hydrogen in dark clouds, as implied by recent observations by Heiles, indicates that ζ must be less than, or of the order of, 10^{-16} sec^{-1} for cosmic rays above about 5 MeV per nucleon. A third limit on cosmic rays at still lower energies was obtained by Steigman and myself. We have set limits on the fluxes of heavy nuclei in cosmic rays between roughly 1 and 2 MeV. These heavy cosmic rays capture electrons by charge exchange with neutral hydrogen, similar to the cosmic-ray protons, as discussed by Sunyaev. The electrons are captured to excited states, and subsequently cascade to the ground state, emitting resonance line radiation at X-ray energies. Below 1 keV the interstellar medium absorbs the X-rays; and above about 6 keV, the abundances of heavy nuclei are too low to produce detectable line emission. However, in the 1 to 6 keV range, nuclei such as neon, magnesium, silicon and iron may produce appreciable line emission along lines of sight in the galactic plane. We have applied our predicted fluxes to a recent rocket observation of an excess diffuse X-ray flux in the galactic plane. Unfortunately the measurement has very poor energy resolution, and we find that it is consistent with $\zeta \leqslant 10^{-15}$ sec^{-1}. Improved energy resolution and increased observing time should enable this limit to be considerably lowered in the near future. In the same way, Steigman and I have also recalculated the cosmic-ray proton flux, as discussed by Sunyaev. With more accurate cross-sections we find that a value of $\zeta \leqslant 10^{-12}$ sec^{-1} would be consistent with the far UV measurements. (Gloeckler, G. and Jokipii, J. R.: 1967, *Astrophys. J. Lett.* **148**, L41; Webber, W.: 1968, *Australian J. Phys.* **21**, 845.)

I should like to consider next the possible role of X-rays in heating the interstellar medium. The diffuse X-ray background extends from MeV energies to below 1 keV. Of importance for the interstellar medium are the soft X-rays. At present there are 5 reported observations of the background at ≈ 0.25 keV. Unfortunately, these measurements differ appreciably in the maximum diffuse flux seen, which varies from ≈ 200 photon cm^{-2} sec^{-1} ster^{-1} keV^{-1} to ≈ 900 photon cm^{-2} sec^{-1} ster^{-1} keV^{-1}. For the present purpose, I will simply take the weighted mean of the observations, which yields ≈ 450 photon cm^{-2} sec^{-1} ster^{-1} keV^{-1}. Werner and I have considered the effect of this diffuse X-ray flux on the interstellar medium. In particular, we find that the effective ionization rate per hydrogen atom is 2×10^{-17} sec^{-1}, about a factor of 20 below the value of ζ required by Field, Goldsmith, and Habing (see Field's Introductory Report, p. 51). Extrapolation to lower energies of the extragalactic component of the diffuse soft X-ray flux does not increase ζ significantly.

Higher ionization rates may be obtained using galactic soft X-ray sources. A recent experiment by the Wisconsin group (Bunner *et al.*, 1969) sets severe constraints on the number of such sources. Observing one twelfth of the sky, they found no source brighter than 1.3 photon cm^{-2} sec^{-1} keV^{-1} at ≈ 0.25 keV. It follows that if soft X-ray sources are to provide a significant heat input to the interstellar medium, there must be either many weak sources or few relatively strong sources. Possible evidence for the first hypothesis is that the same experimenters measured a diffuse flux of ≈ 60 photon cm^{-2} sec^{-1} ster^{-1} keV^{-1} at 0.25 keV in the galactic plane. This cannot be due to an extragalactic flux because of the effects of interstellar absorption. In order to obtain a value of ζ as high as 10^{-15} sec^{-1}, it is necessary to postulate that this flux rises steeply towards lower energies. An effective source temperature of about 50 eV is required; interstellar absorption will turn the spectrum over below about 80 eV. The net result is a diffuse galactic component with spectrum $\propto E^2 \exp(-E/0.05)$ photon cm^{-2} sec^{-1} ster^{-1} keV^{-1}. The number density of sources required is $\approx 10^{-2}$ pc^{-3}, with a luminosity per source of $\approx 10^{32}$ erg sec^{-1} in soft X-rays, giving a total heat input to the Galaxy of 3×10^{-26} erg cm^{-3} sec^{-1}. The corresponding ionization rate ζ, in regions where $n_e/n_H \lesssim 10^{-2}$, is 5×10^{-16} sec^{-1} ionizations per H atom. The distribution of these sources must be that of Population II objects to give the observed degree of isotropy at 0.25 keV when interstellar absorption is taken into account. There is indirect evidence that the spectrum of this galactic component must be extremely soft; otherwise, anisotropy would be detectable above 0.5 keV, where interstellar absorption is less important. The Wisconsin observers note that no such anisotropy is found as they scan in galactic latitude. For comparison purposes, note that the soft X-ray luminosity of the Sun (averaged over the solar cycle) is $\approx 10^{27}$ erg sec^{-1}, whereas the soft X-ray luminosity of Sco X-1 (assumed to be at a distance of 100 pc) is $\approx 10^{35}$ erg sec^{-1}.

The alternative possibility is to have relatively few strong soft X-ray sources. The main constraint then is that only approximately ten such sources can be brighter than about 1 photon cm^{-2} sec^{-1} keV^{-1} at 0.25 keV. This requirement, together with the restriction that these sources not appear abnormally bright above 1 keV, leads to the following model. The source luminosity is $\approx 3 \times 10^{37}$ erg sec^{-1} in soft X-rays below 0.25 keV, and the source density is $\approx 3 \times 10^{-8}$ pc^{-3}. These sources must emit $\approx 3 \times 10^{36}$ erg sec^{-1} at 0.25 keV; therefore, for a thermal bremsstrahlung source, the effective temperature must be ≈ 0.1 keV. Observations of soft X-ray sources are still in a very preliminary state. However, the soft X-ray source Vela X-2 may be considered a possible candidate for our hypothetical strong source. Tentative identification of this source with the Wolf-Rayet star γ Vel implies a luminosity at 0.25 keV exceeding 3×10^{36} erg sec^{-1}. Moreover, three of the X-ray sources detected at ≈ 1 keV in the Sagittarius region (GX349+2, GX5-1, and GX13+1) show evidence of interstellar absorption consistent with their being at a distance of ≈ 10 kpc. Their luminosities are correspondingly estimated at $> 10^{36}$ erg sec^{-1} above 1 keV. The Crab is the only X-ray source for which a reliable distance estimate is available; the luminosity of the Crab in soft X-rays above 0.01 keV is $\approx 5 \times 10^{37}$ erg sec^{-1}. (Silk, J. and Werner, M. W.: 1969, *Astrophys. J.* **158**, 185; Bunner, A. N., Coleman, P. C., Kraushaar, W. L., McCammon, D., Palmieri, T. M., Shilepsky, A., and Ulmer, M.: 1969, *Nature* **223**, 1222.)

Sunyaev: Neutral hydrogen in the peripheries of galaxies will be ionized by intergalactic photons at $\lambda < 912$ Å. The ionization will lead to ionized layers around the galaxies, similar to 'Strömgren-spheres' in the Galaxy. 21-cm observations of galaxies can in principle indicate where at the edge of a galaxy a Strömgren-sphere sets in, and from this one obtains an estimate of the intergalactic ionizing flux. Observations of good quality exist for a dozen galaxies. They yield only an upper limit for the intergalactic ionizing flux, but the upper limit is significantly lower than can be obtained from photon counters on board of space probes. The upper limit on the ionizing flux indicates that the density of intergalactic matter is less than one-third the critical density (if the Hubble constant $H = 100$ km sec^{-1} Mpc^{-1}). More details may be found elsewhere.* [Sunyaev, R. A.: 1969, *Astrophys. Lett.* **3**, 33; Zel'dovich, Ya. B. and Sunyaev, R. A.: 1969, *Zh. eksp. Teor. Fiz.* (1970, *Soviet Phys. JETP*, in translation).]

* See also the criticism by J. E. Felten and J. Bergeron, *Astrophys. Lett.* **4** (1969), 155. (Ed.)

2. The Boundary Layer Between the Stable Gas Phases

Zel'dovich: Field showed a picture (Figure 3, p. 55), which resembles closely an isotherm in a van der Waals gas. In both cases there are two phases, one with low density and high temperature, the other with high density and low temperature. The two phases have the same pressure. In the case of the van der Waals gas only one pressure is stable, namely that pressure for which the two hatched areas in Figure 1 are equal. This follows from thermodynamic principles. But in interstellar

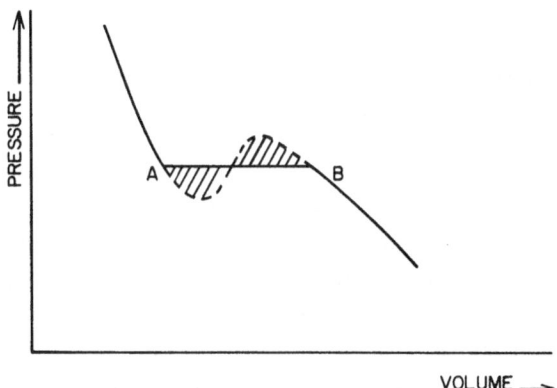

Fig. 1. (See the remark by Zel'dovich.) (*p-V*) curve as discussed by Field (p. 55), in which is indicated (schematically) the situation when there is no net heat production in the intermediate layer.

space we do not have closed systems and we cannot apply thermodynamic principles. Still, Pikel'ner and I discovered that also in interstellar space only one pressure is stable. If there is a dense cloud in pressure equilibrium with the rarefied surrounding, then there should be a sheet between them. In this sheet thermal conductivity occurs and has to be considered in addition to cooling and to cosmic-ray heating. The stable equilibrium is determined by the condition that the three processes balance at each point in the sheet. This condition leads to an integral which should be zero; the integral vanishes for only one pressure.* If we have a different pressure, the border sheet will penetrate into the gas in one way or the other. (Zel'dovich, Ya. B. and Pikel'ner, S. B.; 1969, *Zh. Eksp. Teor. Fiz.*)

Field: Dr. Zel'dovich, I was not aware of the existence of your paper. But just before this Symposium I learned that essentially the same results have been obtained in the United Kingdom by Penston and Brown, apparently independently from you. One thing that emerged from the Penston and Brown paper was that, if the pressure is not the stable one, the conducting boundary layer will move at only an extremely slow rate. In the situation we are considering, Penston estimated a time-scale of 3×10^8 yr, so that in actual situations this equilibrium pressure will not necessarily be found.

---

* In Figure 1 the cosmic-ray heating exceeds the cooling in the hatched area near *B*; in the similar area near *A* the reverse is the case. The vanishing of the integral expresses the requirement that the extra heat produced in one part of the sheet is absorbed in the other part. (Ed.)

Pikel'ner: Indeed, the border moves very slowly through the gas. We concluded that only the initial conditions, and not the physics of the sheet, determine the pressure. For example, suppose that, as in the gravitational theory of spiral structure, gas is condensed into a spiral arm and starts forming clouds (provided that in the interarm region no clouds existed because of the large rarefaction). Then initially the pressure would be the critical pressure (see Field's Figure 3, p. 55 or his Table II, p. 56); but after that the pressure would decrease and the clouds would grow very slowly. Ultimately they would reach the stable pressure, but probably there is not enough time available.

3. Theoretical Aspects of Interstellar Gas Dynamics and the Formation of Clouds

Spiegel: Field mentioned a discrepancy between observed and predicted temperatures. Probably this could be explained if the observed gas is not in its final equilibrium. Let me take Figure 3 of Field's paper (page 55) as describing the instantaneous p-ϱ relation. Now consider the idealized case of a uniform layer of height h, density ϱ and pressure p. ϱ and ϱ_3 are the stable equilibrium densities as described by Zel'dovich (see Figure 1 at p. 81; ϱ_1 corresponds to point B, ϱ_2 to point A), and we have $\varrho_1 < \varrho < \varrho_3$, then the layer is dynamically unstable. If a weight sits on the layer and the system is in a gravitational field, then a slight reduction in h will lead to an initial collapse of the layer. If the loss function adjusts instantaneously, the collapse will maintain the indicated p-ϱ relation. In the phase plane $\dot{\varrho}$-ϱ, the situation is as shown in Figure 2a. The point $(\varrho = \varrho_2, \dot{\varrho} = 0)$ is a saddle point, while $(\varrho = \varrho_1, \dot{\varrho} = 0)$, $(\varrho = \varrho_3, \dot{\varrho} = 0)$ are centers. For a perturbation which keeps the loss function constant, the 'orbits' in the phase plane are closed curves, such as the indicated dashed curve. The figure eight curve is the zero energy orbit. Strictly speaking, the motion does not conserve loss rate; and the system will spiral into the appropriate equilibrium. But, if the parameters are correct, we may treat closed oscillating orbits as instantaneously representing the motion. It is to be expected that, in an orbit such as the one indicated, the system spends most of its time near the ϱ_2 end of the orbit.

Of course, this is an idealization, but this model may indicate the actual behavior much as Baker's one-zone model of stellar pulsation for stars. I would guess that

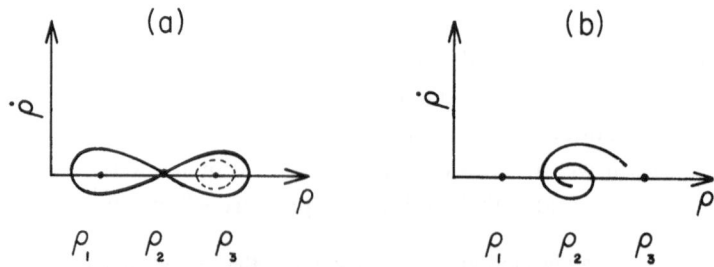

Fig. 2.   (See the remark by Spiegel.) Possible orbits in the phase plane of the system considered.
(a) Undamped motions. (b) Damped motions.

it works for individual clouds, and that the majority of clouds are observed not in equilibrium but rather in the kind of damped oscillation indicated by the diagram of Figure 2b. Whether we get a limit-cycle or spiraling into a stable point depends on the thermodynamic time scales. The question would have to be studied with time-dependent calculations. In this connection, I might mention a thesis by Defouw (1970) at Cal Tech on thermal instability, which shows how thermal instability in a normally stable stratification can lead to convective instability or stability. (Defouw, R. J.: 1970, *Astrophys. J.* **161**, 55.

Field: This is an interesting suggestion. In the solutions by Pikel'ner and Zel'dovich and by Penston and Brown, the boundary layer is so thin that conduction dominates and will stabilize the situation.

Spiegel: Let me raise another subject. The clouds are presumably in the spiral arms. A large number of people believe that those clouds are shocks. What happens when you have a shock in a medium with the equation of state of Figure 1?

Field: That problem has been studied by Goldsmith. He made numerical calculations about a piston ramming into the stable hot phase. If one rams it in with more than a certain critical velocity, there is an instantaneous phase transition to the cold stable phase. The critical velocity is such that behind the shock which forms, the pressure exceeds the critical pressure (see my Report, Figure 3, p. 55). The numerical value is about 10 km sec^{-1} when the ambient pressure ahead of the shock is not far below the critical value. (Goldsmith, D. W.: 1969, *Astrophys. J.* **161**, 41.)

Pikel'ner: In speaking about the formation of clouds due to thermal instability, we should take into account the magnetic field of the Galaxy. This field is strong, and motions can be only along magnetic lines unless the scale is very large. The time for establishing the equilibrium temperature is short, less than 10^6 yr. After that time we have equilibrium temperatures in different points along the field lines, but not a uniform pressure. Pressure gradients set the gas into motion, and compress fluctuations into clouds. If the diameter of clouds are about 10 pc and motions are possible only in one dimension, then the contraction time may attain 10^7 yr, because rarefied gas is collected out of a very large region. But, if the average density is uniform and we consider the formation of very small fluctuations, then the time will be shorter. However, the observations show rather extensive regions of higher density. Since the time of compression in clouds is long, we can observe regions in which the average density is intermediate between the densities of the rarefied and the dense phase.

Parker: We have to be very careful when we talk about the equation of state as shown in Figure 1. There is a finite time needed to reach the thermal equilibrium that shows this equation of state, and we have to distinguish between quasi-stationary situations, where the equation of state can be used, and situations with rapid fluctuations in which the thermal equilibrium cannot be reached.

Field: The transition out of the unstable regime is about 10^6 yr. This is one order of magnitude faster than the Rayleigh-Taylor instability which is due to the gravitational field. So, as the latter instability develops, one can get clumping along the magnetic field on a shorter time scale.

Syrovat-skii: My remark concerns cloud formation. We have heard that the magnetic field is so large that only motion along the magnetic field driven by thermal instability can lead to cloud formation. But if the (strong) magnetic field is not stationary, then a change of the field may produce compression of the gas into a cloud without difficulty. This process may be especially effective near the zero points of the magnetic field, as is shown by examples in the solar atmosphere.

Parker: On a very small scale the magnetic fields inhibit any collapse, just because they are rigid. On a large scale, on the other hand, the magnetic field helps the collapse to proceed. In this context 'large scale' means 'of the order of 100 pc'. 'Small scale' means dimensions of 1 or 2 pc.

Field: It seems to me that Spiegel's suggestion should be studied further, i.e., that the final state of the Rayleigh-Taylor and the thermal instability is in fact a dynamical one. In the Parker model, after the initial instability, the clouds come to rest in the low parts of the magnetic field, unless they are further energized. It is possible that a kind of circulation, such as Spiegel is suggesting, may cause material to evaporate from the dense phase through the unstable phase into the hot phase, and thereby to rise along the field line. Later the material may collapse again to complete the cycle. I raise this point because it seems to me that it is strongly connected with the remark by Weaver (see p. 48) that negative velocities are observed. You see instreaming material; you do not see material going up. Could it be going up in one phase and coming down in another?

Parker: I am intrigued by the suggestion of the continual circulation or dynamical state that Spiegel suggested. There is no ultimate equilibrium that we have been able to discover for the interstellar gas field cosmic-ray system. It seems to me that there is a dynamical balance between the collection of gas into clouds and the disruption of clouds and field by the formation of hot stars and the injection of cosmic rays.

Greenberg: I am trying to connect all this to the discussion on the mass flow into the spiral arms at the Basel Symposium. At the inner edge of the arm, a condensation is produced by a shock wave formed as the mass flows into the potential minimum produced by the density wave. The problem of clouds forming along the spiral arms or along the magnetic fields would then be a separate kind of an instability. In other words, the higher density in the spiral arms is a result of the mass flowing into the potential well, that is the spiral structure, and then the instabilities within the arms produce clouds within the arms. This looks like a consistent picture to me.

Habing: The shock wave that was discussed in Basel is not yet really a physically treated shock. Roberts does not consider what happens perpendicular to the galactic plane. He just considers mass motions in two-dimensional space in the galactic plane.

Pikel'ner: The possibility of formation of a shock wave in the gas flowing into the spiral arms is dependent also on the magnetic pressure. The variation of the transverse velocity of the gas when this gas is flowing into the spiral arms is about 10 to 15 km sec^{-1}. In the rarefied gas, the magneto-sound velocity should be much higher (about 20 or 30 km sec^{-1} if the density is about 0.03 cm^{-3} and the field $B \approx 2\mu G$) and is approximately transverse. Thus the velocity of the gas relative to the spiral arm is lower than the sound velocity, and no shock waves should arise. Even if there is no magnetic field, it is necessary to take into account the fact that the gravitational field, which determined the formation of the arms, accelerates the gas gradually.

4. Observational Aspects of Interstellar Gas Dynamics and the Formation of Clouds

Ozernoi: In making the choice among mechanisms for obtaining the dense and rarefied regions, it is important to know what kind of correlations are established between the densities, temperatures, velocities, and possibly, the magnetic fields inside and outside the spiral arms, respectively, particularly at high latitudes. I would like Dr. Weaver to comment.

Weaver: That subject was a part of the general review which was deleted in the presentation. I have tried to picture the largest complexes of gas within the Galaxy as spiral arms. Within these arms there are fluctuations in density of various sizes and various temperatures.

To understand in greater detail the kinematical properties of H I regions of various sizes remains an important observational problem. In studies made so far, inadequate subdivision of cloud sizes has been made. Adequate division is desirable, since there appears to be a correlation between cloud size and other properties of the clouds. I might mention the studies by Takakubo and van Woerden, involving separation of profiles into Gaussian components. Perhaps van Woerden will comment. Again they obtain a picture of an interstellar medium that is not very homogeneous.

Van Woerden: Our study about the kinematics of neutral hydrogen at intermediate galactic latitudes gave the following results. We distinguished two different types of components in the 21-cm profile: (1) Narrow components, with internal velocity dispersions σ of the order of 2 km sec^{-1}, tentatively identified with normal interstellar clouds of densities n_H of the order of 10 cm^{-3} and of surface densities N_H of 2×10^{20} cm^{-2} (i.e., sizes of about 5 pc). Their average distance from the galactic plane was estimated at about 100 pc. (2) Gaussian components with internal velocity dispersions σ of the order of 10 to 15 km sec^{-1}. It appeared that those were at considerably larger distances ($|z|$ about 150 pc), and we inferred either that they might form a sort of sheath around the spiral arm, i.e., a transition between arm and interarm region, or that they might possibly occupy a transition region between the disk and the halo. The 'random motions' (external motions of clouds) were about 6 km sec^{-1} (r.m.s., radial coordinate only) for both types of component. There was a slight indication for anisotropy in these motions; but the preferential direction varied with galactic latitude, and the relationship of this anisotropy to the local magnetic field was unclear. (Takakubo, K. and van Woerden, H.: 1966, *Bull. Astron. Inst. Netherl.* **18**, 488; Takakubo, K.: 1967, *Bull. Astron. Inst. Netherl.* **19**, 125.)

Verschuur: When van Woerden says he has found narrow components of 2 km sec^{-1}, it is a reflection of the 2 km sec^{-1} bandwidth he used. Correct, Dr. van Woerden?

Van Woerden: The velocity dispersion of the instrumental profile was $\sigma_i = 0.8$ km sec^{-1}; so our narrow components were in general not bandwidth-limited. A small number of components were so narrow that they were hard to resolve; but measurements with a narrower band ($\sigma_i = 0.4$ km sec^{-1}) gave essentially the same distribution of dispersions in the interval $0 < \sigma < 2$ km sec^{-1}, with few if any components having $\sigma < 1$ km sec^{-1}. Therefore, our statistics of velocity dispersions were practically not limited by the bandwidth. It may well be that they were limited by instrumental resolution on the sky. If you have a narrower antenna beam, you may get narrower components.

Verschuur: If you use a narrower beam you see components with smaller velocity dispersions, and we really do not know how narrow these things might become as one increases the resolving power. We must be very careful in interpreting data obtained with a small radio telescope, because the interstellar medium has fine-scale structure; and our present picture is biased entirely by the telescopes we have used.

Van de Hulst: I would like to support the last remark. I have the feeling that, at present, analysis is more limited by too low resolutions of the data than by too little thought.

Verschuur: I wish to show data about one distinct cloud in a position-velocity coordinate diagram (see Figure 3). The observations were made with the 300-foot radio telescope, with a 6.25 kHz bandwidth. Velocity is the horizontal coordinate,

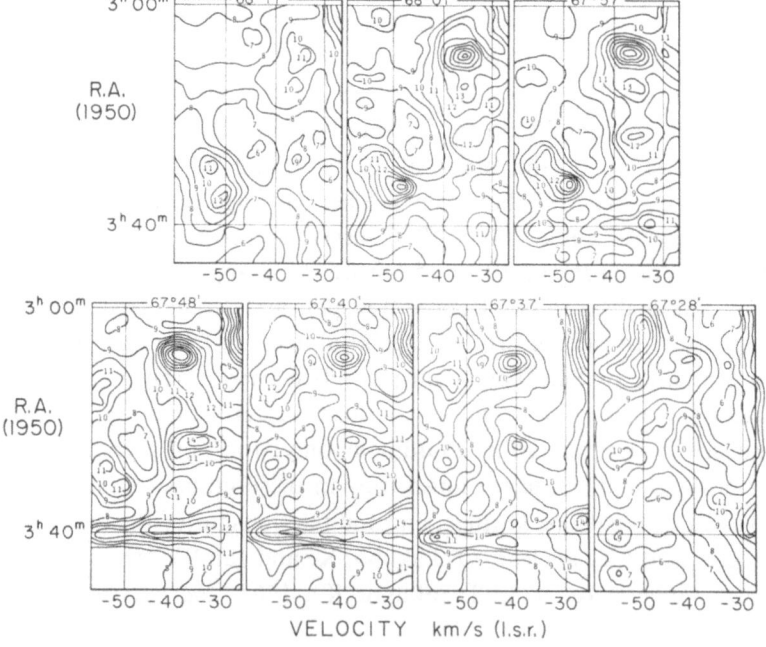

Fig. 3. (See the remark by Verschuur.) 21-cm observations made with the 300-foot radiotelescope at NRAO, Greenbank, W. Virginia (U.S.A.). In each diagram antenna temperatures are plotted in K as a function of right ascension and velocity; the declination is constant for each separate diagram.

and right ascension is the vertical. In Figure 3 the declinations are indicated at the top of each section. There is a cloud at about -40 km sec^{-1} and R.A. 3^h 9^m. It is quite discrete. Its velocity is not associated with either the local or the Perseus spiral arm. The latitude is about $+8°$. Notice also that its velocity at declination 68° 01' is -36 km sec^{-1}; but half a degree lower in declination, the velocity is -42 km sec^{-1}; a change of 6 km sec^{-1}. I do not know whether it is a galactic or an extra-galactic object. It could be either; but, very clearly, it is a rotating 'cloud'. A discussion of another distinct 'cloud' has been published elsewhere. (Verschuur, G. L.: 1969 *Astron. Astrophys.* **3**, 77.)

Habing: I disagree with Verschuur in regard to the very small features. It is probably true that we shall always find smaller features with larger telescopes. But certainly there are intermediate-scale features in the interstellar gas which are difficult to understand on the basis of a model of discrete, but small clouds. In the 21-cm contours, features exist in which there is continuity over 10 or 15 degrees on the sky.

Burke: Can you give us any numbers?

Habing: In the survey I am making together with Carl Heiles, I find structures extending over some 15 to 30 degrees on the sky. I do not know the distances and cannot therefore give real dimensions. But suppose a feature is 15 degrees long and is some 100 pc away; then the scale length is at least 25 pc. You may even have two sheets of ordered motions of gas of 25 pc.

Weaver: There is no question whatever about the size of the telescope used influencing the results. But there are plenty of features of large size that present us with many unsolved problems. The larger the telescope, certainly the smaller the feature observed; and we just add more problems to our list. It would be nice to understand even some of the larger clouds.

Menon: What exactly do we mean by 'random motions' of clouds? How can we distinguish between small-scale, local irregular motions, and large-scale, systematic non-circular motion? The data we get cannot be interpreted in the sense of any theories of the general interstellar medium, because we are averaging over regions with very peculiar motions. And we do not know how to account for this averaging.

Weaver: In reply to Menon it appears to me that reasonable approaches exist to the question of the motion of condensations with respect to each other. First the interstellar medium is broken up into rather large-scale units (perhaps described as sections of spiral arms). Within these sections there is a general structure, which observationally appears as condensations in the velocity cross-section of the arm. A first approximation to the velocity dispersion would be to derive the radial velocities of these condensations with respect to the mean of the velocity for the arm. Such an approach is valid, even though the large-scale (arm) structure appears to have some peculiar motion. Such a simple procedure does not take into account differential rotation within the arm, but that could be allowed for.

Thomas: I want to get closer to a definition of clouds. Pikel'ner said that clouds consist of gas in the high-density phase. Weaver essentially talked also about *density* fluctuations. The picture of two phases goes back to Zanstra at the Second Symposium in 1954. We should also remember Adams and Oort and Burgers. Oort and Burgers, in planning the First Symposium asked whether or not the interstellar clouds could be linked to the 'turbulent eddies', in the aerodynamical sense of correlation distances for velocitity fluctuations discussed in the spectral theories. No definite boundaries are necessary to fix scales, only variation in velocity. Although Adams' picture has been conceptually one of discrete entities, it is reasonably clear that he needs only velocity gradients to interpret the observations. One then has to ask whether such gradients can exist in a continuous medium, or whether we require density boundaries. The new aspect the astronomical investigations brought into the discussion is the necessity of incorporating compressible turbulence. One looks for boundaries, not just correlation lengths. This leads to the discussions along Zanstra's lines of quasi-static equations of state, with several phases. I wish that we would try to be clear about what we mean when we say 'clouds'. Are we talking about scales fixed by velocity gradients, or boundaries fixed by density changes, or both? And what is the interrelation? I think Weaver's film shows the beginning of accumulation of data that will answer these questions and will let us define what we mean, operationally, by clouds. Are we talking about smooth fluctuations in a continuum? Or are we talking about discrete entities moving through a very rarefied substratum (with, perhaps, continuum mechanics not providing a good description)?

Habing: From the theoretical side, we are talking about this discrete picture. I disagree with Weaver that the interstellar gas is characterized by a dense gas in which there are gentle fluctuations. My picture is of a rarefied medium filling almost all of the Galaxy. The medium is, however, more concentrated in the arms and, moving into the arms, the gas becomes unstable and forms discrete clouds.

Thomas: Discrete clouds and discrete boundaries, then. Might there be one or more different kinds of clouds?

Habing: There may be clouds of all sizes.

Thomas: Can one therefore have a continuous distribution of clouds, each with different velocity, each with different density, *but* definite macroscopic velocity for the cloud as a whole?

Habing: Yes. I do not see that this picture is contradicted by the observations, although observationally there is a large-scale ordering, meaning, too, that there are large sheets of clouds.

Weaver: How sharp is the boundary?

Habing: Theoretically, the boundary is determined by conduction, and this would mean a thickness of less than a parsec.

Pikel'ner: We should take into account the magnetic pressures, too. Perpendicular to the magnetic fields, the boundary cannot be really definite; the compression is not very important in this direction. In the direction along the magnetic fields, the boundary should be more definite, perhaps 10^{-3} to 10^{-2} pc. The other question here concerns the different scales of the condensations. The division into clouds and inter-cloud medium is dependent on thermal instability. The large scale condensations should be connected with the Rayleigh-Taylor-type instabilities studied by Parker.

This instability can hardly develop unless it is driven by thermal instability, because the hot gas slides slowly along the magnetic lines. Its height scale is comparable with the height scale of the total layer of the interstellar gas. Thus the thermal instability creates clouds and then these clouds can condense into larger condensations.

Parker: The dynamical theory, involving the magnetic field and the cosmic rays, leads to the prediction that the dimension of clouds in the horizontal direction across the magnetic field may be very small, a few parsecs. It is not clear to me whether or not this effect is observable.

Verschuur: Many clouds are elongated in the direction parallel to the field. I will comment on this in my Report (p. 150).

Mestel: Has anyone considered the effect of galactic differential rotation on the clouds as they form? Other things being equal, the tendency of the differential rotation is to elongate the clouds in the toroidal direction. However, if the arm has a helical magnetic field, the magnetic forces, though not important for the 'grand design' of the spiral structure, may still be strong enough to cancel out effectively the gravitational shear across the spiral arm.

Burke: In this regard I should report one comment by Lin. He maintains that, if the general magnetic field is less than 5 μG, then the large-scale motion is controlled by gravity and not by the magnetic field.

Karpman: The condensation of clouds is usually accompanied by some dissipative progress, which may unfreeze the matter in the clouds from the magnetic field.

Pikel'ner: The magnetic fields are always important in the interstellar gas. Dissipation can occur only in very sharp neutral lines or shocks, but not during the formation of clouds.

Shulman: I should like to add that not only magnetic fields are an important factor in cloud formation, but also rotation. When the density in a cloud increases, the rotation will be accelerated until it controls the contraction process.

Pikel'ner: Observations of Faraday rotation and some other observations show that magnetic fields are generally parallel. That situation is proof, I think, that rotation is retarded by the magnetic tension.

5. Observations of the Rarefied, Neutral Intercloud Medium and of the Interstellar Electron Densities

Van Woerden: Field gave a predicted critical temperature between 7000 K and 9000 K and said that the observations yielded a value of 4000 K. Is this difference significant? As far as the observations go, I doubt whether the velocity dispersion, leading to the 4000 K, has been measured accurately since the intensities are low and the background is broad and irregular.

Field: The difference is significant. On the theoretical side it seems impossible to have a stable phase without Ly-α cooling. That requires temperatures of at least 7000 K. On the observational side, the profiles taken by Heiles do not allow more than 1 km sec^{-1} error in the velocity dispersion. The predicted width is 9 km sec^{-1}; the observed width, 6 km sec^{-1}.

Verschuur: I should like to question the temperature of 2000 K that Weaver obtains for the (neutral) interarm medium. I do not think that one can make a reliable determination from observations with an 85-foot, 35-arc min beamwidth telescope such as he has been using. Observations with a 10-arc min beam at high latitudes indicate much structure – e.g., many small clouds. At low latitudes the number of small clouds must be much higher, and this makes an interpolation method, as has been used by Weaver, very questionable. I myself have some spectra that show very cold clouds superimposed on a broad background. The broad background has a width consistent with a 7000 K gas; the cool clouds have kinetic temperatures of 30 K or less.

Weaver: Certainly there are difficulties in the temperature determination due to an irregular background. You will have noticed that in the diagram I showed (Figure 6, p. 32) the values of the temperature fluctuated considerably. The important point is, however, that nowhere are low temperatures (around 100 K) obtained, nor very high temperatures (around 10 000 K). With respect to the measurement of the broad background that you observed at high latitudes, I think that one has to be very cautious in interpreting the dispersion of the component in terms of temperatures. I frequently found very broad velocity tails, extending to 70 km sec^{-1}, which have nothing to do with temperature, but only with macroscopic gas motions.

Field: Turbulence can increase the dispersion, but not decrease it. Heiles obtains a dispersion of 6 km sec^{-1}, so the temperature cannot be more than 4000 K.

Mills: A comment on the distribution of free electrons in the Galaxy. I plotted the positions of 40 pulsars with observed values of the dispersion measure DM on the plane of the Galaxy. I adopted, as a zero-order approximation, a model with a homogeneous distribution of free electrons with $n_e = 0.06$ cm^{-3}. In the plot I recognize two groups of pulsars: one corresponds to the local spiral arm, the other to the Sagittarius arm. The z-distribution of the pulsars indicates a mean height of about 100 pc and suggests that the electron layer extends beyond this distance. Details may be found elsewhere. (Mills, B. Y.: 1969, *Nature* **224**, 504).

Weaver: What about possible fluctuations in the electron density?

Mills: Eventually one can detect these by statistical means, but 40 is not a large enough number of pulsars. There seem to be too many electrons in the direction of the Vela X pulsar (perhaps associated with the Gum Nebula), whereas in the direction of the Crab pulsar too few electrons are found.

Field: Recently Davies (1969) discussed the relation between the values of DM and of N_H, the total number of hydrogen atoms along the line of sight. Figure 4 shows Davies' data, plotted in a different way. Inside the plot points is written the absolute value of the latitude. If the pulsars were imbedded in a medium with $n_e/n_H = 0.1$ (an upper limit in the model for the interstellar medium, summarized in my Report), they would all lie on a straight line. However, (i) N_H contains a contribution by hydrogen from behind the pulsar, and (ii) the line of sight may pass through cool clouds with considerably lower values of n_e/n_H. Therefore the observed points are to be expected to lie to the right of the straight line; and, indeed, they are found there. It even seems likely that the lower the galactic latitude, the farther the points are removed from the straight line. Figure 4 then shows that the pulsar dispersion measurements are consistent with the model distribution of interstellar matter mentioned before. (Davies, R. D.: 1969, *Nature* **223**, 355.)

Pottasch: Courtès and Monnet have observed general Hα-emission in our Galaxy, confined to within a few hundred pc from the plane. A similar phenomenon was found in other galaxies: a structureless medium on which spiral arms showed up as regions of much higher intensity. Transferring the measurements by Monnet in M51 to our Galaxy (a dangerous thing to do) we have electron densities of 0.1 cm^{-3} or somewhat less. In addition there is interesting information on the electron temperature. Since, besides Hα, the [N II] line is also seen the temperature of the medium is at least 4000 K.

[**Editor's remarks:** At this point a confused discussion started on various observed quantities (mentioned in the introduction to this Discussion) and their relation to densities and temperatures of the interstellar electrons. The confusion arose mainly

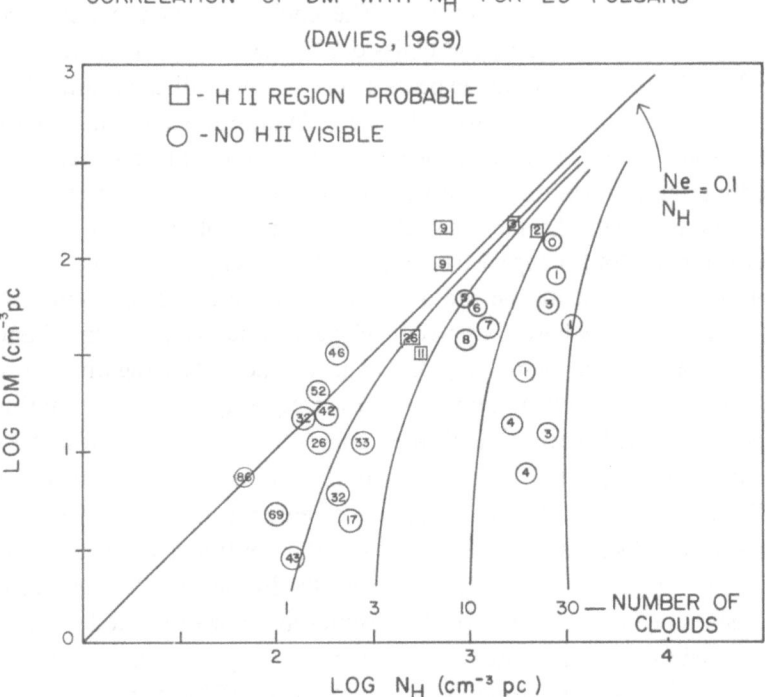

Fig. 4. (See the remark by Field.) Relation between DM, the integral of the electron density along the line of sight, and N_H, the analogous integral of the hydrogen density. The data have been taken from Davies, R. D.: 1969, *Nature* **223**, 355.

because of the complicated, non-linear ways in which most of the observed quantities depend on density and temperature fluctuation. The following conclusions may be extracted from the discussion:

(i) The observed RM- and DM-values indicate that the average interstellar electron density $\langle n_e \rangle$ lies between 0.03 and 0.08 cm^{-3}. This number may be the density in a uniform, rarefied medium; whether this medium is HI or HII is an open question.

(ii) 21-cm observations support the existence of a hot, rarefied HI gas, whose distribution may be somewhat irregular. It seems very attractive to identify this HI-medium with the medium mentioned sub (i), but there is no conclusive evidence for this interpretation.

(iii) Weaver interpreted his 21-cm observations as indicating that in the interarm region the overall H-atom density may be as low as $\frac{1}{40}$ of that inside a spiral arm.

(iv) Mills brought up the paper by Gould (1969), who compared emission and absorption of (radio) bremsstrahlung by galactic interstellar electrons. Gould used mainly data by Wilson (1963) on high-frequency emission and by Ellis and Hamilton (1966) on low-frequency absorption. These data can be explained by a 6000 K medium with $\langle n_e^2 \rangle = 0.06$ cm^{-6}. From the present discussion, two supplemental remarks can be added. First, Gould decided on the temperature of 6000 K by balancing emission and absorption. If T is higher (say 8000 to 10 000 K), there is too much absorption compared to the emission, and a model is required consisting of two components with different temperatures. The cool component would provide the absorption; the hot component, the emission. If the cool component is associated with neutral hydrogen clouds, one expects a very irregular distribution of absorption on the sky. This is in contrast to what Ellis and Hamilton state, but it agrees with a remark by Gould about the coincidence of an intense cloud of HI and a region of unexpectedly large absorption in the survey of Shajn et al. (1961) at 20 MHz. The second remark concerns the fact that the observation of bremsstrahlung emission yields (according to Gould) $\langle n_e^2 \rangle = 0.06$ cm^{-6}, whereas $\langle n_e \rangle = 0.06$ cm^{-3}. This implies that the electrons, producing the radio emission, are confined to a small fraction of space (a limiting case is where six percent of space contains $n_e = 1$ cm^{-3} and the rest of space is empty). Such 'dense' regions (e.g., extended HII regions like the Gum Nebula) also contribute considerably to the low-frequency absorption. In this connection Mezger mentioned that in a high-frequency survey by Altenhoff and Mezger all emission could be attributed to discrete sources (both non-thermal and thermal).

(v) Mills suggested that the z-distribution of electrons may resemble more closely the synchrotron disk with $|z| = 350$ pc, than the neutral hydrogen disk with $|z| = 200$ pc. This point has been brought forward by Bridle and Venugopal (1969). It should be mentioned that on basis of the cosmic-ray ionization model such behavior may be expected (at least qualitatively) since the ratio n_e/n_H increases with decreasing n_H.] (Bridle, A. H. and Venugopal, V. R.: 1969, *Nature* **224**, 545; Ellis, G. R.A. and Hamilton, P. A.: 1966, *Astrophys. J.* **146**, 78; Gould, R. J.: 1969, *Australian J. Phys.* **22**, 189; Shain, C. A., Komesaroff, M. M., and Higgins, C. S.: 1961, *Australian J. Phys.* **14**, 508; Wilson, R. W.: 1963, *Astrophys. J.* **37**, 1038.)

6. The Dynamical Theory of H II Regions

Mezger: I should like to comment on the papers by van de Hulst and Field. It might appear as if the theory and the observations of H II regions agree and that all problems are solved. This is certainly not the case, and I would be very unhappy if theoreticians would stop working on H II regions. First, let me mention the problem of electron temperatures. Theory predicts that electron temperatures increase with increasing distance from the exciting star, because of hardening of the ionizing radiation. However, collisional deexcitation counteracts this increase in regions of high density. As a result, in H II regions with high central densities, we may expect a decrease of temperature in the inner regions, followed by an increase in the outer parts. Observations, however, show only the decrease of temperature. Second, I have observational objections against the theoretical evolution models of H II regions. These theories start with the following initial conditions: an O star is turned on in a homogeneous neutral gas. The ionization front advances very rapidly and establishes the 'initial Strömgren sphere'. Subsequently, the ionization front advances, preceded by a shock front. In later stages, the region is surrounded by a thin, but dense, shell of neutral hydrogen. The theory is well-developed; its only defect is that it considers a situation that has not much relation to reality. Look at some recent observations:

(a) Several independent lines of investigation suggest that the distribution of ionized hydrogen in H II regions is far from smooth. In fact, in young H II regions, clumping factors (defined as: volume occupied by all ionized gas/total volume of dense regions) are found to be typically of the order of 30. A similar value was found some time ago by Osterbrock and Flather. In recent theories such clumping has been taken into account, to my knowledge, only by Dyson.

(b) In all young H II regions observed by us at NRAO, the total mass of ionized hydrogen appears to be only a small fraction (less than ten per cent) of the total star mass. Thus, the dynamics and evolution of an H II region are probably determined by the stars. In fact, it appears as if most of the ionized hydrogen is concentrated around stars.

(c) With only a few exceptions, the ionizing stars of young H II regions are unknown. Therefore these stars are hidden behind a circumstellar dust cloud. How they can ionize the surrounding gas through the dust cloud is still a puzzle.

So, it appears that the evolution of H II regions is very closely connected with the formation and evolution of star clusters. I will discuss this in more detail in my Report (p. 336).

Field: I think that Mezger is correct in stating that, theoretically, the radiation should become harder as one approaches the ionization front. This hardening will tend to increase the temperature of the ionized gas. But there is a counter-effect that will decrease the temperature, and that has to do with the diffuse Lyman-continuum radiation field. This counter-effect was discovered by Hjellming and has been discussed in my Report. I also agree with Mezger that many H II regions are rather irregular, and I admit that density fluctuations pose important theoretical problems. Nevertheless, it may be going too far to say that all H II regions are irregular. I believe that some appear quite regular. I suggest that these be compared with the theory.

[The following part has been taken from the discussion on Tuesday, September 16 after Thomas requested some reactions from the aerodynamicists present. Dr. Goldworthy's comments were inspired by the question: "How can aerodynamicists contribute to and profit from interstellar gas dynamics?"]

Goldsworthy: The dynamic interaction of H II regions with density inhomogeneities presents problems to which the aerodynamicist can, I think, contribute greatly. Field suggested in his Report (p. 51) that everything is known theoretically! However, several real difficulties exist on the theoretical side. The most common types of ionization fronts which occur in the expansion of H II regions are weak R-type and strong D-type. One need not worry too much about nomenclature; but it should be noted that a weak R-type front corresponds to a weak detonation, and a strong D-type to a strong deflagration in normal combustion theory. These types of combustion fronts do not occur under normal laboratory conditions (the Chapman-Jouguet hypothesis). Consequently the expansion of H II regions presents a phenomenon outside the experience of our terrestrial environment and one in which we are very much interested. Ionization fronts do not occur in the terrestrial environment because here, reaction fronts cannot travel relative to the gas at a rate faster than the thermal sound speed. This limitation does not apply in the interstellar medium where radiative transfer plays a dominant role. Both D- and R-types of ionization fronts move with supersonic velocity relative to the gas behind them. The principal difficulty facing the aerodynamicist is that for these types of ionization fronts, the two characteristics of the differential equations involved emerge from the back of the ionization front. This situation is in contrast to a shock wave, which travels subsonically relative to the fluid behind, so that one characteristic enters the shock and the other characteristic leaves it. In order to obtain a unique answer to a problem in which weak R- and strong D-type ionization fronts occur, it is necessary to feed into the calculation the detailed structure of the front. To illustrate the importance of this dictum, I would refer you to the computations carried out by Mathews and referred to by Field in his Report (p. 51). In Mathews' calculations, the shock wave is treated as a diffuse region. In some cases a shock wave occurs within the ionization front. In the numerical treatment this shock wave extends over a major part of the ionization front structure. Clearly, a very drastic approximation is used here and may lead to an incorrect solution to the flow external to the ionization front. If one were to look at the propagation of ionization fronts, considered as discontinuities, one would obtain an infinite number of answers. It is only when you feed in the correct structure that a unique answer is obtained. Thus in undertaking numerical calculations using, say, a total of 100 mesh points, it may be necessary to have perhaps 99 in the structure and 1 elsewhere in the H II region to obtain consistent results. Essentially what one needs is something akin to the Chapman-Jouguet condition, which is applicable in normal combustion theory. That condition has now been obtained for the plane fronts and allows us to solve problems in one space dimension. Inclined ionization fronts constitute a real difficulty, one to which we at the University of Leeds are paying particular attention.

Field: Can you be more detailed in your criticism of Mathews' work?

Goldsworthy: The real difficulty is that, while Mathews does feed structure calculations into his numerical procedure, and everything is all right in the early stages, the results go wrong when the front reaches about twice the thermal sound speed of the H II region. This is precisely the point at which D-type ionization fronts can occur. The problem is aggravated for the higher-powered stars. As I have already mentioned, one must be very sure that the correct structure is being fed. What happens is that at some stage a shock is formed within the ionization front. In Mathews' calculations, he obtains a velocity of a weak R-type front of about the same magnitude as the sound speed; more refined structure calculations show that this is not possible. What is more, if one wants to go further with the calculations, one is in trouble because the range of influence extends from the place where the calculations went wrong. The errors were in the final stages of Mathews' calculations; if he had gone further, he would have obtained completely wrong and meaningless answers.

Field: Would the detached shock have a velocity higher or lower than the velocity that Mathews calculated?

Goldsworthy: I would guess that the shock which goes in front would have a lower velocity, because precisely at the place where the weak R-type front occurs in Mathews' calculations you are more likely to have a strong D-type which is slower moving. The difference would be small, in the H I region, but large in the flow in the H II region.

Field: My second question concerns the discrepancy between the calculations of the shock, and what seems to have been observed by the radio astronomers. The calculations indicate that the thickness of the H I region between the ionization front and the shock front is between 1 and 10 per cent of the radius of the H II region, depending upon the assumed thermal cooling time. On the other hand, the radio observations consistently give 50 per cent for this parameter in six cases. I wonder whether or not you would consider this a serious discrepancy from the theoretical viewpoint, in the light of the uncertainty you have just stated? Later we should ask the observers whether or not these data are reliable.

Goldsworthy: My comments would not make much difference with respect to the distance between the shock in the H I region and the ionization front. No matter what one does aerodynamically, it is difficult to let the shock move away from the ionization front. It is therefore a problem of interpretation of the data; otherwise, we have to think of some other process that will separate shock and ionization front.

Van Woerden: Dr. Field, can you elaborate on the six cases you mentioned?

Field: The technique has been to scan across a number of H II regions in the 21-cm line, using a telescope with a beam of about 1°. The expansion velocity of the H I region is deduced from the splitting of the 21-cm line, along the line of sight through the center of the H II region. The angular extent of the expanding H I shell is judged by the variation of the 21-cm profile across the H II region. This judgment, however, is difficult and problematic according to Riegel (1967). I do not think we can solve the observational difficulties here. I am only asking the observers to re-examine the problem with a view to testing the presented theories. (Riegel, K. W.: 1967, *Astrophys. J.* **148**, 87).

[The following part has been taken from the discussion on Thursday, September 18.]

Toomre: Dr. Mezger, you seem to reject any observational evidence for expanding H I shells surrounding H II regions. Could you give more detail?

Mezger: By correlating the space-velocity distribution of neutral (H I) and ionized (H II) hydrogen one sees immediately that H II regions are located where the H I gas shows a maximum density. In other words, massive stars form in the centers of dense clouds of H I gas. Consequently one would expect to find H II regions surrounded by H I gas of relatively high density and that is what single-dish observations of some H II regions show. However, these are not the relatively thin shells of highly compressed neutral gas which have been predicted to precede an advancing ionization front. Either uniformly expanding H II regions surrounded by dense shells of neutral gas do not exist or the angular resolution of present 21-cm telescopes ($\gtrsim 10'$) is not sufficient to detect them. We shall probably have to wait for interferometer 21-cm line observations to give this answer.

Van Woerden: To name but one example, in the Lacerta association, Sancisi and I (unpublished) have found clear evidence for neutral hydrogen moving away from the stars towards us with a velocity of about 10 km sec^{-1}.

Mezger: Would you interpret it as an expanding shell, or is it just one cloud which is moving away from the star?

Van Woerden: It is not a complete shell around the association. I think it is much too idealized a picture to expect a complete spherical shell. But we do see an extended cloud, or whatever you want to call it, with a velocity of 10 km sec^{-1} coming at us.

Mezger: Is there an H II region observed in this association, e.g., as a thermal radio source?

Van Woerden: I do not know of radio observations, but optical observations have shown an H II region.

Field: Perhaps we cannot settle now this question of the observations of the H I shell. But I am interested in hearing Mezger say that it really is a question of having sufficient resolution. So, there is no evidence against such expansion; rather we might say that there simply is no evidence at all on the question one way or the other.

[The following remark was made at the Tuesday, September 16 discussions.]

Goldsworthy: There is another point I wish to draw attention to. We know that inhomogeneities exist in interstellar gas clouds. We can learn much about the dynamics of ionization fronts by their interaction with such inhomogeneities. I remind you of the work of Dyson (1968) on globules. Suppose one takes a high density globule illuminated by some radiating star (the situation which Mezger accused theoreticians of evading). With low-density gas outside, we imagine that their fronts will move quite rapidly. Probably a fast moving weak R-type front will result. This front will meet high density material. What sort of flow is established? If the density of the 'globule' is not too much higher than the gas outside, the gas inside will be heated to the same temperature as that outside; but the higher density of gas in the globule means a higher pressure. A motion is therefore established which is akin to that in a shock tube. In other words, shocks will be sent out from the object, and rarefaction waves will traverse the high density gas within. This rarefaction wave will weaken the expanding shocks most at the place where the ionization front interacts first with the 'globule'. This weakening of the shock governs the length of time the object is seen in emission within H II regions. The important point to note is that, associated with these objects, there will be sideways shocks carrying the material out with it. If the globule is of high density, the ionization front will be curved, thus running into the difficulties already mentioned of having more than one space dimension involved. Approximative theory is available, however; and in this case the 'hypersonic' approximation is a very useful tool for discussing these problems. (Dyson, J. E.: 1968, *Astrophys. Space Sci.* **1**, 388).

6. DISCUSSION FOLLOWING THE INFORMAL
MEETING ON INTERSTELLAR HI CLOUDS

(Friday, September 12, 1969)

Chairman: S. B. PIKEL'NER

Editor's remarks: On Friday afternoon, September 12, an informal meeting was held to discuss the distribution of neutral interstellar matter. Chairman of the meeting was Thomas. Summaries were presented at the formal discussion meeting, later the same afternoon. No change of importance has been made in the original discussion. The text has been inserted where it seemed most functional, namely closely following the Discussion on the Reports by Weaver and Field.

Pikel'ner: Dr. van Woerden, will you now present to the Symposium your summary of the special discussion meeting held this afternoon on the HI clouds?

Van Woerden: To my surprise, in this afternoon's discussion we have not had any trouble about what a cloud really is. It appeared immediately that all the 21-cm observers define a cloud, or a feature, by its velocity. I have myself added that one should also look at the velocity dispersion of the feature, since that might be another important characteristic. In one line profile there may be two features, with the same velocity but different velocity dispersions, on top of each other, which are due to two separate clouds in two separate parts of space.

We next drew a picture of the distribution of neutral hydrogen in space, in the solar neighborhood, which started out with the map of the Galaxy presented by Weaver earlier in this Symposium (see p. 22). Between the Sagittarius Arm and the Perseus Arm, there is an 'Orion Branch', with an estimated width or thickness of some 500 pc. Weaver thinks within the Orion Branch there is a cucumber-shaped structure around us, with a length of about 300 pc, and an average neutral-hydrogen density of 0.3 atom cm^{-3}. We are not fully agreed on this point; although I am convinced that there is a large structure, as evidenced by the continuity of hydrogen profiles over large areas of sky, I am sure that there are directions where we do not see it, so that the Sun might be at the edge of this cucumber. Then, going down in size, we believe that there is a whole spectrum of sizes reaching down from 100 pc to perhaps 3 to 5 pc and probably even lower. (The uncertainty here is mainly determined by limitations of angular resolution, higher resolution being obtained only in 21-cm line absorption spectra of radio sources.) Within this range of sizes we considered two possibilities. (1) There may be a progression of sizes (Figure 1a): big clouds, smaller clouds, still smaller clouds, which are all separate in space; or (2) there also may be what we call a hierarchy (Figure 1b): a big cloud, a smaller cloud within it, and again a smaller cloud within that one. It seems that both cases occur; I believe we agree on that point.

Habing (ed.), Interstellar Gas Dynamics, 98–107. All Rights Reserved. Copyright © 1970 by IAU

So much for the sizes. The next thing is the spectrum of densities within clouds. There is no good agreement on that. The density within the cucumber shape's large-scale structure was 0.2 atom cm^{-3}; next there are higher densities within the clouds, varying from 1 or 2 to 10 or possibly 100 atom cm^{-3}. The disagreement involves the density contrast between big cloud, small cloud and no cloud. If I understand Weaver

Fig. 1a. (See the summary by van Woerden.) A progression of sizes of interstellar clouds.

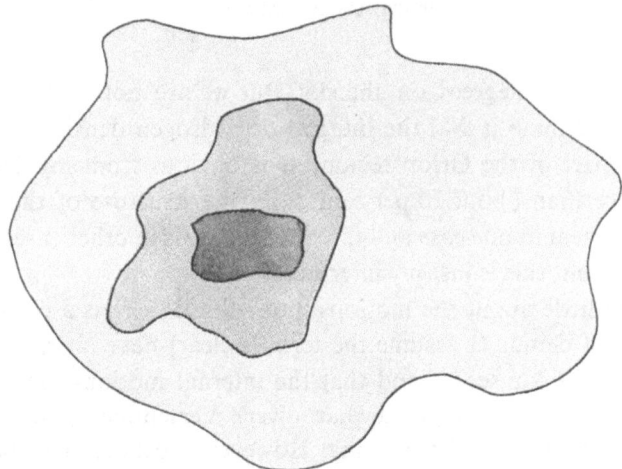

Fig. 1b. A hierarchy of cloud sizes; the smaller clouds are supposed to be inside the larger ones.

correctly, he considers that the density distributions are as shown in Figure 2a: stronger density contrasts for larger scale sizes. I know several cases where the distribution is as shown in Figure 2b: smaller structures having higher density contrast.

In connection with this problem of densities, Stecher has drawn attention to the Ly-α measurements in Orion, which show very slow variation, say 10 per cent over

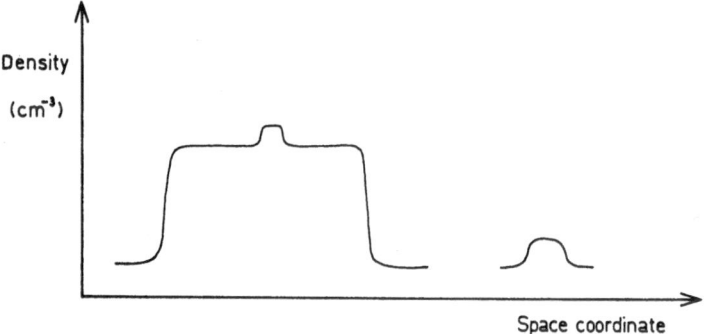

Fig. 2a.   (See the summary by van Woerden.) Density distribution with stronger contrast for larger sizes.

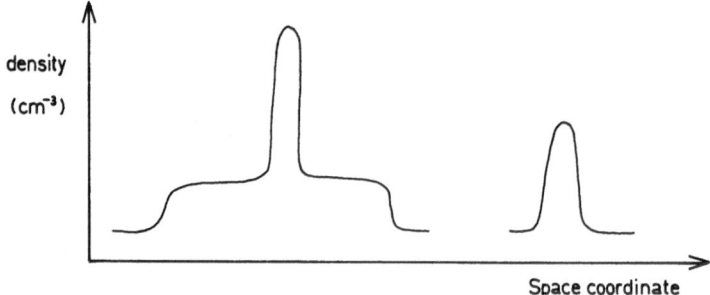

Fig. 2b.   (See the summary by van Woerden.) Density distribution with higher contrast for the smaller structures.

perhaps as much as ten degrees on the sky. But we are not looking at one cloud there; what we determine is N_H, the integral of hydrogen density along the line of sight. In a large part of the Orion region, as is obvious from my 21-cm work, N_H varies by no more than about 10 per cent. Still, the structure of the profile varies: there is one component in one case and three components in other cases. (This was not part of our discussion; this is just my interpretation.)

We have talked little about the motions, but I think there is a consensus that the external motions of clouds (I assume the term is clear) have an r.m.s. value in one coordinate of about 6 km sec^{-1}, and that the internal motions are of the order of 1 or 2 km sec^{-1} (in a few cases somewhat lower). Verschuur thinks the latter value is determined by instrumental broadening. However, I believe that the observations with highest resolution had a sufficiently small bandwidth and that the values for the internal motions are reasonably reliable.

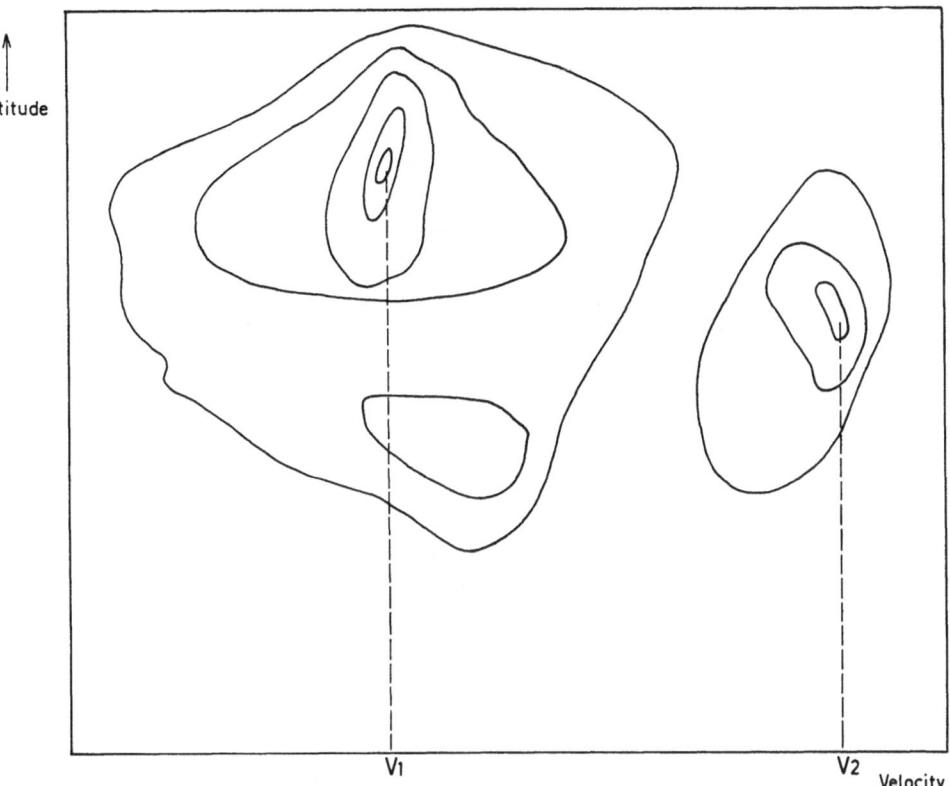

Fig. 3. (See the summary by van Woerden.) Contour diagram showing intensity as a function of latitude and velocity for contrast longitude. There is a 'feature' at velocity V_1, another one at V_2.

That was essentially all of the discussion about interstellar clouds from the observational side.

Thomas: You said we all agree that the clouds are defined by the velocity. How?

Van Woerden: We have not discussed that in detail. However, I think there are two ways. The method followed by Weaver works with contour diagrams: intensity contours in a diagram with velocity and one sky coordinate (longitude l or latitude b) as variables, and the other sky coordinate fixed (Figure 3). If the intensities are high at a particular velocity V, over a range in l or b, one distinguishes a feature (say, a cloud or some big structure) at velocity V and tries to follow it on the sky by considering also the second sky coordinate.

The method used mostly by us at Groningen, but also by other workers (van Woerden, 1967) is as follows: in a line profile $T_b(V)$ obtained at one position (l and b fixed), a few components are recognized (Figure 4), which must mean that in this direction we observe a few groups of atoms, each characterized by an average velocity, by a velocity dispersion around the average, and by the number of atoms in the group. (This is essentially the same thing that Adams and others have done in the analysis of cloud spectra.) We next examine whether similar components (that is, components

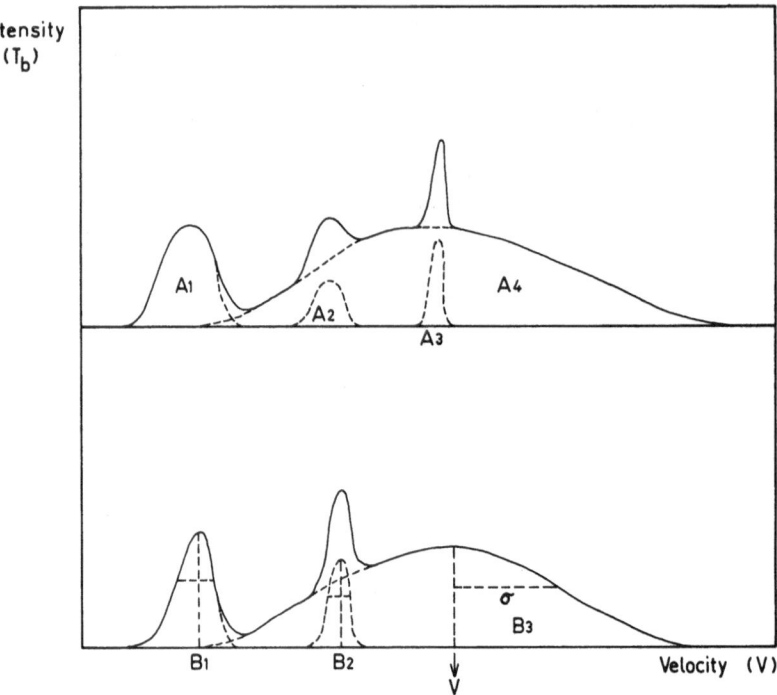

Fig. 4. (See the summary by van Woerden.) Profiles $T_b(V)$ at two neighboring positions on the sky. The top profile has four components: $A1$, $A2$, $A3$, $A4$; the bottom profile three: $B1$, $B2$, $B3$. For the latter three, velocity V and dispersion σ are indicated by the position and half the width of a cross. Components $A1$ and $B1$ are considered to belong to one cloud, $A2$ and $B2$ to another, $A4$ and $B3$ to another; $A3$ belongs to a cloud appearing only at position A.

having similar velocity V and dispersion σ) are present in profiles at neighboring positions; if so, we consider these components as belonging to the same 'cloud', and can then determine the density variations within the cloud and various other cloud properties. Obviously, our procedure requires the variations of V and σ within a cloud to be slow. The two methods described are, I think, not very different in principle. [Van Woerden, H.: 1967, *IAU Symposium No. 31, Radio Astronomy and the Galactic System* (ed. by H. van Woerden), Academic Press, London, p. 11.]

Verschuur: I disagree with about half of what van Woerden has said, but I am not sure yet which half!

Pikel'ner: I now call upon Dr. Field to present a summary of this afternoon's discussions on the theoretical aspects of H I clouds.

Field: The discussion of the theory centered on the presentation by Pikel'ner of a complete, but not yet worked-out, picture of the formation of, first, spiral arms and, second, interstellar clouds. In this picture, the basic mechanism involves a gravitational instability in the stellar disk, which leads to a condensation of stars into spiral arms producing a larger gravitational field in the arm than in the interarm region. This mechanism has not been discussed in detail at this meeting, but we have heard about it several times. It was presented at the IAU Symposium No. 38 in Basel. For our purposes, however, the important point is that the interstellar gas is swept over by this spiral pattern; it experiences an increased gravitational field and consequently a compression in the vertical direction (z-direction). This compression raises the pressure everywhere through the medium and leads to thermal effects which then become important in discussing the formation of clouds. In the (P, ϱ) diagram (see Figure 3 at p. 55) a critical pressure exists, and if within a spiral arm the pressure anywhere exceeds this critical value, a transition must occur to the high-density phase; the results of this transition are identified as clouds. The time-scale for such instability is essentially the cooling time for the material, 10^6 yr. In addition, because of the existence of the magnetic field, this instability leads naturally to a second instability, namely that discussed by Parker in his Report (see p. 168). Basically it is a Rayleigh-Taylor instability. In this second instability the material moving along the lines of force causes a downward depression in these lines and the gas slides further down along them. Cosmic-ray pressure outside the clouds accelerates the growth of the instability. We expect the Rayleigh-Taylor instability to occur behind the leading edge of the spiral arm, along with the development of clouds. The associated time scale is 10^7 yr. As increasing amounts of material stream into massive clouds, their density increases as well, and there may come a point where self-gravitation of the gas becomes important. This can lead to the collapse of clouds and possibly to the formation of stars (including unstable stars such as supernovae), which may accelerate the gas out again and account for the motion of clouds. Moreover, the mere fact that the gas is flowing into the low region in the magnetic field gives it a velocity, the energy source being the cosmic rays which are inflating the magnetic field and tending to drive the gas up to levels of higher gravitational potential.

Pikel'ner mentioned the fact that, the magnetic field being strong, transverse Alfvén waves will be a very important mode of motion. It seems to me that up until this time we have not properly considered Alfvén waves in discussing 21-cm observations. Such motions could be as large as several km sec^{-1}, and therefore could be important in the discussion of both internal and external velocities of clouds. In particular, large-amplitude Alfvén waves may explain the large (probably supersonic) velocities within individual clouds inferred both from 21-cm line and from Ca$^+$-line studies.

It seems to me that we have a splendid opportunity in the next few years to merge theory and observation. On the one hand, the observational situation has greatly improved as a result of the possibilities of accumulating large numbers of profiles and making maps in a limited amount of time. On the other hand, we have developed a fairly clear picture of clouds, and while there is, of course, much work to be done in elucidating the non-linear problems here, still one can already make some predictions. For example, one might expect that clouds would tend to develop toward the rear of the spiral arm, rather than in the front, because of the time delay involved. Among the questions of interest to theoreticians are the ones mentioned earlier: the density contrast, the sharpness of distinction between dense regions and not-so-dense regions, and the velocities within certain structures (are they supersonic or are they subsonic?). Also, can structures be found which stretch along the magnetic field (e.g., along

Mathewson's helix), or are they, as some theories would suggest, compressed into pancakes along the magnetic fields? These questions, I think, are directly relevant for the theoretical developments in the next several years.

Finally I would like to express some thoughts about how we are to use movies that we have seen and those that are still to come. I think the history of solar physics suggests that it is possible to make movies without doing science. Movies can be helpful in giving qualitative ideas about the phenomenon involved; but for comparison with theory we, of course, need numbers. It is absolutely essential that as much of the data as possible be reduced to quantitative numbers.

Thomas: I think it is interesting that we have the same situation in the interstellar medium as we have in stellar atmospheres; viz., turbulence is still just a word of ignorance, associated with very little clear understanding of the physical implications. I got that feeling, after listening to the discussions in the preceding two-hour session on the H<small>I</small> clouds. This is a beautiful field to work in. The concepts are undefined, the interpretation can only be called optimistic. Observers can disagree violently on what they see, then summarize by saying that all agree on definitions. They can argue whether the velocities are subsonic or supersonic, then make a linearized theory for their origin. Also, as an occasional solar astronomer, I have seen many solar movies, and I would take a position 180° from Field's. In a situation such as the interstellar medium or the inhomogeneous solar atmosphere, I think that progress comes first by looking at the most graphic presentation of the greatest possible array of data. I was extremely impressed by Weaver's presentation and his cautious attempt to give what might be a definition of a cloud or concentration. I have the feeling that at present we do not have any more physical feeling for what we mean by 'turbulence', 'cells', 'clouds', or 'concentrations' here than we do when we discuss 'turbulence' and 'inhomogeneities' in the stellar atmosphere. I regard these as optimistic remarks, because there is so much to be done. I hope we can systematize the data and our conceptual thinking about them.

Weymann: How long does it take for a spiral density wave to travel across a given gas element?

Field: Between $(3 \text{ and } 10) \times 10^7$ yr.

Weymann: Do you think it is a trivial or a non-trivial point that only clouds exist in a fairly narrow range of pressure and density on your (P, ϱ) diagram? Can one understand how there could be a whole range of densities?

Field: I do not know a satisfying answer to Weymann's question; I have not dealt with the problem yet. Goldsmith (1969) has considered some linear calculations of stability and to some extent a few non-linear calculations of the development of the thermal instability after being induced by a shock wave. In a few shock calculations Goldsmith found a transition from the unstable to the stable phase in 10^6 yr. One expects that on shorter time scales intermediate phases will indeed be found. This time scale is only a few percent of the time required by a gas element to go through a spiral arm. I think we may well find that the proper resolution of the problem is a dynamical cycle in which we find compression, then condensation; then, following the passage of the arm, expansion again, only to be followed by recompression. And the dynamical aspect of this problem, when fully explored, will show that 10 to 30 percent of the gas is in intermediate phases. That is one new problem to study which I will take home from this Symposium. (Goldsmith, D. W.: 1969, Thesis, University of California, Berkely.)

Habing: I would like to point out a possible difficulty in the cycle that Field described. The difficulty is with the expansion phase. The option is that at the rear of a spiral arm the clouds evaporate, because the surrounding pressures become so low.

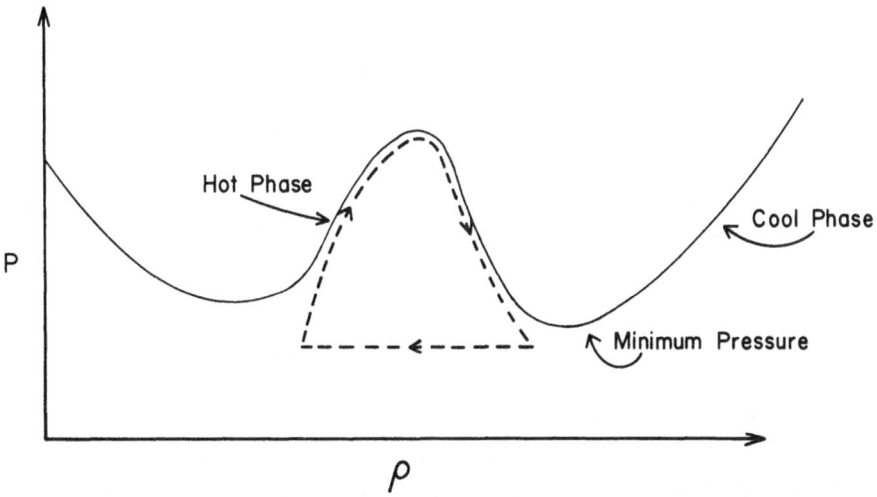

Fig. 5.   (See the remark by Habing.) (P, ϱ) curve similar to that of Figure 3 at p. 55, but extended (schematically) to lower densities and higher temperatures. There is a second minimum in the curve at the lower densities. If this minimum is less deep than the other one, phase transitions at minimum pressure cannot occur in the direction indicated by the broken line. The broken line indicates a possible cycle, as discussed by Field.

One would guess then that evaporation takes place at constant pressure. But if one calculates the extension of the (P, ϱ) curve toward higher temperatures, it turns out that there is no thermal equilibrium point available at the high temperature end (see Figure 5). The (P, ϱ) curve has to increase (for decreasing ϱ) because of bremsstrahlung-cooling. Figure 5 is only schematical; we still have to make more detailed calculations.

Field: In addition to Habing's remarks, I would like to say that the situation at the high temperatures is still a little unclear. The reason is that in order to study the thermal equilibrium in this high-temperature region, one must have the ionization equilibria (in the presence of cosmic rays) of C, N, O, and perhaps Fe, through many stages of ionization. Since all these elements contribute to the cooling at the high temperatures, I think it is too early for us to say that we understand that branch of the curve.

Sunyaev: There exist calculations by Cox and Tucker (1969) of the cooling rate for a plasma with normal cosmic abundances. I know that several people present calculated the thermal equilibrium in the intergalactic medium for H and He. Why can you not use these calculations? (Cox, D. P. and Tucker, W. H.: 1969, *Astrophys. J.* **157**, 1157.)

Field: I think those calculations are not directly applicable to this situation, since we are talking about cosmic-ray heating and cosmic-ray ionization. The ionization equilibrium in Cox and Tucker was for thermal collisions only and did not include cosmic-ray ionization.

Sunyaev: I understand. But if the thermal electron collisions in the plasma give a degree of ionization higher than the cosmic-ray flux, the cooling rate is the same as in the collisional case. And the degree of collisional ionization increases very rapidly with temperature.

Field: I do not think that you are necessarily right. In thermal equilibrium the only ions present in significant numbers are those with ionization potentials of the order of 10 kT. However, in the presence of cosmic rays, ionization can take place to a very much greater degree, and it works out that the ionization equilibrium can in fact include all the stages more or less equally. The ionization equilibrium with cosmic rays may be completely different from that without cosmic rays.

Pikel'ner: Field mentioned that the Rayleigh-Taylor instability collects the gas and that it is one of the mechanisms of formation of the denser clouds. After the formation of hot stars in these complexes, H II regions appear and push the remnants of the gas complex out into space. Very probably part of the gas will not flow out at all, but will stay at the same place. Ultimately the magnetic field will remain slightly curved there. After the H II region disappears, the gas clouds can collect again in this place. We should therefore observe the formation of stars at the same place in time intervals of 10^7 yr. For example, in the Orion region one finds older stars with ages of about 10^{10} yr and very young stars with ages much less than 10^6 yr. I think we observe here the recurrent appearance of hot stars within time lapses of about 10^7 yr.

Field: Dr. Parker, there is an interesting study by Kippenhahn and Schlüter (1957) about confinement of cool solar matter in a magnetic well. They studied a two-dimensional situation. You said earlier that you expect the Rayleigh-Taylor instability to develop with sharp variation in the third dimension. Can you explain to us the difference between this Kippenhahn and Schlüter picture, which ended up with what is observed in the Sun (namely, a long filament suspended in the magnetic field perpendicular to the field), and your picture, where you have sheets along the field? (Kippenhahn, R. and Schlüter, A.: 1957, *Z. Astrophys.* **43**, 36.)

Parker: I will draw it on the board (Figure 6). The model was meant for solar prominences in which material is supported on magnetic fields coming up from the Sun. The material is in its cool, dense phase due to thermal instability. I was stating a different situation for the galactic field. When it becomes unstable, it tends to slice itself in the third dimension, giving sheets parallel to the plane of the paper. The situation studied by Kippenhahn and Schlüter was different since in their case the lines of force are rooted in the Sun, whereas you have no such stabilizing effects in the Galaxy.

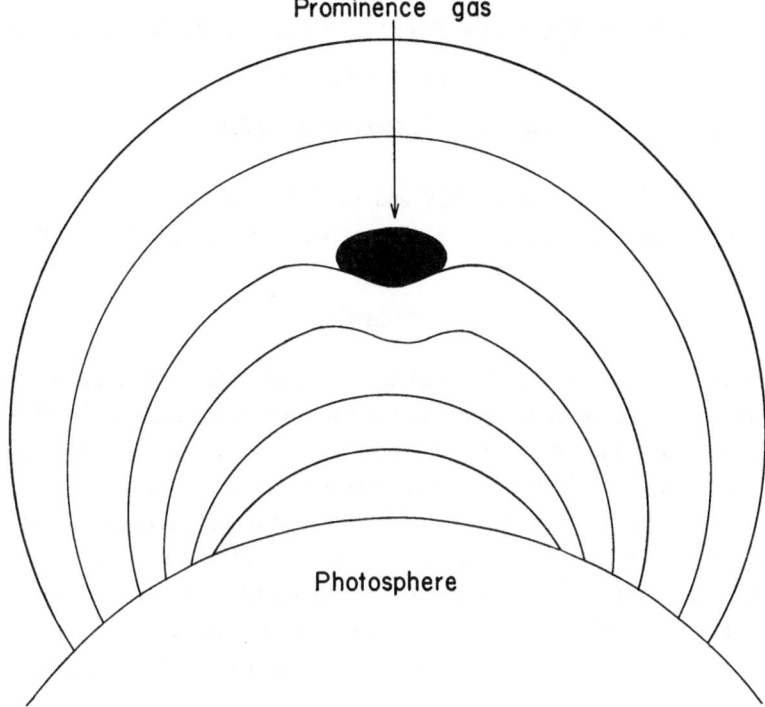

Prominence gas

Photosphere

Fig. 6.   (See the remark by Parker.) Schematic drawing of the Kippenhahn-Schlüter model for cool
solar material suspended in a magnetic well in the solar corona.

Lüst: To me the rooting of the magnetic lines in the Sun is an essential feature of the Kippenhahn-Schlüter model.

Mestel: Kippenhahn and Schlüter found that their model was unstable if the gas is supported by a local dipolar field. The gas tends to slide down into the Sun if the field is given a slight asymmetric disturbance. For stability they required a local quadrupole field.

7. COLLECTIVE PLASMA PHENOMENA AND THEIR ROLE IN THE DYNAMICS OF THE INTERSTELLAR MEDIUM

Introductory Report

(Wednesday, September 10, 1969)

B. B. KADOMTSEV and V. N. TSYTOVICH

Fizicheskij Institut P. N. Lebedev, Akademiya Nauk S.S.S.R., Moskva, S.S.S.R.

1. Introduction

The First Symposium on cosmical gas dynamics clearly demonstrated that magnetic fields play an essential role in the dynamics of the interstellar matter. The Symposium emphasized that the behavior of interstellar gas should be described not by traditional gas dynamics but with the help of magneto-gas dynamics. As we now realize this was only the first step, since in plasmas a very broad range of phenomena can occur which are coupled not with large-scale magneto-gas dynamic motions but rather with more subtle collective plasma motions. To describe properly the dynamics of an ensemble of charged particles we should use not only average characteristics such as density, pressure and mean velocity, but also more detailed features of the particle distribution function.

At present, plasma physics is becoming a well-developed branch of modern physics. It contains an elaborate system of concepts, based both on theoretical investigations of simplified models and on many experimental studies. These investigations have led to the general picture of a plasma as a very unstable medium, in which many types of oscillations and noises can be easily excited. Such oscillations are not only the large-scale Alfvén and magnetosonic waves, which are already known from magneto-gas dynamics, but also the pure plasma oscillations. These oscillations can interact with plasma particles leading toward energy and momentum exchange. If the amplitudes are sufficiently high, nonlinear interaction between these oscillations is significant for the behavior of the plasma.

It is the purpose of the present paper to discuss the essential role that collective processes play in cosmic plasmas. The first part (Sections 1 through 5) deals in a general way with the concepts developed in recent years of collective phenomena in plasmas. Various types of instabilities, which are associated with the excitation of waves by particles (for example by cosmic rays), are considered. A picture is presented of nonlinear interactions and wave transformations that produce the spectra of oscillations in a weakly turbulent plasma. In the second part of this Report (Sections 6 through 10) acceleration of particles in a turbulent plasma is considered, and the role of this process in the generation of cosmic rays is discussed. The role of Langmuir turbulence is emphasized particularly. The turbulence together with the electromagnetic radiation that it excites and with the induced Compton scattering, gives a power-type spectrum of relativistic particles.

Habing (ed.), Interstellar Gas Dynamics, 108–132. All Rights Reserved. Copyright © 1970 by IAU

The first part of the report is written largely by B. B. Kadomtsev, the second part largely by V. N. Tsytovich.

2. Waves in a Plasma

The presence of electromagnetic forces in an ensemble of charged particles leads to an elastic behavior. For example, if electrons are displaced relative to ions, an electric field is produced which tends to return the electrons to their initial position. When a magnetic field is present the quasi-elastic forces occur even if the electrons and ions are displaced together. In contrast to usual hydrodynamics, where small velocities can give rise to large displacements, a weak motion in plasmas does lead to small oscillations around some stationary state. It appears that such oscillations play a significant role in plasma dynamics. First we consider three cases with no external magnetic field present. Then we turn to the more complex case of plasma oscillations in the presence of an external field.

A. LANGMUIR OSCILLATIONS

Consider first the 'Langmuir' or 'plasma' oscillations which arise when the electrons are displaced relative to the ions, causing small perturbations in the state of quasi-neutrality. This type of oscillation is the simplest example, in which the long-range Coulomb forces are essential. Let the electrons be cold and let their density n_0 be constant in space. Give the electrons a small displacement ξ. For simplicity take the magnetic field zero. Then

$$\ddot{\xi} = -\frac{e}{m_e} \mathbf{E}, \tag{1}$$

where e is the charge of an electron, m_e its mass, and $\mathbf{E}$ the electric field strength. Such displacement leads to a perturbation of the charge density

$$n_e' = n_0 \operatorname{div} \xi. \tag{2}$$

Since the ion mass is much larger than the electron mass, we may consider the ions to be at rest and, therefore,

$$\operatorname{div} \mathbf{E} = -4\pi e n_e' = 4\pi e n_0 \operatorname{div} \xi. \tag{3}$$

Since $\operatorname{curl} \xi$ and $\operatorname{curl} \mathbf{E}$ both vanish, it follows from Equations (1) and (3) that, in an infinite plasma,

$$\ddot{\xi} = -\omega_{pe}^2 \xi \tag{4}$$

where the 'Langmuir frequency' $\omega_{pe} = \sqrt{(4\pi e^2 n_0/m_e)}$. Equation (4) holds for any displacement $\xi(\mathbf{r})$. As we shall see below it is very useful to expand $\xi(\mathbf{r})$ in a Fourier series, i.e. to consider $\xi(\mathbf{r})$ as a superposition of plane waves

$$\xi(\mathbf{r}) = \int \xi_k \exp(i\mathbf{k} \cdot \mathbf{r}) \, d\mathbf{k}. \tag{5}$$

Each harmonic ξ_k in this expression corresponds to a plane wave, which propagates in the direction of the wave vector $\mathbf{k}$ with the phase velocity ω/k. If the electron temperature T_e (the dimension of T_e is energy, it is expressed in eV (Ed.)) is not zero, the gradient of the electron pressure should be taken into account in Equation (1). But as this gradient is proportional to k, it may be neglected for $k \ll \omega_{pe}\sqrt{(m_e/T_e)}$. So far we have assumed the absence of an external magnetic field. If such a field is present, the equation of motion (Equation (1)) is more complicated. But in practical astrophysical cases the cyclotron frequency $\omega_{Be} = eB/m_ec$ is much less than the plasma frequency ω_{pe}, and the magnetic field does not influence the frequency of the Langmuir oscillations.

B. ION-SOUND WAVES

If the electron temperature T_e is at least three times as great as the ion temperature, then in a plasma in the absence of a magnetic field another branch of longitudinal oscillations occurs, the so-called 'ion-sound waves'. In such waves the ions oscillate inertially and we have

$$\xi_i = \frac{e}{m_i}\mathbf{E} \tag{6}$$

where m_i is the ion mass. Since the electrons are more mobile, they can come to an equilibrium. Hence, we have

$$T_e\nabla n_e = -e\mathbf{E}. \tag{7}$$

In the equation for $\mathbf{E}$, we must take into account the perturbations in both the ion and the electron densities. Using, again, $n_e' = n_e - n_0$ we get

$$\text{div}\,\mathbf{E} = -4\pi e n_0\,\text{div}\,\xi_i - 4\pi e n_e'. \tag{8}$$

From these equations, it is easy to find the expression for the frequency ω of a plane ion-sound wave

$$\omega^2 = \frac{k^2 V_s^2}{1 + k^2 d_e^2}. \tag{9}$$

Here $V_s[=\sqrt{(T_e/m_i)}]$ is the sound velocity and $d_e(=\sqrt{[T_e/(4\pi e^2 n_0)]})$ is the Debije length. The wave propagates for small k with the sound velocity V_s, whereas for large k, the frequency is near the ion Langmuir frequency $\omega_{pi} = \sqrt{(4\pi e^2 n_0/m_i)}$.

C. ELECTROMAGNETIC WAVES

If there is no external magnetic field, the transverse waves (for which $\text{div}\,\mathbf{E} = 0$) are completely separated from the longitudinal waves. They are described by Maxwell's equations, in which the current density $\mathbf{j}$ is expressed as a function of $\mathbf{E}$ by the equations of motion. The calculations give for the frequency ω

$$\omega^2 = \omega_{pe}^2 + k^2 c^2, \tag{10}$$

For large k this is the usual electromagnetic wave. For small k the frequency ω does not become arbitrarily small but approaches ω_{pe}, and if $\omega < \omega_{pe}$ the electromagnetic waves cannot penetrate in the plasma.

D. PLASMA OSCILLATIONS IN A MAGNETIC FIELD

If an external magnetic field is present, but if it is sufficiently weak (i.e. if $\omega_{Be} \ll \omega_{pe}$), the above expressions for Langmuir and electromagnetic waves remain valid, but a new qualitative feature appears – the possibility of penetration of low frequency waves in a plasma. In the language of magnetohydrodynamics, these are known as Alfvén waves and as (fast and slow) magnetosonic waves. Alfvén waves correspond to the displacement of plasma across **B** and **k** without compression. The frequency is $\omega = k_\parallel V_A$ where $V_A [= B / \sqrt{(4\pi n_0 m_i)}]$ is the Alfvén velocity and where $k_\parallel$ is the wave vector in the direction of the magnetic field. In magnetosonic waves, plasma compression occurs, which (for a fast wave) is in phase and (for a slow wave) is in antiphase with the enforced magnetic field. (The slow wave is, in some sense, a modified ion-sound wave.) The high-frequency behavior of the oscillations is strongly influenced by the dispersion, i.e. by the dependence of the phase velocity ω/k upon the wave number k. It is necessary to distinguish between waves of different circular polarization (in the case of propagation along the magnetic field lines) or between waves of different elliptical polarization (in the general case). Alfvén waves have a strong dispersion and become polarized in the direction of the ion Larmor rotation when the frequency of the waves ω is near the ion cyclotron frequency ω_{Bi}. An Alfvén wave can propagate only if $\omega < (k_\parallel/k)\omega_{Bi}$. A magnetosonic wave can propagate at much greater frequencies up to the electron

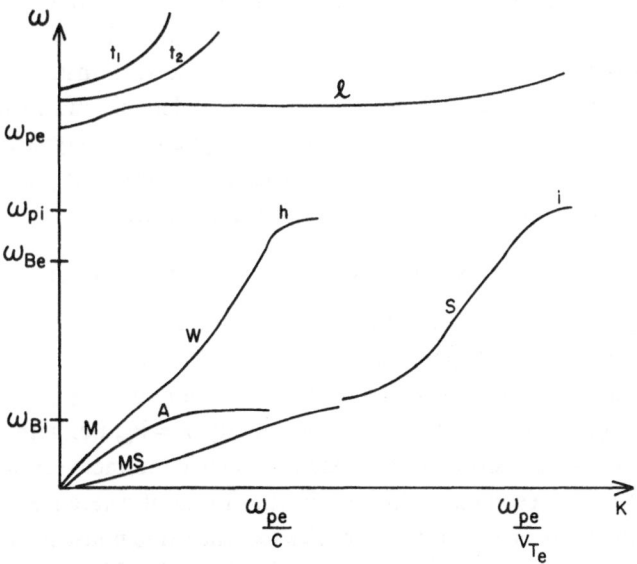

Fig. 1. Wave number k vs frequency for various oscillation modes. MS = slow magnetosonic, S = ion-sound, A = Alfvén, M = fast magnetosonic, w = whistlers, h = hybrid modes, l = Langmuir waves, and t_1 and t_2 are the two polarizations of the transverse waves.

cyclotron frequency ω_{Be}. If $\omega \gg \omega_{Bi}$ the ions are at rest and the oscillations are known as 'whistlers'. These conclusions follow from the equation for the frequency of a wave that propagates along the magnetic field:

$$\frac{k_\parallel^2 c^2}{\omega^2} = 1 - \frac{\omega_{pe}^2}{\omega(\omega \pm \omega_{Be})} - \frac{\omega_{pi}^2}{\omega(\omega \mp \omega_{Bi})}. \tag{11}$$

Here the upper and lower signs correspond to the polarization in the direction of ion and electron rotation, respectively. From Equation (11) one can see that for very low frequencies $\omega^2 = k^2 V_A^2$, as found before. For the whistlers (upper sign) we have

$$\omega \approx \frac{k_\parallel^2 c^2 \omega_{Be}}{\omega_{pe}^2}. \tag{12}$$

The qualitative dependence of the frequency ω upon the wave number k is shown for different branches in Figure 1.

3. Plasma Instabilities and Excitation of Oscillations *

The importance of oscillations in plasmas is determined by the question how special the conditions are by which the oscillations are excited. In this respect, it is amazing that many types of oscillations can be excited so easily. By intuition we connect this with the great mobility of a plasma. The interaction with the magnetic and electric fields gives the particles a collective behavior that very easily entails energy exchange between waves and particles. Spontaneous wave excitation in plasma is a result of its instability – i.e. a large number of theoretical investigations show that various different, infinitesimally small deviations from a stationary equilibrium will grow exponentially. We shall not consider configurational instabilities produced by large scale inhomogeneities in the magnetic fields and in the plasma, because such instabilities can be well described by magnetohydrodynamics and are not called plasma instabilities in the proper sense. But outside the framework of magnetohydrodynamics there exist kinetic instabilities, in which oscillations are excited by resonant particles and which are very sensitive to the form of the particle distribution functions.

A. AMPLIFICATION AND DAMPING OF LANGMUIR OSCILLATIONS; CHERENKOV RESONANCE

The simplest example of the kinetic instabilities is the excitation of Langmuir oscillations by an electron beam. Suppose that on the tail of the electron distribution function $f(V)$ (where V is the electron velocity) there exists a 'hump' so that $(\partial f/\partial V)_{V=V_0} > 0$. Now consider the Langmuir wave with the phase velocity $V = V_0$, i.e. with the wave number $k = \omega_{pe}/V_0$. It can be easily seen that such a wave will interact very strongly with the particles near $V = V_0$ (resonance). Indeed, in the frame of reference that moves with such particles, the wave is at rest, the particles feel the same phase for a long time, and the wave can change the particle energy considerably. If $(\partial f/\partial V)_{V_0} > 0$ there are more

---

* The authors use the terms 'oscillation' and 'wave' without distinction. (Ed.)

particles faster than the wave than there are slower than the wave. So the transfer of energy from the particles to the wave is greater than the transfer in the reverse process and, consequently, the amplitude of the wave grows with time. However, if $(\partial f/\partial V)_{V_0} < 0$ the amplitude decreases with time ('Landau damping'). The amplification of waves by a particle beam corresponds to the emission of induced Cherenkov radiation. Particles moving in a medium with a velocity greater than the phase velocity $V_0 = \omega/k$ emit Cherenkov radiation. The condition $(\partial f/\partial V)_{V_0} > 0$ means that near $V = V_0$ the population of the energy levels is inverted, i.e. the temperature is negative, which is a necessary condition for the maser effect to occur. From this point of view the expansion of the oscillations in plane waves is not only a formal mathematical procedure but it also has a real physical sense – each wave has its own phase velocity and interacts with a separate group of resonant particles. For large wave number k the phase velocity becomes small and for $k > d_e^{-1}$, where d_e is the Debije length, the damping of the wave by the thermal electrons becomes very large. Excitation of oscillations by a beam causes energy losses for the resonant particles. As a result, a 'plateau' is created in the distribution function, i.e. $\partial f/\partial V$ vanishes in vicinity of $V = V_0$. More exactly, the 'plateau' exists only around those velocities where the possibility for the induced emission is satisfied, i.e. $\partial f/\partial V > 0$ initially. Theoretically this effect has been investigated in the so-called quasi-linear approximation. The creation of a plateau and anomalously large losses of kinetic energy were found in a great number of experiments dealing with the interaction of particle beams and tenuous plasmas. In the preceding paragraph, V represented the longitudinal component of the velocity. Using the same letter for $|\mathbf{V}|$ we can write the condition for Cherenkov resonance

$$V \cos \Theta = \omega/k, \tag{13}$$

where Θ is the angle between $\mathbf{V}$ and $\mathbf{k}$. From Equation (13) it follows that if the phase velocity ω/k is fixed, a larger value of V gives a smaller interval for Θ in which interaction is possible.

B. ANOMALOUS RESISTIVITY

The instability described in the preceding paragraph can also occur when a current exists in a plasma. The instability is most pronounced when the drift velocity of the electrons $U = |\mathbf{j}|/en_0$ ($\mathbf{j}$ is the current density) is greater than the ion-sound velocity $V_s = \sqrt{(T_e/m_i)}$. In this case the Joule heating makes the electron temperature larger than the ion temperature and so the necessary condition for the existence of ion-sound waves is fulfilled. A similar condition as before ($U > V_s$) leads to amplification of the ion-sound oscillations and the current density saturates even if the electric field applied to the plasma continues to increase. Usually this fact is called 'anomalous resistivity' or 'turbulent heating'. The role of the fast electrons in this process is not yet completely clear but the existence of anomalous resistivity has been proven experimentally. In cosmic conditions anomalous resistivity will probably be important near surfaces with opposite magnetic fields where a high current density can exist. Such regions can at the same time be the sources of the plasma waves.

C. CYCLOTRON RESONANCE

In the presence of an external magnetic field the charged particle rotates with the cyclotron frequency

$$\omega_B = eBc/\varepsilon, \tag{14}$$

where $\varepsilon = \sqrt{(m^2c^4 + p^2c^2)}$ is the energy of the particle. This gives the possibility of cyclotron resonance when, in the frame of reference moving along the field, the frequency of plasma oscillations equals the cyclotron frequency or its harmonics. (The harmonics of the cyclotron frequency are effective only if the transverse wave length is large compared to the Larmor radius.) Taking the Doppler shift into account, the resonance condition is of the form

$$\omega - k_\| V_\| = l\omega_B \tag{15}$$

where ω is the plasma frequency and where $k_\|$ and $V_\|$ are the projections on $\mathbf{B}$ of the wave vector $\mathbf{k}$ and the particle velocity $\mathbf{V}$, respectively. l is an integer (positive or negative). If $l > 0$, we have the normal and if $l < 0$, we have the anomalous Doppler effect. The meaning of this is more easily understood if one interprets Equation (15) in terms of energy and momentum conservation laws by means of a quantum mechanic analogue. Consider the charged particle as a harmonic oscillator with equidistant energy levels (Landau spectrum). If a quantum is emitted with energy $\hbar\omega$ and momentum along the magnetic field $\hbar k_\|$, we have

$$\begin{aligned} \hbar\omega + V_\| \Delta p_\| - l\hbar\omega_B &= 0 \\ \hbar k_\| + \Delta p_\| &= 0 \end{aligned} \tag{16}$$

where l represents the change in the orbital quantum number. If $l > 0$ the transverse energy of the particle decreases, but if $l < 0$ (the anomalous Doppler effect) the emission of waves is accompanied by an increase in the transverse energy. Using this analogy it is easy to find the condition for induced emission, i.e. for instability: we require that the population of the energy levels be inverted, i.e. the value of the distribution function $f(p_\|, p_\perp)$ in the initial state is larger than that of final state $f(p_\| + \Delta p_\|, p_\perp + \Delta p_\perp)$. In the nonrelativistic case $\Delta p_\perp = -l\hbar\omega_B m/p_\perp$, i.e. for small $\Delta p_\|$ and $\Delta p_\perp$

$$\int_{-\infty}^{+\infty} \left(k_\| \frac{\partial f}{\partial p_\|} + \frac{l\omega_B m}{p_\perp} \frac{\partial f}{\partial p_\perp} \right) \times \delta(\omega - k_\| V_\| - l\omega_B)\, dp_\|\, dp_\perp > 0. \tag{17}$$

Here the δ-function insures that only resonant particles are able to emit or absorb the wave. For relativistic particles this condition is of the same form, but ω_B depends on the particle energy and $\omega_B m = eB/c$. Using Equation (15) we rewrite Equation (17) as follows

$$\int\limits_{-\infty}^{+\infty} \left[\frac{\omega}{p_\parallel} \frac{\partial f}{\partial p_\parallel} + l\omega_B \left(\frac{1}{p_\perp} \frac{\partial f}{\partial p_\perp} - \frac{1}{p_\parallel} \frac{\partial f}{\partial p_\parallel} \right) \right]$$

$$\times \delta(\omega - k_\parallel V_\parallel - l\omega_B) \, dp_\parallel \, dp_\perp > 0. \qquad (18)$$

The second term in the integral is important only if the distribution function is anisotropic $[(1/p_\parallel)(\partial f/\partial p_\parallel) \neq (1/p_\perp)(\partial f/\partial p_\perp)]$. If the energy in the transverse direction is larger than that in the longitudinal direction, $(1/p_\perp)(\partial f/\partial p_\perp) > (1/p_\parallel)(\partial f/\partial p_\parallel)$ and the instability is due to the normal Doppler effect; while in the opposite case it is due to the anomalous Doppler effect.

D. INSTABILITY OF COSMIC RAYS

The instability of plasma oscillations due to normal and anomalous Doppler effects may be responsible for the observed high degree of isotropy of cosmic rays (Ginzburg, 1965; Tsytovich, 1965, 1966a; Lerche, 1967; Wentzel, 1968). Indeed, even for a small anisotropy Equation (18) shows that instability can be found for waves with frequencies much less than the cyclotron frequency, i.e. for Alfvén and for magnetosonic waves. When the transverse pressure is greater than the longitudinal one, waves polarized in the direction of proton rotation (cosmic rays) are unstable due to the normal Doppler effect. If the transverse pressure is smaller, the wave with the opposite polarization is excited but, in this case, since the particles move faster than the phase velocity, the resonant particles rotate in the same direction as before. The instability can also be due to the presence of a drift velocity (i.e. absence of right-left symmetry along the magnetic field). Estimates show that for a very small anisotropy of about 1 percent as well as for a small longitudinal disturbance in the distribution function, the instability can develop with a growth time of the order of 10^2 years, which is negligibly small compared to the characteristic diffusion time of 10^6 years for cosmic rays in the Galaxy. Isotropization of cosmic rays may be due also to the presence of fluctuations in the electric and magnetic field that exist independently of the cosmic rays. However, the instability described here is the result of a direct response of the plasma to anisotropy in the distribution function and, therefore, the isotropization mechanism will be more effective when the density of the cosmic rays is sufficiently high.

E. POSSIBLE MECHANISMS OF WAVE EXCITATION IN THE COSMIC RAY PLASMA

Global anisotropy of the cosmic rays is only one of the possible causes for kinetic instability of the cosmic ray plasma. The instability may appear also because the cosmic rays flow out of the region of their origin. Moreover, in magnetohydrodynamic waves (i.e. low frequencies and large scales) regions of low magnetic field are produced which act as magnetic bottles. In such bottles, as many laboratory experiments indicate, the nonequilibrium distributions of fast particles seem to produce fast oscillations. This example shows that a low frequency oscillation of the magnetic field can cause an excitation of high frequency waves. In addition, excitation of high frequency waves, e.g. Langmuir waves, can take place at shock fronts, which occurs, for example, when beams are injected into regions having large magnetic field variations. Also, high frequency electromagnetic radiation may excite waves of other branches by non-linear decay processes (see *e.g.* Figure 1 for such branches). In other words, in the presence of violent plasma motions, it is not surprising that oscillations are excited in all possible branches.

4. Nonlinear Interaction of Waves

If waves with finite amplitudes are excited, nonlinear wave interaction (*i.e.* anharmonicity) becomes essential, for which several different mechanisms exist. If the excitation of the waves is weak all mechanisms are of a resonant nature: the interaction is important only when frequency and wave number fulfill certain conditions. To the lowest order of the oscillation amplitude such processes are known either as three-wave processes of coalescence and of decay of waves or as induced wave scattering by plasma electrons and ions. The frequency and wave number conditions for the first process (decay and coalescence of waves) are of the form

$$
\begin{aligned}
\omega - \omega' &= \omega'' \\
\mathbf{k} - \mathbf{k}' &= \mathbf{k}'',
\end{aligned}
\tag{19}
$$

where ω, ω', ω'' are the frequencies, and $\mathbf{k}$, $\mathbf{k}'$, $\mathbf{k}''$ are the wave vectors of the interacting waves. In the second process (scattering of an ω-wave into an ω'-wave) we have

$$
\omega - \omega' = (\mathbf{k} - \mathbf{k}') \cdot \mathbf{V},
\tag{20}
$$

where $\mathbf{V}$ is the velocity of the particles on which the scattering occurs. In the decay process the total energy and momentum of waves are conserved, while in the scattering process the total number of plasmons* is preserved. The scattering process causes energy transfer into the region of lower k and the decay processes result in the broadening of the spectrum. In some cases it is necessary to take into account four-wave processes, corresponding to the scattering of plasmons by one another. The waves participating in the interaction may belong to different branches of plasma oscillations. In this case the nonlinear interaction is an induced transformation of one type of oscillation into another. (For example, plasma waves are transformed into electromagnetic waves, or high-frequency waves into low-frequency waves.) The characteristic times of the nonlinear interactions are very small on an astrophysical scale. This means that the plasma oscillation spectra are determined by nonlinear interactions. For example, the time τ characteristic for transformation of the energy of Langmuir waves due to the scattering on ions is about

$$
\frac{1}{\tau} \approx \omega_{pi} \frac{V_s}{V_{Te}} \frac{W}{n_0 T_e}.
\tag{21}
$$

* Somewhat similar to the association of photons with electromagnetic waves, one associates so-called 'plasmons' with plasma waves. The energy of a plasmon is $\hbar\omega$, the momentum $\hbar k$. Equation (19) expresses conservation of energy and momentum when two plasmons merge to give a third plasmon. Equation (20) may be written as $\Delta E = \Delta p \cdot V$ where ΔE and Δp are the difference in the energy and the momentum before and after scattering. (Ed.)

Here W is the turbulent energy per cm^3, $n_0 T_e$ is the thermal energy, ω_{pi} is the ion plasma frequency, and V_{Te} is the thermal velocity of the electrons. If W is of the order of $10^{-6} n_0 T_e$, then for $n_0 \approx 0.01$ cm^{-3} τ is of the order of 10^3 sec. A comprehensive summary on nonlinear interactions of waves in plasmas may be found in the monographs by Kadomtsev (1964), by Tsytovich (1967), and on linear properties by Stix (1962) and by Spitzer (1962).

5. Weak Turbulence

It is customary to call a plasma state 'weakly turbulent' if waves are excited in large numbers but with comparatively small amplitudes. As in usual turbulence, a transfer of turbulent energy occurs from the region where it is created to the dissipation region. This transfer takes place because of nonlinear interactions between the turbulent oscillations. On the one hand, turbulence in a plasma is more complicated than that in liquids because so many types of motions and oscillations are possible. On the other hand, it is easier to deal with in the sense that if the amplitudes are low, the oscillations can be well described by means of expansion methods with a small expansion parameter – the ratio of the energy of the oscillations to the energy of the thermal particles. The corresponding methods have been recently developed and applied successfully in investigations of different types of plasma turbulence in which only one branch of oscillations is important – magnetohydrodynamic, ion-sound, Langmuir turbulence and others. However, usually oscillations in one branch also excite to some extent the other branches, *i.e.* transformation occurs of one type of the waves into another and then it is necessary to take many types of oscillations into account.

6. Turbulence of the Interstellar Medium and the Acceleration of Cosmic Ray Particles

Many observations show that apart from ordinary thermal plasma the interstellar medium also contains a large number of high-energy particles. First of all, there are the 'cosmic rays' consisting of ultrarelativistic particles (protons and heavier nuclei) with energies from 10^9 to 10^{19} eV. Second, there are particles with energies of the order of 10^7 to 10^8 eV, called 'subcosmic rays'. In the third place, the existence of electrons of energies between 10^8 and 10^9 eV is inferred, because they are responsible for the cosmic radio emission. It is possible that electrons exist with both higher and lower energies. Although the number density of high-energy particles is not large, their total energy density is quite comparable with the energy density of the thermal plasma. This implies that the high-energy particles are able to interact significantly with the thermal plasma. The fact that the cosmic rays are isotropic indicates that such interaction actually occurs. Presumably a number of local sources exist which produce fast particles that subsequently are isotropized by the plasma. For this process to take place the plasma should probably be turbulent. Therefore, we shall consider here a simple model of the interaction between high-energy particles and turbulent plasmas. It should be kept in mind that the interstellar turbulence is expected to be inhomogeneous, by which we mean that there exist both active regions with a high level of turbulence and comparatively quiet regions. For example, intensive turbulence may exist in supernovae shells, in active nuclei of galaxies, in the vicinity of pulsars, neutron stars and other active stars. Therefore, we shall consider a wide range of the temperatures, densities, and turbulence levels in the plasma.

There is one outstanding observational fact that should find a theoretical explanation, namely that the spectra of relativistic electrons and ions are usually of the type

$$f(\varepsilon) = \text{const} \times \varepsilon^{-\gamma}. \tag{22}$$

Here $f(\varepsilon)$ is the number of fast particles per cm^3 and per energy interval dε, so that $\int_0^\infty f(\varepsilon)\mathrm{d}\varepsilon = n$ is the total number density of the particles. It could be assumed that this spectrum is produced by statistical superposition of separate spectra arising from different active regions. In practice, however, the value of γ varies only slightly from source to source and it is within the range $1 < \gamma < 3$ with an average value $\gamma = 2.7$. Therefore it is more probable that the acceleration mechanism itself produces this type of spectrum. The modern theory of turbulence provides such a mechanism.

7. Nature and Rate of Particle Acceleration in a Turbulent Plasma

A. ACCELERATION RATE

The rate of particle acceleration is characterized by an average energy gained by a particle in unit time. Let dε/dt depend on ε as

$$\mathrm{d}\varepsilon/\mathrm{d}t = \beta\varepsilon^\mu. \tag{23}$$

There is a qualitative difference between the two cases $\mu < 1$ and $\mu > 1$. When $\mu < 1$, the high-energy particles are accelerated at a slower rate and they determine the time required for reaching a stationary spectrum. When $\mu > 1$, the acceleration time is determined by the minimum energy, the injection energy. If we had $\mu < 1$, spectra would be expected with sharp cut-offs at some high energy. Since these have not been observed, it seems that the case $\mu > 1$ is preferable. In this case, one and the same mechanism could be responsible for acceleration of cosmic rays up to the highest energies and the usual argument, that the particles of highest energies have an origin outside of our Galaxy, would not work.

B. INFLUENCE OF RESONANCE OF PARTICLE-TURBULENT WAVE INTERACTION ON THE ACCELERATION RATE

A possible acceleration mechanism with $\mu > 1$ is predicted by modern theory (see Section 2). According to Equation (13) Cherenkov resonance will occur if

$$\omega = kV \cos \Theta \tag{24}$$

from which follows

$$V > V_p \equiv \omega/k. \tag{25}$$

This condition shows that when the particle energy ε increases there is also an increase in the range of phase velocities at which the interaction works. Therefore, a larger number of turbulent oscillations participate in the acceleration and the acceleration rate increases with ε. In the presence of a magnetic field this remains true. To see why, consider the case when the phase velocity V_p is much less than the particle velocity, i.e. $\omega \ll k_\parallel V_\parallel$. According to Equation (15)

$$- k_\parallel V_\parallel \approx l \frac{ZeBc}{\varepsilon} \tag{26}$$

where Z is the charge of the accelerated particle.

From $| k_\parallel V_\parallel | < kV$ we obtain

$$\varepsilon > \frac{ZeBlc}{kV}. \tag{27}$$

When ε increases, smaller values of k are allowed and the wave number interval of interacting waves is broadened. Equation (27) assures that the acceleration rate increases with ε even in the ultrarelativistic limit $\varepsilon \gg mc^2$; this is not true in the case of Equation (25). Equation (27) implies that the Larmor radius of the particle $r_\lambda = \varepsilon V/(ZeBc)$ be larger than the wavelength $\lambda = 1/k$. When the energies are high, the Larmor radius of the particles becomes large, and only for small k will the acceleration rates depend appreciably on the energy.

It should be kept in mind that if k is very small not every type of wave is possible. But, as is seen in Figure 1, at least two types of waves with small k exist, the Langmuir and the Alfvén (or magnetosonic) waves. However, as k decreases, the phase velocity V_p of the Langmuir waves increases and becomes greater than the velocity of light for $k < \omega_{pe}/c$, so that Cherenkov resonance and cyclotron resonance are no longer possible. Corresponding to the wavenumber ω_{pe}/c there is a critical energy ε_c such that for $\varepsilon > \varepsilon_c$ the acceleration rate no longer increases with ε. Requiring that $\varepsilon_c \gg mc^2$ and by substituting $k = \omega_{pe}/c$ and putting $Z = l = 1$, we find from Equation (27)

$$\frac{B^2}{4\pi n_0 mc^2} \gg \frac{m}{m_e} \tag{28}$$

or, what is the same,

$$\frac{V_A^2}{c^2} \gg \frac{m^2}{m_e m_i} \tag{29}$$

Even when the accelerated particles are electrons $(m = m_e)$ Equation (29) requires very strong magnetic fields $(V_A \gg c/40)$. For nonrelativistic particles, i.e. subcosmic rays, Equation (25) gives an increase in acceleration rate with increase of ε. For magnetohydrodynamic and for Alfvén waves the phase velocity V_p does not depend on k and is less than V_A if $V_A > V_s$. Therefore, if the particle velocity V is much larger than V_A the condition $V \gg V_p$ that is required by Equation (27) is fulfilled for any small k. This means that if only very long waves are present (k very small) resonance and acceleration can take place only for particles with large ε, i.e. with large Larmor radii.

This restricts the acceleration to high-energy particles. However, it is known that the frequencies of magnetic oscillations (with wavelengths larger than the Larmor radii of thermal electrons and ions) are proportional to the cosine of the angle between the direction of wave propagation and the magnetic field, and this offers the possibility of interaction of the waves with particles with $V < V_p$. Writing $\omega = kV_A \cos \Theta$ and using the fact that $k_{\parallel} V_{\parallel} \ll \omega_B$, we obtain from Equation (15)

$$\frac{ZeBl}{mcV_A k} < 1. \tag{30}$$

Since Equation (30) does not contain the particle velocity it allows the injection of low energy particles. This is very essential for the explanation of the chemical composition of the cosmic rays because it gives a preferential injection of multicharged, heavy ions.

C. THE INVERSE COMPTON EFFECT ON PLASMA WAVES

In addition to the effects considered above, resonance with electromagnetic waves emitted by the turbulent oscillations is possible for high-energy particles. One such mechanism is the Compton effect, in which the particle oscillates in the turbulent field and emits an electromagnetic wave. From the point of view of elementary processes the effect corresponds to a Feynman diagram with absorption of a turbulent wave σ, and emission of an electromagnetic wave t. The resonance condition states that in the rest frame of the particle the frequency does not change:

$$\omega^{\sigma} - \mathbf{k}^{\sigma} \cdot \mathbf{V} = \omega - \mathbf{k} \cdot \mathbf{V}. \tag{31}$$

Here ω^{σ} and $\mathbf{k}^{\sigma}$ are the frequency and the wave vector of the turbulent wave, respectively; ω and $\mathbf{k}$ are those for the electromagnetic wave and $\mathbf{V}$ is the velocity of the particle. Let us consider the case of electromagnetic waves of very high frequencies and particles with ultrarelativistic energies. Then $\omega = kc$ and, if Θ is the angle between $\mathbf{k}$ and $\mathbf{V}$,

$$\omega = \frac{\omega^{\sigma} - \mathbf{k}^{\sigma} \cdot \mathbf{V}}{1 - (V/c) \cos \Theta}. \tag{32}$$

From the condition $\cos \Theta < 1$ we have

$$\omega < \frac{\omega^{\sigma} - \mathbf{k}^{\sigma} \cdot \mathbf{V}}{1 - (V/c)} \approx 2 \frac{\varepsilon^2}{(mc^2)^2} (\omega^{\sigma} - \mathbf{k}^{\sigma} \cdot V). \tag{33}$$

This means that the higher the energy of the particle, the higher the frequency of the electromagnetic waves that can interact with it.

The Langmuir oscillations with $V_p \gg c$ have been mentioned above. Then $\omega^{\sigma} \gg k^{\sigma}c$ and, a fortiori, $\omega^{\sigma} \gg k^{\sigma}V$. We may therefore approximate Equation (33) by

$$\omega < 2 \frac{\varepsilon^2}{(mc^2)^2} \omega_{pe}. \tag{34}$$

This implies that the acceleration process does not depend on the wavelength distribution of the turbulent oscillations but only on the density of the turbulent energy. This is very important because in such a situation we can expect a universal spectrum of accelerated particles. But the theory should answer the question whether the main part of the turbulent energy can be concentrated in Langmuir oscillations with $V_p > c$; the theory should also predict the frequency distribution of electromagnetic radiation. We note that for other types of turbulent oscillations Equation (33) shows that the resonance depends appreciably on the wavelength of the oscillation. For instance, if $V \gg V_p = \omega^\sigma / k^\sigma$ we have

$$\omega < 2 \frac{\varepsilon^2}{(mc^2)^2} k^\sigma c \, |\cos \Theta| \, . \tag{35}$$

The problem of electromagnetic wave emission by turbulent oscillations has been discussed by Kaplan and Tsytovich (1969).

Plasma mechanisms of emission of electromagnetic waves is a subject of great interest for the interpretation of cosmic radio emission. But at the basis of such an interpretation is the spectrum of the accelerated particles, and first the mechanisms that produce such a spectrum should be understood. In radiation processes oscillations of essentially different kinds can operate, but for acceleration processes the Langmuir oscillations are of greatest interest. Indeed, transparent waves are important for cosmic radio emission and nontransparent waves for particle acceleration, due to the fact that the radiation that can be reabsorbed on the fast particles can interact with them most intensively.

D. ACCELERATION OF RELATIVISTIC PARTICLES BY THE WHOLE TURBULENT SPECTRUM

The acceleration depends upon the particle energy only if the wave numbers of the turbulent oscillations cover a wide interval, so that for particles of higher energy more oscillations can take part in the acceleration. When the whole wave number spectrum is involved (*i.e.* all the turbulent oscillations interact with the particles), the acceleration, which has a stochastic nature, will be determined by an energy diffusion coefficient that does not depend on the particle energy ε, so that

$$\varepsilon^2 = 4 \, Dt, \tag{36}$$

or

$$d\varepsilon/dt = 2 \, D/\varepsilon, \tag{37}$$

and according to Equation (23) $\mu = -1$. This statement is true only at ultrarelativistic energies. For nonrelativistic particles, even if waves of all available wave numbers are interacting, the diffusion coefficient depends on V and therefore on the energy ε, because with increasing velocity there is a decrease in the fraction of turbulent oscillations that produces acceleration. As can be easily seen from the Cherenkov condition [Equation (13)], the extra factor V^{-1} gives

$$d\varepsilon/dt = \beta \varepsilon^{-3/2} . \tag{38}$$

This situation occurs for the ion-sound waves whose phase velocities V_p do not exceed the sound velocity V_s. Therefore, if $V_p < V_s$, $\mu = -\frac{3}{2}$. Since the acceleration rate increases rapidly with the decreasing particle energy, it is very high at low energies, *i.e.* Equation (38) gives the possibility for an efficient injection mechanism.

These arguments show that the acceleration rate will increase with particle energy only for energies small enough, so that the particles interact with part of the turbulence spectrum. From Figure 1 it can be seen that for the whistlers the values of k are limited to a narrow interval $\omega_{pi}/c < k < \omega_{pe}/c$ and that their phase velocities are small. The ion-sound and magnetic-sound waves yield an acceleration rate given by Equation (38). Therefore, it is of primary interest to know the distribution of wave numbers for Langmuir and for magnetohydrodynamic turbulence.

8. Magnetohydrodynamic Collisionless Turbulence

A. TURBULENT SPECTRA

The recent development of the concept of plasma turbulence opens a new quarter in our knowledge of magnetohydrodynamic motions. Basically there exist two types of magnetohydrodynamic oscillations: the first (Alfvén and magnetohydrodynamic waves) are waves whose frequencies are much lower than the ion collision frequency v_i, and whose wavelengths are much larger than the (particle) free path l_i (this is the traditional region of magnetohydrodynamics); the second are waves whose frequencies are much greater than v_i and for which $kl_i \gg 1$ (this is the collisionless region). For collisionless waves the Landau damping and induced scattering on electrons and ions result in a transformation of turbulent energy to long wavelengths. For isotropic turbulence the distribution of the turbulent energy as a function of k may be characterized by W_k – the energy per cm^3 of the plasma in the wave number interval dk, so that

$$W = \int_0^\infty W_k \, dk. \tag{39}$$

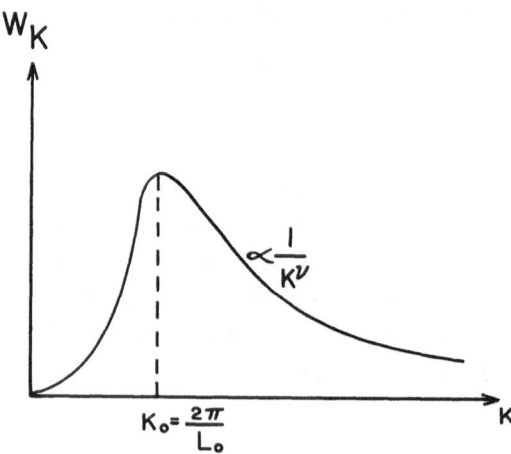

Fig. 2.   Magnetohydrodynamic turbulence spectrum showing energy W_k vs k. The decrease of W_k at large wave number behaves as $1/k^\nu$.

For magnetohydrodynamic turbulence W_k differs only by a normalization factor from the distribution of turbulent energy of frequencies W_ω because of the linear dependence of ω on k.

If the magnetohydrodynamic waves have an anisotropic distribution, the turbulent spectrum characterizes their distribution integrated over all angles, and the acceleration is practically independent of angle.

The energy flow in the spectra of collisionless magnetohydrodynamic and Alfvén waves is in the direction of small values of k $(k \approx l_i^{-1} = v_i/V_{Ti})$ where the maximum of the turbulent energy will be found. Therefore in this case the basic scale of the turbulence has the order of magnitude $L_0 = 2\pi/k_0 = l_i$ at $V_A \gg V_{Ti}$. This spectrum is shown schematically in Figure 2. In the region of $k \gg k_0$, $W_k \propto k^{-\nu}$. If the characteristic dimension of the active region is less than the mean free path, as in the case of the solar wind, k_0 is defined by the dimension of the active region. For $V_{Ti} \ll V_A \ll V_{Te}$, Landau damping of magnetohydrodynamic waves is the primary cause of energy loss of turbulent oscillations because of the energy exchange between Alfvén and magnetohydrodynamic oscillations; $1 < \nu < 2$. This spectrum corresponds to that in the interplanetary magnetic field.

B. ACCELERATION OF PARTICLES

The rates of particle acceleration due to particle interaction with magnetohydrodynamic oscillations have been studied by Tsytovich (1963) and by Tverskoy (1967) (see Figure 3). For $\nu = 2$, we obtain $\mu = 1$, and the acceleration is similar to Fermi acceleration

$$\beta = \frac{V}{L_0} V_A^2 \frac{8\pi W}{B^2}. \tag{40}$$

Here $L_0 = 2\pi/k_0$ is the main scale length of turbulent spectra shown in Figure 2; this length and the Alfvén velocity, V_A, correspond respectively to the distance between

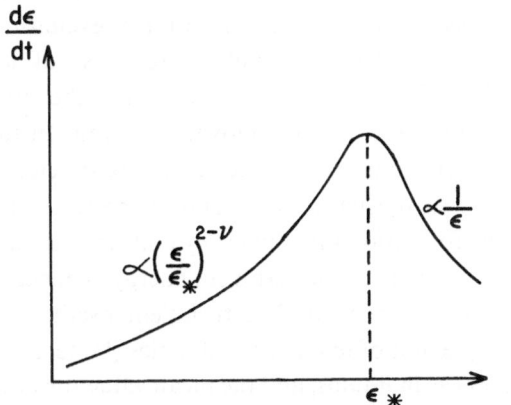

Fig. 3. Rate of energy gains, $d\varepsilon/dt$, vs energy of particles accelerated in magnetohydrodynamic turbulence. ε_* is the energy of a particle whose Larmor radius is the same as the main scale of the turbulence.

clouds and to the cloud velocity in the Fermi acceleration mechanism. In Equation (40) a new factor appears: the square of the ratio of the amplitude of Alfvén waves at the maximum of the energy-spectrum to the constant external magnetic field. According to polarization measurements this factor is of the order of 10^{-2} (Pikel'ner, 1968). We can conclude, therefore, that the acceleration due to magnetohydrodynamic oscillations is an equivalent of the Fermi acceleration in the framework of the modern concepts of plasma turbulence. In Figure 3, ε_* corresponds to the energy of a particle at which its Larmor radius becomes equal to the main scale length of the turbulence. This corresponds to the maximum wavelength for which the criterion of Equation (27) is fulfilled. The physical reason for the decrease of the acceleration rate for $\varepsilon > \varepsilon_*$ is that the particles interact with the whole turbulent spectrum and the diffusion coefficient becomes constant. It was frequently pointed out that the efficiency of the Fermi acceleration is very small (see e.g. Ginzburg and Syrovat-skii, 1963). Equation (40) is even more pessimistic because of the additional factor $8\pi W/B^2 \approx 10^{-2}$ discussed above. L_0 corresponds to the original value that was accepted for the Fermi acceleration and cannot be decreased. Moreover, $v=2$ corresponds only to $\mu=1$, but not to $\mu > 1$, and is reached only when $V_A \gg V_{Te}$. For $V_A \ll V_{Te}$ (and this is more probable under astrophysical conditions) $1 < v < 2$ and, therefore, $\mu < 1$. On the other hand, the injection of heavy multicharged ions can be done very effectively by Alfvén oscillations. The diffusion coefficient for such particles is proportional to $m^v Z^{2-v}$.

9. Langmuir Turbulence

It is convenient to characterize the spectrum of isotropic turbulence by W_k as was done in the preceding section. Just as in liquids, where the turbulent spectrum is defined by only one parameter, i.e. the turbulent energy flow, one single parameter determines the whole turbulent spectrum in a plasma. Let Q be this parameter denoting the oscillation energy generated per cm^3 per sec. Spectra of Langmuir turbulence have been calculated by Pikel'ner and Tsytovich (1968) for $T_e = T_i$ and by Tsytovich (1969) for $T_e > T_i$. The spectra are formed as a result of excitation by different kinds of instabilities. The most effective instabilities are those of low phase velocities close to the mean thermal velocity of electrons. Then the nonlinear interactions transform the phase of the waves in the following fashion. In the first stage, where $kd_e > (m_e/m_i)^{1/5}$, the transformation is due to scattering on electrons; in the second stage, $kd_e < (m_e/m_i)^{1/5}$, to scattering on ions; and in the third stage, for still greater phase velocities, the transformation takes place by collisions of plasmons. It is important to realize that the main part of the turbulent energy is concentrated in the region where the oscillations are not excited. The turbulent oscillations are very quickly transformed through the region of low phase velocities (large k). This transformation is due to a cascade process as in liquids, and the mean value for one step in this cascade is:

$$k_* = \sqrt{\left(\frac{m_e}{m_i}\right)\frac{1}{d_e}}. \tag{41}$$

If the source of turbulence (*e.g.*, the beam of particles) has phase velocities near V_{Te}, the spectrum for $k > k_{**} = 1/d_e(m_e/m_i)^{1/5}$ is $W_k \propto 1/k^{5/2}$ (see Figure 4). In the region $k_* < k < k_{**}$, the spectrum is flat and $W_k = $const (Liperovskii and Tsytovich, 1969). If the phase velocities of the source are higher than $V_{Te}(m_i/m_e)^{1/5}$, the spectrum begins at the k number where the generation occurs. For $k \ll k_*$, collisions between the plasmons play an important role and form a 'Maxwellian' type distribu-

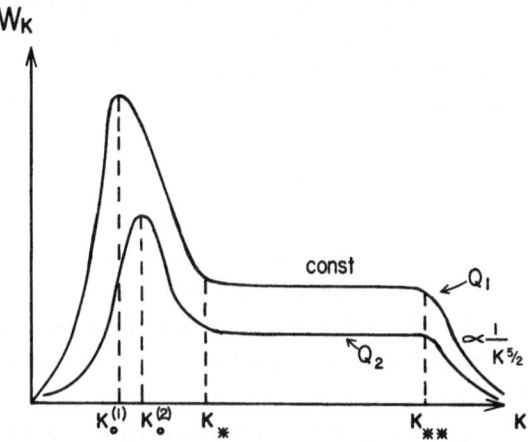

Fig. 4.   The turbulent energy W_k for Langmuir turbulence shown as a function of k for two different rates of excitation Q_1 and Q_2.

tion. The maximum, $k = k_0$, of the spectrum corresponds to the energy-containing region, where the oscillations are damped by the usual collisions. In the asymptotic region, where $k \gg k_0$, Pikel'ner and Tsytovich found a spectrum $1/k^\nu$ where $2.84 < \nu < 4$. The existence of such spectra with a maximum at $k = k_0$ requires that $k_0 \ll k_*$. k_0 depends on Q (Liperovskii and Tsytovich, 1969)

$$k_0 = k_* \left(\frac{8 n_0 T_e \nu_e^2}{\omega_{pe} Q} \right)^{1/(2\nu - 2)} \tag{42}$$

where ν_e is the collision frequency of electrons. In Figure 4, the turbulent Langmuir spectra for $T_e = T_i$ are shown for two different values of the power generation, Q. For small Q the maximum in the spectrum disappears. This means that the power of turbulence generation is so low that the turbulent energy can be absorbed while transforming from the region of generation up to k_*. Since the slowest transformation takes place at large wave numbers, the growth rate of the instability must be close to threshold. Even for low Q's the maximum in the spectrum exists, and as a rule, k_0 is such that the phase velocity at k_0 is of the order of $V_p^0 \approx 10\omega_{pe}/k_* \approx 10^3 V_{Te}$. In the majority of circumstances of astrophysical interest, this value is of the order of light velocity or higher which is important for the process of resonance emission.

10. Acceleration and Isotropization of Fast Particles by Langmuir Turbulence

A. ACCELERATION BY OSCILLATIONS WITH $V_p < c$

The subsequent events of induced emission and induced absorption of turbulent oscillations by fast particles give rise to particle diffusion both in angle and in energy. The diffusion in angle results in isotropization and that in energy leads to acceleration. The isotropization and the acceleration are produced by the same processes, but the isotropization is faster by a factor $(V/V_p)^2$. Therefore, all oscillations with low phase velocities such as ion-sound and magnetohydrodynamic waves lead more rapidly to isotropy than to acceleration. For the acceleration and isotropization by Langmuir turbulence, an essential parameter is

$$\eta = k_c/k_*. \tag{43}$$

Here the wavenumber $k_c (= \omega_{pe}/c)$ corresponds to a phase velocity equal to c, whereas k_* is the minimum value of k on the plateau of Figure 4 (p. 125). If $\eta \ll 1$ (that is $T_e < 10$ to 20 eV if $T_e = T_i$) an interval exists for $V_p < c$ where the spectrum is proportional to $\propto k^{-\nu}$; this interval is important for the interaction with subcosmic rays. The rates of isotropization and acceleration are then equal. If the plasma is hot ($\eta \gg 1$), the isotropization is faster than the acceleration by a factor of approximately $\ln [(c/V_{Te})(m_e/m_i)^{1/5}]$. Figure 5 shows schematically the acceleration rates for electrons and ions under conditions $\eta \gg 1$ and $\eta \ll 1$. If $\eta \ll 1$, the acceleration is effective and $\mu = \frac{1}{2}(\nu - 1)$, i.e. $\mu = 0.94$ for $\nu = 2.84$ and $\mu = \frac{3}{2} > 1$ for $\nu = 4$. The coefficient β is proportional to $\sqrt{Q}$. The main part of the Langmuir turbulence energy near the maximum is not effective in this mechanism. Nevertheless, very effective isotropi-

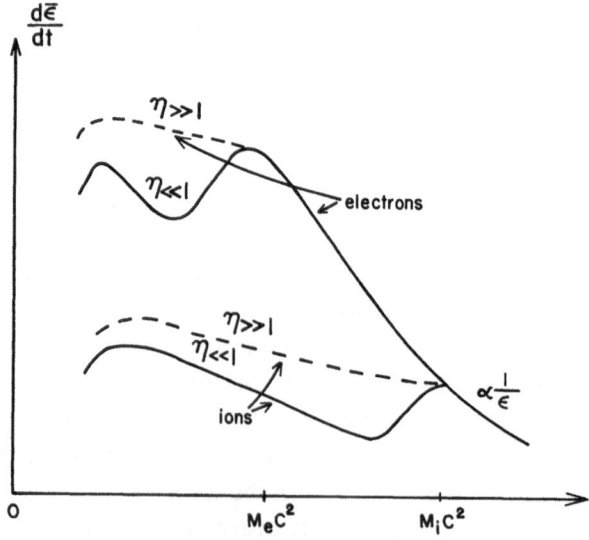

Fig. 5.   Rate of energy gain for electrons and ions in Langmuir turbulence. The parameter η determines the ratio of acceleration to isotropization.

zation and acceleration of subcosmic rays can occur. According to the calculations given by Pikel'ner and Tsytovich (1969), the isotropization of subcosmic rays can be brought about by Langmuir turbulence in a heterogeneous model of the Galaxy, in which there is a mixture of active and passive regions.

B. PARTICLE ACCELERATION BY LANGMUIR OSCILLATIONS WITH $V_p > c$ DUE TO INVERSE
   COMPTON EFFECT

Now let us consider the acceleration arising from the interaction of relativistic particles with waves having wavenumbers at the maximum of the Langmuir turbulence spectrum. This interaction can lead to an effective acceleration only by the reabsorption of electromagnetic waves. In addition, the spectrum of the electromagnetic

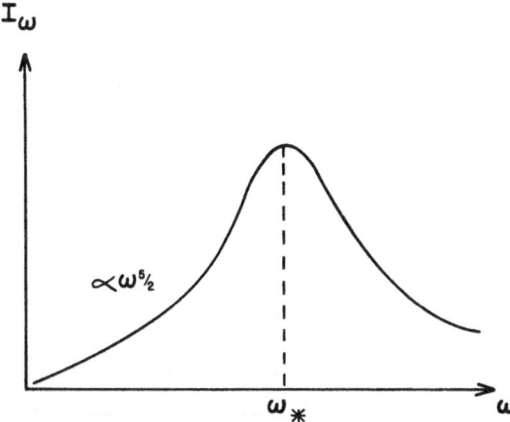

Fig. 6.   Power spectrum of electromagnetic waves, I_ω, vs frequency ω in equilibrium with particles
and Langmuir turbulence.

waves should have a form as sketched in Figure 6. In that figure, ω_* is the maximum frequency for which the active region is opaque for electromagnetic radiation emitted by fast particles.

Common for ultrarelativistic particles and well known for synchrotron radiation is the fact that the intensity $\propto \omega^{5/2}$. This behavior ($\propto \omega^{5/2}$) can be easily found from Equation (35) if we consider that $T_{\text{eff}} \approx \varepsilon \approx mc^2 \sqrt{(\omega/2\omega_{pe})}$. Reabsorption of synchrotron radiation also gives an effective acceleration (Tsytovich, 1966b). We consider the sum of the synchrotron and plasma mechanisms of acceleration since the two mechanisms give the same energy dependences and differ from one another only by numerical factors. This can easily be seen if we compare Equation (35) with a corresponding condition for emission of synchrotron radiation

$$\omega < \left(\frac{\varepsilon}{mc^2}\right)^2 \frac{eB}{mc}. \tag{44}$$

From Equation (44) the maximum energy ε_*, for which effective acceleration can be expected, is found to be

$$\varepsilon_* = \sqrt{\left(\frac{2\omega_*}{\omega_{pe}}\right)} mc^2 \,. \tag{45}$$

For large active regions or for regions with a large number of relativistic particles, the value of ε_* can be very high. From Equation (35) it is clear that if $\varepsilon < \varepsilon_*$ the whole spectrum of radiation cannot take part and that the higher the energy of the particle the more waves can accelerate the particle. This growth of acceleration ceases at $\varepsilon \approx \varepsilon_*$, because practically all the waves take part in acceleration, and for $\varepsilon \gg \varepsilon_*$ the acceleration rate decreases as $1/\varepsilon$. Figure 7 shows the acceleration rates that correspond to these qualitative considerations (Tsytovich and Chikhachev, 1969). The relation

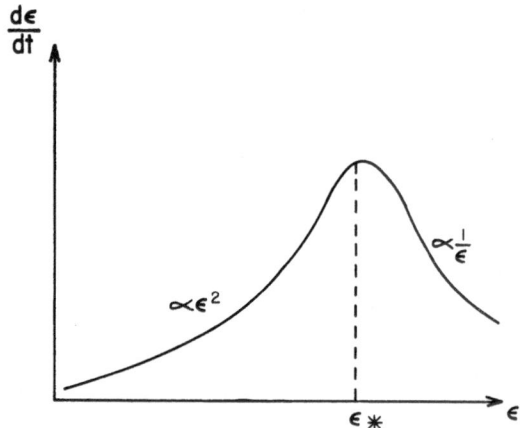

Fig. 7. Rate of energy gain of particles due to 'Compton' reabsorption of electromagnetic waves.

$d\varepsilon/dt \propto \varepsilon^2$, i.e. $\mu=2$, corresponds to the above requirements of effective acceleration. It does not depend on the turbulent oscillation distribution and is therefore universal. The coefficient β is given, order-of-magnitude-wise, by

$$\beta = \frac{\omega_{pe}^4}{n_0^2 m}\left[W + a\sqrt{\left(\frac{eB}{cm\omega_{pe}}\right)}\frac{B^2}{8\pi}\right] \tag{46}$$

(Kaplan and Tsytovich, 1968) where a is a numerical factor of order unity. We can consider the Cherenkov acceleration with $V_p < c$ as an injection mechanism. Hence, the injection energy is determined by intersection of curves in Figures 5 and 7. Since both acceleration mechanisms are coupled with the same turbulent spectrum, the acceleration time depends only on the parameter Q:

$$\frac{1}{\tau} = bZ^{10/3}\left(\frac{m_e}{m_i}\right)^{4/3}\left(\frac{Q}{v_e n_0 T_e}\right)^{5/6} \,. \tag{47}$$

The parameter $Q/(v_e n_0 T_e)$ is the ratio of the turbulent energy to the energy of thermal motion and has a maximum value of about 1. Table I shows the coefficient b as a function of plasma density and temperature. This table shows that acceleration is extremely rapid in dense and hot turbulent plasmas. Applications to such objects as pulsars, quasars and supernova remnants can be made using the table.

TABLE I

Values of the coefficient $1/b$ for particle acceleration
[see Equation (47)] (dimension of $1/b$: sec)

n_0 (cm^{-3})	T (eV)			
	10	10^2	10^3	10^4
10^{20}	10^{-3}	1.5×10^{-5}	3×10^{-6}	2×10^{-7}
10^{17}	0.6	10^{-2}	2×10^{-3}	10^{-4}
10^{14}	3×10^2	50	1	6×10^{-2}
10^8	10^8	1.5×10^6	3×10^5	2×10^2
10^2	3×10^{13}	4×10^{11}	10^{10}	6×10^9

11. Spectra of Accelerated Particles

A. THE ENERGY LOSSES OF FAST PARTICLES IN A TURBULENT PLASMA

To find the particle spectrum it is necessary to know not only the rate of acceleration but also the energy losses, because the balance between the acceleration and energy losses determines the stationary distribution of accelerated particles. In addition to the usual forms, a new type of energy loss exists in a turbulent plasma – the Compton effect on the plasma turbulent oscillations. It is analogous to the well-known cosmic ray energy losses from collisions with the black body radiation. In general, the energy loss on the Langmuir turbulence is largest, in which case the spectrum of radiation for a power law particle energy distribution [Equation (22)] is also a power law

$$Q_\omega = \text{const} \times \omega^{-\nu}, \qquad \nu = \tfrac{1}{2}(\gamma - 1). \tag{48}$$

In this sense, the plasma mechanism of radiation is similar to the synchrotron radiation. All the energy losses due to both the Compton effect on the turbulence and the synchrotron radiation are

$$\frac{d\varepsilon}{dt} = -\frac{16\pi\, e^4 Z^4}{3m^2 c^3} \left(\frac{\varepsilon}{mc^2}\right)^2 \left(\frac{B^2}{8\pi} + \frac{W}{6}\right). \tag{49}$$

The other types of energy losses, for example, the ionization or nuclear collision ones, are the same as in a quiescent plasma.

B. THE SPECTRA OF SUBCOSMIC RAYS

The stationary spectrum of accelerated particles arises as a balance of acceleration and energy losses. For low energy particles the most important losses are those of ionization and nuclear collisions. Similar to the Fermi mechanism, the acceleration by the magnetohydrodynamic and Alfvén oscillations results in a power law spectrum $\varepsilon^{-\gamma}$ only for a turbulent spectrum $\nu = 2$. For $1 < \nu < 2$ the spectra do not follow a power law because the energy losses are due to processes that are very different from the acceleration, *i.e.* ionization and nuclear collisions. Therefore, even for $\nu = 2$, there is a difference between acceleration and energy losses and the parameter γ can vary over wide intervals from 1 to ∞. Since this is contrary to the observations, it has been for a long time the principal argument against the Fermi acceleration mechanism. Therefore, this type of acceleration is probably more important as an injection mechanism, especially for heavy multicharged ions (Melrose, 1967).

Similarly, the acceleration by Langmuir oscillations with $V_p < c$ for $\varepsilon \gg mc^2$ does not result in a power law spectrum. The explanation again is due to the difference between the acceleration and energy loss mechanisms. For $\varepsilon \gg mc^2$ the distribution is Maxwellian with the effective temperature depending on the parameter Q

$$T_{\text{eff}} = \frac{mc^2}{2\gamma_0}; \quad \gamma_0 = \sqrt{\left(\frac{Q_c}{Q}\right)}; \quad Q_c = \frac{2m^2 v_e^2 n_0 T_e T_i}{27\pi m_i m_e \omega_{pe}(T_e + T_i)}. \tag{50}$$

$T_{\text{eff}} \gg mc^2$ at $\gamma_0 \ll 1$, i.e. $Q \gg Q_c$. The quantity Q_c is very small. As an example, for a plasma beam instability one has $T_{\text{eff}} > mc^2$ even near the threshold of the instability.

Therefore, even the main part of the turbulent energy has $V_p > c$ the energy of the oscillations with $V_p < c$ is sufficient to give an effective injection for acceleration by radiation. For nonrelativistic subcosmic rays the distribution has the form

$$f(\varepsilon) = \frac{\text{const} \times \sqrt{\varepsilon}}{(\varepsilon + \gamma_0 T_e)^{\gamma_0}} \tag{51}$$

i.e. is a power law only if $\gamma_0 > 1$, $\varepsilon \gg \gamma_0 T_e$.

C. THE SPECTRA OF COSMIC RAYS

In the case of acceleration by radiation due to the Compton effect on the turbulence, the acceleration and energy losses are coupled by the same process (induced and spontaneous emission). The spectrum always is a power law for energies $\varepsilon \ll \varepsilon_*$. This is due to the fact that both the energy gain and energy losses are proportional to ε^2. Since β in Equation (23) depends on γ and γ is found independently from the equation for the particle energy distribution, we have an equation (Tsytovich and Chikachev, 1969) which shows that γ depends on two parameters $\kappa = W/(n_0 mc^2)$ and $\xi^{-1} = \omega_{pe} mc/(eB)$. The result of numerical solution of this equation is shown in Figure 8. It is interesting that γ is within a very narrow interval $1 < \gamma < 3$, that corresponds to the observed interval of spectra of radio sources. There exists also an extra point $\gamma = 2.7$ for $\xi = 10^{-2}$ to 10^{-3} that corresponds to the observed spectrum of cosmic rays. If Q and B are varied over wide limits, γ remains very close to 2.7. Therefore, the possibility exists here of determining the parameters of the turbulence from the spectral index of radio sources.

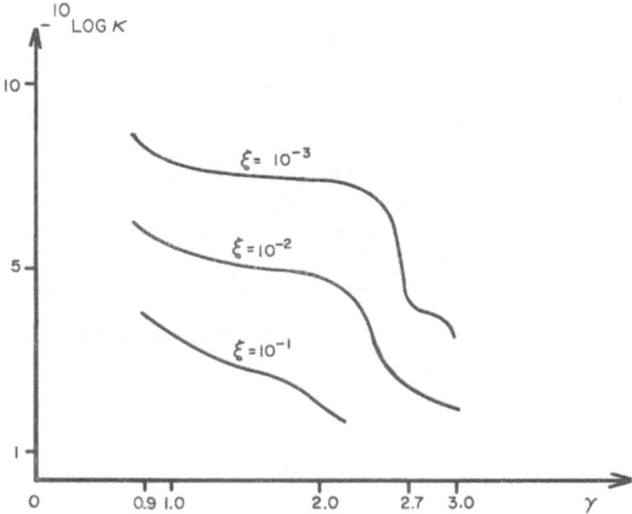

Fig. 8.   The ratio of wave energy density to rest mass ($\kappa = W/n_0 mc^2$) vs energy exponent γ for various values of the ratio of plasma to cyclotron frequency, $\xi^{-1} = mc\,\omega_{pe}/eB$.

The above analysis shows that at present radiative acceleration in a turbulent plasma may be the best mechanism for the explanation of the spectrum and the acceleration of cosmic rays.

12. Conclusion

At present, a broad system of ideas and conceptions concerning collective processes in rarefied plasmas has been developed. The application of this system to astrophysics has only begun, but even now it has led to an understanding of many processes which take place in cosmic conditions. It seems that the further development of these ideas and their wide application to astrophysics will make it possible to understand more clearly the essence of cosmic phenomena.

References

Ginzburg, V. L.: 1965, *Astron. Zh.* **42**, 1129 (1966, *Soviet Astron.* **9**, 877).

Ginzburg, V. L. and Syrovat-skii, S. I.: 1963, *The Origin of Cosmic Rays*, Izd. Akad. Nauk SSSR, Moscow (translation 1964, Pergamon Press, Oxford).

Kadomtsev, B. B.: 1964, *Plasma Turbulence*, Izd. Akad. Nauk SSSR, Moscow (translation 1965, Academic Press, New York).

Kaplan, S. A. and Tsytovich, V. N.: 1968, *Astron. Zh.* **45**, 777 (1969, *Soviet Astron.* **12**, 618).

Kaplan, S. A. and Tsytovich, V. N.: 1969, *Usp. Fiz. Nauk* **97**, 77 (1969, *Soviet Phys. Usp.*, under translation).

Lerche, I.: 1967, *Astrophys. J.* **147**, 689.

Liperovskii, V. A. and Tsytovich, V. N.: 1969, *Zh. Eksp. Teor. Fiz.* **57**, 301 (1969, *Soviet Phys. JETP*, under translation).

Melrose, D. B.: 1967, Preprint Belfer Graduate School, New York.

Pikel'ner, S. B.: 1968, *Ann. Rev. Astron. Astrophys.* **6**, 165.

Pikel'ner, S. B. and Tsytovich, V. N.: 1968, *Zh. Eksp. Teor. Fiz.* **55**, 977 (1969, *Soviet Phys. JETP* **28**, 507).

Pikel'ner, S. B. and Tsytovich, V. N.: 1969, *Astron. Zh.* **46**, 8 (1969, *Soviet Astron.* **13**, 5).

Spitzer, L.: 1962, *Physics of Fully Ionized Gases*, 2nd ed., Interscience, New York.

Stix, T.: 1962, *Theory of Plasma Waves*, McGraw-Hill, New York.

Tsytovich, V. N.: 1963, *Zh. Eksp. Teor. Fiz.* **44**, 946 (1963, *Soviet Phys. JETP* **17**, 643).

Tsytovich, V. N.: 1965, *Astron. Zh.* **42**, 33 (1965, *Soviet Astron.* **9**, 24).

Tsytovich, V. N.: 1966a, *Astron. Zh.* **43**, 528 (1966a, *Soviet Astron.* **10**, 419).

Tsytovich, V. N.: 1966b, *Usp. Fiz. Nauk* **89**, 89 (1966b, *Soviet Phys. Usp.* **9**, 370).

Tsytovich, V. N.: 1967, *Nonlinear Effects in Plasmas*, Izd. Akad. Nauk SSSR, Moscow.

Tsytovich, V. N.: 1969, Lebedev Physical Institute, preprint.

Tsytovich, V. N. and Chikhachev, A. S.: 1969, *Astron. Zh.* **46**, 486 (1970, *Soviet Astron.*, under translation).

Tverskoy, B. A.: 1967, *Zh. Eksp. Teor. Fiz.* **53**, 1417 (1968, *Soviet Phys. JETP* **26**, 821).

Wentzel, D. G.: 1968, *Astrophys. J.* **152**, 987.

8. DISCUSSION FOLLOWING THE REPORT BY
KADOMTSEV AND TSYTOVICH

(Wednesday, September 10, 1969)

Chairman: S. A. COLGATE

Editor's remarks: The following discussion was long and chaotic, in spite of the energetic efforts of the Chairman. Apparently the subject was unfamiliar to many participants. I did quite some reshuffling. A long contribution by Ozernoi on photon whirls in the primeval Metagalaxy has been transferred to Chapter 13.

Kaplan: I would like to suggest some possible observations of plasma turbulence in the interstellar medium. Before I do this, it is necessary to note that all plasma wave modes are confined in relatively small regions of interstellar or circumstellar space, because only there the conditions are met for excitation of turbulence. Plasma waves cannot travel long distances. It is also very improbable that, in any given volume of the interstellar medium, the conditions for excitation of plasma turbulence will exist for a long time. Streams of charged particles, which are the most effective agent for producing high-frequency modes in the plasma turbulence, usually undergo a fast isotropization. The shock which can induce low-frequency turbulence passes through any given point in a short time. Plasma waves without continuing excitation decay soon by collisions or by Landau damping. I would therefore guess that plasma turbulence in the usually quiet interstellar medium will be encountered in small volumes and for short periods only, but quite frequently. I will call such volumes 'plasma-turbulence pockets'. Of course, in quasars and in pulsars, plasma turbulence is much more common; but here we are discussing the interstellar medium.

Can we see these 'pockets'? It is well known that Langmuir waves can easily be converted into electromagnetic waves of nearly the same frequency. This process explains many features of sporadic solar radio noise. But the Langmuir frequency in the interstellar medium is approximately 10^4 Hz and cosmic radio noise of that frequency cannot be detected. Direct observations of this electromagnetic radiation are therefore impossible. However, there are some ways to observe the interstellar plasma turbulence pockets.

(1) Plasma waves with frequency ω_p can be transformed into electromagnetic waves with frequency $\omega \approx 2\omega_p (c/V_p)(\varepsilon/mc^2)^2$ by scattering on relativistic cosmic-ray electrons. Here V_p is the phase velocity of plasma waves, c is the velocity of light, and ε is the energy of the relativistic electrons. In this way, not only Langmuir waves but also very low-frequency ion-sound or magneto-sound waves can be transformed into electromagnetic waves with observable frequencies.

The radiation so produced has the same spectrum as the synchrotron radiation; and the intensities of both are comparable, if the energy density of plasma-wave turbulence is of the same order of magnitude as the magnetic energy density. However, since the pockets of plasma turbulence are rather small, the turbulent radiation is rarely seen against the continuous synchrotron background radiation in the Galaxy.

(2) The conditions for the observation of plasma-turbulence pockets are much better, if the relativistic electrons in these regions have an anisotropic velocity distribution; for example, if there are streams of relativistic electrons. The induced conversion of plasma waves into electromagnetic waves by scattering on the streams leads to an instability in the electromagnetic waves (interstellar maser-effect in continuous spectra). The amplitude of the electromagnetic waves increases until nearly all the energy of the stream of relativistic electrons is transformed into radiation. The plasma waves function as an ignition mechanism. In the ideal case of a very well-collimated stream, the frequency of converted radiation is the same as mentioned sub (1), but more often it is of the order of $\omega \approx 2\omega_p c/(\langle\theta^2\rangle V_p)$. Here θ is the angle between the electron velocity and the direction of the stream. The spectrum of this radiation also resembles the synchrotron radiation, and in addition there is almost complete polarization. Therefore, the interstellar plasma pockets in which the velocity distribution of relativistic electrons is anisotropic can be seen on the synchrotron background as bright spots with different polarization and with an intensity that is changing rapidly with the time because of isotropization of the stream.

(3) The anisotropy of relativistic particles and their concentrations may be increased by the turbulence pocket itself, since the plasma turbulence can serve as an effective accelerator of charged particles. But in this case, the output of electromagnetic energy is not so high, because the energy of the fast particles is taken from the energy of the plasma turbulence in the same pocket. I believe, incidentally, that most of the energy of the cosmic rays does not derive from the interstellar medium, but from violent events (supernovae).

(4) Under appropriate conditions, plasma turbulence pockets can enhance spectral-line radiation of background ions and atoms and in this way form interstellar masers. I am not sure that this mechanism can really explain the anomalous OH emission, but the possibility deserves further study. I cannot give the details here, but refer you to a recent paper (Kaplan and Tsytovich, 1969a).

(5) When an electromagnetic wave passes through a plasma turbulence pocket, its frequency is changed by the non-linear fusion (coalescence) with plasma waves, $(\omega'=\omega\pm\omega_p)$. This process is stochastic and therefore leads to an overall increase of the width of the emission line. In the paper cited above, the following formula has been established for the increase of the line width by Langmuir turbulence:

$$\langle(\varDelta\omega_*)^2\rangle = \frac{\omega_p^4}{\omega_*^2}\frac{W}{n_e mc^2}\frac{r}{\lambda_p}.$$

Here ω_* is the frequency of the line, λ_p is the wavelength of the plasma waves, W is their energy density, and r is the path length through turbulent regions. I shall use this formula for some estimates.

The observed emission lines of interstellar OH are very narrow. As a matter of fact the increase of their width due to turbulence in the interstellar medium cannot be more than $\Delta\omega_* = 10^2$ Hz. We have $\omega_* = 10^{10}$ Hz and if we assume $\omega_p = 10^4$ Hz and $n_e = 0.1$ cm^{-3}, we have $W \lesssim 10\lambda_p/r$ erg cm^{-3}. The value of the ratio λ_p/r is unknown, but we may estimate $\lambda_p/r \approx 10^{-18}$, so that W is less than 10^{-17} erg cm^{-3}. Of course, in small pockets the energy density of plasma turbulence may be much higher. Let me add that this process of fusion of electromagnetic and plasma waves leads to an apparent increase in the angular diameters of OH sources.

I would like to finish my remarks by referring to a review paper by Tsytovich and myself on the plasma mechanisms of radiation (Kaplan and Tsytovich, 1969b). (Kaplan, S. A. and Tsytovich, V. N.: 1969a, *Astrofiz.* **5**, 21; Kaplan, S. A. and Tsytovich, V. N.: 1969b, *Usp. Fiz. Nauk* **97**, 77.)

Colgate: Dr. Kaplan, you mentioned the width of an OH line. At what distance was this source? Can we estimate the probability of the existence of regions of at least lighter turbulence in our Galaxy?

Kaplan: I took the distance to the OH source as 300 pc and assumed $\lambda = 10^3$ cm.

Syrovat-skii: I have a remark concerning the excitation of high-frequency plasma turbulence. To elucidate the question, consider the main channels of energy transformation in astrophysical situations (see Figure 1). It seems likely that the primary form of energy is the kinetic energy of macroscopic plasma motion. The causes of this motion may be nuclear burning, stellar explosions, gravitational contraction, rotation, convection, and stellar wind, as in the case of Earth's magnetosphere. These processes supply kinetic energy to the plasma motion and lead to the distortion and amplifica-

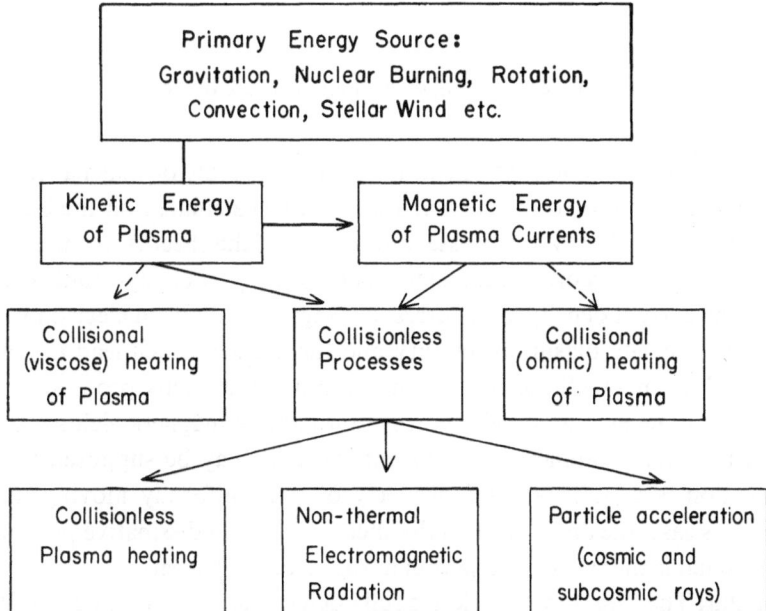

Fig. 1.   (See the remark by Syrovat-skii.) Main channels of energy transformation in astrophysical situations.

tion of the magnetic field. Because of the low collision rate between particles in a rarefied plasma, the dissipation by viscosity and finite conductivity is slow; and the main channel of energy transformation is seemingly collisionless dissipation through the excitation of plasma turbulence and acceleration of particles. As main causes of plasma turbulence, Kadomtsev mentioned the collisionless shock wave and the anisotropy of velocity distributions produced by magnetic contraction of a plasma.

I should like to add another process, which may be even more effective than shock waves. This is the formation of current sheets (see Figure 2). Such a sheet develops,

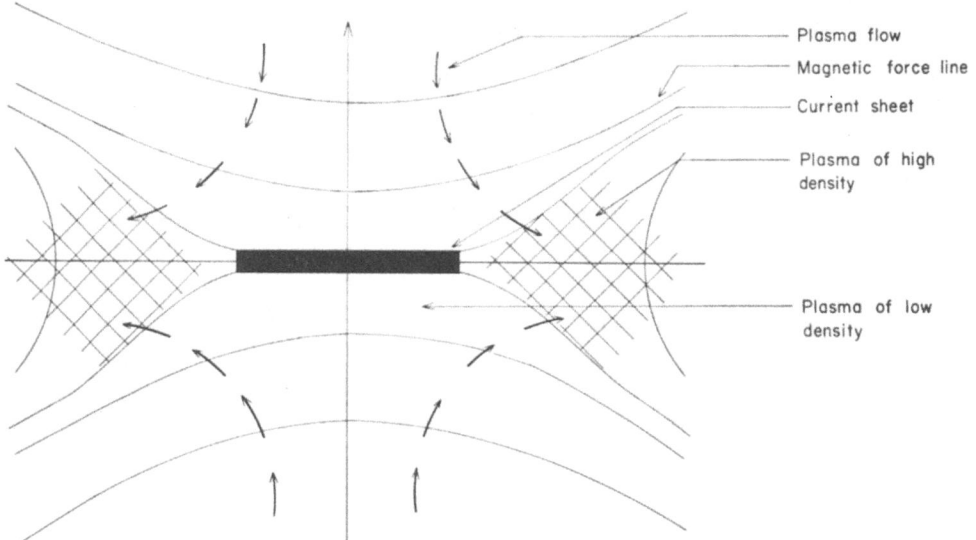

Fig. 2.   (See the remark by Syrovat-skii.) Structure of a current sheet.

for example, near magnetic zero (neutral) lines as a result of plasma motion in its vicinity. The plasma flow to the current sheet can be calculated on the basis of magnetohydrodynamics. However, the essential feature of the sheet is its thickness, which is determined by microscopic parameters such as the particle gyroradius. It may be shown that the current density and, correspondingly, the directed electron velocity are large enough to excite plasma instability and turbulence, which lead to collisionless dissipation of the magnetic energy in the current sheet. This process is especially effective because the energy of neighboring, antiparallel magnetic fields is completely released in the small volume of the current sheet. It may be supposed that current sheet formation is a quite general property of non-uniformly moving magnetized plasma. In this case, the dissipation in the sheet is the main dissipative process.

Colgate: What is the rate of dissipation in such a current sheet?

Syrovat-skii: The rate of dissipation is fully determined by the velocity of the influx of magnetic energy into the sheet, which depends on the plasma flow at large distances.

Colgate: It seems to me that the dissipation in these current sheets is always limited by the rate at which the plasma in the sheet can emit the energy produced. In some cases, this may be only cooling by radiation; in others, it may be the emission of waves. Have you considered the ratio of the wave emission to heat emission in the form of light and X-rays? Which is more important?

Syrovat-skii: In quasi-stationary cases, the energy released in the sheet may be estimated using only the external parameters of the flow. The effective resistivity is such that it produces the required dissipation. It is a very complicated problem to calculate the ultimate energy losses.

Dubov: Dr. Syrovat-skii, I think you should place shock waves in a separate rectangle in your figure because it represents a different mode of energy dissipation.

Colgate: To me the difference between the current sheet and the shock wave is that in a shock wave the fluid passes through the discontinuity and has only a finite energy density behind the shock wave; whereas, in the current sheet, matter continuously, flows into a discontinuity, with no escape. Therefore a fundamental difference exists between the properties of a shock wave and of a current sheet composed of opposite magnetic fields. These different properties lead to rather different final conditions of the plasma.

Tsytovich: I have two comments. The first is about a possible mechanism of excitation of turbulence in the presence of non-thermal electromagnetic radiation. This particular mechanism is decay of an electromagnetic wave into another electromagnetic wave and a turbulent wave. We know that energy flows out from radio galaxies, quasars, and the like. From the observed output of radiation by such sources, it is possible to estimate the energy of the turbulence in the source. One finds that there is much less excitation of low-frequency turbulent motions than of high-frequency motions. The most important mode is excitation of Langmuir turbulence. From the amount of radiation one can estimate that in the case of quasars, for example, the energy of Langmuir turbulence must be of the order of 1 erg cm^{-3}. It is of the same order of magnitude as the thermal energy of the particles. Therefore the turbulence must be very high in quasars. For supernovae, one can estimate that the energy of the turbulence from the observations is about 10^{-4} to 10^{-6} of the thermal particle energy, but much greater ($\sim 10^4$ to 10^5) than the thermal plasmon energy. For the interstellar medium the energy of the turbulence excited by the radiation is so low that it is much less than the thermal energy of the plasmons. Thus the radiation in the radio-frequency range cannot excite turbulence in the interstellar medium.

My other remark concerns the possible channels of the energy dissipation of the turbulence. Figure 1 of the Report by Kadomtsev and myself shows the dispersion of the various types of plasma waves. Because of non-linear effects, a high degree of isotropization exists. In isotropic turbulence the non-linear effects transform waves downward in frequency. Therefore the transfer of turbulent energy is downward along the curves of Figure 1 at p. 111. Consider first ion-sound turbulence. When $\omega = \omega_{Bi}$, ion cyclotron damping occurs, which heats the ions. In the outer regions, there is Landau damping on electrons, and this damping heats the electrons. Thus two channels exist for the dissipation of the ion-sound turbulence. For Langmuir turbulence, the situation is different. Landau damping can occur only for large values of k, and for isotropic turbulence the energy transfer is away from the region of Landau damping. Therefore only collisional damping occurs in the region of small k. However, if the velocity distribution is anisotropic, or if some magnetic bottles exist, the direction of energy transfer can be reversed, and we will have a high rate of collective heating of electrons. For all low-frequency waves, efficient Landau damping on electrons occurs because the phase velocity is less than V_{Te}. Therefore low-frequency turbulent motion will heat the electrons and the plasma. Langmuir turbulence is better adapted for acceleration of particles, although it too shows Landau damping and produces the tail of the Maxwell distribution. In summary, the first channel of dissipation of turbulent energy is collective dissipation in the form of Landau damping or cyclotron damping. The second channel is collisional dissipation, similar to the channel in usual hydrodynamical turbulence. The third channel is particle acceleration and the fourth is radiation. Langmuir turbulence can be dissipated efficiently and transformed into radiation. For a certain critical dimension of the plasma all the energy of the turbulence goes into radiation.

Colgate: Rosenbluth and McCrea have been calculating the transformation of the Langmuir turbulence into the transverse mode of the electromagnetic radiation. The reason this transformation becomes so important is the following: the usual cross-section for scattering photons off electrons is the Compton cross-section. But when Langmuir turbulence is excited and plasma oscillations occur, then the cross-section is increased by a factor that can be as large as the number of electrons per cubic Debije length $(n_e \lambda_D^3)$. In addition, this factor is multiplied by $W_k/(n_e kT)$. For a supernova ejection, a total factor can result as large as 10^{18}. Therefore one has to multiply the Compton cross-section by a very large number to get the cross-section for the scattering of electromagnetic radiation by the cooperative modes in a plasma oscillation. The result is a diffusion of photons in k-space; and just as in the preceding remark by Tsytovich, turbulence decay is represented by a diffusion of phonons in k-space.

Syrovat-skii: Tsytovich discussed emission of radiation by plasma turbulence. What about absorption? Do we have some indication of absorption of radio waves from discrete sources?

Tsytovich: Reabsorption of radio waves emitted by the source can be negative, i.e., the emitted waves can be amplified. The necessary condition for such a plasma maser effect is very easy to fulfill, and therefore the efficiency of the plasma mechanism of emission is very high. There is evidence of scattering of radar waves on high-frequency turbulence in the Sun. Will Dr. Gordon say a few words about the observations?

Gordon: I would like to discuss an example of nonlinear interaction of electromagnetic waves and plasma waves which occurs in the solar corona. My discussion is based on the study of radar echoes of the Sun carried out since 1961 by J. C. James at 38 and 25 MHz (James, 1964, 1966). In the conventional theory the radar echo by the Sun is explained by the scattering of the beam in the vicinity of the layer where the refraction index tends to zero and the laws of geometrical optics are violated. In the framework of this theory the only parameters on which the effective cross-sections depend are the global distributions of the electron concentration n_e and the electron temperature T_e. The variations in these distributions are small, even in the course of the cycle of solar activity. We therefore expect that the effective cross-sections will remain between 1 and 3 times $\pi R_\odot^2$. However, the observed radar echoes from the Sun behave quite differently. First, the effective cross-sections vary over a large range, from $\approx 100\pi R_\odot^2$ down to a few times $10^{-2}\pi R_\odot^2$. Second, the frequency displacements have a symmetrical distribution of components, and if the radar reflections originate on inhomogeneities 'frozen' in the outflowing solar wind the displacements must be predominately to the 'violet'. A third difficulty is that reflections originating in very high layers of the corona, up to $3R_\odot$ from the center, demand a 100-fold increase of the local density, which is highly improbable even if we are dealing with strong shocks there. Also, on days when reflections are strong, sporadic radiation too is strong and fluctuating. This peculiar interdependence cannot be explained in the framework of the classical theory. But the results of the radar experiments can be explained self-consistently by the assumption that the reflected signal arises from the back-scattering from the turbulence excited in a coronal plasma situated above the plages (Gordon, 1968, 1969). The frequency displacements are therefore not Doppler shifts, but result from the induced processes of wave coalescence and decay of the type $t^l \rightleftarrows t + S$, where t and S represent photon and ion acoustic waves, respectively. The analysis of the reflected signal enables us to ascertain that the reflections do not arise in the quiet corona. (James, J. C.: 1964, *IEEE Trans. Ant. Propag.* **AP-12**, 876, 1966, *Astrophys. J.* **146**, 356; Gordon, I. M.: 1968, *Astrophys. Lett.* **2**, 49, 1969, *Astrophys. Lett.* **3**, 181.)

Meyer: I would like to comment on the physical picture underlying these radar observations. H. U. Schmidt and I (Meyer and Schmidt, 1969) have investigated hydrodynamic flows along magnetic flux tubes in the low viscosity solar atmosphere. We considered flux tubes that arch back to the photosphere and found that very small pressure differences between the two footpoints must lead to the following characteristic flow pattern. The flow rises slowly at the base rapidly near the top of the flux tube and reaches the sonic velocity at the summit of the magnetic arch (or very near the summit, depending on the cross-sectional changes along the flux tube). The flow descends supersonically on the other side, passes through a shock front by which it adapts to the pressure condition on the low pressure side, and moves subsonically down to the other end. Since the viscosity in the solar atmosphere is small, even small pressure differences drive such flows. Such small pressure differences are to be expected between the two feet of the arched tube where the magnetic flux density is different, as will in general be the case.

On the basis of such a model one expects sonic and supersonic velocities along magnetic flux tubes that arch back to the solar surface. This situation is in striking contrast to all such flux tubes that lead into the solar wind. There the flow is limited by Parker's solar wind solution, which reaches the sound velocity only at the distance of several solar radii and accordingly leads to velocities of only 10 to 30 km sec^{-1} in the corona.

The suggestion that sonic and supersonic flows in magnetic arches are responsible for the features of the radar observations of the solar corona seems to be in general agreement with those observations. The disappearance of high-velocity signals above 1.5 $R_\odot$ agrees with the indication from eclipse pictures that there are no magnetic arches above this level but only outgoing flux tubes, belonging to the solar wind. On these arches one will find both ascending and descending velocities up to and larger than the velocity of sound, 100 to 200 km sec^{-1}. The appearance of these flows depends on the existence of the magnetic arches, and the radar signals therefore will change with the rotation of the Sun and the solar activity. One might add that plage regions especially are regions of possible footpoints of such coronal magnetic arches. (Meyer, P. and Schmidt, H. U.: 1969, *Z. Angew. Math. Mech.* **48**, T218.)

Field: If plasma turbulence is important in the interstellar medium, we are interested in the sources of such turbulence. Dr. Tsytovich, do you expect the sources of the turbulence to be localized around energetic sources such as supernovae explosions, shock waves from various sources, etc.? And if so, can this turbulence then propagate into the quiescent interstellar medium? Can it do so at all, particularly if Langmuir turbulence is involved, since plasma oscillations have a very small phase velocity?

Tsytovich: I think that the region in which plasma turbulence exists must be much larger than the sources of the turbulence itself, e.g., supernovae. Many processes exist that can transport turbulent energy: ejection of fast particles; high power radiation, and other processes.

Thomas: Can we be more specific on the source of turbulence, which means some motion decaying? You talk about the hot stars as sources; long ago turbulence was believed to come from differential galactic rotation.

Colgate: I want to make sure that we know what we are discussing. In electromagnetic or magnetic turbulence, a charged particle, if displaced, will oscillate back and forth through some equilibrium point. This oscillation is a unique property of all turbulences associated with charged particles and a magnetic or electrostatic field. In gas dynamic turbulence, there is no such thing as an equilibrium value, unless you should be speaking of sonic turbulence. But almost always one speaks of turbulence in gas as being an arbitrary displacement of a particle, with only a small chance that the particle returns to the same point. This interpretation primarily means, in terms of physics, that you can have a persistent function like an eddy, which has no a priori reason to unwind. The distinction is decisive for the description of the two forms of turbulence. The aerodynamic turbulence is only poorly described by waves; but in the plasma case, we speak of k-values and waves and the modes of waves. Now first we have the sort of turbulence discussed so far, which has very short wavelengths, of the order of the Debije length (meters, or kilometers). Second there is hydromagnetic turbulence with energy input at $1/k$ of the order of 100 pc, as discussed by Parker.

Parker: I think it is an oversimplification to say that the main input is the Rayleigh-Taylor instability at 100 pc. It is certainly one of the inputs. I think non-uniform rotation of the Galaxy, plus the stirring of hot stars and sporadic local production of cosmic rays must be added, too. The Rayleigh-Taylor instability, together with the thermal instability, causes the gas to condense into clouds; and the hot stars and cosmic ray production tend to blow the clouds apart.

Colgate: I disagree with the differential rotation idea which leads to the concepts of two-dimensional turbulence. In two-dimensional turbulence, the eddies always progress to larger size.

Parker: You can develop three-dimensional turbulence in a one-dimensional shear, such as the non-uniform rotation. Therefore, I see no justification for saying that the non-uniform rotation cannot produce turbulence.

Colgate: Do you think that pure differential rotation alone with no heat input, as in the Earth's atmosphere, gives rise to turbulence?

Parker: I do not know. But of course, we do not have the shear alone; we have all the other heating, stirring, and agitation, which takes energy out of the non-uniform rotation in the form of turbulence.

Field: From the point of view of gas dynamics, can one characterize the turbulence spectrum of the plasma oscillations by some mean values which can then be put into the magnetohydrodynamic equations? For example, we might use a turbulent pressure which would be important in discussing shock propagation. [This question remained unanswered during part of the discussion. Finally Syrovat-skii brought it up again. Ed.]

Syrovat-skii: I should like to make a remark on the question by Field. The small-scale plasma turbulence gives us a new method of dissipation of large-scale motions. A good example is given by the solar flares. If we calculate the time of dissipation by the usual collisional processes, we obtain, say, a hundred or a thousand years for the dissipation of sunspot magnetic fields. But when we observe a solar flare, we see the field changing radically in a few minutes. Such a short time scale is brought about by small-scale plasma processes. Field's question is therefore how one effectively includes the small-scale plasma processes in the macroscopic description of interstellar gas dynamics.

Tsytovich: I can answer this question. Let us consider, for instance, ion-sound turbulence, or anomalous resistivity. One can introduce effective collisions that depend on the energy of the turbulence. They are very similar to normal collisions, in the sense that the dependence on the velocity is the same. This leads to a description of turbulent resistivity. However, we have no generalization of this procedure available. In some cases you cannot introduce effective collisions because the dependence on the velocity is not the same. Another question is: can one treat the turbulence as an external force in the magnetohydrodynamic equations? Work on this problem has been started, and it should be continued, I think.

Kaplan: There is a direct answer to Field as regards shock waves. All plasma turbulence processes in shock waves have dimensions less than the mean free paths. Thus these processes are beyond the scope of hydrodynamics and are therefore not included in hydrodynamic equations. Also, the energy density of plasma turbulence is usually much smaller than the density of thermal energy. But, if effective acceleration of particles in shock-wave turbulence is really taking place, these particles remove energy from the shock. This phenomenon must be taken into account in the Rankine-Hugoniot equation. The energy of flow of plasma across the shock front is not conserved, and such shocks resemble the well-known shock with radiation.

Mestel: Do I take it correctly that acceleration of fast particles is not an important energy sink for dissipation of large-scale turbulent energy?

Parker: If you look at the damping processes both for very large-scale, hundred-pc disturbances in interstellar space and for small-scale things, which you might call plasma turbulence, it seems to me that the major dissipation known is not cosmic rays. It is diffusion and Landau damping. That is, the energy does not seem to go into the cosmic rays. The other side of the coin is that the energy which goes into cosmic rays is five or ten times as much as is available to the interstellar medium. If anything, it is the cosmic rays which drive the interstellar medium, rather than vice-versa.

Tsytovich: We know that the Galaxy is moving through the intergalactic medium. The velocity of this motion is about 100 km sec^{-1}, which is of the order of the thermal velocity of the electrons. Can this motion of the Galaxy be a source of turbulence excitation?

Colgate: Is not the density of the Metagalaxy $n \approx 10^{-5}$ cm^{-3}? With n for the Galaxy of the order of 1 cm^{-3}, the ratio of densities excludes, I think, the excitation of either waves or turbulence of a very sizeable magnitude in the Galaxy.

Tsytovich: The excitation depends on the nature of the instabilities. For a plasma instability, the density ratio of 10^{-5} must be multiplied by the plasma frequency (10^4 sec^{-1}); thus the growth time is 10^{-1} sec^{-1}. Only 10 sec, therefore, would be needed to excite turbulence. The kinetic energy of the inflow is enough to produce the cosmic rays.

Menon: From the observer's point of view the energy input is in the form of high velocity motions of small bits of matter and it goes ultimately into random motions of 20 to 100 pc clouds. The fact that the kinetic energies agree is not enough. We really must consider the question of the energy transformation.

Colgate: That comment is probably correct. But Tsytovich's question implied that the plasma turbulence is excited by the wave friction of the Galaxy moving through metagalactic space. One should ask first: is there sufficient energy in the collision alone? Then, after that question is answered, we should look at the wave field that would be excited. And the wave field that might be excited might have the fine-structure scale that you are asking for.

Pikel'ner: If the Galaxy moves in the intergalactic gas with a rather high velocity, it would form a stationary shock wave in front of it. The thickness of this shock would depend on the plasma instability. Behind the front, there should be a layer with excited plasma turbulence. This front can also accelerate cosmic rays. [In his Report Spiegel, p. 201, discusses a related problem. Ed.]

Colgate: We now direct the discussion toward the subject of collisionless shock waves.

Kaplan: One of the most important points in the theory of the collisionless shock is the width of the shock. I would like to present a new argument that the width is of the order of an ion Larmor radius. The instability of plasma inside the shock front is usually connected with an ion-distribution function with two maxima, as was demonstrated earlier by Mott-Smith and by Tidman. This leads to excitation of plasma turbulence with low-frequency modes as follows. As in strong shocks, the inequality $T_e \gg T_i$ is fulfilled (for instance, by heating the gas in front of the shock by electron conductivity or by radiation); and we may expect powerful ion-sound turbulence (see the Report by Kadomtsev and Tsytovich, p. 108). Usually the first excited waves are those with wave numbers of the order of V_s/ω_{pi} where V_s is the velocity of the shock and ω_{pi} is the ion-plasma frequency. Then, by the non-linear process of scattering on ions, the ion-sound waves decrease their wave numbers and frequencies until they reach their lower limit. If the frequency of ion-electron collisions is really very small, the lowest possible frequency is the gyrofrequency of ions ω_{Bi}, since at that frequency strong damping of ion-sound waves occurs. The shock front thickness will be of the order of the free path of the wave with smallest wave number; in this case, it is approximately $10V_s/\omega_{Bi}$. In the interstellar medium this quantity is about 10^{10} or 10^{11} cm. If in a particular situation there is no magnetic field at all, or if the ion-electron collisions are somewhat more frequent, the smallest possible wave number is determined by other factors. One of the models appropriate for solar conditions was elaborated by a former student of mine (Zaitsev, 1967). He found the value for the thickness of the shock front to be $n_e^{1/2}\lambda_D^{5/2}$, where n_e is the electron density and λ_D is the Debije radius for the plasma after the shock. By coincidence, the same value is found for the shock thickness in the interstellar medium. Therefore, it seems to me that the value of 10^{10} to 10^{11} cm is a realistic estimate of the width of the shock front in the interstellar medium. This value is many orders of magnitude smaller than the value of 10^{15} to 10^{17} cm found if only the mean free path of ions and atoms is used. (Kaplan, S. A. and Pikel'ner, S. B.: 1963, *The Interstellar Medium*, Moscow (translation: 1970, Harvard University Press, Cambridge); Zaitsev, V. V.: 1967, *Astron. Zh.* **44**, 490 (translation: *Soviet Astron.* **11**, 392).)

   Colgate: We have heard from the theoreticians that collisionless shock waves exist in plasmas and in the interstellar medium. I want now to bring up the question: is there any evidence, in the Galaxy or in the laboratory, that the collisionless shock really exists?

   Davis: Space observations of the bow shock of the earth and of other shocks in free space should remove completely any doubt about the existence of collisionless shock waves.

   Colgate: Is there any estimate of shock thickness or of the dissipation mechanism in the satellite experiments of the bow shock?

   Davis: One can make estimates of the thickness, but it is not easy because the shock is not stationary. In principle you sail your spacecraft through the shock and see how long it takes. The problem is that the shock is moving rapidly relative to the spacecraft. The thickness is clearly small. The experiments cannot easily distinguish between a thickness of the order of the proton gyroradius and a thickness of the order of the geometric mean of the proton and electron gyromagnetic radii. The dissipation mechanisms are not clear, but strong electric fields are involved,

Parker: We have much detailed evidence on the behavior of collisionless shocks in the solar system, and there are observations showing that fast particles are often generated copiously in these shock waves. One would conclude, scaling in a very crude way, that the much more vigorous blast waves from novae and supernovae must certainly generate a great many fast particles through plasma turbulence mechanisms.

Van de Hulst: I want to repeat that spatial resolution in astrophysical observations is not good enough to observe widths of less than 10^{15} cm anywhere in interstellar space. Only by indirect evidence can one conclude that collisionless shocks exist.

Podgornii: In the Institute of Space Research in Moscow, Dubinin, Managadze and I have tried to construct collisionless shock waves in the laboratory. In our experiments the parameters were: velocity V: 3×10^7 cm sec^{-1}; density n: 10^{13} cm^{-3}; electron temperature T_e: 15 to 20 eV; ion temperature T_i: 5 eV; and B (frozen in the plasma) 40 G. These parameters were chosen to make the shock similar to those in the solar wind. For example, this laboratory plasma flow has the same Mach number as the solar wind, with respect to the Alfvén velocity and the magneto-sound velocity. The free path of the particles is 10 times larger than the size of the experiments. The ratios $\beta = 8\pi nkT/B^2$ and ω_{pi}/ω_{Bi} are comparable in the laboratory and in the solar wind, where ω_p and ω_B are the plasma frequency and the cyclotron frequency, respectively. The magnetic Reynolds number was about 10^4. In the experiment we obtained collisionless shocks with a length of about c/ω_{pi}, where ω_{pi} is the plasma-ion frequency and c the velocity of light. We tried to determine how the energy was dissipated. One of the theories of collisionless shocks assumes instability of Alfvén and magneto-sound waves in the shock front. Under this assumption one should observe magnetic field fluctuations at a frequency near the ion Larmor frequency and the dimension of the fluctuations should be about c/ω_{pi}. We measured such fluctuations by means of magnetic probes. One probe was placed upstream where there were almost no magnetic fluctuations. But in the shock and behind the shock, the magnetic fluctuations increased. By placing two probes in or behind the shock, we can estimate the lengths of the fluctuations from the measured correlation function.

A typical oscillogram of the fluctuations is presented in Figure 3. The upper trace shows the change of the magnetic field strength in the upstream probe. The second trace shows the output of the probe placed in the shock front. Figure 3 shows oscillograms from two downstream probes, when the distance between the two probes was less than 3 cm. A rather good correlation of signals is seen. The correlation coefficient of the signals from the two probes as a function of their distance is shown in Figure 4. The half-width of the curve gives the correlation length of the fluctuations, which turns out to be, indeed, c/ω_{pi}. In the region of interaction between the plasma flow and the magnetic field, c/ω_{pi} equals the ion Larmor radius. Measurements of the magnetic fluctuations have not been made previously. It is an interesting point that the spectra of the fluctuations around ω_{pi} are the same in our experiments and in space [$P(\omega) \propto \omega^{-3}$]. The experiments indicate that dissipation of the energy of the plasma flow arises from the instability of magneto ion sound or Alfvén waves, induced probably because of the

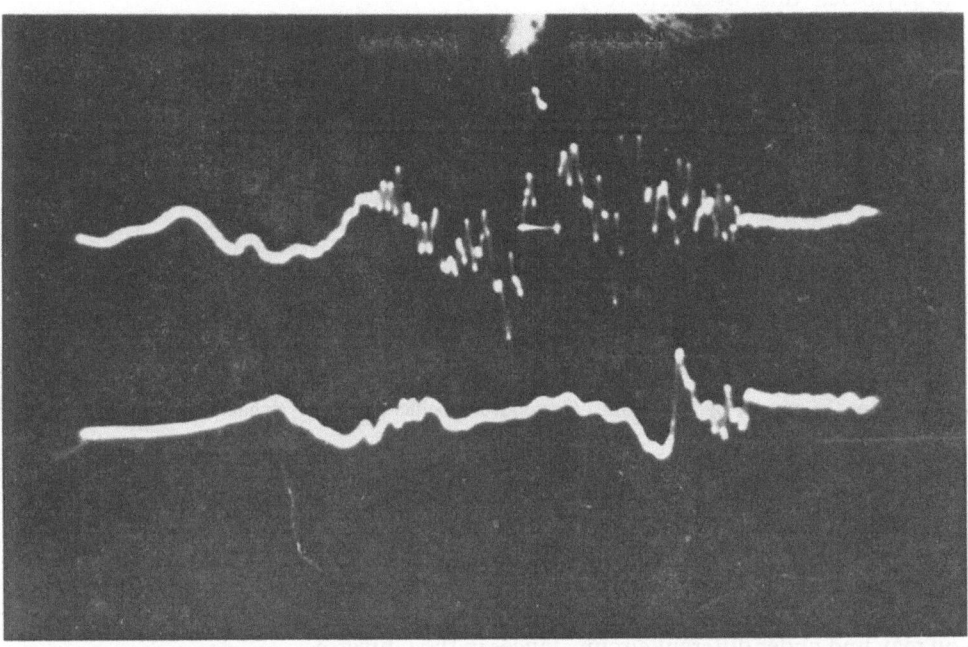

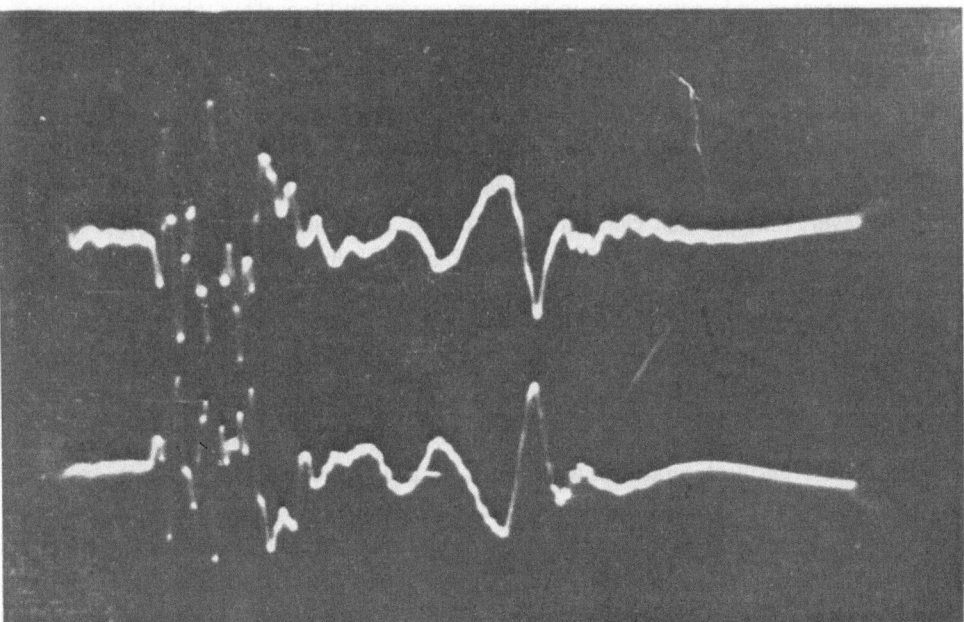

Fig. 3. (See the remark by Podgornii.) Oscillograms of magnetic-field fluctuations observed in laboratory experiments on the solar wind.

anisotropy of the plasma pressure in the magnetic field, as discussed by Kadomtsev and Tsytovich (this volume, p. 108).

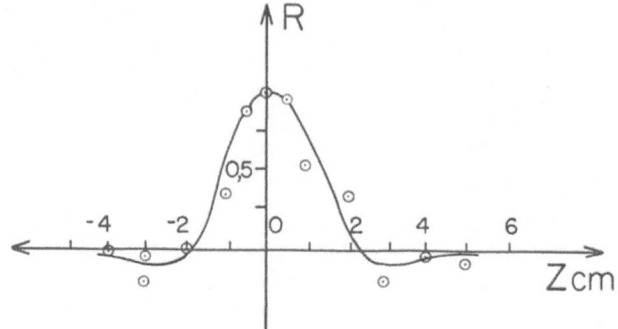

Fig. 4.   (See the remark by Podgornii.) Correlation coefficient of the signals from two probes as a function of their distance.

Parker: There is one point which has been demonstrated in many investigations of the problem. Depending on the various dimensionless numbers that characterize any particular collisionless shock, there are many different interactions which may dominate the scene; and therefore, there are many different shock structures which you may find under different circumstances in the Universe.

Colgate: The dissipation mechanisms behind the shock determine whether the entropy is generated in ion waves or in electron waves, and therefore whether the energy goes into high energy electrons or into high energy ions. If enough energy goes into electrons, we might hope to see radio emission.

Kaplan: I have said previously that the plasma turbulence pockets can be seen, if there is an anisotropic distribution of fast electrons. In very strong shocks we may have such a distribution due to the acceleration of particles inside the shock. I am not sure, however, that the fast electrons can achieve relativistic energy when crossing a shock front thickness of small dimensions. Finally, I hope that the law of changing of the velocity of the shock during the passage through a medium with varying electron density ($V \propto n_e^{-1/4}$) (Kaplan, 1967) is also valid for collisionless shocks. (Kaplan, S. A.: 1967, *Astron. Zh.* **44**, 384 (translation: *Soviet Astron.* **11**, 302).)

Field: Is it true that the final result of the generation of plasma turbulence is to increase the temperature far downstream from the main shock front? Therefore, are the Rankine-Hugoniot relations applicable as long as one is far enough from the shock front, because complete thermalization has occurred?

Colgate: Yes, the Rankine-Hugoniot relations are always fulfilled. The decay lengths behind the shock are associated with the excitation of additional degrees of freedom as the plasma approaches thermal equilibrium. What are the widths of waves coming out of the neutron star in the Crab?

Woltjer: The width of the waves in the Crab is of the order of 0.01 pc.

Colgate: If the particles are relativistic, the wavelength is small compared to the thermalization length.

Zel'dovich: To answer Field's questions: At large distances there will be an equilibrium due to the relaxation of all the established processes. But near the front, there are several velocities at any one point. Instead of one Maxwellian distribution, there are several δ-functions. However, for pressure and density this effect is not very important; therefore, a very exact Rankine-Hugoniot equation exists at a distance which is not necessarily very large. Very good laboratory experiments on the widths of shocks with transition from a wide shock to a normal shock are made by Sagdeev, who is not only a theoretician, but also has recently become an experimentalist. I would like Dr. Karpman, who has worked with him, to comment.

Karpman: There was a series of experiments in Novosibirsk, performed by the Sagdeev group in the last few years in which detailed structures of collisionless shocks were investigated. If in a rather cold plasma a shock front exists perpendicular to a magnetic field, then, as was predicted theoretically and observed experimentally, the structure of the collisionless shock is of an oscillatory kind, with the value c/ω_{pe} for the dimension of one oscillation. The decay length depends on the nature of the dissipation. If the dissipation from the turbulence is of the collective type, then several large-scale oscillations appear. The oscillatory structure appears only if the angle θ between the normal to the shock front and the direction of the magnetic field is less than $\sqrt{(m_e/m_i)}$. If the angle is larger, the size of one oscillation is $\lesssim (c/\omega_{pi})\theta$. All this applies to cases with rather small Mach numbers. If the Mach number is large, the structure becomes non-oscillatory, aperiodic, and like a usual [collisional [?] Ed.] shock structure. The length or thickness in this case is approximately the Larmor radius. The ions in the front are rotating around the magnetic field. All these types of shocks were observed experimentally in special tubes with different geometry for the magnetic fields.

Podgornii: I want to compare the Sagdeev experiment and the experiments I described before (p. 144). The principal difference consists in the conditions of the experiments. In the Sagdeev experiment, a low-density plasma was used; the ratio $\beta = 8\pi nkT/B^2$ was much less than unity. But in our experiments, where we tried to simulate the solar wind, a high-density collisionless plasma was employed. We have β somewhere near unity. In our case, there is no regular magnetic structure, only microfluctuations. In a recent issue of *Usp. Fiz. Nauk*, Sagdeev and I have described all these cases and compared the results of all different experiments.

Colgate: I should like to mention a problem associated with the power law spectrum of cosmic rays at low energies. Observationally we obtain the low-energy cosmic-ray spectrum by correcting for the interplanetary modulation. Theoretically we have predictions based on the various models for a supernova explosion. In both cases we seem to obtain a smooth curve down to very low energies. The difficulty, from the standpoint of collisionless shock wave theory and the interaction with the interstellar medium, is that this power law does not give you anywhere a sharp discontinuity in the velocity spectrum. All the excitation mechanisms of instabilities along the lines of force require such a discontinuity and are therefore excluded. Only the component of the explosion which is across the local magnetic field should lead to a collisionless shock and then only when the energy densities of the particles have been reduced to a value of β of the order of unity. This implies that the explosion will blow a hole in the magnetic field of the Galaxy, of the order of 100 pc in dimension, before the stress of the particles in the hole becomes comparable to the field. These predictions, both the one about the direction perpendicular to the field and the one parallel, disagree with the observations. This forms a problem that has worried me for many years, and I wonder if anyone here knows the solution.

Field: McKee at Berkeley has been considering collisionless shock-fronts in a relativistic twinstream situation. Thus far his computer studies indicate that an instability will develop and, presumably, also a shock; I cannot speak, however, on your specific point that the gradual build-up of particles may preclude such a shock.

I want to bring up a question connected with collisional shock fronts in neutral gas. I wonder if the presence of a small ionization of, say, one per cent will significantly change the properties of the shock waves. I assume that the shock fronts move so slowly that they do not ionize the gas. Will there be a major change? It does not seem so to me.

Pikel'ner: In a shock front in a partially ionized gas, ions have a collective instability and form a sharp front. Neutral atoms do not have a collective instability, and they can form only collisional shocks. Fifteen years ago I calculated the structure of shocks in a partially ionized gas (Pikel'ner, 1954). There are several layers, a sharp shock for ions and a thick layer in which there are interactions (especially charge exchange interactions) between ions and neutral atoms. In this thick layer the densities of ions and atoms change gradually. But under the usual conditions in the interstellar gas, even the thickness of the thick layer should be less than can be observed.

I should like to add another possible observation on plasma phenomena. For instance, there are some filamentary nebulae which are shock waves. To calculate their emission, you have to know the electron temperature immediately behind the fronts; but the distribution of energy between electrons and ions is dependent on the kinds of instability which transform the energy of shock waves into the thermal energy of particles. From a classical point of view, this shock energy should go to the ions and from a plasma point of view, it should go to the electrons. But the emission of the filament depends very strongly on whether the ions or the electrons happen to receive the shock energy. Quite generally, plasma effects are important for dissipation of magnetic fields in neutral lines. Usually in cold interstellar space, plasma turbulence will be very weak. Only in some very violent places in the Galaxy, for instance in the central parts or in shock waves going out from supernovae, are there conditions favorable for high energy plasma turbulence and for the acceleration of fast particles. (Pikel'ner, S. B.: 1954, *Izv. Krymsk. Astrofiz. Observ.* **12**, 94.) (A discussion in English of this work may be found in Sections 9 and 10 of *Interstellar Gas Dynamics* by S. A. Kaplan, Pergamon Press, Oxford, 1966.)

Spiegel: Since you are really trying hard to find an observable collisionless shock, may I raise the possibility that we look for such things in stellar systems? For example, if the spiral arms are waves, they are narrower than a stellar mean-free path; that result would be suggestive of a collisionless shock.

Colgate: The term collisionless shock pretty well implies the concept that there exists an enhanced dynamic friction over and above the r^{-2} force law, independent of whether it is gravitational or electrostatic force. The collisionless shock occurs when you take a large number of particles of one kind of charge and have them interact in a cooperative way with a large number of particles of the opposite sign. The collective interaction may be either through electrostatic forces or through magnetic forces. But in any case, the idea is that the collective interaction is stronger. The shocks that you get in stellar systems ought to be called collisional shocks.

Spiegel: I am speaking of waves which involve only collective interaction.

Parker: The difficulty with the stars is, as Colgate said, that you do not have opposite charges to give overall neutrality. Therefore, whenever you have two systems falling through each other, there is always a large potential well, and the characteristic length for instability (the interaction between the two oppositely streaming components of stars) is always comparable to the dimensions of the components of the system itself. You never can get evidence for a clean-cut shock.

Field: I think Spiegel is quite correct. There is a plasma-like mode in a stellar system for which the dispersion relation is: $\omega^2 = \omega_g^2 + k^2 c^2$ with $\omega_g^2 = 4\pi G\varrho$. The point is that, instead of having a real plasma frequency, you have an imaginary one corresponding to the attractive forces of gravitation. The analogue of a plasma oscillation is then the Jeans instability.

The characteristic length, $\lambda_J (= 2\pi c/\omega_g)$ associated with such a shock might be the Jeans length, corresponding to the Debije length in the plasma shock case. However, it must be kept in mind that the dispersion relation is derived from moment equations. One finds from the Boltzmann equation that waves with $k \gg k_J = \omega_g/c$ are heavily Landau damped and therefore do not propagate as readily as the dispersion relation would suggest. I am not sure what the effect of the damping would be on the formation of a shock. I would take issue with Parker. While it is rigorously true that the Jeans length tends to be of the order of the size of the system, one can think of a special system that is quite interesting, for example, a rotating disk supported by stellar motions (pressure) in the vertical direction and by centrifugal force in the plane. The radius can be far greater than the Jeans length. A spiral galaxy is such a system. The Jeans length in the stars of the Galaxy is about 100 pc; it is conceivable, therefore, that we could have a shock with a thickness of about 100 pc, still much smaller than the radius of the system, 10^4 pc.

Pikel'ner: If the stellar system is rotating, then the coriolis forces are important and play the role of the magnetic field in the plasma. This was studied in detail by Marochnik.

9. OBSERVATIONAL ASPECTS OF GALACTIC
MAGNETIC FIELDS

Introductory Report

(Friday, September 12, 1969)

G. L. VERSCHUUR

National Radio Astronomy Observatory*, Green Bank, W.Va., U.S.A.

1. Introduction

I will review the impressive advances in the observations of the galactic magnetic field made since the time of van de Hulst's review at the 1966 Noordwijk Symposium (van de Hulst, 1967). Most of these observations are so recent that the consequences have not yet been worked out very well and are in need of discussion. Luckily the emphasis in the present Symposium, unlike that in many others, is on discussion. For this reason I will not hesitate to include in my review provocative speculations. I will base my discussion on Mathewson's (1968) elegant magnetic field model, consisting of a local field in the form of a sheared helix, superimposed on a large scale longitudinal field. I will include the criticism of this model by Gardner *et al.* (1969b). Mathewson has succeeded in accounting for such data as the distribution of background polarized radiation from the Galaxy, the distribution of rotation measures of extragalactic radio sources and even the spurs and ridges in galactic continuum emission. There are critics who inherently distrust models that account for too many things at one time, but I feel that we should try to account for as many things as possible with the least number of models. Mathewson has succeeded in uniting much data and I will only add a few pieces to his model.

The progress since 1966 in the number, variety, and quality of the observations is indeed impressive. Gardner *et al.* (1969a) have presented data on the radio polarization of 366 sources at three frequencies. The observations have been discussed by Gardner and Whiteoak (1969) and Gardner *et al.* (1969b). Mathewson has largely filled in the gap in optical polarization data in the southern hemisphere by measurements on 2000 stars, so that his model could be based on 7000 stars in the northern and southern skies together. Enormous progress has also been made in the area of direct measurements of the field strength. First, measurements of the Zeeman splitting at 21 cm has shown the existence of fields of a few μG up to 50 μG. Second, the discovery of pulsars has made it possible to measure both the (Faraday) rotation measure (RM) and the signal-dispersion measure (DM); the ratio of these two yields directly an estimate of the mean value of the line-of-sight component of the magnetic field. These results give average field strengths between a few times 0.1 μG to 3 μG.

I start the review with Table I which is my adaptation of a table originally presented

---

* Operated by Associated Universities, Inc., under contract with the National Science Foundation.

Habing (ed.), Interstellar Gas Dynamics, 150–167. All Rights Reserved. Copyright © 1970 by IAU

TABLE I

Observational data about the galactic magnetic field
(modified from the original version by Van de Hulst, 1967)

Category		Magnitude	Direction
A	Optical interstellar polarization	q	
	Polarization of non-thermal radio emission	–	e
	Faraday effect on radio sources	f	
B	Elongation of interstellar clouds	–	f
C	Cosmic-ray energy, density and confinement	q	q
D	Cosmic-ray anisotropy	–	–
E	Zeeman effect, H	e	f
F	Zeeman effect, OH	–	–
G	Pulsars	f	e

Key: e = excellent; f = fair or fine; q = questionable or marginal; – = no data, or do not believe.

by van de Hulst (1967). The present table (which makes clear what my biases are) groups several fields of research together in a more optimistic way than the analogous display of 1966. One should appreciate the amount of data included in category A which encompasses the results of radio-source polarization measurements on hundreds of sources, continuum surveys, background polarization data, and optical polarizations for thousands of stars. Category B includes the shapes of filamentary nebulae (following a proposal by Shajn, 1955). Van de Hulst left this category out, but, as we will explain in Section 7, we disagree with him on this point.

2. Mathewson's Model and Associated Work

Mathewson (1968) has combined his southern polarization data on 1400 stars (mostly within 500 pc of the Sun) with the northern hemisphere results (a total of 7000 stars), in order to deduce a magnetic field model for the solar neighborhood. Figure 1a shows the polarization data, as projected on the sky. As is well known, the polarization vector gives the average of the direction of the magnetic field component perpendicular to the line of sight. Mathewson then took a helical field structure, and with a trial-and-error method he chose the parameters of this helix such that they predicted maps (those of Figures 1b and 1c) as similar as possible to Figure 1a. The best-fitting model has the field lines form tightly wound (pitch angle 7°), right-hand helices which lie on the surfaces of tubes having elliptical cross-sections of axial ratio 3 with major axes parallel to the galactic plane. The helical pattern is sheared by 40° on the plane of the Galaxy, in an anti-clockwise sense when viewed from the galactic North Pole. The Sun is 100 pc toward the galactic center from the magnetic axis, and 10 pc below the galactic plane. Two predicted maps are shown in Figures 1b and 1c. Figure 1b shows the component, perpendicular to the line of sight, of the tangents to helices with a semi-major axis of 250 pc, and Figure 1c represents in the same way the families of helices with semi-major axes of 115 pc (thick lines) and 100 pc (thin lines). The differ-

ences between Figures 1b and 1c are due mainly to the offset in the Sun's position with
respect to the axes of the helices. It is seen that the superposition of Figures 1b and 1c
very well duplicate Figure 1a. Already a first look at Figure 1a shows that the field
lines are curved. The picture does not seem to agree at all with parallel lines converging in the distance. This disagreement, it seems, is the basis for Mathewson's assertion
that the observations are not compatible with a longitudinal field. I therefore differ
in opinion from Gardner *et al.* (1969b) who claimed that the data shown in Figure

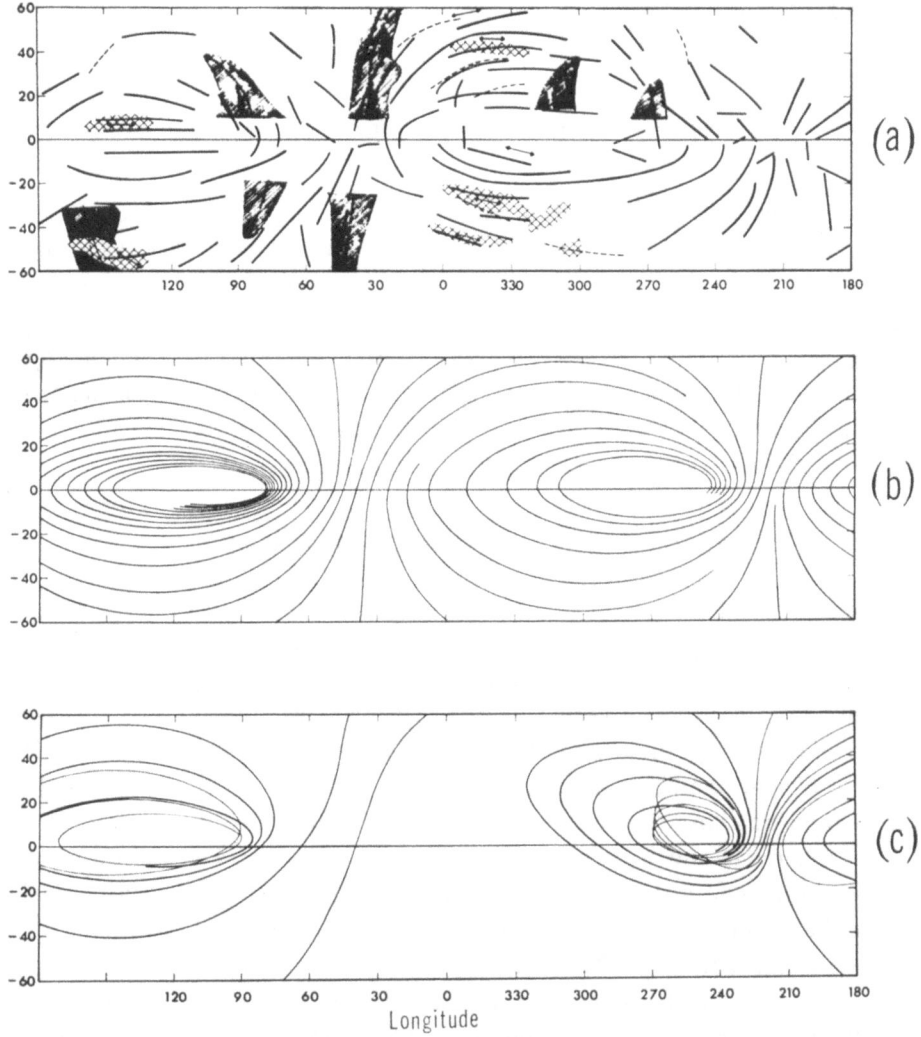

Fig. 1. Optical polarization data and field models from Mathewson (1968). – (a) Heavy lines
represent 'flow patterns' formed by *E*-vectors of optical polarization. Radio spurs are shaded. Cross-
hatched areas are strongly polarized at 408 MHz. Arrowed lines give directions of magnetic field
obtained from radio measurements. – (b, c) Projection on sky of several families of helices (see text).

1a "... are also consistent with an axial field toward $l^{II} = 50°$". Still, such a field would have been indicated if a smaller sample of stars had been taken, especially if the stars were near the plane. Therefore I do not give much weight to Seymour's (1969) analysis of 550 stars which yields an axial field. It would be of interest if Seymour were to apply his analyzing method to all the polarization data now available.

Also included in Mathewson's 'Unified Field Theory' are data by Appenzeller (1968) on 308 stars near the poles and (in the plane) near the anomalous region around $l^{II} = 140°$. Gardner, Morris, and Whiteoak claim that "... Appenzeller found a low altitude field aligned parallel to Gould's belt in Perseus ($l^{II} = 145°$)" and then state "this orientation defies explanation in terms of the magnetic helix". These are strong words, first since Appenzeller's data do fit Mathewson's model, in particular in the North Pole region (these data are not shown in Figure 1 but may be found in Mathewson, 1968); and second since Mathewson (1969) himself had already stated that the helical field is probably associated with Gould's belt. Appenzeller does not, in so many words, invoke Gould's belt and admits that more data are needed in any case before a comprehensive model can be made. Much needs to be said on the association of the magnetic field structure and the distribution of stars in the Gould belt system, but I should perhaps leave that for the discussion. Clube (1968) has been rather specific on this point but he unfortunately used the contour map of Gardner and Davies (1966) which has been superseded by later work.

Before I move to the modification of the helical model I would like to stress the impressive property of this model in that it provides a simple explanation for the spurs and ridges of continuum radio emission which may be tracers of the field lines. Non-thermal electrons are injected into the helix, probably from sources nearer the plane, and these electrons then spiral away from the plane along the field lines and appear as enhanced spurs of emission. Bingham (1967) had already shown that field lines do indeed lie along the galactic spurs. Hall (1958) had noted that vectors from $l^{II} = 350°$ to 40° seemed to converge to $l^{II} = 35°$ and he speculated that this might have a 'particular significance'. This point is at the base of the North Polar spur and Figure 1 shows what the significance is.

In a more recent paper Mathewson and Nicholls (1968) propose a modification of the helical model in order to account for Faraday rotation data. Consider first the contribution by the local helical field. As is known, the Faraday rotation measure (RM) equals $0.81 \int n_e B_{\parallel} \, dl$ rad m^{-2} (if n_e is in cm^{-3}, l in pc, and $B_{\parallel}$ in μG), where l is the distance along the line of sight. Because of the condition div $\mathbf{B} = 0$, the geometry determines the field strength along the field lines in terms of the strength at, say, the minor axis of the elliptical cross sections of the helix. This makes it possible to construct contour maps of $B_{\parallel}$, one of which is shown in Figure 2. The contours are in units of 0.025 times the field strength on the minor axis of the helical cross section. The semi-major axis is 250 pc, but any helix wound around a tube of larger cross section would give a similar pattern; smaller tubes would not, because the Sun is off the axis of the family of tubes. (Note the difference between this Figure 2 and Figures 1b and 1c: in the two latter figures the curves indicate only the direction of the field

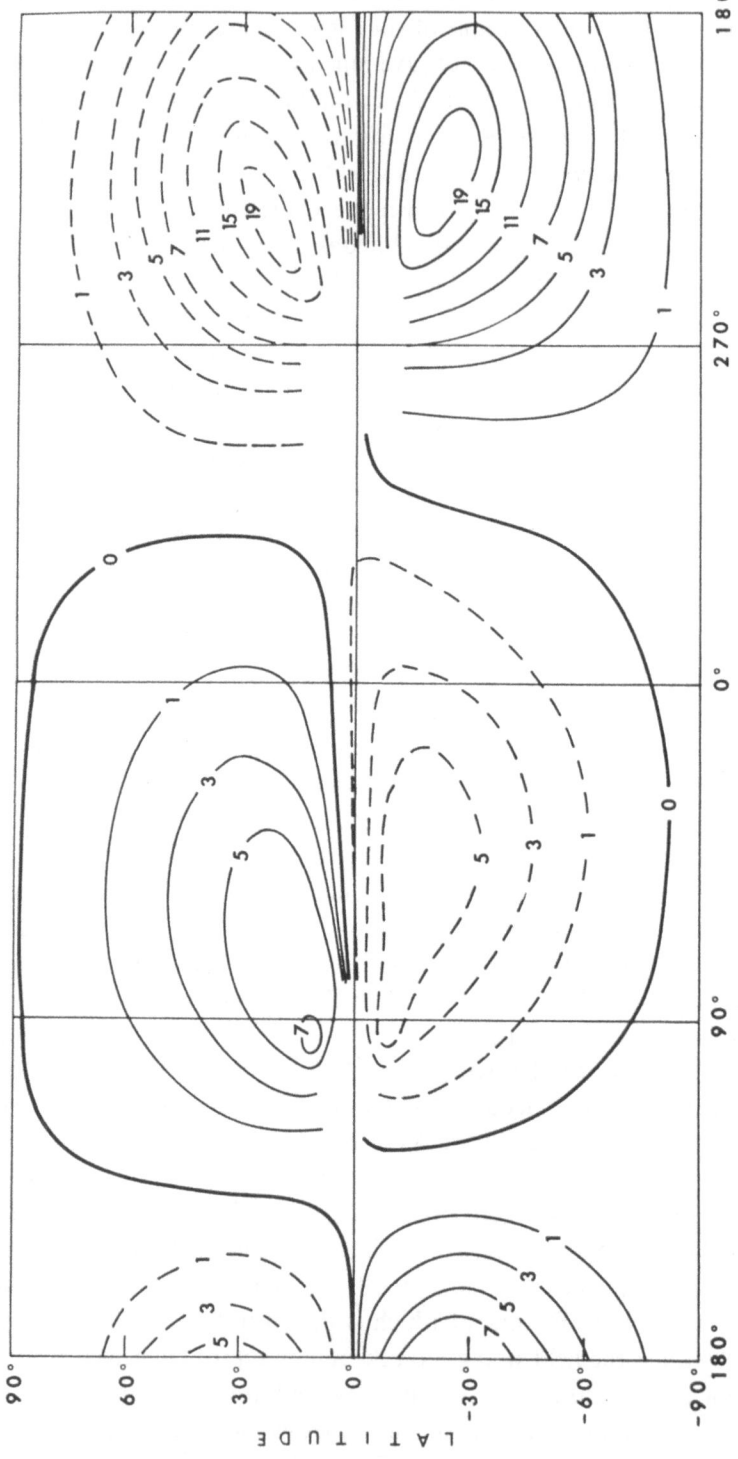

Fig. 2.   Contours of the line-of-sight component of the magnetic field for a family of helical magnetic lines of force lying on the surface of an elliptical tube with semi-major axis of 250 pc. Full lines – field toward the observer. Broken lines – field away from observer. Contour unit is 0.025 of the strength on minor axis of the tube (Mathewson and Nicholls, 1968).

component perpendicular to the line of sight, *i.e.*, Figures 1b and 1c are not contour diagrams.) Mathewson and Nicholls (1968) claim that one can account for the RM distribution as found by Gardner *et al.* (1969a) by adding to the helical field a larger-scale longitudinal field in the direction $l^{II}=90°$ and $b^{II}=0°$, i.e., the helix is just a superimposed local phenomenon.

I have a few critical remarks regarding this modified model. First, since the longitudinal field only exists beyond several hundred pc, one wonders how it can contribute to the rotation measure at high latitudes where one expects to find very few electrons. Second, Mathewson and Nicholls suggest that south of the galactic plane the longitudinal field dominates between $l^{II}=320°$ to $0°$ and $l^{II}=130°$ to $160°$, whereas in between these longitudes the combination of the longitudinal field, directed to $l^{II}=90°$, and the helical field (Figure 2) accounts for the RM-signs. But north of the plane it has to be the helical field that determines the RM-sign, either by virtue of its strength or as a result of higher local electron density. Since the pulsar data do not suggest a systematic difference in integrated electron content between the northern and southern hemispheres, there must be a stronger field present in the north. This of course destroys the symmetry in the model and takes away some of its elegance. In Section 6 I will return to the difficulties associated with the RM-distribution, but first I consider in the following sections the suggestion that the Galaxy has a basic longitudinal field, directed along the spiral arms and distorted by local structures which we will call 'field pockets'. I will show that, observationally at least, there is evidence for motion and structure of interstellar clouds which can be associated with our local field pocket.

3. The Distance at Which Optical Polarization is Produced

To test whether the helical field is indeed a local phenomenon we may investigate whether there is a correlation between the polarization of a star and its distance. Behr (1959) initially found that the polarization increased with distance but neither Hall (1958) nor Hiltner (1956) found such an effect. One pertinent observation indicating that most of the observed polarization is produced rather nearby is that by Krzeminski and Serkowski (1967). They noted that two clusters in the same area in the sky (around $l^{II}=134°$) had about the same degree of polarization, although their distances differed by large amounts, the first (Stock 2) being at 300 pc, the second (h and χ Persei) at 2000 to 3000 pc. Therefore, they concluded that probably most of the net observed polarization takes place in the solar neighborhood.

I have studied this problem using the large amount of data available on polarization of (open) clusters and associations. By taking groups of stars instead of individual field stars we hopefully reduce the 'noise' in the polarization data and in the distance determinations. One may study the magnitude of the polarization, the mean position angle and the scattering around this mean, all as a function of distance and as a function of *l* and *b*. I collected data on 56 groups of stars containing five or more members with polarization data available (see Table II). I will not give details here but hope to publish these elsewhere. In Figure 3a the mean polarization is plotted as a

TABLE II

Mean polarization as function of distance for clusters and associations

Longitude range	Number of objects	Correlation coefficient
0° –30°	10	0.25
60° –89°	10	−0.12
90° –121°	11	0.38
122°–180°	14	0.60
181°–359°	11	0.39

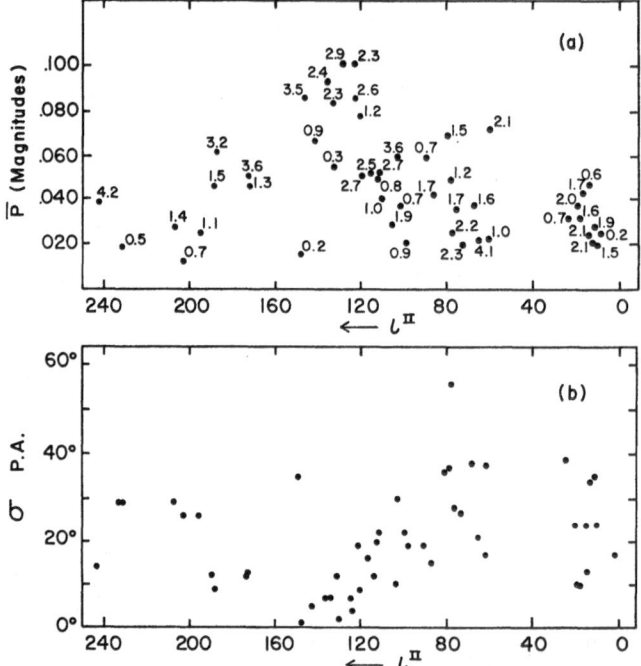

Fig. 3. Polarization properties of clusters and associations. – (a) Mean polarization as a function of longitude. Numbers indicate the distance to the cluster of association in kpc. – (b) The scatter in position angle of member stars (one standard deviation) as a function of longitude.

function of l and b with the distance to each cluster indicated next to the various points. It is clear that the effect noted by Krzeminski and Serkowski is evident elsewhere. Take, for instance, the group of objects near longitude 15°. The mean polarizations are all of about the same value, whereas their distances vary from 0.6 to 2.1 kpc. To test this further, I estimated the correlation coefficient between mean polarization and distance in various longitude intervals. The results in Table II show that polarization is not a function of distance.

The scatter in polarization position angle for stars within the individual clusters and associations is shown as a function of longitude in Figure 3b (σ equals one standard deviation value). The minimum scatter is around $l^{II} = 140°$ which is consistent with the helical model. Figure 3b looks similar to a plot given by Hiltner (1956). The combination of Figures 3a and 3b is not inconsistent with the suggestion that the polarization is produced relatively locally, *i.e.*, within about 500 pc of the Sun.

4. The Zeeman Effect

The Zeeman splitting of the 21-cm line has now been observed in a considerable number of neutral hydrogen clouds (Verschuur, 1968, 1969a, b, c; Davies *et al.*, 1968). It often remains difficult to estimate the magnetic field strength due to the complexity of most of the spectra. It is interesting that in all the Zeeman observations made so far the directions of the fields fit Mathewson's helical-plus-longitudinal field model. There are, however, large problems with the field strengths. For example in Orion A ($l^{II} = 209$, $b^{II} = -19°$) I find a dense cloud with a field of -50 μG*, whereas at the same longitude, but at the other side of the plane I was able to make one successful Zeeman observation (at $l^{II} = 238.25$, $b^{II} = +38.05$), which yields a magnetic field of 4.5 ± 5.3 μG. These values cannot be reconciled with the predictions of Figure 2; addition of a longitudinal field directed to $l^{II} = 90°$ would make the disagreement even worse.

A twofold explanation for this seems to be, first, that the strongest fields are found in the densest clouds, which are gravitationally stable at least against internal motions, and second that the field strength is the result of amplification by contraction of the clouds (Verschuur, 1969d).** More data recently obtained (Verschuur 1969b, c) allow us to examine this suggestion further.

Table III lists the neutral hydrogen features in which fields have been detected, whereas Table IV contains information on a number of negative results. Column 4 gives the velocity of the feature with respect to the local standard of rest (l.s.r.); column 5, the field strength, from Verschuur (1969a, b, c); and column 6 gives the density in the clouds. These have either been taken from previously published works, listed in column 7, or estimated here. Column 8 indicates the reliability of the density estimates and (in some cases) of the field estimate. 'Good' indicates that little further improvement can be expected on the data given. 'Fair' indicates that further observations might help to improve the estimates, whereas 'Poor' implies that further observations are essential to establish the best values for these features.

For M17 I have measured an accurate absorption line profile and have used the observed dispersion (σ) of the apparent Zeeman pattern, together with an estimate of the optical depth, to derive n_H. This value differs a little from that suggested by Clark (1965). In general in Table III I have assumed a cloud diameter of 10 pc whenever Clark did not give interferometer data. The data of Table III are shown in Figure 4, where the values of $B_{\parallel}$ (in μG) are plotted against n_H (in cm^{-3}). The solid

---

* I have adopted the convention to call a value of $B_{\parallel}$ negative if the field is directed toward the observer. This is in agreement with the generally-accepted definition of negative radial velocities. It is not in agreement with the less generally used definition of rotation measure.
** See also the remarks by Mestel during the following discussion, p. 187 (Ed.).

error bars indicate well-determined field values and the dashed lines indicate less certain values or cases where only estimates of the field or density could be made.

For an isotropically contracting cloud with a 'frozen-in' magnetic field we expect the density to be $\propto R^{-3}$ (R=radius) whereas the magnetic field will be $\propto R^{-2}$. Therefore the field, B, will be proportional to $n_H^{2/3}$. I have drawn in Figure 4 a line

TABLE III

Magnetic fields in clouds of known density

Direction	l^{II}	b^{II}	Velocity (km sec^{-1})	Field estimate (μG)[a]	n_H(cm^{-3})	References	Integration time (hr)	Significance [b]
Tau A	185	− 6	+10	− 3.5 ± 0.7	14 (16)	1 (2)	75	G
			+ 4	− 1.5 ± 0.9	27	2		F
Cas A	112	− 2	−38	+18.0 ± 1.9	193	1	34	G
			−48	+10.8 ± 1.7	87	1		G
Cyg A	76	+ 6	+ 3	+ 3.0 ± 2.2	1–4	3	26	F
			−84	+ 4.0 ± 2.2	2.5	3		F
M17	15	− 1	+14	+25.0 ± 10.0	60–100	Estimate (see text)	9	P
Orion A	209	−19	+ 7	− 50.0 ± 15.0	680	2	8	G
			+ 2	− 70.0 ± 20.0	350	2		P

1. Verschuur, 1969a.
2. Clark, 1965.
3. Shuter and Verschuur, 1964.
[a] A negative sign indicates a field toward the observer.
[b] G = Good, F = Fair, P = Poor. See text.

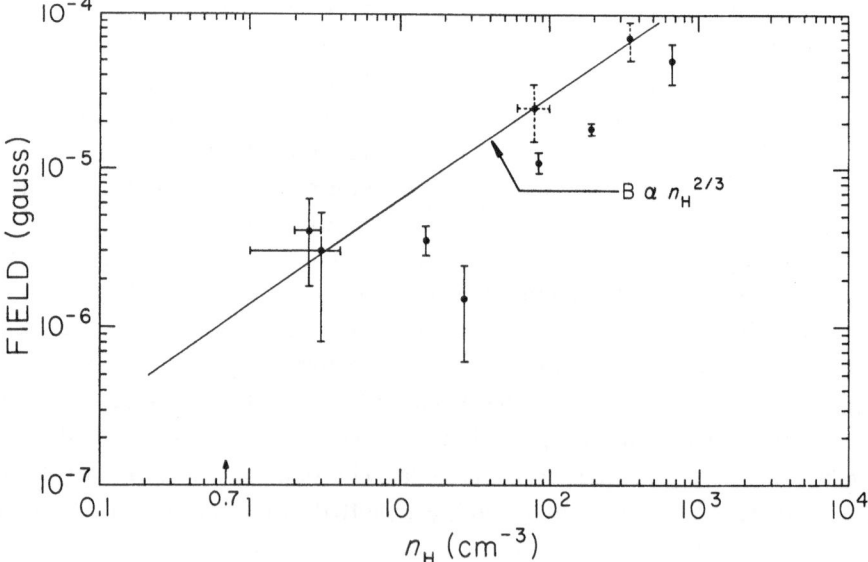

Fig. 4. Magnetic fields (B) in neutral hydrogen clouds as a function of their density (n_H). The solid lines indicate well-determined values. The broken bars indicate less certain values. The line with slope $\frac{2}{3}$ shows the expected form for the case of magnetic fields frozen into contracting clouds.

with a slope of $\frac{2}{3}$. It is clear that the line represents the observations very well. This suggests strongly that contracting clouds of neutral hydrogen do in fact carry 'frozen-in' fields with them, without dissipation, when they contract over nearly three orders of magnitude in density.

Since one is only seeing one component of the field in the Zeeman data the upper envelope of the points in Figure 4 should describe the history of the field strength in a contracting cloud. If this argument is valid one can estimate the value at the onset of cloud formation by extrapolating back to the average interstallar hydrogen

TABLE IV

Magnetic fields in other clouds

	l^{II}	b^{II}	Velocity (km sec^{-1})	Field estimate (μG) [a]	Integration time (hr)
Sag A	0	0	-53	-8.0 ± 8.1	6
Sag A	0	0	0	$+3.0 \pm 5.0$	6
M8	6	-1	$+9$	$+3.0 \pm 6.7$	9.5
HD 142096	12	$+31$	0	$+2.2 \pm 2.6$	30
M16	17	$+1$	$+1$ and $+6$	-4.5 ± 12.5	6
W43	31	$+1$	$+24$	-2.0 ± 21.0	6.5
IV-cloud	108	$+71$	-29	-2.5 ± 7.4	50
Cas A	112	-2	0	$+0.2 \pm 0.7$	21
3C 123	170	-11	$+5$	$+1.0 \pm 11.2$	10
Dust cloud	174	-14	$+6$	-4.7 ± 4.5	26
Cloud C	238	$+38$	$+6$	$+4.5 \pm 5.3$	23
CP 1133	240	$+69$	0	-2.0 ± 4.9	24
HD 147550	350	$+25$	0	$+7.0 \pm 11.0$	13

[a] A negative sign indicates a field toward the observer.

density, which will be the value of the undisturbed "interstellar" magnetic field. Kerr (1969) has given a value of 0.7 cm^{-3} for the mean density in the region of 4 to 11 kpc from the galactic center, whereas in the direction of Scorpius, Jenkins *et al.* (1969) find 2 cm^{-3} as the mean value from Ly-α data. Considering these values as extremes I find from Figure 4 that the mean interstellar magnetic field is between 1 and 3 μG. Taking a value of 2 μG I find the energy density of the field to be 1.6×10^{-13} erg cm^{-3} and, if I assume a radius of 15 kpc and a thickness of 200 pc, I derive a total magnetic field energy of 6×10^{53} erg in the Galaxy.

5. Pulsar Data

The rotation measure is given by RM $= 0.81 \int n_e B_{\|} \, dl$. The dispersion of the pulsed radiation yields a 'dispersion measure' DM $= \int n_e \, dl$. When both quantities are measured, we readily obtain $B_{\|}$. Clearly this is the mean field in the line of sight weighted by the electron density.

TABLE V

Mean fields in the direction of pulsars

Pulsar	l^{II}	b^{II}	DM (pc cm^{-3})	RM (rad m^{-2})	References	Field strength [a] (μG)
AP 2015 + 28	68.1	− 4.0	14.2	− 30	a	+ 2.0
CP 0328	145.0	− 1.2	26.8	− 95	a	+ 3.5
				+ 63	d	+ 2.8
NP 0532 (Crab)	184.5	− 5.8	56.9	− 25	c	+ 0.6
NP 0527	183.8	− 6.9	49.3	+ 36	d	+ 0.9
CP 0950	228.9	+ 43.7	2.98	< 4	b	< 0.2
PSR 0833–45	263.5	− 2.8	63	+ 42	e	− 0.8
				+ 33	f	− 0.7
CP 0808	139.9	+ 31.6	5.8	− 11	g	+ 2.3

[a] Rotation measures are negative for a field away from the observer. However, in the Zeeman experiment a negative sign is taken to mean a field toward the observer.
References: a = Smith, 1968b; b = Smith, 1968a; c = RM from Morris and Berge, 1964; d = Staelin and Reifenstein, 1969; e = Radhakrishnan, 1969; f = Ekers *et al.*, 1969; g = Smith, F. G. (private communication).

Table V shows data for seven pulsars together with the reference for the rotation measures. Values of DM may be found in many papers (e.g., Taylor, 1969; Davidson and Terzian, 1969; Bridle and Venugopal, 1969; Davies, 1969). The average value for the mean field in Table V is about 1.6 μG. This agrees well with the Zeeman data discussed before, although the statistics are not yet good.

The interpretation of the Zeeman data in the previous section suggested that high fields can only occur in relatively dense H I clouds. CP 0328 has two dense absorbing clouds in front of it (Gordon *et al.*, 1969), and Hjellming *et al.* (1969) have derived a mean density of 8 cm^{-3} in these clouds, assuming a kinetic temperature of 50 K and a distance to the pulsar of 1 kpc. According to Figure 4 the line-of-sight field component is about 4 μG in clouds of this density. The pulsar data alone give approximately 3 μG, but this value is an average weighted by the electron density. A combination of a path length of 50 pc in cold clouds, having $n_e = 4.4 \times 10^{-2}$ cm^{-3} and a field of about 4 ($\pm$2) μG, with a pathlength of $\approx$ 1 kpc of intercloud medium, having a field of 2 μG and $n_e = 2.6 \times 10^{-2}$ cm^{-3} gives RM-values of 8 and 42 rad m^{-2} respectively, which compares well with the observed total value of 63 rad m^{-2} (Staelin and Reifenstein, 1969). For this pulsar the intercloud medium dominates in producing the rotation measure. Radhakrishnan (1969) has pointed out that in the direction of PSR 0833-45 there is no dense H I cloud. The rotation measure is +42 rad m^{-2}, which must then be due to the intercloud medium alone unless a significant amount is due to the Vela-X supernova remnant and the associated H II regions.

In the case of the Crab Nebula pulsar the picture is somewhat complicated. The RM data indicate a mean field toward the source, the Zeeman data show a cloud with a field in the opposite direction. The line of sight to the source probably goes through parts of two spiral arms and the H I absorption data indicate at least four to six absorbing clouds in the line of sight. The Zeeman splitting gave 3.5 μG in one of these, but the pulsar data give a very low average field of 0.6 μG. Clearly, field reversals must be occurring, or the contribution to the RM from the Crab Nebula itself dominates the total RM.

6. Some Problems in the Interpretation of Faraday Rotation Measurements

In the distribution on the sky of the Faraday rotation measures (RM) several large irregularities are found. While it is true that the largest RM-values are found within 10° from the galactic plane, it seems hard to justify the statement that the RM-values decrease with increasing latitude for $|b| > 10°$. For example, a radio source that is usually ignored is 3C 287 at $b^{II} = 81°$ with RM $= +67$ rad m^{-2}. Admitting that some sources have a large intrinsic RM weakens all subsequent discussion! Other exceptions may be found among the five new RM-values obtained by Berge and Seielstad (1969). One is CTA 21 (or PKS 0316+16) with a RM of $+240$ rad m^{-2} which lies within 3° of the radio sources PKS 0300+16 and PKS 0307+16; the latter two, according to Gardner *et al.* (1969a), have RM-values of -17 and -15 rad m^{-2}. In this context

TABLE VI

Orientation of H I clouds with respect to magnetic field

l^{II}	b^{II}	Axial ratio	Position angle	Direction of magnetic field	Difference
Intermediate velocity clouds					
A 136	+54.5	2:1	105°	90° O, T	15°
B[a] 146.7	+39.5	⩾3:1	100°	95° T	5°
C 104	+69	2:1	55°	60° T	5°
D (1) 86.5	+59.5	2:1	50°	60° T	10°
(2) 89.0	+56.5	5:1	50°	60° T	10°
E 102	+36	2:1	90° (+10)	70° T (98° O)	20°
RII 136	+51.5	3:1	110°	90° T	20°
RI 148	+38.0	3:1	110°	100° T	10°
High velocity clouds					
MI 160	+66	4:1	102°	105° T	3°
MII 185	+65	5:1	120°	130° T	10°
CI 90	+45	Large	50°	65° T	15°
CIII 120	+52	3:1	80°	80° O, T	0°
Sm 43	−15	3:1	130°	150° O, T	20°
Sp 50	−80	3:1	135°	130° O (160° T)	5°
A 150	+35	7:1	40°	100° T	60°
B 168	+38.5	2:1	140°	110° T	30°
ACI 183	−12	3:1	120°	45° T	75°
ACII 192	−22	7:1	140°	50° T	90°

[a] H I data on this cloud are of poor quality.

TABLE VII

Orientation of dust clouds (and reflection nebulae in their neighborhood)

Region (1)	Cloud numbers [a] (2)	R.A. (1950) (3)	Dec. (4)	l^{II} (5)	b^{II} (6)	Position angle of cloud (7)	Position angle of magnetic field (8)	Comment (9)
A	2, 5, 6, 8, 9	4^h20^m	+25°00'	172°	−16°	70°	70°	(1)
B	10	4 37	+25 30	174	−14	110	60	(2)
C	3, 4	4 19	+54 48	150	+4	60	130	
D	12	5 02	+25 29	178	−9	80	60	
E	13	5 15	+7 17	195	−17	35	45	(3)
F	16, 17, 19 (9, 10)	16 25	−19 15	358	+19	130	120	
G	18	16 36	−14	4	+21	130	130	(4)
H	20 (12, 13)	16 44	−10	8	+21	70	130	
I	21, 22	16 44	−14	5	+19	130	130	(5)
J	25, 26	17 44	−4 30	21	+12	135	140	
K	27, 28, 29	18 14	−4 33	25	+5	100	150	
L	32	23 35	+48 15	111	−13	60	90	
M	(1)	2 04	+75 54	128	+14	75	100	
N	(11)	16 24	−24 30	353	+9	110	130	(6)
O	(16)	16 47	−15 18	4	+18	100	140	
P	(18)	17 01	−22	0	+11	125	130	

[a] Numbers from Heiles (1969). Bracketed numbers are from his Table II.

(1) These are regions discussed by Shajn (1955).

(2) This is cloud #2 of Heiles (1968) in which H_2CO, OH, etc. have been found. The nearby HV-cloud ACI had position angle of 120°.

(3) The position angle quoted corresponds to nearby emission nebulosity.

(4) The position angle quoted corresponds to distinct striations in the region (cloud P.A. = 45°).

(5) The position angle quoted corresponds to distinct striations in the region (cloud P.A. = 170°).

(6) Much striation. One large feature has P.A. = 130°, however.

reference is made to the suggestion by Gardner, Morris, and Whiteoak of the existence of 'magnetic loops' and to the suggestion by Bologna *et al.* (1969, see also Bologna *et al.*, 1965) of the presence of small-scale depolarizing structures. Davies (1968) has also proposed small-scale depolarization but this analysis has been criticized by Gardner *et al.* (1969b).

Sofue *et al.* (1969) have suggested that the RM of an extragalactic radio source depends on the redshift. From this they derive an intergalactic magnetic field of 10^{-9} G. I have plotted recently obtained RM-values in the diagrams of Sofue *et al.*, and the result is not an overwhelming support for their case, but the plots warrant further investigation.

There is obviously much to be done in the interpretation of RM-data. We must not be too disappointed if a simple picture of our galactic field pattern is not derived.

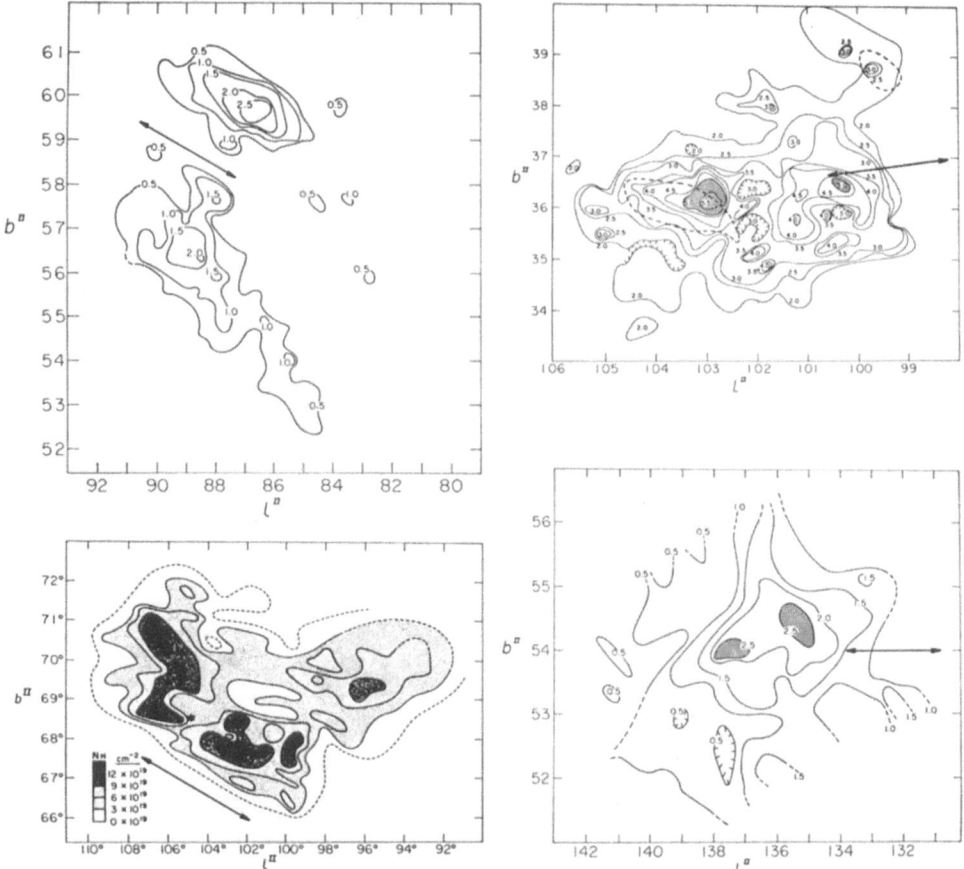

Fig. 5. Four intermediate velocity clouds of neutral hydrogen with heavy arrows representing the magnetic field vectors in that direction from the model (or data) in Figure 1. Contours are integrated hydrogen density in units of 10^{20} atom cm^{-2} except for lower left cloud.

7. Evidence for Cloud Elongation Parallel to the Magnetic Field

In the study of a cloud at high latitudes it was noted (Verschuur, 1969e) that the cloud is elongated in the direction of the magnetic field, as found from the polarization of background radiation. To see whether such an alignment is a more common phenom-

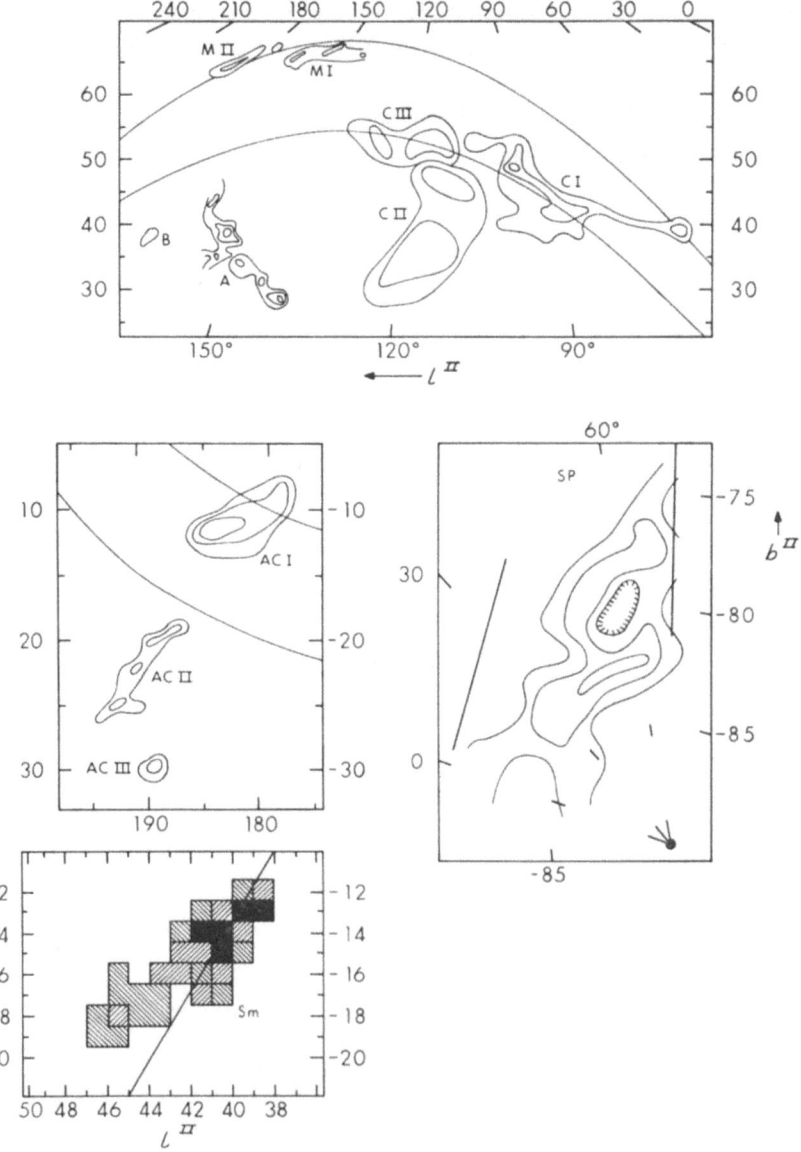

Fig. 6.   The magnetic field lines in the region of the high velocity clouds taken from Hulsbosch (1968) and Smith (1963). Heavy lines indicate the field vectors taken from Figure 1.

enon I have collected in Table VI data on eight intermediate velocity (IV) clouds ($|V| < 70$ km sec^{-1}) and on ten high-velocity (HV) clouds. The data are from Verschuur (1969e, f, g), from Rickard (1969), from Hulsbosch (1968), and from Smith (1963). Column 6 contains the position angle of the magnetic field read from Figures

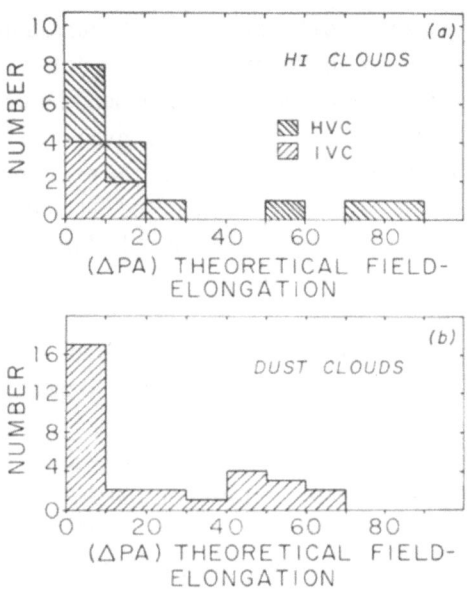

Fig. 7. Histograms of the difference in position angle of elongated clouds and magnetic field in their direction. – (a) Neutral hydrogen clouds compared with the model magnetic field of Figure 1b. – (b) Dust clouds compared with the same model.

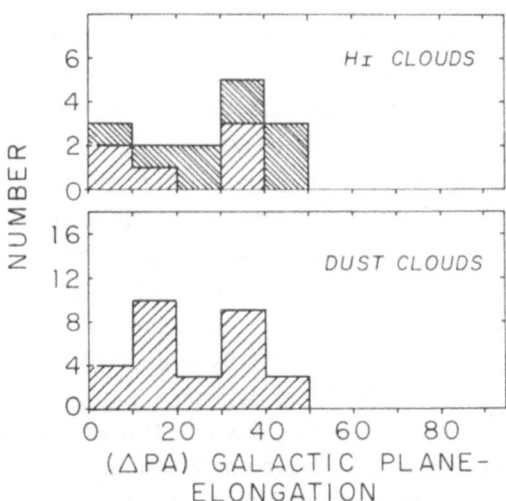

Fig. 8. Histogram of the difference between cloud axes and the galactic plane.

1 a or 1 b. Figure 5 shows maps of 5 of the IV-clouds, and in Figure 6 some HV-clouds are presented. All the IV-clouds appear elongated in the direction of the (helical) magnetic field and 6 out of the 10 HV-clouds do the same. The HV-clouds as a group do not give the impression of parallel alignment. It is clear that high resolution observations of the HV-clouds are essential to decide whether small-scale features exist within them, which might be indicative of the presence of magnetic fields. This, in turn, would argue strongly for these clouds to be inside our Galaxy.

Shajn (1955) has already noted that a number of dust clouds were elongated parallel to the optical polarization vectors of stars in their neighbourhood. We have briefly followed this up by taking the list of dust clouds given by Heiles (1968) and comparing the elongations, if any, of these clouds with the direction of the magnetic field read of Figures 1a or 1b. Table VII shows the data used. The position angle of the cloud (column 7) was obtained by inspecting the Palomar Sky Survey Plates. The histograms of the difference in the expected field direction and cloud position angle (column 7 minus column 8) is also shown in Figure 7. In Figure 8 we have plotted the histogram of position angles of both the hydrogen and dust clouds compared to the galactic plane to illustrate that the clouds are more closely aligned parallel to the field direction than the galactic plane.

It is worth reviving Meaburn's (1965) model for HV-clouds in which he noted that they occur near the spurs of galactic continuum emission. This is still true for the HV-clouds discovered since then, as well as for the IV-clouds. If, as Mathewson suggests, the spurs are the result of explosions in the plane injecting relativistic electrons along the field lines then perhaps these explosions have injected H I clouds along these lines as well. This, however, does not explain the predominance of negative velocities.

8. Summary

The local magnetic field appears to be helical in shape. This is consistent with optical polarization data, with the existence of spurs of galactic continuum emission and with their polarization. A basic longitudinal field may exist beyond the helix, which may be regarded as a local field pocket in which the net optical polarization is produced. Many such field pockets probably exist in the Galaxy. Spurs of continuum emission are the result of injection of non-thermal electrons into parts of the helical field. Zeeman-effect data show that the magnetic field in neutral hydrogen clouds is proportional to their density and that the basic interstellar field has been amplified in gravitationally contracting clouds. Extrapolation to a mean intercloud density suggests a mean interstellar field of 2 ± 1 μG. This is partially supported by the fact that 10 other clouds have fields less than 10 μG. Pulsar data show that the mean field, weighted by electron density, is also of the order of a few μG. It is not yet possible to derive the field strength in the local helical field region unambiguously.

Many interstellar clouds, both H I and dust, are elongated parallel to the helical field lines. The overall motion of these clouds may also be determined by the field.

References

Appenzeller, I.: 1968, *Astrophys. J.* **151**, 907.
Behr, A.: 1959, *Veröff. Univ.-Sternw. Göttingen*, No. 126.
Berge, G. L. and Seielstad, G. A.: 1969, *Astrophys. J.* **157**, 35.
Bingham, R. G.: 1967, *Monthly Notices Roy. Astron. Soc.* **137**, 157.
Bologna, J. M., McClain, E. F., Rose W. K., and Sloanaker, R. M.: 1965, *Astrophys. J.* **142**, 106.
Bologna, J. M., McClain, E. F., and Sloanaker, R. M.: 1969, *Astrophys. J.* **156**, 815.
Bridle, A. H. and Venugopal, V. R.: 1969, *Nature* **224**, 545.
Clark, B. G.: 1965, *Astrophys. J.* **142**, 1398.
Clube, S. V. M.: 1968, *Observatory* **88**, 243.
Davidson, K. and Terzian, Y.: 1969, *Astron. J.* **74**, 849.
Davies, R. D.: 1968, *Nature* **218**, 435.
Davies, R. D.: 1969, *Nature* **223**, 355.
Davies, R. D., Booth, R. S., and Wilson, A. J.: 1968, *Nature* **220**, 1207.
Ekers, R. R., Lequeux, J., Moffet, A. T., and Seielstad, G. A.: 1969, *Astrophys. J. Lett.* **156**, L21.
Gardner, F. F. and Davies, R. D.: 1966, *Australian J. Phys.* **19**, 129.
Gardner, F. F., Morris, D., and Whiteoak, J. B.: 1969a, *Australian J. Phys.* **22**, 79.
Gardner, F. F., Morris, D., and Whiteoak, J. B.: 1969b, *Australian J. Phys.* **22**, 813.
Gardner, F. F. and Whiteoak, J. B.: 1969, *Australian J. Phys.* **22**, 107.
Gordon, C. P., Gordon, K. J., and Shalloway, A. M.: 1969, *Nature* **222**, 129.
Hall, J. S.: 1958, *Publ. U.S. Naval Observ.*, 2nd series, **17**, No. VI.
Heiles, C.: 1968, *Astrophys. J.* **151**, 919 (#2 list).
Heiles, C.: 1969, *Astrophys. J.* **156**, 493.
Hiltner, W. A.: 1956, *Astrophys. J. Suppl.* **2**, 389.
Hjellming, R. M., Gordon, C. P., and Gordon, K. J.: 1969, *Astron. Astrophys.* **2**, 202.
Hulsbosch, A. N. M.: 1968, *Bull. Astron. Inst. Netherl.* **20**, 33.
Jenkins, E. B., Morton, D. C., and Matilsky, T. A.: 1969, *Astrophys. J.* **158**, 943.
Kerr, F. J.: 1969, *Ann. Rev. Astron. Astrophys.* **7**, 39.
Krzeminski, W. and Serkowski, K.: 1967, *Astrophys. J.* **147**, 988.
Mathewson, D. S.: 1968, *Astrophys. J. Lett.* **153**, L47.
Mathewson, D. S.: 1969, *Proc. Astron. Soc. Australia* **1**, 209.
Mathewson, D. S. and Nicholls, D. C.: 1968, *Astrophys. J. Lett.* **154**, L11.
Meaburn, J.: 1965, *Nature* **207**, 179.
Morris, D. and Berge, G. L.: 1964, *Astrophys. J.* **139**, 1388.
Radhakrishnan, V.: 1969, *Proc. Astron. Soc. Australia* **1**, 6.
Rickard, J. J.: 1969, preprint.
Seymour, P. A. H.: 1969, *Monthly Notices Roy. Astron. Soc.* **142**, 33.
Shajn, G. A.: 1955, *Astron. Zh.* **32**, 381.
Shuter, W. L. H. and Verschuur, G. L.: 1964, *Monthly Notices Roy. Astron. Soc.* **127**, 387.
Smith, F. G.: 1968a, *Nature* **220**, 892.
Smith, F. G.: 1968b, *Nature* **218**, 325.
Smith, G. P.: 1963, *Bull. Astron. Inst. Netherl.* **17**, 203.
Sofue, Y., Fujimoto, M., and Kawabata, K.: 1969, *Publ. Astron. Soc. Japan* **20**, 388.
Staelin, D. H. and Reifenstein, E. C.: 1969, *Astrophys. J. Lett.* submitted for publication.
Taylor, J. H.: 1969, *Astrophys. Lett.* **3**, 205.
Van de Hulst, H. C.: 1967, *Ann. Rev. Astron. Astrophys.* **5**, 179.
Verschuur, G. L.: 1968, *Phys. Rev. Lett.* **121**, 775.
Verschuur, G. L.: 1969a, *Astrophys. J.* **156**, 861.
Verschuur, G. L.: 1969b, *Nature* **223**, 140.
Verschuur, G. L.: 1969c, *Astrophys. J.* Sept. 1970.
Verschuur, G. L.: 1969d, *Astrophys. J. Lett.* **155**, L155.
Verschuur, G. L.: 1969e, *Astron. Astrophys.* **1**, 473.
Verschuur, G. L.: 1969f, *Astron. Astrophys.* **3**, 77.
Verschuur, G. L.: 1969g, in preparation.

10. THE ORIGIN AND DYNAMICAL EFFECTS OF THE MAGNETIC FIELDS AND COSMIC RAYS IN THE DISK OF THE GALAXY

Introductory Report

(Friday, September 12, 1969)

E. N. PARKER

Dept. of Physics, University of Chicago, Chicago, Ill., U.S.A.

1. Introduction

The topic of this presentation is the origin and dynamical behavior of the magnetic field and cosmic-ray gas in the disk of the Galaxy. In the space available I can do no more than mention the ideas that have been developed, with but little explanation and discussion. To make up for this inadequacy I have tried to give a complete list of references in the written text, so that the interested reader can pursue the points in depth (in particular see the review articles Parker, 1968a, 1969a, 1970). My purpose here is twofold, to outline for you the calculations and ideas that have developed thus far, and to indicate the uncertainties that remain. The basic ideas are sound, I think, but, when we come to the details, there are so many theoretical alternatives that need yet to be explored and so much that is not yet made clear by observations.

2. The Galactic Field

Consider first what is presently known about the magnetic field in the disk of the Galaxy. (See Verschuur, this Volume, p. 150.) Observations of the polarization of starlight (Hiltner, 1956) indicate that the magnetic field in the disk of the Galaxy is oriented generally in the azimuthal direction around the disk. Observations of Faraday rotation of distant polarized radio sources indicate that the sense of the field may be in either direction at various places in the disk (Morris and Berge, 1964; Davis and Berge, 1968). In addition to the changes in sign of the field, there are large local fluctuations in direction and in magnitude (Hiltner, 1956; Jokipii and Lerche 1969; Jokipii *et al.*, 1969) so that if we write the field as the sum of the mean field $\mathbf{B}_0$ plus a fluctuating component $\Delta\mathbf{B}$,

$$\mathbf{B} = \mathbf{B}_0 + \Delta\mathbf{B}, \tag{1}$$

we have

$$\langle \Delta\mathbf{B}^2 \rangle = O(\mathbf{B}_0^2). \tag{2}$$

The mean strength of the field is subject to some uncertainty, but a few μG is suggested by Faraday rotation measurements of polarized radio signals from both extragalactic sources and from pulsars (Davis and Berge, 1968; Jokipii and Lerche, 1969) and appears not to be contradicted by measurements of Zeeman splitting in the dense cold HI regions where the effect is observable. What is more, the observed

Habing (ed.), Interstellar Gas Dynamics, 168–183. All Rights Reserved.

dynamical behavior of the gas and field in the galactic disk suggests a strength of a few μG, on which we will have more to say below.

The galactic field is 'frozen' into the gas in the disk of the Galaxy. In H$\textsc{ii}$ regions the electrical conductivity σ is about 10^{13} e.s.u. so that the resistive diffusion time over small dimensions of 1 pc is large compared to the age of the Galaxy. In H$\textsc{i}$ regions the field is frozen to the electrons and ions, which have densities of 10^{-2} cm^{-3}. The neutral gas is tied to the electrons and ions by collisions, so that the characteristic (ambipolar) diffusion coefficient is of the general order of magnitude of 10^{21} cm^2 sec^{-1} or less. Diffusion over dimensions of 1 pc requires 3×10^8 yr or more. Thus the field is frozen into the H$\textsc{i}$ regions for most purposes even on scales as small as 1 pc.

The cosmic rays, which we consider to be a gas when viewed on a galactic scale, are tied to the magnetic field. Altogether, then, the interstellar medium is a composite fluid, made up of a thermal gas and a cosmic-ray gas, bound to the lines of force of the galactic field. Each of the three constituents has about the same energy density and pressure. The energy density of a field of 5 μG is 1.0×10^{-12} erg cm^{-3}, and the pressure is the same, in dyne cm^{-2}. This is to be compared with the energy density 1.5×10^{-12} erg cm^{-3} and pressure of 0.5×10^{-12} dyne cm^{-2} of the cosmic rays in the disk (Parker, 1966a), and the energy density 1.0×10^{-12} erg cm^{-3} and turbulent pressure of 0.7×10^{-12} dyne cm^{-2} of the interstellar gas (assuming a mean density of two hydrogen atoms cm^{-3} and an r.m.s. small-scale velocity of 7 km sec^{-1}).

The first question we might ask is what is the origin of the magnetic field. There are three possibilities that spring to mind. For instance, the field may be primordial, trapped in the original matter in the universe, and carried along with the matter into the present form of the Galaxy. The non-uniform rotation of the Galaxy would stretch the field out in the azimuthal direction, whatever the cause of the field, and it is an easy matter to show that the field is trapped in the gas in the disk for periods in excess of 10^{10} yr. I can think of no fundamental objection to this idea except, perhaps, that the field would be wound too tightly by galactic rotation.

We note, however, that the primordial field would be irrelevant if there were active generation of magnetic flux at the present time, which would simply obliterate the initial primordial field, replacing it with fields of more recent origin.

Turning then to active generation of magnetic flux, we might conjecture that the field has been generated by the observed random turbulent motions of the interstellar gas. It has been speculated (Alfvén, 1947; Biermann and Schlüter, 1951) that turbulence in a highly conducting fluid will build up any initial magnetic fields to a level where the field energy is comparable to the kinetic energy of the turbulence

$$\langle B^2/(8\pi) \rangle \approx \langle \tfrac{1}{2}\varrho v^2 \rangle. \tag{3}$$

The other possibility, that the magnetic energy does not build up to the kinetic energy, based on the similarity of the vorticity equation and the hydromagnetic equation for the field (Batchelor, 1950) has been suggested, too. In this case the galactic field would *not* be the result of random turbulence because the field energy builds up until it is equal to the kinetic energy in the small eddies, which possesses only a tiny fraction of the total kinetic energy density, $\langle \frac{1}{2} \varrho v^2 \rangle$ of the turbulence. Recently Kraichnan and Nagarajan (1967) have examined in detail the dynamical equations for a turbulent conducting fluid in the presence of magnetic fields. They point out that the generation of magnetic field by a turbulent flow depends upon several terms in the equation, all of the same order of magnitude. They show that the question of whether the field does, or does not, build up to equipartition of energy with the velocity field, depends upon the mean values (over wave number) of the various terms, which can be determined only by formal solution of the equations, and cannot be determined by any of the qualitative physical arguments proposed so far. Thus the question of the magnetic fields in a turbulent fluid is still an open question, after almost twenty years of debate.

It is possible to make a formal calculation of the magnetic field in a turbulent flow if the magnetic field, produced by the turbulence up to the time t, is statistically independent of the turbulent velocity at time t (Parker, 1969b). This condition is satisfied so long as (a) the field is too weak to affect the velocity significantly and (b) the correlation time of the turbulent flow is very short. In real turbulence the correlation time of the motion v on a scale l is of the order l/v i.e., during its lifetime an eddy turns through an angle of the order of one radian, giving a change $\delta \mathbf{B}$ in the field, which is comparable to $\mathbf{B}$, and hence is strongly correlated with the final field $\mathbf{B} + \delta \mathbf{B}$. Only if the correlation time is very short compared to l/v is $\delta \mathbf{B}$ so small that it is essentially uncorrelated with $\mathbf{B} + \delta \mathbf{B}$. On the other hand if (a) and (b) are satisfied, the field $\mathbf{B}$ and the velocity $\mathbf{v}$ can be treated as independent random variables. It can then be shown, by applying the theory of random functions to the hydromagnetic equations, that the magnetic field grows at all wave numbers. If dissipation destroys the field at large wave numbers, the net result of the turbulence is the generation of field at small wave numbers (large scale). One could imagine, then, that at some time in the distant past, turbulence in the interstellar gas may have generated a weak magnetic field (perhaps some fraction of a μG) which was then sheared by the non-uniform rotation of the galaxy into the presently observed field of several μG with the lines of force predominantly in the azimuthal direction (Parker, 1969c). It must be kept in mind, however, that application of the formal theory for short correlation times to real turbulence, in which the correlation time is l/v, is a conjecture of the same order as the earlier equipartition conjecture and the analogy with vorticity.

Observations give no particular support for any of the theoretical ideas. For instance, the solar photosphere is turbulent, as a consequence of the convective zone beneath, with a kinetic energy density of the order of 10^2 erg cm^{-3}, corresponding to the energy density of a field of 50 G. But the magnetic fields in the solar photosphere are observed to range from 1 G in very extensive regions, to 3000 G in sunspots. One may note the approximate equality of the energy density of the interstellar (galactic) field and the interstellar turbulence. But, of course, this example begs the question, because we are concerned with the problem of whether the interstellar field is, in fact, the result of the interstellar turbulence. And, in any case, it must be remembered that the interstellar field is dominated by the non-uniform rotation of the Galaxy rather than by the local turbulence. It is the non-uniform rotation which stretches the field into the observed azimuthal orientation and greatly intensifies the field in the process. We shall argue later on that the magnetic field strength is controlled by the cosmic-ray production rate.

It seems to me, therefore, that the source of the galactic field need not be very strong. Any mechanism which can produce a field of $1\mu G$ is entirely adequate. Non-uniform rotation does the rest. The approximate equality between $B^2/(8\pi)$ and the kinetic energy density of the interstellar gas is to be understood, it seems to me, from the dynamical instability of the gas-field-cosmic ray system, in which the energy of the field and cosmic rays is converted into kinetic energy of the gas. The gas is driven into clouds, leading to star formation, and then the clouds are disrupted to newly formed hot stars (Oort, 1952, 1954; Oort and Spitzer, 1955; Savedoff, 1956; Spitzer, 1968; Parker, 1967b). The equality of field energy and turbulent energy is then only a very loose relation. So, altogether, it cannot be shown yet whether the magnetic field in the disk of the Galaxy owes its origin to the turbulence there.

Consider, then, the third possibility, that the galactic field has been generated by motions which contain some order, such as might result from the Coriolis forces of galactic rotation. It was pointed out many years ago that cyclonic convective motions, together with non-uniform rotation (both the result of convection and Coriolis forces) are able to regenerate the dipole field of the Earth (Parker, 1955) and to produce the migratory fields which lead to the sunspots on the Sun (Parker, 1955, 1957; Leighton, 1969). What might such motions generate in the Galaxy? As in the Sun and Earth, the non-uniform rotation of the Galaxy stretches out the magnetic fields so that the dominant component is in the azimuthal direction. In the Sun and Earth the cyclonic convective motions move upward across the azimuthal field, locally lifting and twisting the lines of force into loops with non-vanishing projection on the plane perpendicular to the azimuthal field (i.e., with non-vanishing projection on the meridional planes). These loops coalesce to give large-scale loops of field in the meridional planes. For the Earth this leads to the observed dipole field. The question is whether some similar process might be operative in the Galaxy. The general idea of a galactic dynamo would begin with the non-uniform rotation. The non-uniform rotation of the Galaxy generates a toroidal field – presumably the observed azimuthal field – from whatever poloidal fields are present. The missing link is the generation of the poloidal fields from the existing toroidal fields. There is considerable turbulence and convection in the disk, and there are the Coriolis forces due to galactic rotation, so the turbulence must be slightly cyclonic. If cyclonic turbulence produces loops of flux in meridional planes with predominantly one sense, then the loops coalesce to give an overall poloidal field. For a given sense of rotation, rising and sinking cells produce loops of opposite sense, so that if rising and sinking cells occur in equal numbers, there is no net production of poloidal field. In the core of the Earth we believe that rising cells dominate. The sinking fluid is spread out over the broad regions between and has but little cyclonic rotation. Hence there is net production of poloidal field in the core of the Earth. There may be some such concentration of rising or sinking cells in the galactic disk. If there is, then the gas operates as a dynamo and we have the explanation for the galactic magnetic field. Indeed, we might point to the preponderance of negative radial velocities of interstellar gas at high latitudes and consider the matter resolved. But there is another possibility. Steenbeck *et al.* (1966) have made the general point that if fluid motions extend over one or more scale heights in an atmosphere, then rising cells of fluid are diverging laterally and sinking cells are converging laterally. Hence the Coriolis force produces retrograde rotation in the rising cells and direct rotation in the sinking cells (contrary to rising and sinking cells in an incompressible fluid such as the core of the Earth where both undergo direct rotation). The sense of the loop of field produced (in the meridional plane) by the cyclonic cells of fluid is determined by the product of the vertical motion and the rotation. Hence rising and falling cells extending over a scale height or more produce meridional loops with the same sense. If we apply this point to the Galaxy it raises the possibility that the galactic field is generated by the combination of non-uniform rotation and the cyclonic gas motions perpendicular to the disk. On the other hand, we must not be too hasty in asserting that this is the explanation of the galactic field, because a simple sketch of the cyclonic rotation, the non-uniform rotation, etc. shows that the basic dynamo mode is migratory, as in the Sun, with the direction of migration perpendicular to the disk (and, incidentally, in the direction opposite to the direction of progress of a right-hand thread turning with the Galaxy i.e., in the direction of the galactic North Pole). Such migration is blocked by the boundary of the disk. The question is whether there are higher modes of the dynamo which can regenerate the galactic field. I do not know the answer to this question yet. The origin of the galactic field is still a matter of speculation.

3. Equilibrium in the Galactic Disk

Whatever may be the origin of the galactic magnetic field, consider the conditions necessary for the dynamical equilibrium state in which we find the field today. It is readily shown from the virial equations that a magnetic field tends to expand (Chandrasekhar and Fermi, 1953). So the first question is what keeps the galactic field confined to the disk of the Galaxy? We showed some time ago (Parker, 1966b) that the field cannot be confined by the galactic nucleus (with a force-free configuration in the disk) unless the field increases at least as fast as r^{-3} toward the center of the Galaxy. This seems to be contrary to observation. So the only theoretical possibility remaining is that the field is confined to the disk of the Galaxy by the weight of the gas in the disk. The field necessarily penetrates through the gas if the gas is to hold down the field. This is just what observations would suggest, of course. There is a horizontal magnetic field (in the mean) through interstellar space, and the interstellar gas is distributed along and across the field. The scale height Λ of the gas is observed to be of the order of 100 to 200 pc (Schmidt, 1956; see also the remark by van Woerden, p. 184). Equilibrium in the vertical direction (the z-direction perpendicular to the disk) requires that

$$\frac{d}{dz}\left(p + \frac{B^2}{8\pi}\right) = -\varrho g \tag{4}$$

where g is the gravitational acceleration perpendicular to the disk, ϱ is the gas density, and p the gas pressure. We use smeared-out average values of ϱ, p and B here, because both the gas and the field are subject to small-scale fluctuations which are not of interest to the large-scale equilibrium.

At this point we recall that the cosmic-ray pressure P is comparable to both $B^2/(8\pi)$ and the turbulent pressure of the gas, so it too should be included in the equation for equilibrium,

$$\frac{d}{dz}\left(\frac{B^2}{8\pi} + P + p\right) = -\varrho g. \tag{5}$$

Now write $p = \varrho u^2$ where u^2 is the mean-square small-scale gas velocity (thermal plus turbulent) in the vertical direction. The mean-square velocity must be suitably averaged over both HI and HII regions, which we do not distinguish in this large-scale consideration of equilibrium. Then if the total pressure in intergalactic space is small compared to the total pressure within the disk, it is evident that $B^2/(8\pi)$ and P must vanish with z at least as rapidly as the density ϱ. For if they do not decrease with height as fast as ϱ, then above some height there is not sufficient weight $\int \varrho g \, dp$ to confine the pressure $P + B^2/(8\pi)$ at that height. For simplicity suppose that $B^2/(8\pi)$, P and $p = \varrho u^2$ all decrease in proportion, so that

$$\frac{B^2}{8\pi} = \alpha\varrho u^2, \qquad P = \beta\varrho u^2, \tag{6}$$

where α and β are constants. Suppose too that u^2 is independent of z. Then Equation (5) can be written

$$\frac{1}{\varrho u^2}\frac{d}{dz}\varrho u^2 = -\frac{g}{u^2(1 + \alpha + \beta)}. \tag{7}$$

The scale height Λ is, accordingly,

$$\Lambda = \frac{u^2(1 + \alpha + \beta)}{\langle g\rangle_\Lambda} \tag{8}$$

where $\langle g\rangle_\Lambda$ denotes the mean value of $g(z)$ over the scale height. If we take $\Lambda = 160$ pc, then $\langle g\rangle_\Lambda \cong 2 \times 10^{-9}$ cm sec^{-2}. Observation suggests that the small scale r.m.s. velocity u is of the order of 7 km sec^{-1}, from which we conclude that $\alpha + \beta \cong 1$. The reader may use his own favorite numbers for Λ, g, and u if he wishes. The point here is that with $\alpha + \beta$ of the order of unity, as suggested by observations, there is no reason to doubt that the distribution of gas and field in the disk is in large-scale hydrostatic equilibrium. The gas density necessary to confine the field and cosmic rays follows from

$$\frac{B^2}{8\pi} + P \equiv \varrho u^2(\alpha + \beta). \tag{9}$$

The cosmic-ray pressure is one-third the energy density, or 0.5×10^{-12} dyne cm^{-2}. A field of 3.5 μG has the same pressure, so that with $\alpha + \beta = 1$ and $u = 7$ km sec^{-1}, we find $\varrho = 2 \times 10^{-24}$ gm cm^{-3}. Thus the weight of one or two hydrogen atoms cm^{-3} is required to confine a galactic field of 3 to 4 μG to the galactic disk. The mean hydrogen density in the disk appears to be about two atoms cm^{-3}, or even a little more (Schmidt, 1956, 1963), though both higher and lower figures are sometimes quoted. We note that again observation is not inconsistent with the simple equilibrium condition of Equation (5). And inasmuch as the gas and field appear to be in a quasi-steady equilibrium, we shall assume that Equation (5) is satisfied whatever varieties of u, Λ and ϱ may be suggested by various observations.

It is important to note that, unless $\Lambda \varrho g$ is much larger than present observations suggest, the Galaxy cannot accommodate a galactic field in excess of about 5 μG. There simply would not be enough gas to confine a stronger field to the disk. We suggest, then, that the r.m.s. magnetic field in the disk of the Galaxy is bounded by

$$\langle B^2 \rangle^{1/2} \lesssim 5\mu\text{G}. \tag{10}$$

4. Internal Dynamics of the Galactic Disk

Before we inquire into the dynamical behavior of the magnetic field in the disk of the Galaxy, a few words should be said on the properties of cosmic rays, whose pressure P plays a role in shaping the dynamics. As already noted, the energy density of the cosmic rays is $U \approx 1.5 \times 10^{-12}$ erg cm^{-3} (1 eV cm^{-3}), half of which is carried by particles above 10 GeV per nucleon. The cosmic rays are highly relativistic, forming a gas with pressure

$$P \approx \tfrac{1}{3}U. \tag{11}$$

The number density of relativistic particles is $n \approx 10^{-10}$ cm^{-3}. The cyclotron radius R of a typical 10 GeV proton in a field of 3 μG is 10^{13} cm. This is a small fraction 3×10^{-8}, of the thickness of the galactic disk. Thus the average cosmic-ray particle is tightly bound to the lines of force of the galactic field and can drift across the lines of force only at a very slow rate as a result of field gradients or scattering. (For a detailed discussion of the kinetic properties of the cosmic-ray gas see Lerche and Parker, 1966; Lerche, 1967c; Scargle, 1968.)

The cosmic-ray gas is isotropic to better than one percent at the present time in the neighborhood of the Sun. Lerche has shown that anisotropies in which the cosmic-ray pressure parallel to the field differs from the perpendicular pressure are rapidly destroyed by unstable collective plasma interactions with the thermal gas and magnetic field. When $P_{\parallel} > P_{\perp}$, an unstable magnetosonic mode is excited (Lerche, 1967a).

When $P_\perp > P_\parallel$, the cosmic-ray electrons are unstable, producing a transverse mode propagating perpendicular to the field (Lerche, 1966, 1968; see Kadomtsev and Tsytovich, this volume, p. 108). These instabilities play the role of collisions, maintaining isotropy and permitting the cosmic-ray gas to be treated as a hydrodynamic fluid in many cases where the changes are slow (the fluid equation overlooks Landau damping and resonances). The bulk cosmic-ray streaming velocity follows as

$$\left(\delta + \frac{P}{c^2}\right)\frac{d\mathbf{w}}{dt} \cong \nabla P \tag{12}$$

where δ is the cosmic-ray gas density (n multiplied by the relativistic mass of the particles, see Tolman, 1946; Lerche and Parker, 1966; Lerche, 1967b; Scargle, 1968). If magnetic fields and a thermal gas are present too, the equations of motion become

$$\left(\delta + \frac{P}{c^2}\right)\frac{d\mathbf{w}_\parallel}{dt} = -\nabla_\parallel P \tag{13}$$

for the cosmic rays and

$$\varrho\frac{d\mathbf{v}_\parallel}{dt} = -\nabla_\parallel p_\parallel + \varrho\mathbf{g}_\parallel \tag{14}$$

for the thermal-gas motion parallel to the field (neglecting the gravitational forces on the cosmic-ray gas) and

$$\varrho\frac{d\mathbf{v}_\perp}{dt} = -\nabla_\perp(p + P) + \frac{(\nabla \times \mathbf{B}) \times \mathbf{B}}{4\pi} + \varrho\mathbf{g}_\perp \tag{15}$$

for the perpendicular thermal-gas motion. Since both the cosmic-ray gas and the thermal gas are tied to the lines of force, we have $\mathbf{w}_\perp = \mathbf{v}_\perp$. We neglect the inertia of the cosmic-ray gas. The hydromagnetic equation for the magnetic field can be written

$$\frac{\partial\mathbf{B}}{\partial t} = \nabla \times (\mathbf{v}_\perp \times \mathbf{B}). \tag{16}$$

The equations of continuity are

$$\frac{\partial\varrho}{\partial t} + \nabla\cdot(\varrho\mathbf{v}) = 0, \tag{17}$$

$$\frac{\partial\delta}{\partial t} + \mathbf{w}\cdot\nabla\delta + \left(\delta + \frac{P}{c^2}\right)\nabla\cdot\mathbf{w} = 0. \tag{18}$$

For small variations the pressure and density in each gas are related by the effective speed of sound in that gas. It is readily shown from these equations that there is an additional hydromagnetic wave mode as a consequence of the cosmic-ray gas. There is also some modification of the familiar fast mode (Parker, 1958, 1965b).

It has been shown recently (Wentzel, 1968, 1969; Kulsrud and Pearce, 1969; see Kadomtsev and Tsytovich, this volume, p. 108) that plasma turbulence is generated if the cosmic-ray gas streams along the magnetic field with a velocity relative to the thermal gas which is greatly in excess of the Alfvén speed in the thermal gas. The instability producing the turbulence involves scales of the same order as the cyclotron radius of the cosmic-ray particles, which is, in most cases, small compared to the collision mean-free path of the ions in the thermal gas. Thus the neutral atoms do not usually participate directly and the Alfvén speed is to be computed for the ionized component of the thermal gas alone, and may be 50 to 200 km sec^{-1} in typical H$_I$ regions. The net effect of the plasma turbulence is to introduce a frictional coupling between $\mathbf{w}_\parallel$ and $\mathbf{v}_\parallel$. Suitable friction terms should be introduced on the right-hand sides of Equations (13) and (14) to take account of this coupling when $|\mathbf{w}_\parallel - \mathbf{v}_\parallel|$ exceeds the critical value. The existence of such coupling was first suggested by Ginzburg (1965), who stressed its importance in regulating and controlling the escape of cosmic rays from the disk of the Galaxy, on which we will say more later.

Now consider the stability of the simple hydrostatic equilibrium described in the previous section. Let $p = \varrho u^2$ in the equilibrium state again and write

$$\delta p = \gamma u^2 \, \delta \varrho \tag{19}$$

for small variations in density and pressure. The coefficient γ can be as high as $\frac{5}{3}$ for rapid variations in which radiative transfer and thermal conduction are negligible. For the long term variations (10^7 to 10^8 yr) with which we shall be concerned, it is well known (Savedoff and Spitzer, 1950) that radiative transfer causes the temperature to decline with increasing density, so that $\gamma < 1$, and in some cases $\gamma < 0$ (see the recent work by Pikel'ner, 1967; Spitzer, 1968; Spitzer and Tomasko, 1968; Field et al., 1969; Goldsmith et al., 1969 and the references therein).

Now it is well known that a uniform thermal gas in free space is subject to instability if $\gamma < 0$. In this case the pressure decreases with increasing density, so that a slight compression leads to collapse as a consequence of the pressure of the surrounding gas. It is also well known that an isothermal atmosphere in equilibrium in a uniform gravitational field is unstable provided only that $\gamma < 1$. In this case the temperature and the scale height decline with increasing density so that a slight compression tends to collapse as a consequence of the weight of the overlying gas. Some of the thermal effects are discussed in detail by Field at this Symposium (see p. 51), who considers not only small perturbations to the equilibrium, but treats the final highly inhomogeneous equilibrium state of the gas. For the present discussion, illustrating the dynamical properties of the magnetic field and cosmic rays, it is sufficient to continue with the crude representation of the thermal effects in terms of γ.

The perturbation of an equilibrium atmosphere of thermal gas, horizontal magnetic field, and cosmic-ray gas in a gravitational field g, representing large-scale conditions in the disk of the Galaxy, can be carried out in a straightforward manner from Equations (13) through (19). We note that the inertia of the cosmic-ray gas is negligible so that Equation (12) contributes nothing. The cosmic-ray pressure remains uniform along each line of force and the volume of each tube of flux remains contant to first order. Hence

$$\frac{dP}{dt} \equiv \frac{\partial P}{\partial t} + v_z \frac{dP}{dz} = 0.$$ (20)

We suppose that α and β are constant in the initial equilibrium. A simple normal mode analysis, with solutions of the form exp $(t/\tau + i\mathbf{k} \cdot \mathbf{r})$ leads to an adequate instability criterion. It is sufficient for the present discussion to restrict attention to perturbations involving variations and motions in the vertical z-direction and in the horizontal y-direction along the magnetic field. Then instability occurs for

$$\gamma < \frac{(1 + \alpha + \beta)^2}{1 + \beta + \alpha[\frac{3}{2} + 8(k_y^2 + k_z^2)\,\Lambda^2]}.$$ (21)

For long wavelengths and $\alpha = \beta = 0.5$, this criterion is $\gamma < \frac{16}{9}$, which is to be compared with the criterion $\gamma < 1$ in the absence of magnetic field and cosmic rays. The characteristic growth times are of the order of the free-fall time over one scale height, or the time in which the speed of sound or Alfvén speed traverses one scale height, typically $(1$ to $3) \times 10^7$ yr. This is of the same order as the thermal instability time in the absence of field and cosmic rays. The characteristic scale of the instability is comparable to Λ, say 200 pc.

The magnetic field and cosmic rays both enhance the instability, each contributing about equally, as may be seen from Equation (21) above. The unstable effect of the magnetic field can be understood if we remember that the thermal gas is constrained to motion along the lines of force, like beads on a string. Then if the horizontal lines of force are perturbed, by raising them in some places and lowering them elsewhere, the gas tends to slide from the high places along the lines into the low places, further burdening down the low places, and unloading the high places where the field is then free to expand upward. The cosmic-ray gas contributes to the instability because its pressure remains constant along the perturbed line of force. Hence the cosmic-ray pressure is higher than the surrounding pressure on the raised portions of a line of force, tending to inflate and buoy up the raised portion further. On the depressed portions the cosmic-ray pressure is below the surrounding pressure, thereby permitting compression and sinking. Altogether the effects are much like the well-known Rayleigh-Taylor instability, in which the light fluids – the field and cosmic rays – try to bubble up through the heavy thermal gas.

The pressure of the thermal gas resists the accumulation of gas in the low places, but unless γ is larger than the value given by Equation (21), the resistance is ineffectual and the gas clumps into clouds.

We have explored the dynamical behavior of the interstellar gas-field-cosmic ray system in some detail (Parker, 1966b, 1967a, b, 1968b, c, d, e, 1969a; Lerche and Parker, 1967; Lerche, 1967c, d) carrying through calculations with the boundary conditions appropriate to the disk of the Galaxy, etc. The calculations show that introduction of the third dimension, with wave number k_x, enhances the instability somewhat. The most unstable modes have $k_x^2 \gg k_y^2, k_z^2$, suggesting a tendency for the interstellar gas to break up into relatively thin vertical sheets lying along the field. There appears to be no final inhomogeneous equilibrium state into which the gas can settle, suggesting that the gas clouds are shifting forms, rather than stable entities. It is, then, the combined effects of the cosmic rays, the galactic magnetic field, and the small γ (thermal instability, see Field, this volume, p. 51) which are responsible for clumping the interstellar gas into clouds and compressing those clouds toward compact masses in which star formation occurs. The overall dynamical picture of the interstellar gas which emerges from these theoretical considerations is one in which there is continual and vigorous competition between the dynamical instability, which tends to form the gas into shifting patterns of concentration, and the usual disruptive effects of hot stars and local cosmic-ray production, which tend to disperse the concentrations (Parker, 1968e). Self-gravitation (which we have omitted from the equations to keep the algebra as simple as possible) is relatively unimportant in most cases until the individual gas clouds become very dense and massive. The scale along the field is typically a few hundred pc, though the gas from such a region may then collapse into a much smaller volume. The scale across the field may be only a few pc, according to our simple linearized calculations. It will be interesting to see what turns up in future detailed observations of cloud structure. And of course it would be desirable to carry out more detailed and complete theoretical calculations of the behavior of the gas when the perturbations have grown to large amplitude and the cloud structure really begins to take shape. Some attempts have been made in these directions in the references cited above, but only restricted examples have been dealt with and the problem is still relatively unexplored.

5. The Role of Cosmic Rays in the Galactic Disk

It is interesting to inquire into the origin and evolution of the cosmic-ray gas in the context of the above dynamical considerations. It is generally presumed that the cosmic rays are generated by energetic phenomena in the galactic disk, such as novae, supernovae, pulsars, etc. There are more extravagant ideas available, of course, such as universal cosmic rays, originating in radio galaxies, quasars, etc. But for the present we pursue the conservative idea that most of the cosmic rays in the disk of the Galaxy are generated somewhere in the disk (see the general discussion of Ginzburg and Syrovat-skii, 1964; Parker, 1968a).

It is known from studies of radioactive nuclei in meteoritic material that the cosmic-ray density has been approximately steady over the past 10^5, 10^7 and 10^9 yr (see discussion and references in Parker, 1968a). The mean density over these periods has not varied significantly. So presumably the cosmic rays are in an overall steady state in the disk (apart from local outbursts and instabilities).

Studies of the abundances of such rare nuclei as Li, Be, B, He^3, etc. indicate that the heavier cosmic-ray nuclei have passed through a significant amount of matter, approximately 3 g cm^{-2}. It is the breakup of the more common heavier nuclei in passage through matter that produces the rare nuclei. Translated into distance through interstellar space, 3 g cm^{-2} is about $2 \times 10^{24}/n$ cm. A mean interstellar density of $n = 2$ atoms cm^{-3} gives 3×10^5 pc or 10^6 light-years, suggesting that the cosmic-ray particles which are observed today have spent about 10^6 yr in the disk of the Galaxy. It appears, then, that cosmic rays escape from the Galaxy after having spent about 10^6 yr in interstellar space.*

Now if cosmic rays are generated within the disk and escape out of the disk after a residence of $t = 10^6$ yr, then the cosmic rays have a mean streaming velocity S of the order of L/t where L is the mean distance from the source to the exit. A streaming velocity S leads to an anisotropy such that an observer looking upstream sees a cosmic-ray intensity (measured at a particular particle energy) which is $1 + \varDelta$ times the mean, where

$$\varDelta = (2 + \varGamma) \, S/c \qquad (22)$$

where $\varGamma \approx 2.6$ is the exponent of the differential cosmic-ray energy spectrum, $E^{-\varGamma}$. Hence there is a cosmic-ray anisotropy which is a direct measure of the streaming of cosmic rays. If the mean distance for escape is $L = 10^4$ pc or more, corresponding to distances along the galactic arm, or to distances to and from the nucleus of the Galaxy, we have $S \approx 10^9$ cm sec^{-1} and $\varDelta \approx 10^{-1}$. But, as we have already noted, plasma turbulence prevents streaming of cosmic rays in excess of a few hundred km sec^{-1}, and observations indicate that $\varDelta$ is very small, probably $\varDelta < 10^{-3}$ ($S \leqslant 60$ km sec^{-1}) in the neighborhood of the Sun at the present time, showing that the cosmic rays in the vicinity of the Sun today are streaming at a very leisurely pace. Of course we are near the central plane of the Galaxy, which, as a plane of symmetry, may be a stagnation surface. And the streaming velocity may be larger nearer the surface of the disk. But whatever the situation, it appears that the cosmic rays do not escape along the length of the magnetic field, but must escape out the surface of the disk within a few hundred pc of their place of origin.

We suggest that the large-scale dynamical instability of the interstellar gas-field-cosmic ray system discussed in the section above, is essential to the escape of cosmic rays from the disk. The only means by which cosmic rays can escape is by inflation of the raised portions of the galactic field (Parker, 1965a). They must literally push their way out. The cosmic-ray pressure observed at the position of the Sun is 0.5×10^{-12} dyne cm^{-2}, equal to the pressure of a magnetic field of 3.5 μG. So there should be no difficulty for the cosmic rays to push their way out from some region where the field is raised and expanded to values well below 5 μG, the r.m.s. field strength given by Equation (10). This argument puts an upper limit on the field strength. But the field cannot be much smaller either, for if it were less than 3 μG deep in the disk, the raised portions of the field would offer no resistance whatever to the cosmic-ray gas. The

---

* It is conjectured by some that the cosmic rays may spend up to 10^8 yr in a galactic halo during which time they make occasional brief excursions into the disk where they accumulate 10^6 yr of residence before escaping from the Galaxy altogether. This idea does not affect the present discussion.

only limitation would then be the 'friction' of the plasma turbulence with the thermal gas when the streaming exceeds a few hundred km sec^{-1}. (Parenthetically we remark that indeed this offers the fascinating prospect of pushing interstellar gas to positions high above the disk ($z \gg \Lambda$) from where it falls back in clumps and is observed with high negative velocity at high galactic latitudes (see van de Hulst, this volume, p. 3). For instance, we might then expect to see a larger anisotropy in the cosmic rays here at the position of the Sun. We would also have some difficulty in explaining the storage of cosmic rays for 10^6 yr in the disk, unless, of course, the sources are quite densely distributed through the disk, because the cosmic rays would tend to burst directly out of the disk wherever they are produced, rather than spreading out along the field through the disk before escape. Finally, if the field is weak, then $B^2/(8\pi) \ll \frac{1}{2}\varrho v^2$ and the field would be completely disrupted and distorted by the turbulent motions of the interstellar gas. As pointed out by Pikel'ner, one could not then understand the large-scale order of the galactic field, with a well-defined reversal of sign across the galactic plane. Hence, all things taken together, it appears that

$$B \gtrsim 3\mu G. \tag{23}$$

This lower limit, together with the upper limit given by Equation (10) confine the field strength rather closely if we are to understand the dynamical behavior of the interstellar gas-field-cosmic ray system in a simple way in terms of the present observational estimates of the gas density, turbulent velocity, scale height, etc.

To pursue the problem of escape of the cosmic rays a little further, recent theoretical considerations, together with observational studies of solar fields, indicate that the lines of force of the magnetic fields in nature are stochastic (Jokipii and Parker, 1969a, b). Pick any two lines of force which are neighbors, with separation $h(0)$ at some position along the field. Following along the lines of force a distance s we find that their separation $h(s)$ undergoes a random walk, so that on the average $h(s)$ is larger than $h(0)$.

The observed dispersion in the direction of the galactic field (Hiltner, 1956; Jokipii *et al.*, 1969) indicates that the lines of force random walk to the surface of the galactic disk (say to $z = 150$ pc) in distances of only 500 pc (Jokipii and Parker, 1969b). This stochastic property of the galactic field, in which each line of force comes close to the surface at various places, appears to be an essential property of the field accounting for the 10^6 yr cosmic-ray life in the disk. Were the field not stochastic, the cosmic-ray pressures in the disk would increase the scale height Λ a little, and would produce more violent instabilities of the interstellar gas-field-cosmic ray system.

We should not fail to note that the escape of cosmic rays through inflation of the fields at the surface of the disk must extend inflated bubbles of field out from the disk for some distance, producing a thick boundary layer of field and cosmic rays over the surface of the disk (Parker, 1965a). Presumably the inflated bubbles of field eventually free themselves from the Galaxy through the diffusive and dissipative effects of plasma turbulence. We have no way of computing how far out they extend before this occurs.

In summary, then, the escape of cosmic rays from the disk of the Galaxy appears to be from the surface of the disk. The streaming of cosmic rays along the field as they move toward escape is limited to speeds of a few hundred km sec^{-1}, or less, by the friction of plasma turbulence in the interstellar gas. Access to the surface of the disk is facilitated by the stochastic character of the lines of force of the field. The ultimate escape from the surface of the disk involves disengaging the cosmic-ray particles from the lines of force of the field of the disk. We have suggested that the particles disengage by inflating the field at the surface to form bubbles of extended field and cosmic rays, which presumably are eventually freed from the Galaxy by the dissipative effects of plasma turbulence. The cosmic rays which inflate them are then free of the Galaxy.

Now to comment in a broader context; the mechanism by which cosmic rays escape from the Galaxy is one which has been treated lightly for too long. I have proposed that cosmic rays escape by inflating the field and pushing their way out because I can think of no alternative mechanism. There may be alternatives. And if there are, they should be formulated and explored. Scattering out of the galactic field by plasma turbulence has been suggested as the means of escape, but upon close examination it does not seem adequate. As was already mentioned the cyclotron radius R of a 10 GeV proton in a field of 3 μG is 10^{13} cm. The cyclotron period is 2×10^3 sec. If we suppose that the 10 GeV proton is scattered n times through a large angle, the proton may random walk a distance of the order of $n^{1/2}R$ across the field. Escape from the disk implies diffusion across the field for some 100 pc, requiring $n = 10^{15}$. Even if the proton were scattered through a large angle as often as once each cyclotron period, the escape time is 2×10^{18} sec or 0.7×10^{11} yr!

The problem of escape becomes particularly acute in the universal theory of cosmic rays, which explains cosmic rays in the Galaxy by supposing them to be the dominating phenomenon of the universe, filling all space to the high density which we observe in the disk, and originating in colossal releases of energy in distant galaxies. In this theory the cosmic-ray density is more or less uniform throughout all space, with cosmic rays entering the disk of the Galaxy from outside, remaining for not longer than 10^6 yr, and departing. The uniform cosmic-ray pressure precludes cosmic rays pushing their way either in or out. And it has not yet been shown how they can penetrate across the fields of the disk of the Galaxy and back out again in only 10^6 yr. It has been suggested instead that the galactic fields lie 'open' to the outside in some way, with the lines of force in the rotating disk maintaining a direct connection into the intergalactic field. The proposal is contrary to the usual ideas of hydromagnetism, that lines of force move with the background plasma and can break and reconnect in a changing pattern only in the characteristic dissipative diffusion time. On the other hand, it cannot be ruled out that sufficient plasma turbulence is present to maintain connection of the galactic field with an intergalactic field. The problem is one of fundamental importance to the universal theory and merits serious inquiry.

The problem of cosmic-ray escape is of particular interest too when we consider the strength of the galactic field under the assumption that cosmic rays are generated within the disk of the Galaxy. We have indicated in the discussion (Parker, 1965a) that the cosmic-ray gas inflates the interstellar gas-field system until the cosmic-ray pressure becomes comparable to the magnetic pressure, whereupon the cosmic rays begin to escape by inflating the surface fields. Now suppose for the moment that there is some easier means of escape, such as direct connection into the intergalactic field. Then it is only plasma turbulence which limits the escape along the lines of force. If, as we have supposed, this is an easier escape than inflation of the fields, then the cosmic-ray pressure does not build up to the magnetic pressure. But observations suggest that the cosmic-ray pressure is, in fact, comparable to the magnetic pressure. The rough equality of the cosmic-ray pressure and magnetic pressure suggested by observations can be interpreted, then, as an indication that there is not, in fact, an easier way out than inflation of the fields. But, as noted above, there is still considerable uncertainty in the mean field strength in the disk, so this conclusion must be considered tentative for the time being.

It is evident from these qualitative considerations that our understanding of cosmic-ray escape would be greatly increased theoretically if a quantitative treatment of the limitations on cosmic-ray streaming by plasma turbulence could be applied to the dynamics of the inflation of a bubble of field from the surface of the disk, with the simultaneous downward streaming of the thermal gas and upward streaming of the cosmic rays.

Acknowledgement

This work was supported by the National Aeronautics and Space Administration under grant NASA-96-60.

References

Alfvén, H.: 1947, *Monthly Notices Roy. Astron. Soc.* **107**, 211.
Batchelor, G. K.: 1950, *Proc. Roy. Soc. London* **A201**, 405.
Biermann, L. and Schlüter, A.: 1951, *Phys. Rev.* **82**, 863.
Chandrasekhar, S. and Fermi, E.: 1953, *Astrophys. J.* **118**, 116.
Davis, L. and Berge, G. L.: 1968, *Nebulae and Interstellar Matter* (ed. by L. Aller and B. Middlehurst), University of Chicago Press, Chicago, Ch. 15.
Field, G. B., Goldsmith, D. W., and Habing, H. J.: 1969, *Astrophys. J. Lett.* **155**, L149.
Ginzburg, V. L.: 1965, *Astron. Zh.* **42**, 1129 (translation: 1966, *Soviet Astron.* **9**, 877).
Ginzburg, V. L. and Syrovat-skii, S. I.: 1964, *Origin of Cosmic Rays*, Pergamon Press, New York.
Goldsmith, D. W., Habing, H. J., and Field, G. B.: 1969, *Astrophys. J.* **158**, 173.
Hiltner, W. A.: 1956, *Astrophys. J. Suppl.* **2**, 389.
Jokipii, J. R. and Lerche, I.: 1969, *Astrophys. J.* **157**, 1137.
Jokipii, J. R. and Parker, E. N.: 1969a, *Astrophys. J.* **155**, 777.
Jokipii, J. R. and Parker, E. N.: 1969b, *Astrophys. J.* **155**, 799.
Jokipii, J. R., Lerche, I., and Schommer, R. A. 1969, *Astrophys. J. Letters* **157**, L119.
Kraichnan, R. H. and Nagarajan, S.: 1967, *Phys. Fluids* **10**, 859.
Kulsrud, R. and Pearce, W. P.: 1969, *Astrophys. J.* **156**, 445.
Leighton, R. B.: 1969, *Astrophys. J.* **156**, 1.
Lerche, I.: 1966, *Phys. Fluids* **9**, 1073.
Lerche, I.: 1967a, *Astrophys. J.* **147**, 681.
Lerche, I.: 1967b, *Astrophys. J.* **147**, 689.
Lerche, I.: 1967c, *Astrophys. J.* **149**, 395, 553.
Lerche, I.: 1967d, *Astrophys. J.* **150**, 651.
Lerche, I.: 1968, *Phys. Fluids* **11**, 1720.
Lerche, I. and Parker, E. N.: 1966, *Astrophys. J.* **135**, 106.

Lerche, I. and Parker, E. N.: 1967, *Astrophys. J.* **149**, 559.
Morris, D. and Berge, G. L.: 1964, *Astrophys. J.* **139**, 1388.
Oort, J. H.: 1952, *Astrophys. J.* **116**, 233.
Oort, J. H.: 1954, *Bull. Astron. Inst. Netherl.* **12**, 177.
Oort, J. H. and Spitzer, L.: 1955, *Astrophys. J.* **121**, 6.
Parker, E. N.: 1955, *Astrophys. J.* **122**, 293.
Parker, E. N.: 1957, *Proc. Nat. Acad. Sci. Am.* **43**, 8.
Parker, E. N.: 1958, *Phys. Rev.* **109**, 1328.
Parker, E. N.: 1965a, *Astrophys. J.* **142**, 584.
Parker, E. N.: 1965b, *Astrophys. J.* **142**, 1086.
Parker, E. N.: 1966a, *Astrophys. J.* **144**, 916.
Parker, E. N.: 1966b, *Astrophys. J.* **145**, 811.
Parker, E. N.: 1967a, *Astrophys. J.* **149**, 517.
Parker, E. N.: 1967b, *Astrophys. J.* **149**, 535.
Parker, E. N.: 1968a, *Nebulae and Interstellar Matter* (ed. by L. Aller and B. Middlehurst), University of Chicago Press, Chicago, Ch. 16.
Parker, E. N.: 1968b, *Astrophys. J.* **154**, 49.
Parker, E. N.: 1968c, *Astrophys. J.* **154**, 57.
Parker, E. N.: 1968d, *Astrophys. J.* **154**, 515.
Parker, E. N.: 1968e, *Astrophys. J.* **154**, 875.
Parker, E. N.: 1969a, *Space Sci. Rev.* **9**, 651.
Parker, E. N.: 1969b, *Astrophys. J.* **157**, 1119.
Parker, E. N.: 1969c, *Astrophys. J.* **157**, 1129.
Parker, E. N.: 1970, *Astrophys. J.* **160**, 383.
Pikel'ner, S. B.: 1967, *Astron. Zh.* **44**, 1915.
Savedoff, M. P.: 1956, *Astrophys. J.* **124**, 533.
Savedoff, M. and Spitzer, L.: 1950, *Astrophys. J.* **111**, 593.
Scargle, J. D.: 1968, *Astrophys. J.* **151**, 791.
Schmidt, M.: 1956, *Bull. Astron. Inst. Netherl.* **13**, 247.
Schmidt, M.: 1963, *Astrophys. J.* **137**, 758.
Spitzer, L.: 1968, *Nebulae and Interstellar Matter* (ed. by L. Aller and B. Middlehurst), University of Chicago Press, Chicago, Ch. 1.
Spitzer, L. and Tomasko, M. G.: 1968, *Astrophys. J.* **152**, 971.
Steenbeck, M., Krause, F., and Radler, K. H.: 1966, *Z. Naturforsch.* **21**, 369.
Tolman, R. C.: 1946, *Relativity, Thermodynamics, and Cosmology,* Clarendon Press, Oxford.
Wentzel, D.: 1968, *Astrophys. J.* **152**, 987.
Wentzel, D.: 1969, *Astrophys. J.* **156**, 303.

11. DISCUSSION FOLLOWING THE REPORTS BY PARKER AND VERSCHUUR

(Friday, September 12, 1969)

Chairman: S. B. PIKEL'NER

Editor's remarks: The Discussion has been rearranged into four sections: 1. Various Questions; 2. The Galactic Halo; 3. The Origin of the Magnetic Field; 4. Summary. Section 4 is the outcome of another meeting (on September 17). One very extensive remark has been condensed. The latter part of the discussion on September 12 dealt with the informal meeting on H I gas and is presented separately as Chapter 6.

1. Various Questions

Van Woerden: Parker uses a value of 0.2 cm^{-3} kpc for the product $n_H \Lambda$. This product is equivalent to the integral $\int_0^\infty n_H dz$. I estimate 0.1 cm^{-3} kpc as an average over the ring between 9 and 11 kpc. The estimate comes from unpublished work, based on integrations in the diagrams published by Westerhout (1957). This estimate may be too low because of molecular hydrogen and because of concentrations of atomic hydrogen in dense clouds. (Westerhout, G.: 1957, *Bull. Astron. Inst. Netherl.* **13**, 201.)

Burke: A model-independent value for the integral can be derived for the solar neighborhood from observations in the direction of the pole. What is this number averaged over all latitudes above, say, 40°?

Van Woerden: The high-latitude survey by Tolbert (1970) gives a value close to 0.1 cm^{-3} kpc. (Tolbert, C. R.: 1970, in preparation.)

Field: I am surprised that neither of our speakers discussed the strong magnetic field (i.e., a field of 10 μG or more) which has been the subject of controversy at many previous symposia. Evidence for a strong field was derived from the intensity of synchrotron radiation, together with estimates of the flux of cosmic-ray electrons near the Earth. A recent study by Anand *et al.* (1969) concludes that at present such arguments yield 6 μG as the best value. The difference is due to the postulate of a very significant solar modulation of electrons below about 1 GeV. This means that the number of radiating electrons is considerably larger in interstellar space than close to the Earth. Consequently, the magnetic field required to explain the synchrotron emission is lower than the previous estimate. It is interesting that the value of 6 μG is comparable to the numbers which were derived from other lines of evidence and were mentioned by Verschuur in his report. (Anand, K. C., Daniel, R. R., and Stephens, S. A.: 1968, *Proc. Ind. Acad. Sci.* **67**, 267.)

Davis: Is the information on modulation sufficiently certain so that in another two years the field may not have dropped a bit more?

Field: The modulation derived by Anand, Daniel, and Stephens agrees with that

Habing (ed.), Interstellar Gas Dynamics, 184–200. All Rights Reserved. Copyright © 1970 by IAU

assumed for cosmic-ray protons. To change the subject, may I draw attention to the elongated clouds discussed by Verschuur in his report. I am worried about the existence of supersonic velocities, or rather, velocity gradients over very small distances. The observations are done with finite beam widths and therefore one observation includes information on regions quite large distances apart. Consequently, the problem may not be too great, but, on the other hand, the calcium-line profiles indicate a large turbulent velocity in a very small angular region. Is it possible that one can have large amplitude hydromagnetic waves, basically transverse waves, which could be supersonic and not dissipate at the same rate as compression waves? Nevertheless, for a cloud with 10 cm^{-3} and $B=5$ μG, the Alfvén velocity $V_A = 3.0$ km sec^{-1}, which is not much more than the thermal speed.

Pikel'ner: According to a paper by Heiles (1967), the average density in clouds is between 2 and 4 cm^{-3}. These values will increase V_A by a factor of about 2. (Heiles, C.: 1967, *Astrophys. J. Suppl. Ser.* **15**, 97.)

Weaver: I agree with Pikel'ner on the densities he quotes. Such values seem to be indicated by the observations.

Van Woerden: I think there is still evidence for hydrogen densities of 10 or 20 cm^{-3} in clouds.

Woltjer: I have a question to Parker. If, as Verschuur has discussed, there is some evidence for a helical field in the neighborhood of the Sun, you could think of a configuration where the field is force-free in the z-direction. Have you considered such fields and would they affect your conclusion concerning the strength of the field?

Parker: I have considered such fields, but they do not offer possibilities to increase the magnetic field strength. The point is, if you twist a helix until it really becomes force-free, and external pressures are no longer required to confine it, you will have increased the pressure in the direction along the helix by an equal amount. And there is still an instability which tends to make things bulge, i.e., you have the same problem as before.

Woltjer: Except that in the plane of the Galaxy you have to take into account differential rotation, and a very small change in rotational velocity gives you a comparatively large dynamical effect.

Parker: That is true. But, if you have pressures within the disk of the Galaxy, the gas-field system is unstable through loops bulging out in the z-direction without significant change in the instability criterion. And you lose your force-free configuration as soon as it becomes unstable.

Van de Hulst: I have a question to Parker. Parker said in his lecture that sheets of matter might be envisaged which are oriented along the gravity vector and the magnetic field lines. Could you suggest any observational program to check that prediction?

Parker: That is a question that I am interested in pursuing here. The linear analysis of a disk with a uniform field shows that the most unstable modes are vertical sheets. I have from time to time asked various observers whether or not this is something that you could see. I don't know whether the sheets would have enough relative velocity to be separated by Doppler shifts or not; and it is not obvious to me how one would go about detecting the sheets. Nor is it obvious that vertical sheets still persist in the ultimate, very non-linear dynamical state just because the linear analysis for small perturbations indicates their existence.

Weaver: In the velocity-latitude diagrams of the kind that I have been showing in my report (see Figure 3 at p. 27), isolated velocity peaks are seen extending over a considerable interval in b (or in l). This is very suggestive of a sheet, but one may easily be deceived since the nature of the diagram does not give direct information on the hydrogen distribution in depth along the line of sight.

Van de Hulst: Dr. Parker, is the dust also tied to the magnetic field so that we can say that the dust must follow roughly the same spiral pattern? Also, you mentioned the possibility of a magnetic field of 1 μG. Is that not ruled out by the spectrum of cosmic rays at very high energies, or do you think that at very high energies the cosmic rays are intergalactic?

Parker: I will answer your last question first. The reduction of the magnetic field from, say 4 μG to 1 μG, would have very little effect on the precise value of the energy at which you can no longer contain particles in the Galaxy; roughly speaking the energy goes down by a factor of 3 or 4. In answer to the first question, the dust is tied to the magnetic field if photoelectric effects give each dust grain a charge of at least one electron. In fact, without a magnetic field one cannot understand striations, and one can put a lower limit of about 1 or 2 μG in the case of fine striations in the Pleiades. I imagine that one could put lower limits in a similar way over a good part of the Galaxy.

Van de Hulst: Parker mentioned that the ratio of cosmic-ray pressures, $|P_{\parallel} - P_{\perp}|/P_{\parallel}$, is 10^{-2} or less. He suggested that isotropizing instabilities bring any anisotropy down to 10^{-2} in 10^2 yr and down to 10^{-3} in 10^6 yr. If the observations are good enough to determine an anisotropy of 10^{-3} or less, this measurement is very important for the origin of cosmic rays. A few years ago it was claimed that anisotropy of this order had been detected and at succeeding cosmic-ray conferences this claim has been repeated. But can Dr. Parker tell us what the practical limitations are? Is the value unknown just because people have not spent enough effort, or is it too difficult, or what?

Parker: The difficulty in observing the anisotropy of cosmic rays is that we are sheltered in the solar system by a field of 50 μG over scales of 10^{13} cm. We are in a shelter with the wind blowing by outside, if you want to look at it that way. You cannot expect to see the anisotropies for particles whose cyclotron radius in a 50 μG field is less than 10^{13} cm, and therefore you must observe particles above about 0.5×10^{12} eV. The cosmic-ray spectrum drops off steeply with energy; the number of cosmic-ray particles above an energy E declines as $E^{-1.7}$. There are many measurements up to about 0.2×10^{12} eV. Unfortunately the interpretation of these measurements in terms of anisotropies is complicated by the effects of the solar magnetic field. But some equipment exists, built for neutrino experiments, which may have a sufficiently high counting rate at 10^{12} eV to look for anisotropies of the order of 10^{-3} or 10^{-4}; the equipment has not yet been used for that purpose. At the present time, based on measurements at 0.2×10^{12} eV it looks as though the anisotropy is less than or equal to 10^{-3}. It will probably be a couple of years before there is better information. I quite agree with van de Hulst that from anisotropy information one might learn a great deal about the origin and behavior of cosmic rays in the Galaxy.

Tsytovich: Dr. Parker, have you estimated these isotropization times of 10^2 and 10^3 yr from the linear approximation to the instability theory?

Parker: My estimate is based on the calculations of Lerche, who used a purely linear theory, not a quasi-linear theory. I would be most interested to hear your comments.

Tsytovich: There is experimental evidence that in some cases a linear theory does not yield a reliable estimate of the relaxation time. For example, let us consider isotropization and relaxation of a beam in a plasma. The plasma particles have a Maxwellian velocity distribution, and the beam provides a second hump in the velocity distribution at the beam velocity. As Kadomtsev has pointed out in his Report (see Section 3a, p. 112), the second hump develops rapidly into a velocity-plateau; the energy of the beam is converted into turbulent and thermal energy and the beam is isotropized. Now if one applies this theory to solar bursts of type III, one finds that the beam extends over a distance three orders of magnitude larger than follows from the linear approximation. Therefore non-linear stabilization exists. It can be shown that non-linear stabilization occurs for those beams for which $V > V_{crit} \approx 3 (m_i/m_e)^{1/4} V_{Te}$, where V_{Te} is the thermal velocity of the electrons; V_{crit} is the velocity at which non-linear coupling between electrons and ions occurs. Observations of solar bursts of type III confirm this predicted behavior. If the temperature is not too high $V_{crit} < c$. Since the velocity of cosmic rays is close to that of light, non-linear stabilization occurs for cosmic-ray instabilities. Therefore the isotropization of cosmic rays may take a much longer time, up to 3 orders of magnitude, than predicted from the linear theory.

Pikel'ner: May I bring up a question about the Rayleigh-Taylor instability? When this instability develops, there are two possibilities. First, compression of the magnetic field may push out gas and field at the other side of the galactic plane. Therefore, the magnetic field will not be increased considerably in massive clouds. But consider a second case in which the gravitational forces, associated with the gas, are so strong that they 'attract' the perturbation at the other side of the galactic plane. Then we have compressed gas together with a compressed magnetic field. To find out which of these possibilities is realized we probably have to rely on observations.

Mestel: I would like to enlarge on Pikel'ner's remark about self-gravitation. Consider a region with a large-scale galactic magnetic field $\mathbf{H}_i$ and an associated density ϱ_i. Now let the Field-Pikel'ner instability start locally, so that a massive blob begins to contract, causing a local distortion in the magnetic-field structure. For simplicity we assume the blob is spherical and cool enough for thermal pressure to be ignored. From the virial theorem we have a simple criterion relating the mass M and the magnetic flux F in order that indefinite gravitational collapse can ensue:

$$\frac{9\pi^2}{5} GM^2 > F^2. \tag{1}$$

This may be written alternatively as

$$\eta \equiv \frac{8\pi^2}{3} \frac{GR^2\varrho_c^2}{H_c^2} > 1,$$ (2)

where R is the radius of the sperical blob, H_c the central field-strength and $(\varrho_c + \varrho_i)$ the central density (see Figure 1). Thus with $\eta > 1$, indefinite collapse can occur, and the

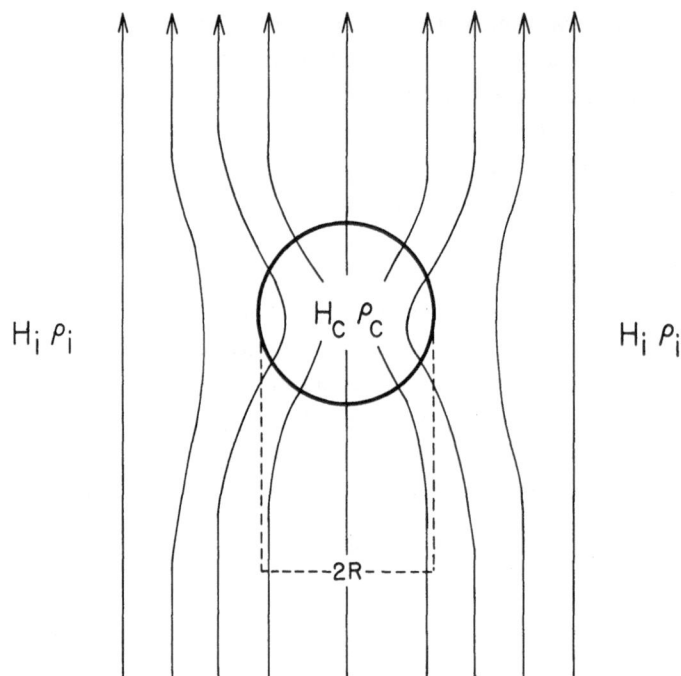

Fig. 1.   (See the remark by Mestel.) Magnetic-field strength in a contracting cloud.

cloud can ultimately break up into a proto-star cluster (by flow down the field, if flux-freezing continues to hold). But if $\eta < 1$, the collapse must be halted when the distorted field exerts forces strong enough to balance its self-gravitation. This magneto-gravitational equilibrium is achieved when

$$\frac{\varrho_c + \varrho_i}{\varrho_i} = \frac{1}{1 - \eta^{1/3}}$$ (3)

and

$$\frac{H_c}{H_i} = \left(\frac{1}{1 - \eta^{1/3}}\right)^{2/3}$$ (4)

I suggest that this model describes approximately what Verschuur has discussed in his Report (p. 157). An increase of 5 in the field-strength requires a density increase (in this spherically-symmetric model) of $(5)^{3/2} \approx 11.25$. The mass required to maintain this local field distortion is ≈ 0.87 of the initial value for indefinite collapse. Verschuur drew my attention to the paper by Clark (1965) that discussed the parameters of the first two clouds in which later the Zeeman effect was found. It is gratifying to see that one cloud has an estimated mass of $2500 \, M_\odot$. The other one has only $800 \, M_\odot$, and this mass is too small to maintain a distorted field. However, Schmidt pointed out to me that since Clark has only lower limits on his cloud diameters, the masses are correspondingly lower limits also. I would like to see some detailed models computed of such clouds, including thermal pressure, especially to clarify the conditions under which the clouds are oblate or prolate about the field direction. (Clark, B. G.: 1965, *Astrophys. J.* **142**, 1398.)

Field: In the equation for the critical mass, presented by Mestel [his Equation (2)], the term $R\varrho_c$ represents mass per unit area. This parameter is observed directly by the 21-cm observers. One can therefore re-examine Verschuur's Table III and see whether Mestel's criterion applies. In a preliminary list of 14 clouds measured by Verschuur, roughly half had large values of the magnetic field. In every case $R\varrho_c$ exceeds the value corresponding to 3 μG in Mestel's equation. On the other hand, when the mass per unit area is small, the observed magnetic field is about 3 μG. I therefore support Mestel's suggestion that in the clouds of low surface-density one is observing essentially the intercloud magnetic field; in the cases of high surface-density we witness contraction of the magnetic field. I have also considered a little bit the problem of expanding Mestel's theory by inclusion of the gas pressure. The effect of gas pressure will be to expand the gas along the field line. I find that there is a critical radius of about 4 pc (given the conditions of thermal balance based on cosmic-ray heating). If across the magnetic field a cloud has a radius less than 4 pc, gravitational forces are less than the pressure forces along the field. Such a cloud cannot contract, and it will be stretched along the magnetic field into a prolate shape. If, on the other hand, the radius across the field is bigger than 4 pc, the radius along the field depends on the surface density but it has a maximum of 4 pc, which is attained when the surface density, is becoming so large that contraction across the field can occur. It follows that clouds extending more than 4 pc across the field, and having surface density sufficient for gravitational contraction to occur, will be oblate, with the minor axis along the field. Hence observers may want to study the shapes of the clouds which have large magnetic fields.

2. The Galactic Halo

Mestel: During the third Symposium in this series, in 1957, there was a long debate between Spitzer and Pikel'ner on the nature of the galactic halo. I would like to ask the observers: what has happened to the idea of a $10^6 \, \text{K}$ halo? One of the reasons for introducing it was, I think, to maintain the disk of neutral hydrogen at a constant thickness over the galactic radius, in spite of the variation of the z-component of the gravitational field. Do the new 21-cm observations of spiral arm thicknesses destroy this argument?

Verschuur: There *is* no halo!

Mills: There *is* a halo!

Burke: Perhaps I can try to summarize the discussion on the interpretation of our data, that Mills and I have had running over several years. The data of Mills supported the idea of a halo, but came only from the Southern Hemisphere observatories. Turner and I made a 200-MHz survey of the sky from the Northern Hemisphere (Burke, 1967) which showed no evidence for the original concept of a halo, that is, an object more or less spheroidal in shape with a radius of 16 or 20 kpc. However, a flattened, spheroidal distribution some one or two kpc thick, is allowed by the data. Perhaps Dr. Mills would like to comment? (Burke, B. F.: 1967, in *Radio Astronomy and the Galactic System, IAU Symposium No. 31* (ed. by H. van Woerden), Academic Press, London, p. 361.)

Mills: Yes. Since the earlier work of about ten years ago, only one set of systematic southern observations has been made. They have just been completed by Hamilton using the Parks 210-foot reflector at various frequencies (principally 150 MHz and lower). Hamilton's paper is at present in press and I have no detailed information about it. But it does seem that the isophotes at the southern galactic latitudes around the galactic center show evidence for a rather flattened, small halo or corona. I certainly agree with Burke that there is no evidence for a very large halo. If it exists, it is fairly small. In the disk it has a radius to the point of half emissivity of the order of 10 kpc; perpendicular to the disk this radius is about 5 kpc. The axial ratio is thus about 2:1.

Menon: If the concept of a halo, especially that of an extensive halo, is rejected, then the original problem of Spitzer (1956) is unsolved, namely that of keeping the gas down in a layer of constant thickness. In addition we know now that the cosmic rays also exert pressure, and Spitzer's problem becomes even more important. (Spitzer, L.: 1956, *Astrophys. J.* **124**, 20.)

Field: I have done some calculations on the back of an envelope, which indicate that a halo of 10^6 K and a number density of 10^{-3} cm^{-3} is detectable by current X-ray techniques. Such experiments should be done. In addition to the (historical) argument mentioned by Menon, I want to add another of Spitzer's arguments, that of the confinement of clouds at high latitudes. There are clouds known at z-distances larger than 1 kpc (Münch and Zirin, 1961) and the original Spitzer proposal was to confine these clouds by outside pressure produced by the hot, rarefied halo gas.[*] (Münch, G. and Zirin, H.: 1961, *Astrophys. J.* **133**, 163.)

Syrovat-skii: For a long time one of the arguments in favor of a halo has been that one needs a place to store all cosmic rays. I want to point out that this argument is no longer valid as can be shown from age determinations of cosmic rays. At the recent Budapest Conference on Cosmic Rays, Shapiro and others reported that the age of cosmic rays is less than 50×10^6 yr; the age may be 1×10^6 yr, perhaps 3×10^6 yr. These results were based on measurements of the abundance ratios Li/B and Be/B in the cosmic-ray flux. A time of 1×10^6 yr or 3×10^6 yr is needed by the cosmic rays

---

[*] 21-cm observations of three of these clouds (Habing, H. J.: 1969, *Bull. Astron. Inst. Netherl.* **20**, 177) show internal velocities so large that the clouds cannot be confined by a halo of the proposed properties. (Ed.)

to get out of the galactic disk. Therefore it seems unlikely that the cosmic rays spend a long time in the halo.

In addition I should like to point out that the existence or non-existence of the halo has no influence on the estimate for the energy output U (dimension: erg sec^{-1}) of cosmic-ray sources in the Galaxy. If the energy density of the cosmic rays W_{cr} (for which we take the local value, 1 eV cm^{-3}), and if the age of the cosmic rays is t yr, then $U = W_{cr} V t^{-1}$, where V is the volume of the Galaxy in which the rays are found. But this expression is equal to $W_{cr} M_g c x^{-1}$, where M_g is the total amount of interstellar gas in the Galaxy, c the velocity of light, and x the amount of matter, traversed by the cosmic rays during their existence ($x = c\varrho_g t$, where ϱ_g is the mass density of the interstellar gas). From the abundances in the cosmic-ray gas it follows that x is between 3 and 10 g cm^{-2} and if we take $M_g \approx 10^{10} M_\odot$ then we find $U = 3 \times 10^{40}$ erg sec^{-1}. This estimate is independent from the existence or non-existence of the halo.

Weymann: In this connection may I ask Dr. Parker the question whether or not the escape of cosmic rays from the disk of the Galaxy leads to an extensive halo?

Parker: I cannot tell you the extent of the halo produced by the cosmic rays, because I do not know how to compute the extent of the cosmic-ray inflated bubbles at the moment when they are cut loose from the Galaxy by a plasma instability. I suggest that the inflated field will extend up at least a few hundred pc.

Woltjer: How does the bubble cut itself from the plane? Don't you run the risk that, once you have the bubble, you have a permanent hole?

Parker: The topology of the magnetic field, I think, prevents any permanent hole, at least with any scheme that I have been able to think of. Perhaps you have something in mind?

Davis: After the cosmic rays have escaped, does one not arrive at a configuration like that in the Sun after the solar wind has blown out a weak field loop? You get streaming lines of force that go out forever and probably have enough wiggles in them so that the cosmic-ray gas cannot flow out very fast.

Verschuur: Dr. Parker, how could one observe those bubbles?

Parker: They might emit some synchrotron radiation, but I am not sure. Presumably the thermal gas has slid down the lines of force, and there is not much thermal gas remaining in the bubbles. Let me think first about the escape of cosmic rays in a weak field ($B \lesssim 1 \mu$G). In this case the cosmic rays are confined more by friction with the thermal gas than by magnetic field pressure. The friction might possibly lift some gas up into the halo; and then (coming back to a remark that Field made) you might see gas raining down upon you in some condensed state, as you see in the solar corona. One could imagine processes like that, even if the field were several μG. There is an infinite number of alternatives at this point, and I think we should be very cautious in pursuing any one of them.

Field: Would one not expect filaments of neutral gas raining down on the galactic plane along the magnetic field? And is there not evidence consistent with this?

Verschuur: I cannot picture a bubble configuration that accounts for the data. Some relevant questions are: given the distribution of intermediate- and high-velocity clouds, what is the three-dimensional picture of the magnetic field and where are we situated? Besides these questions, there is the problem that one might not get sufficiently large velocities during the streaming downwards.

Weaver: Certainly the idea of gas flowing into the local gas structure is completely consistent with the observations. But the observed velocities are very high, often up to 50 and 60 km sec^{-1}, not infrequently 100 km sec^{-1}.

Field: I think you are quite right about high-velocity or intermediate-velocity clouds. But is there not a general inflow at about 7 to 10 km sec^{-1}, both above and below the plane, observed by Dieter and, I believe, the southern observers?

Weaver: I do not know about a specific inflow velocity. The point is that there is always a negative velocity tail in the velocity distribution, and perhaps this tail represents the falling down into the galactic plane.

Van de Hulst: If the matter rains down somehow in the form of filaments, then you can see one of those filaments in projection and see an elongated cloud. If, on the other hand, it moves down in the form of a plane sheet, containing the direction of gravity and the direction of the magnetic field, then we may observe this sheet from two directions. Looking along the galactic plane perpendicular to the field, you don't see the sheet elongated. Looking along the galactic plane and along the magnetic field you see it elongated, but then there is not a projected magnetic field component. And so, in either case, you cannot see what Verschuur mentioned in his Report. Therefore it is not clear to me what the actual interpretation is of the elongated clouds which Verschuur has demonstrated.

3. Origin of the Magnetic Field

Woltjer: I have a question referring to the primordial field. It sounds very reasonable that the present field is in the azimuthal direction, just because of galactic differential rotation. But during the history of the Galaxy an originally uniform field will have undergone a large number of windings and, therefore, if you look in the neighborhood of the Sun, between 10.0 and 10.1 kpc, you would have the field in one direction; between 10.1 and 10.2, in the opposite direction and so on. This would create very conspicuous effects in the Faraday rotation pattern, which are not in fact observed. Does Dr. Verschuur agree?

Verschuur: You are correct, in the plane there are only a few radio sources, which are not polarized.

Syrovat-skii: Parker mentioned three possible mechanisms for the origin of the magnetic field. I want to add a fourth – supernova explosions. During the lifetime of the Galaxy there have been 3×10^8 supernova explosions, if the frequency of this event is one per 30 yr. Let us take 10^{67} cm^3 for the total volume of the Galaxy. Then we have one supernova event per 3×10^{58} cm^3, which is the volume of a sphere with a radius of 6 pc. The Crab nebula (assuming it can serve as a prototype) has, at present, a volume with a diameter of 1 pc and a magnetic field of 300 μG. If it expands to a diameter of 10 pc the field will be 3 μG, a value equal to that of the interstellar field.

So it seems possible to obtain the average flux density of the magnetic field of the Galaxy from independent supernova explosions. But there is the observation of the regularity of the magnetic field. Why is the field regular if individual supernovae produce it? Perhaps there is some correlation between the orientations of stellar angular momenta, reflecting stellar formation in the differentially rotating Galaxy. Then we can expect correlation between partial magnetic fields and the existence of some mean field.

Pikel'ner: But if there is such a correlation, then you would expect the total efficiency of this process to be much lower than you assumed.

Syrovat-skii: Yes, but only for the mean field.

Kardashev: I should like to draw attention to the nucleus of the Galaxy. It probably has a regular field which, near the nucleus, may be dipolar. At large distances from the nucleus the field will be twisted. Perhaps it extends out as far as the 3 kpc arm. The field may be similar to the magnetic field generated in a supernova explosion, as described in papers by Piddington (1966) and by myself (Kardashev, 1964). The main process is twisting of the magnetic flux that connects the fast-rotating stellar remnant with the envelope. Near the remnant we have a dipole field and $B \propto R^{-3}$, where R is the distance to the remnant. Farther out there is a radial component $B_r \propto R^{-2}$, and a tangential component $B_t \propto R^{-1}$. B_t increases linearly with the number of revolutions made by the stellar remnant. [Kardashev, N. S.: 1964, *Astron. Zh.* **41**, 807 (1965, *Soviet Astron.* **8**, 643); Piddington, J. H.: 1966, *Monthly Notices Roy. Astron. Soc.* **133**, 163.]

Ozernoi: First, if the magnetic fields arise in supernova explosions, then we need very large magnetic fields for the pre-supernova stars. Then we are left with the question where this embryonal field comes from. Second, if galactic magnetic fields arise from explosions in the nucleus, then irregular galaxies without nuclei cannot have magnetic fields. But they do have magnetic fields, because, for instance, they have large amounts of non-fragmented gas.

Zel'dovich: Dr. Ozernoi, indirect evidence is not enough in such an important problem! Let me present my opinion. There is a problem with the time scale for the production of the magnetic field in the Galaxy. Together with Sagdeev and Pikel'ner I made some calculations on the turbulent dynamo proposed by Steenbeck *et al.* (1966). The characteristic rotation time of the Galaxy is about 2×10^8 yr and the characteristic turbulence time is 10^7 yr. Since the efficiency of the dynamo is low (less than 0.1) there is little amplification. But in the nucleus of a galaxy the time scales become more favorable. So I suggest that in the nucleus of a galaxy a primordial field is created (e.g., by thermo-electric effects, or by rotational friction of electrons against the 3 K radiation). This field is amplified by nuclear rotation and flows out of the nucleus, together with matter. This suggestion would imply that galaxies without a nucleus have no magnetic field. (Steenbeck, M., Krause, F., and Radler, K. H.: 1966, *Z. Naturforsch.* **21**, 369.)

Verschuur: I think that observations by Mathewson (unpublished) indicate polarization of starlight in the Magellanic Clouds. If I am not mistaken, he finds a correlation of polarization with spiral arms in the Large Cloud. And the Large Magellanic Cloud has no nucleus. (Polarization measurements of stars in the Magellanic Clouds have been published by Th. Schmidt: 1970, *Astron. Astrophys.* **6**, 294.)

Menon: In addition, the Magellanic Clouds, as well as other irregular galaxies, are non-thermal radio sources. So they contain magnetic fields.

Grebinskii: I should like to point out that dynamos are unlikely to work on the galactic scale, because the galactic rotation is so slow. For example, Brajinsky's investigation of a terrestrial dynamo shows that it takes an extremely long time – of the order of 1000 revolutions – to reach a stationary state.

Tsytovich: There is the possibility of generating magnetic fields by plasma turbulence. The time scales are very short. The generated fields can be regular if the original turbulence field was regular. If somewhere there is turbulence (e.g., Langmuir turbulence; see the Introductory Report by Kadomtsev and Tsytovich, p. 108) the currents in the plasma produce certain forces, which react on the electrons and therefore amplify the currents. This generates a magnetic field. Then, an additional possibility exists: turbulent plasma may become unstable and separate into (a) regions with higher plasma densities and low turbulent-energy densities and (b) with lower plasma densities and higher turbulent-energy densities. The same thing can happen with respect to regions of different magnetic energy densities. The growth rate of the instability is proportional to $k \approx 1/L$, where L is a dimension of the created magnetic field. Therefore smaller scales grow faster.

Pikel'ner: But Dr. Tsytovich, this produced field will be of small scale and so the problem is: how do we get a general field?

Tsytovich: I do not know. I only want to point out that in a turbulent plasma magnetic fields can be produced in a short time. But I admit, they have small dimensions.

Woltjer: In the amplification of the magnetic field there are two time scales. You can increase the magnetic energy of the field on the kinematical time scale, L/v. You can increase the magnetic flux only on the time scale on which effects of finite conductivity become important. Unless you can make this time scale short, I cannot see how you can increase the flux of the galactic magnetic field appreciably, and, therefore, I cannot quite see how all these models and explosions from galactic nuclei and similar effects can give you a magnetic field of any sizeable scale.

Zel'dovich: I am not sure that you are right. Of course we know that axisymmetric dynamos do not work, and neither do plane dynamos. But there exist three-dimensional motions, which are able to act as a dynamo. And in addition, we always speak about collisional conductivity, but might not turbulence increase the resistivity and decrease the conductivity and so decrease the time scale for creation of magnetic flux?

Woltjer: In answer to Zel'dovich: the three-dimensional dynamo will not work if you have infinite conductivity. It is a dynamo obtained by balancing terms that involve conductivity.

Parker: Woltjer is right, except that the growth time of the dynamo refers to the smallest participating turbulent element. So in effect there is no limitation on the time.

Woltjer: But you get very small scales.

Parker: I think that reasonable growth times of less than 10^8 yr can be obtained for scales of 1 pc, if one puts in ambipolar diffusion and everything else.

Csada: I want to discuss a three-dimensional, periodic solution of the magneto-hydrodynamic equation of motion, which solution may act as a model for the dynamo of the interstellar magnetic field. The method of solution is a continuation of the 'eigenvalue' theory of Elsasser (1946) and Bullard (1955). Jayanthan's work (1968) differs basically from mine. Part of my results have been published previously (Csada, 1966).

The mathematical problem is to solve the equation

$$\frac{\partial \mathbf{H}}{\partial t} - \mathrm{curl}(\mathbf{v} \times \mathbf{H}) = \kappa \Delta \mathbf{H} \tag{1}$$

by the series

$$\mathbf{H} = \sum_{n=1}^{\infty} B_n(t) \, \mathbf{H}_n(x, y, z).$$

The velocity $\mathbf{v}$ is a known function of the coordinates, κ is the resistivity ($\kappa = c^2/4\pi\sigma$) and $\mathbf{H}_n$ is a certain given set of orthogonal functions. I have been able to show that a velocity model $\mathbf{v}(x, y, z)$ can be found such that the functions $\mathbf{B}_n(t)$ are periodic and stable, provided that the conductivity is infinite. For finite conductivity, dissipation of magnetic energy occurs. In this way a dynamo is obtained, in which the velocity field and the magnetic field are consistent with Equation (1). For a detailed analysis, suitable for discussion of actual observations of interstellar matter, a much more complicated field pattern may be required. [This remark has been condensed (Ed.).] (Bullard, E. C.: 1955, *Proc. Roy. Soc. London Ser.* **A233**, 285; Csada, I. K.: 1966, *Bull. Astron. Inst. Csl.* **17**, 321; Elsasser, W. M.: 1946, *Phys. Rev.* **69**, 106; Jayanthan, R.: 1968, *Monthly Notices Roy. Astron. Soc.* **138**, 477.)

Burke: So far everybody seems to agree that there is a general magnetic field. But an important aspect of Verschuur's report was that there is very good evidence for a magnetic field with a scale of the order of a few hundred parsecs, but not for a field on a considerably larger scale.

Verschuur: Let me extend Burke's argument. Why are magnetic fields permanent? Can't they be transient? We now believe that the once permanent HI-clouds are also transient.

Parker: Once you have magnetic fields, it is hard to get rid of them, in fact equally as hard as to create them.

Greenberg: In M31, the Andromeda galaxy, optical polarization follows the spiral arms. This shows the existence of large-scale magnetic fields.

Woltjer: Optical polarization does not allow you to determine what the magnetic flux is. You have exactly the same polarization, whether all your field-lines are parallel or whether half of them are opposite to the other half.

Pikel'ner: I do not like the idea of an ensemble of independent local fields. I am afraid that it would be difficult to explain the diffusion of the cosmic rays and the uniformity of the spectra of relativistic electrons. In addition cosmic rays will leave the Galaxy in the neutral surfaces between the local fields.

I would prefer the hypothesis of the primordial origin for the magnetic field. As an argument in favor of this hypothesis I would propose the existence of bridges between galaxies. These bridges cannot be understood without a magnetic field between galaxies, and a quasi-regular field of that order cannot have been produced by a dynamo mechanism or any other mechanism. There is one more point: If pulsars are indeed neutron stars with a very strong magnetic field, then this proves that the main part of the magnetic field of the star was in the nucleus and was condensed when collapse took place. Quasi-regular magnetic fields which go through the nucleus of the star argue for the assumption that a field already existed in the gas clouds out of which the pre-supernova star was formed.

Burke: First, with respect to the extra leakage of cosmic rays through the neutral sheets, I would like to answer Pikel'ner that, if the neutral sheets are separated by something of the order of the width of the arm, then you are increasing the surface through which cosmic rays can escape by a geometrical factor of the order of 2 or 4. There will not be a serious effect on the required production rate of cosmic rays. The second point concerns the intergalactic bridges, which are not necessarily evidence of magnetic fields. They could be caused by gravitation as well. In Prendergast's numerical experiments (unpublished) short-living bridge-like structures occur; intergalactic bridges are short living, too, since not all pairs of galaxies show bridges.

Zel'dovich: Is there polarized light in bridges? Are they made out of stars?

Pikel'ner: The bridges are somewhat similar to spiral arms, in that they contain young, hot stars and HII-regions. Without a magnetic field the gas would turn into stars in a short time.

Zasov: There are some arguments in favor of magnetic connections between galaxies. First, Arp has observed polarization in one bridge. Second, to keep gas inside a thin tube for a long time you need a strong magnetic field.

Van Woerden: How do you know the gas is kept together for a long time?

Zasov: The frequency of bridges is very high. Several percent of the galaxies show bridges or tails.

Burke: As far as I remember intergalactic bridges are observed between close pairs of galaxies. A. Toomre (unpublished) has shown that a close encounter between the Magellanic Clouds and our Galaxy can draw a string of material out of our Galaxy. This process requires only gravitational forces.

4. Summary

[On September 16 a group gathered to discuss informally galactic magnetic fields. On September 17 Dr. E. N. Parker presented a summary to the formal meeting. His summary and the ensuing discussion appear here in the Proceedings instead of in their chronologically correct place (Ed.).]

Parker: Several participants interested in galactic magnetic fields met to discuss this subject. We hoped to express and to clarify our views, to argue, and perhaps to arrive at a consensus. The first question that came up was the old one of whether or not the Galaxy really has a general field; the consensus was that it has. The evidence has been accumulating rather steadily in the past and now it accumulates fairly rapidly. The polarization of starlight and the Faraday rotation measurements would certainly be extremely difficult to understand if there were not a large-scale, *i.e.*, a general field in the Galaxy; I might add that cosmic rays would also be difficult to understand. Their persistence in the solar system for so many (up to 10^9) years, would also be difficult to understand without making rather grandiose assumptions. I think, therefore, that everyone would agree that there is apparently a general field in the Galaxy. When one comes to the more detailed question of the structure of that field, however, some points are debatable. But there seems to be agreement that the simplest picture one can form for the galactic field, or for the observable fields in the neighboring arms of the Galaxy, is that, to a first approximation, there is a longitudinal field along the direction of the arms. The field may not always have the same sense but the lines of force seem to lie along the arm. As Weaver mentioned (this volume, p. 22), the Sun seems to reside in a local spur. Some have suggested that in this Spur the field is a tightly wound and somewhat stressed helix or, perhaps, a uniform field through that spur with a local loop. We made no attempt to hammer out which of these pictures fits most closely. At the present time, we seem to be in a spur which has a field of a somewhat more complicated form, perhaps, than the neighboring arms. But, on the other hand, one will have to see what future observations show in the details of the fields in the local arms. An important point, when we begin to think about the significance of the magnetic field, is that the energy density of the magnetic fields, based on the observational indications of a few μG is comparable to the turbulent energy density of the interstellar medium. That is, if you multiply cloud velocities times mean, smeared-out densities, you get 10^{-12} erg cm^{-3}, as van de Hulst emphasized in his Report (p. 3). And these two energies, the magnetic energy and the turbulent energy, are comparable to the cosmic-ray energy. The thickness of the disk of the Galaxy seems to be determined by the sum of these energies, plus the cosmic-ray energy density. All three inflate the disk of the Galaxy. Very roughly, there is an upper limit on the strength of the field of 5 or 6 μG, if we are to understand the observed thickness of the Galaxy in terms of the observed densities of material, etc. That number may change somewhat when observational values change, but it is the right order of magnitude.

Then we spent considerable time discussing a lower limit on the field, i.e., the question of how weak the field could be and still perform the duties that we think it performs. There are a number of arguments that one can give. In our discussion, I took a point of view that I do not normally hold. I argued that the field in the Galaxy could be very weak, with a field strength of only 1 μG. It is clear, however, that with such a weak field present it is difficult to understand the more or less steady, stagnant nature of the cosmic ray gas; although by appealing to plasma turbulence, one might be able to do so. Other members of the group added some very sound arguments which probably make the weak, 1 μG field an untenable hypothesis. Among the points made was that you would have difficulty explaining the non-thermal radio emission, if the field were weaker than about 3 μG. Another point was that if the field were only 1 μG, it would be completely at the mercy of the turbulence in the interstellar medium. It would be much too weak to resist the stresses; and the turbulence would therefore very quickly tangle up the field, so that there would be no recognizable pattern. We know, however, that in fact there is some tendency for the field to lie along the arms, and that there are sharp reversals of fields. Reversals in the sense of the direction of the field would be quickly obliterated by the turbulence in the interstellar medium. Roughly speaking, the field must therefore be 3 μG or more, if it is to maintain its integrity in the face of the turbulence. Unless I am mistaken, then, the consensus of the group was that, for a variety of reasons, the field lies between 3 and 5 or 6 μG, and that, if it should lie outside that range, we must make complicated assumptions about what we see.

Van Woerden: Where does this field of 3 to 6 μG apply? Apparently not in the densest clouds as Verschuur (p. 150) stated, nor in the interarm regions. Should it apply anywhere within the Orion branch, or possibly within a normal spiral arm?

Verschuur: Yes, it should.

Van Woerden: If it is correct that the field between the arms is much smaller than in the arms, the non-thermal radiation between the arms should also be much smaller, a conclusion which fits rather well with the older ideas of Mills.

Pikel'ner: The field between the arms should be less than in the arms. It can be shown by the equilibrium conditions that the density of the gas between the arms is low and no equilibrium is possible for strong fields, especially if the density of cosmic rays is more or less uniform in the galactic disk.

Field: It is still not clear to me whether or not we have information about the field in other spiral arms, other than that it is preferentially along the galactic plane. It seems to me that the information from the Faraday rotation pertains almost entirely to the nearest 500 pc or so and does not extend to other spiral arms.

Verschuur: The Zeeman data which I have on other spiral arms (and there are about four or five clouds which I know must be in other spiral arms) are not inconsistent with a longitudinal field running in a clockwise direction as viewed from the galactic north pole.

Parker: To what distances does the polarization of starlight establish a direction of the field? As I recall, the stars that were observed ranged up to at least one kpc. That observation dictates that the mean distance which you are seeing is of the order of 500 pc.

Verschuur: The stars observed extend up to 3 or 4 kpc, but the helical model applies to within 500 pc. And it appears, as I tried to point out in my lecture, that stars beyond that may be predominantly influenced by the local field and its polarizing material.

Mestel: Does the field run along the spiral arm, or does it actually run in the toroidal direction? Remember that the angle between the Lin wave and the toroidal direction is small. Non-uniform rotation acting steadily on a field with a toroidal component will make the field point in a nearly toroidal direction; whereas the inclination of the gravitational spiral arm is presumably determined by the large-scale structure of the stellar background. There is no obvious reason why the direction of the field and of the arm should coincide; perhaps in a spiral galaxy with less tightly wound arms, the two directions would be markedly different.

Verschuur: If you indeed have a magnetic field which is inclined relative to the gravitational spiral arms and you inject relativistic electrons, they are going to follow the field lines. The non-thermal emission would not be concentrated in the arms; it would be spread throughout the disk.

Parker: As Pikel'ner has pointed out the field in the interarm region can hardly be as strong as in the arm region. The electrons would not radiate when they were in the interarm region; they would radiate only when they were inside the arm.

Burke: I have a question for someone more skilled in cosmic rays than I. The Sun is not going to stay in a spiral arm for very long; it must wander from spiral arm to interarm space and back to spiral arm. If there should be a weaker magnetic field in the interarm region, there should be fewer cosmic rays. One should, therefore, over a period of $(50 \text{ to } 100) \times 10^6$ yr actually see changes in the cosmic-ray intensity. Is such a change observable?

Parker: The following data are available: you look for radioactive nuclei in meteorites, which can be there only because they were produced by cosmic rays. You pick various nuclei with various half-lives, and you find that in every case the number of those radioactive nuclei is what you expected if that meteorite had been exposed to the present cosmic intensity for a period in excess of the half-life of that specific isotope. This finding, of course, gives you only an average of the cosmic-ray intensity over the half-time considered. Now you ask, what happened over the last 10^8 yr? The trouble in answering is that we know the average intensity over the last 10^7 yr (from some isotope which I cannot recall at this moment) and an average over the last 10^9 yr (from an Argon isotope). The 10^7 yr isotope does not go back far enough, and the 10^9 yr isotope averages over all the 10^8 yr ripples. Therefore to answer your question we need an isotope with a half-life of 10^8 yr.

Field: Dr. Burke, would you tell us again why you would expect the cosmic-ray flux to vary as the spiral arm went by?

Burke: I was merely following up the consequences of assuming that the magnetic field between the arms is much lower than in the arms. I wanted to examine this question because Pikel'ner has struck on a really very crucial point. If there is magnetic field running through the interarm regions and into the spiral arms, then Robert's picture of a shock wave along the arm may be open to serious questioning.

Pikel'ner: If the magnetic field between the arms is less than within the arms, then the cosmic ray pressure may or may not be less there, because it is dependent on the isotropy of the cosmic rays. If the cosmic rays between the arms are isotropic, they should have the same density in interarm space as in the arms, if, as is more likely, the magnetic lines connect both regions.

The second point I wish to discuss is the importance of the magnetic field for the shock wave in front of the arms. According to the Lin theory the jump in velocity is about 10 km sec^{-1}. Does such a jump lead to a shock in the rarefied interarm gas? And can this shock be responsible for very massive clouds at the inner edge of the arms? If between the arms $B \approx 2$ μG, $n_H \approx 0.05$ cm^{-3}, and $T \approx 10\,000$ K, then the magnetosound velocity is about 20 km sec^{-1}, and the jump in velocity is not enough to produce a shock wave. As for the second question, let us assume that the shock exists. Then it is difficult to see how we could have the frequently discussed intercloud gas ($n = 0.1$ cm^{-3}, $T = 7000$ K) behind the shock and very low density interarm gas ahead of the shock ($n = 0.05$ cm^{-3}). For the interarm gas moves into the shock with a speed of 10 km sec^{-1} and the shock relation $p + \varrho V^2 =$ constant cannot be fulfilled, since $p_2 > (\varrho V^2)_1$. If the gas behind the shock is not the intercloud medium then it is difficult to see how massive clouds can be expected at the inner edge of the spiral arm.

12. THE GAS DYNAMICS OF ACCRETION*

Introductory Report

(Thursday, September 18, 1969)

E. A. SPIEGEL

Dept. of Astronomy, Columbia University, New York, N.Y., U.S.A.

1. Introduction

The term accretion originally referred, in astronomical contexts, to the capture of mass by stars, and, later, to mass capture by other centers of gravitational force. As such, the process has not proved to be of general importance, in spite of early hopes. However, there are other aspects of the problem which may yet prove worthy of attention in interstellar gas dynamics. In particular, the effects of stars, galaxies, or even clusters of galaxies, on ambient matter streaming by them may be detectable, directly or indirectly, and it is on such possible effects that I shall concentrate here. These effects are related to the original accretion problem but may be thought of separately; nevertheless, I retain the use of the term accretion to refer to all aspects of the motions induced in an ambient medium by a gravitating object.

In a certain sense, a disturbing gravitating object can be thought of as a probe which may be of use in the diagnostics of the ambient medium. However, the distinction from plasma probes is that we concentrate here more on the effects produced in the medium than on the response of the probe. Anyone who wishes to go deeply into the subject would do well to study the literature on plasma probes. In the list of references, I give a sampling of seven papers.[†] But here, I shall only go into the astronomical literature, which provides a suitable introduction.

2. Traditional Theory

To provide a background to the fluid dynamical problems, I should like to mention briefly some of the early work on the traditional accretion process. A detailed review has been given by Lyttleton (1953) and it is not necessary to go into details. The simplest case to consider is that of a stationary star embedded in a uniform medium. Bondi (1952) gave a gas dynamical treatment of this problem, the equations of which are now familiar since they are essentially those of Parker's solar wind theory (1958). Bondi found the rate of accretion

$$A = \lambda \pi \left(\frac{2GM}{c^2} \right)^2 \varrho_0 c \tag{1}$$

where M is the mass of the star, c is the speed of sound in the medium whose density

---

* During the Symposium this Report was presented later than its present location suggests. (Ed.)

† These are marked with asterisks in the bibliography.

infinitely far from the star is ϱ_0, and where λ is a pure number which varies from 0.25 to 1.1 as the ratio of specific heats, γ, varies from $\frac{5}{3}$ to 1. For $\varrho_0 \approx 10^{-24}$ g cm^{-3}, $M = M_\odot$, $c = 1$ km sec^{-1}, we find that $A \approx 5 \times 10^{-12} M_\odot$ yr^{-1}, which indicates that under normal conditions, spherically symmetric accretion is not a significant process. Of course, one can imagine extreme situations where accretion is more important, but they may also involve rather different conditions than contemplated here. Moreover, the Bondi theory does not delve into the physical conditions at the surface of the star, which can make a serious difference. Some additional work along these lines was done by Mestel (1954) who found that unless the ambient medium is dense enough, the H II region around the star will prevent any rapid accretion; he finds the critical density to be 10^3 cm^{-3} for $M \leqslant 1.75 M_\odot$. This would seem to preclude accretion by early-type stars. Late-type stars, on the other hand, would be expected to have strong winds, thus they too would not be likely to accrete. However, we might still consider a small accumulation of matter in a shell around the star, where the wind and accretion inflow bring one another to rest. Such a shell is probably unstable and I do not know whether there would be any observable effects.

A more extensive literature exists on the capture of mass by stars moving with respect to the ambient medium. The simplest case is that in which the medium is cold, uniform, collisionless, and non-gravitating, and the star moves uniformly through it. In the frame of the star a gas particle has the orbit

$$\frac{1}{q} = \frac{R_A}{2s^2}(1 + \cos\alpha) + \frac{1}{s}\sin\alpha, \tag{2}$$

where q is distance from the star, α is the angle measured from the downstream symmetry axis, or accretion axis, s is the impact parameter, and

$$R_A = 2GM/V_0^2, \tag{3}$$

where V_0 is the velocity at upstream infinity (see Figure 1). It is easy to see that the

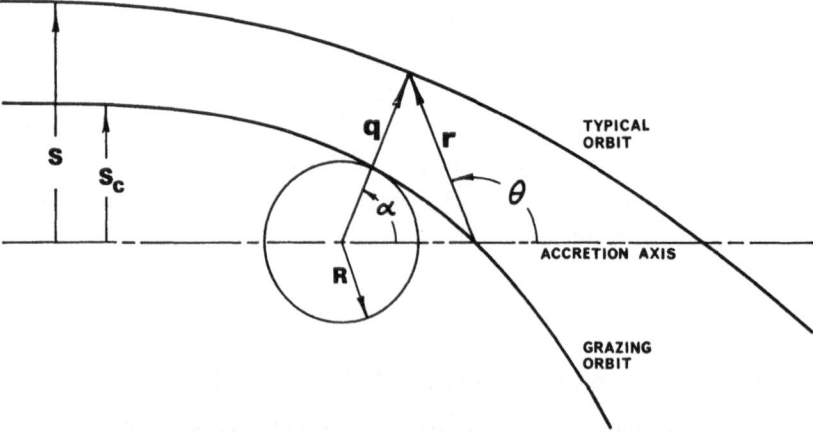

Fig. 1.   Illustrating the notation. s is impact parameter; s_c is the critical impact parameter of Equation (5). Coordinates (q, α) are used in this section. Coordinates (r, θ) are introduced in Section 3.

smallest value of q for given s is

$$q_{\min} = 2s^2/[R_A + (R_A^2 + 4s^2)^{1/2}].\tag{4}$$

This formula can be used to determine whether the particle can strike the surface of the star, whose radius is R. We readily find that particles for which

$$s \leqslant s_c \equiv [R(R + R_A)]^{1/2}\tag{5}$$

strike the star. If we assume that all such particles are captured, we then obtain the accretion rate

$$A = \pi s_c^2 \varrho_0 V_0.\tag{6}$$

With the same assumptions as above, and with $R = R_\odot$ and $V_0 = 10$ km sec^{-1} we obtain $R_A = 7 \times 10^{14}$ cm and $A = 10^{-18} M_\odot$ yr^{-1}.

Lyttleton (1953) has argued that inelastic collisions among the streaming particles could lead to an enhanced rate of accretion. The argument is based on the recognition that if the incident gas is cold, a large density develops on the accretion axis. In the limit of zero temperature, the density distribution becomes

$$\varrho(q, \alpha) = \tfrac{1}{2}\varrho_0 \csc \frac{\alpha}{2} \left(\frac{R_A}{q} + \sin^2 \frac{\alpha}{2} \right)^{1/2}$$
$$\times \left[\frac{R_A}{2q} + \sin^2 \frac{\alpha}{2} + \sin \frac{\alpha}{2} \left(\frac{R_A}{q} + \sin^2 \frac{\alpha}{2} \right)^{1/2} \right],\tag{7}$$

where the contribution of particles which have already crossed the axis is neglected. (For a detailed discussion of the density distribution see Danby and Camm, 1957.) We see that for $\alpha = 0$, $\varrho = \infty$; the fact that in this flow rings of gas squeeze down into points on the axis is associated with the singular behavior of ϱ. Hoyle and Lyttleton argued that in this high density region, inelastic collisions will occur and annihilate the transverse momentum of the particles. The downstream velocity is approximately conserved, and particles which cross the axis close enough to the star move too slowly to escape and are accreted. By conservation of angular momentum we find that if no collisions occur, the α-velocity of a particle is

$$v_\alpha = - sV_0/q,\tag{8}$$

while energy conservation implies that its q-velocity is

$$v_q^2 = V_0^2 \left(1 + \frac{R_A}{q} - \frac{s^2}{q^2} \right).\tag{9}$$

From Equation (2) we see that the particle crosses the accretion axis ($\alpha = 0$) at a distance $q = s^2/R_A$ from the star so that at $\alpha = 0$, $v_q = V_0$, for all particles. Thus particles which cross the axis at distances less than about R_A from the star, on having their α-momenta destroyed in collisions, will be below escape velocities. From Equation

(2) we see that a particle for which $q = R_A$ at $\alpha = 0$ has $s = R_A$, so that we obtain an accretion rate

$$A \approx \pi R_A^2 \varrho_0 V_0, \tag{10}$$

which exceeds the estimate of Equation (6) by a factor of about R_A/R_0, but still does not give encouragement.

The theory has been further elaborated in the particle picture by Bondi and Hoyle (1944; see also Lyttleton, 1953) and a picture of the flow in a wake of free particles has been developed. However, the estimated capture rate is not appreciably modified. Various details of these calculations have been cause for debate, and a list of some of the more important papers is given in a brief review by McCrea (1955).

Here I want to mention what appears to be the most serious objection, which has been raised by Danby and Camm (1957) and Danby and Bray (1967). These authors have considered the effect of a finite temperature on the density distribution in the wake of a star and have shown how the singular behavior near the axis is averted. Under typical interstellar conditions, they contend, the density is lowered sufficiently so that collisions will not be important, and they conclude that the mechanism of Hoyle and Lyttleton will not operate. On the other hand, they do not include self-interaction of the gas either through plasma effects or gravitation, and it seems likely that such effects will be of importance. Just what the details of the motion become then is not clear, but if we consider that different 'streams' interpenetrate as they cross the axis, we might speculate on the possibility of collective instabilities which perhaps act like collisions. Whether the collective energy resulting from instabilities is carried away in waves or results in plasma instabilities probably depends on the Mach number, but there does seem to be a case for treating the region near the accretion axis as a continuum to gain some impression of the general nature of the flow. At any rate, with only these vague assurances, I shall devote the rest of this discussion to a gas dynamical discussion of accretion processes.

3. Linearized Gas Dynamics

The treatment of the gas dynamics of a flow past a center of gravitational force is sufficiently difficult that some drastic approximations have been made in this subject. One quite tempting, but not necessarily justifiable, approach is linearization, in which the perturbing body is assumed to produce small deviations from uniformity. This approach was used by Dokuchaev (1964) who considered the uniform motion of a point mass through a uniform medium; this problem is closely similar to that of the motion of a charged satellite moving through a plasma as treated by Kraus and Watson (1958). Dokuchaev also considered the effects of a uniform magnetic field along the motion and the effects of mass loss from the star; but let us discuss here the purely gravitational case, as it stems from the classical problem of the previous section. (I shall also mix into the discussion some unpublished results on this problem obtained by Prendergast, Ruderman, and me.)

Let the density in the ambient medium be $\varrho = \varrho_0 + \delta\varrho$ where ϱ_0 is a constant and $|\delta\varrho| \ll \varrho_0$. Similar remarks apply to the other state variables and it is assumed that the gas velocities are small in the stationary frame. We also assume isentropic motion. Then, on linearizing the equations of motion and performing the usual manipulations of acoustics (Ward, 1955), we obtain the wave equation (Appendix A)

$$\frac{\partial^2 \Psi}{\partial t^2} - c^2 \nabla^2 \Psi = 4\pi GM \, \delta(\mathbf{r} - \mathbf{V}_0 t), \tag{11}$$

where

$$\Psi = \delta\varrho / \varrho_0, \tag{12}$$

and where c is the speed of sound and δ is the Dirac function. This kind of problem is very like that of Cherenkov radiation and procedures for solving it are standard (Landau and Lifshitz, 1960). If we ignore the homogeneous solutions of Equation (12) and consider the forced solution, we obtain for $M \equiv V_0/c < 1$

$$\Psi = \frac{R_A M^2}{q(1 - M^2 \sin^2 \alpha)^{1/2}} ; \tag{13}$$

while for $M > 1$,

$$\Psi = \begin{cases} \dfrac{R_A M^2}{q(1 - M^2 \sin^2 \alpha)^{1/2}} & \text{for} \quad \alpha < \arcsin(1/M) \\[2mm] 0 & \text{for} \quad \alpha > \arcsin(1/M). \end{cases} \tag{14}$$

These solutions are expressed in the frame of the star, where they are time-independent. The supersonic solution has a discontinuity on the Mach cone, which indicates the formation of a shock. However, on the cone the density is singular which is a standard difficulty in the linear theory of supersonic motion of point sources (Ward, 1955). It can be removed by considering a mass of finite size (Huebner and Herring, unpublished report).

If we have the density, we can then find the velocity from the equations of motion. In the star's reference frame this gives

$$u_\| = - V_0 - \tfrac{1}{2} R_A V_0 \left[\frac{1}{q} - \frac{2}{q(1 - M^2 \sin^2 \alpha)^{1/2}} \right] \tag{15}$$

$$u_\perp = - \tfrac{1}{2} R_A V_0 \left[\frac{1 + \cos\alpha}{q \sin\alpha} - \frac{2 \cot\alpha}{q(1 - M^2 \sin^2 \alpha)^{1/2}} \right]$$

behind the shock and

$$u_\| = - V_0 - \tfrac{1}{2} R_A V_0 q^{-1}, \qquad u_\perp = \tfrac{1}{2} R_A V_0 \frac{1 + \cos\alpha}{q \sin\alpha} \tag{16}$$

ahead of the shock, where $u_\|$ and $u_\perp$ are the velocities parallel and perpendicular to the original uniform flow, $\mathbf{V}_0$. We see that behind the shock $u_\| = 0$ on the surface

$$(q + \tfrac{1}{2} R_A)^2 = \frac{1}{1 - M^2 \sin^2 \alpha}. \tag{17}$$

Between the shock cone and this 'accretion' surface, $u_\parallel$ is starwards; while downward of the surface, $u_\parallel$ is downstream. This is in line with what one would expect, but $u_\perp$ has an unpleasant feature. Behind the shock $u_\perp$ is negative, whereas ahead it is positive. This means that the velocity is into the shock from both sides, and this does not seem to be physically meaningful. However, mass is conserved to the accuracy of linear theory and to get an idea of the flow we must consider the linearized mass flux: $\mathbf{V}_0'(\varrho_0 + \delta\varrho) + \varrho_0\mathbf{u}$. This gives the flow pattern shown in Figure 2. Mass capture does not occur in this

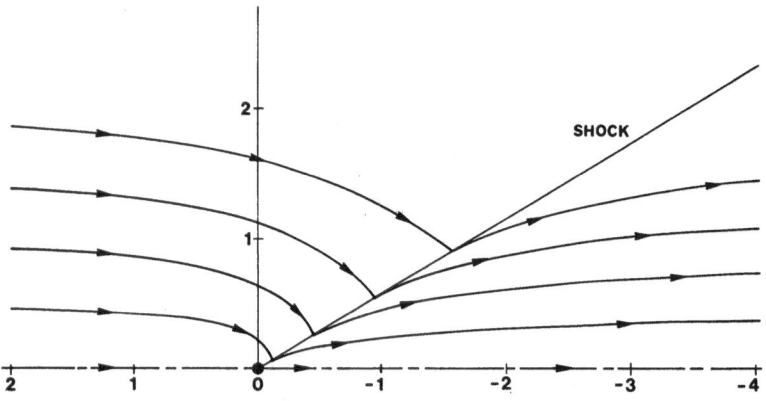

Fig. 2.   The mass current as given by linear theory. Unit of length $= R_A$.

model unless a sink is introduced in the continuity equation; the strength of such a sink would have to come from aspects not yet considered, such as conditions on the stellar surface.

Another result of interest from linear theory is the drag, F_D, on the body. This results from the gravitational force of the disturbed gas on the star. If we calculate this drag by Fourier transform techniques we find that the Fourier inversion gives a divergent integral, as is usual with Coulombic potentials; indeed the calculations are much like those encountered in Cherenkov radiation. As in that problem, the limits of integration must arbitrarily be cut off at large and small distance. The long range cut-off is not needed, however, if the self-interaction of the gas is included, and a sort of gravitational shielding occurs. On the other hand, the introduction of self-gravity of the gas brings Jeans instability into the problem so that the results with self-gravity are only suggestive of what happens. For $M = V_0/c > 1$ and with

$$k_J^2 = 4\pi G\varrho_0/c^2 \tag{18}$$

we find, for $M > 1$,

$$F_D = \pi \ln\left[\frac{\pi}{Rk_J}\left(\frac{M}{\sqrt{(M^2 - 1)}}\right)\right] R_A^2\varrho_0 V_0, \tag{19}$$

where R is the stellar radius, introduced as the short range cut-off.

4. Galactic Wakes*

The linear theory gives some inkling of the difficulties involved in the gas dynamics of accretion, but it is rather unsatisfactory in many ways, and it seems worthwhile to attempt a nonlinear theory. In doing this it is helpful to distinguish two cases $R \gtrsim R_A$ and $R < R_A$, where R is the radius of the moving body. For our Galaxy, $M \approx 2 \times 10^{44}$ g, and if we assume $V_0 \approx 200$ km sec^{-1}, we find $R_A \approx 20$ kpc which is about the radius of the Galaxy. Thus, galaxies belong typically in the class where $R \gtrsim R_A$. This means that gas focused into the wake of a galaxy has already begun to leave the zone of gravitational influence, and this permits a simplification of the problem which is not usually possible in the stellar case.

We treat the Galaxy as spherical and its motion as hypersonic so that the upstream gas is effectively cold. The gas is also collisionless upstream. We assume that particles which strike the Galaxy are absorbed so that no pressure builds up ahead of the Galaxy and we may ignore bow shocks. An orbit which just grazes the Galaxy crosses the accretion axis at a distance, R_0, from the Galaxy (Figure 1), where

$$R_0 = R(1 + \delta^{-1}), \qquad \delta = R_A/R, \tag{20}$$

so that $R_0 > 2R \geqslant 2R_A$. If no collisions occurred, we could completely describe the density and velocity fields by means of orbit theory; for our purposes such a description is most effectively expressed in spherical polar coordinates with origin at $q = R_0$, $\alpha = 0$. Orbits which miss the Galaxy converge conically onto the axis about an apex at this point. We introduce spherical polar coordinates (r, θ) and express quantities in this system. Then density [Equation (7)] is quite complicated in these new coordinates, but for $r \ll R_0$ it simplifies to

$$\varrho = \varrho_0 R(1 + \delta)^{1/2}/(2r \sin \theta), \tag{21}$$

while the velocity becomes

$$
\begin{aligned}
u &= V_0 [1 + \delta(1 + \delta)^{-1}] \cos(\theta + \theta_i), \\
v &= V_0 [1 + \delta(1 + \delta)^{-1}] \sin(\theta + \theta_i),
\end{aligned}
\tag{22}
$$

where u and v are the r- and θ-components and

$$\tan \theta_i = \frac{\delta}{(1 + \delta)^{1/2}}. \tag{23}$$

This describes the flow near the apex of the convergence onto the axis. We assume that this flow encounters a (collisionless) shock and that behind the shock the motion obeys the equations of gas dynamics, but that by now we are far enough downstream to neglect the gravity. We also treat the flow as adiabatic and steady in the Galaxy's frame of reference. Equations (21) and (22) suggest that we look for similarity solutions

* This section is based on an unpublished MS of M. A. Ruderman and the author, as is Section 5C.

of the form

$$\varrho = \tfrac{1}{2}\varrho_0 (R/r) f(\theta), \qquad p = \tfrac{1}{2}\varrho_0 V_0^2 (R/r) g(\theta),$$
$$u = V_0 U(\theta), \qquad\qquad v = V_0 V(\theta). \tag{24}$$

For adiabatic flow the equations admit such solutions (Appendix B) and we obtain a set of ordinary differential equations for U, V, g, and f. These equations can be solved numerically but I will not discuss such solutions here. It suffices to point out that near the axis $(\theta \ll 1)$ the solutions behave like

$$U = W_0 \left[1 - A\theta^{2(\gamma-1)/(2\gamma-1)} + \cdots\right], \qquad V = -\frac{2\gamma-1}{2\gamma}\theta + \cdots,$$
$$f = f_0 \theta^{-2(\gamma-1)/(2\gamma-1)} + \cdots, \tag{25}$$
$$g = g_0 W_0^2 \left[1 + \frac{f_0}{2g_0\gamma}\left(\frac{2\gamma-1}{2\gamma}\right)^2 \theta^{2\gamma/(2\gamma-1)} + \cdots\right],$$

where W_0, f_0, g_0, and A are constants with the constraint

$$A = \frac{g_0}{f_0}\left(\frac{\gamma}{\gamma-1}\right),$$

and where we have assumed $V(0)=0$. We see that even in the gas dynamical case the density is singular on the axis, as it is in the collisionless case. (For $\gamma=1$, f is not singular, but V is.)

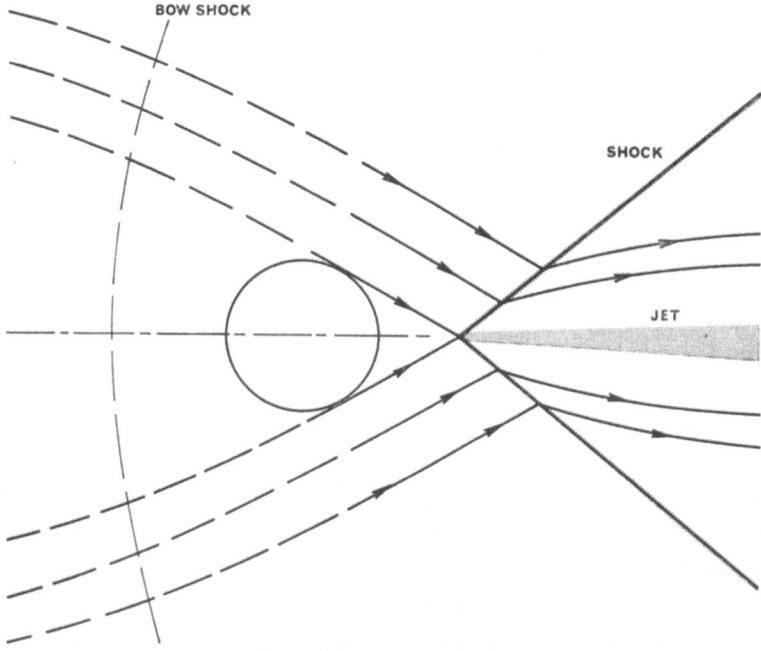

Fig. 3.   Solid flow lines indicate the flow sketched from the theory. Dashed lines are the orbits which these should join onto; the bow shock is not in the theory but is indicated.

The gas flow of Equations (25) must now be matched to the incoming flow with the usual conservation conditions for strong shocks, and from this matching we obtain

$$A = \left(\frac{2\gamma - 1}{2\gamma}\right)^2 \frac{1}{2\gamma} \theta_s^{2\gamma/(2\gamma-1)}, \qquad \theta_s = \frac{2\gamma(\gamma - 1)}{3\gamma - 1} \theta_i,$$

$$W_0^2 = 1 + \delta^2 (1 + \delta)^{-1}, \qquad f_0 = \left(\frac{\gamma + 1}{\gamma - 1}\right) (1 + \delta)^{1/2} \theta_s^{-1/(2\gamma-1)}, \tag{26}$$

where θ_s is the shock angle. Thus the solution is determined for given δ. Figure 3 shows a typical flow pattern. The picture we have here is only schematic since many simplifying assumptions have been built into the calculation, but it shows how one might begin to get an idea of which way the flow goes. To do better one should probably try to integrate the equations numerically, but in view of the uncertainties in the plasma effects, perhaps the time for this is not yet ripe.

5. Some Applications

Accretion has at times been suggested as the cause of a variety of phenomena including heating of the solar corona, X-ray sources, galaxy formation, spiral arm formation, comet formation, and novae. I would not like to try to evaluate all of these, and instead, will compound the list a bit.

A. JETS

We know that a variety of galaxies and quasars reveal jet-like appendages, and we might wonder whether these could be accretion wakes. After all, whether you treat the accretion flow as collisionless or gas dynamical, you seem to get evidence of jets. However, it is unlikely that the ambient medium is anywhere dense enough to give rise to an observable galactic jet. On the other hand, the flow we considered in Section 4 need not be the result of accretion, but simply a portion of a non-spherical collapse or implosion. To this extent, the accretion work points a moral. If we want to make a jet we must focus momentum. We would not find this easy to do with ordinary explosions, but the possibility of jet formation in asymmetric collapse does arise. Thus some observed galactic jets might be produced by asymmetric implosions, just as collapsing bubbles often produce prolonged jets (Benjamin and Ellis, 1966), and the kind of convergent flows considered in accretion theory may be quite relevant to this problem, but in a rather different context.

B. THE TAIL OF BETELGEUSE (α ORIONIS)

The star α Orionis shows a jet-like appendage with what looks like a dust core (Morgan *et al.*, 1955). Very little is known except for the picture shown in Figure 4; the tail just barely shows up on the Sky Survey, but it is highly unlikely that even this faint object could be produced by accretion alone. However, the accretion wake might be rendered visible if it were seeded by dust escaping from the star, and the following picture seems possible.

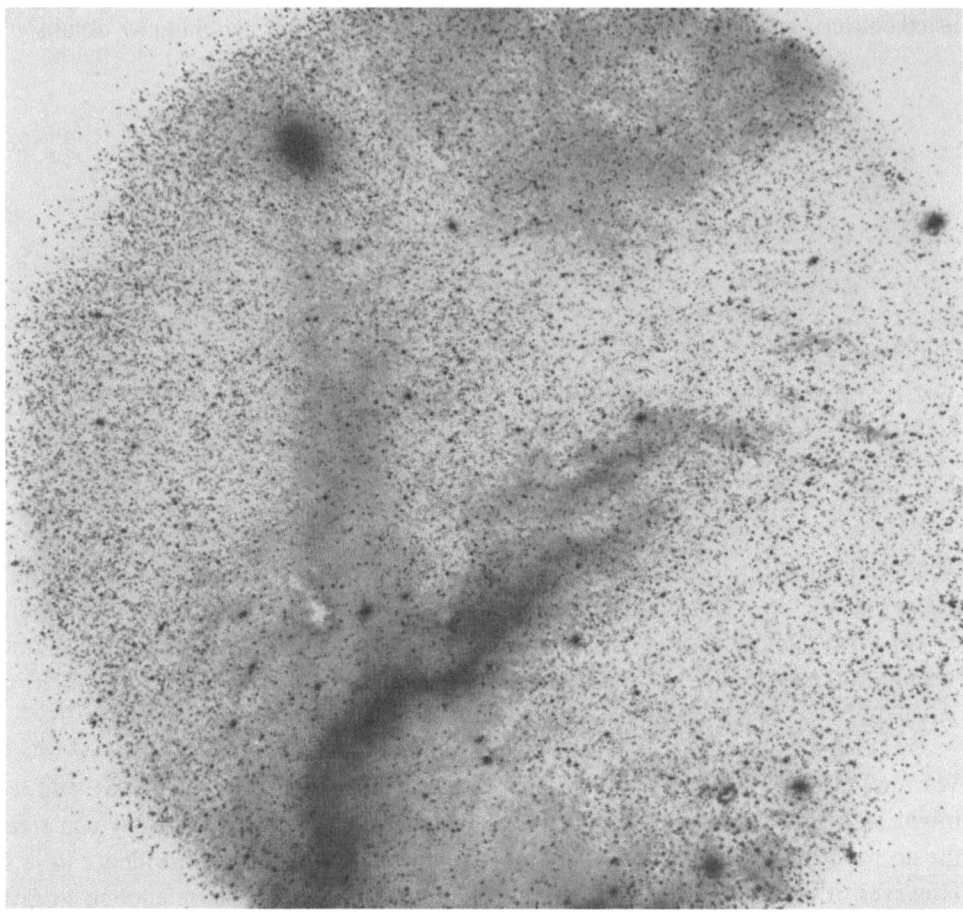

Fig. 4. The Tail of α Orionis: Yerkes Observatory photographs by H. M. Johnson with Meinel 8 in. $f/1$ Schmidt camera and Hα interference filter; diameter of field, 12° (cf. Morgan *et al.*, 1955). Prof. Morgan, who kindly provided the plate reproduced here points out that it was made from a lantern slide, as the original is missing. The tail is nevertheless visible, especially when the plate is held at arm's length.

Suppose that grains form in the atmosphere of α Orionis at some constant rate and that they are driven off by radiation pressure. Inevitably, they must drag some gas with it and thus produce a stellar wind. The extensiveness of such a wind is a matter of debate (Weymann, 1962; Wickramasinghe *et al.*, 1966) and depends very much on the nature of the dust particles. However, that dust particles do exist around α Orionis seems strongly indicated by its infrared excess. The dust wind would behave much like an ordinary wind and have a sonic transition just as in Parker's theory. We would then expect the wind to be arrested in a shock. The shock occurs in the gas which is kept cool by the dust, so that the gas comes virtually to a stop. The gas then drags the dust to a stop and this gives rise to a dusty shell around the star. This shell probably has a radius larger than R_A, so that the theory outlined above is relevant. But it must be modified to take account of the fact that material flowing by the star pulls dust into it so that the flow is enriched. This dust is then squeezed onto the downstream axis to produce the observed tail.

Of course, only detailed observations can decide the correctness of this picture, but at least it provides us with one instance where we may be witnessing an interaction of a star with the interstellar medium.

C. HEATING OF INTERGALACTIC GAS

The preceding example depended on an understanding of the details of the accretion flow. But there are gross considerations which are relevant such as the heating of an ambient gas by supersonically moving objects. The rate of working by such an object is $F_D V_0$, where F_D is the drag force. If the radius of the object exceeds R_A, this formula must be amended to include the geometrical drag which would result from a bow shock. Such corrections for galaxies are not appreciable if V_0 is not in excess of 500 km sec^{-1}.

The heating of the intergalactic gas is important in setting limits on detectability and perhaps the most promising place to look for intergalactic gas is in clusters of galaxies. Typical clusters do not seem gravitationally bound and it has been speculated that the binding is accomplished by intergalactic matter. This is by no means a reliable prediction, but it might be testable if the gas is heated by shock waves generated by the galaxies (Ozernoi and Zasov, in preparation). For example, in the Coma cluster the number of galaxies $N \approx 10^3$ and the dissipation rate by accretion shocks is

$$\varepsilon \approx N\pi C_A R_A^2 (nm_H) V_0^3, \tag{27}$$

where $V_0^2 = 2GM_c/R_c$ and C_A is the logarithmic factor in Equation (19). Subscript c denotes values of R and M for the whole cluster. With $M_G = 4 \times 10^{44}$ g, $M_c = NM_G$, $R_c \approx 5 \times 10^{24}$ cm, we find $V_0 \approx 300$ km sec^{-1} and

$$\varepsilon \approx 5 \times 10^{47} n \text{ erg sec}^{-1}. \tag{28}$$

Now the cluster has a gravitational energy $GM_c^2/(2R_c) \approx 10^{63}$ erg and this would be dissipated in a time of $2 \times 10^8 \, n^{-1}$ yr. If, as seems plausible, we assume that the clusters are at least 10^{10} yr old we must have $n < 10^{-2}$ cm^{-3}, which is not inconsistent with the densities required for binding ($n \lesssim 3 \times 10^{-3}$). The gas would be raised to a temperature somewhat less than

$$T_0 = \frac{m_H}{k} \frac{GM_c}{R_c} \approx 10^8 \text{ K} \tag{29}$$

so that temperatures on the order of 10^7 K or a little greater would be expected. In these circumstances we would expect to see thermal X-rays generated at a rate $1.4 \times 10^{-27} \, n^2 T^{1/2}$ erg sec^{-1} cm^{-3}. The emission comes from the wakes of galaxies which occupy a volume which may be estimated from the theory of the accretion shock; but crudely it is of the order of a few times N times a galactic volume, so that about

10^{-5} of the gas radiates the X-rays. We obtain then an X-ray luminosity of $\approx 10^{48} \, n^2$ erg sec^{-1}. If we compare this with Equation (28) we see that the X-ray efficiency is fairly low at intergalactic densities. (Of course, the formulae cannot be used at high densities where self-absorption enters.) If we take $n \approx 3 \times 10^{-3}$ cm^{-3}, we can get an X-ray luminosity for the Coma cluster of about 10^{43} erg sec^{-1}. This should be readily detectable and would seem to militate against the assumption that there is intergalactic matter in clusters sufficient to bind them.

Similar examples can be worked out for other clusters, individual galaxies, clusters of galaxies moving through the intergalactic medium, and even star clusters in dense galactic nuclei and perhaps quasars.

Appendix A

The equations of isentropic motion are

$$\varrho \left(\frac{\partial u}{\partial t} + \mathbf{u} \cdot \nabla \mathbf{u} \right) = - \nabla p + \varrho \nabla \varphi, \tag{A1}$$

$$\frac{\partial \varrho}{\partial t} + \nabla \cdot (\varrho \mathbf{u}) = - A, \tag{A2}$$

$$\nabla^2 \varphi = - 4\pi G (\varrho + \varrho_*), \tag{A3}$$

$$\frac{dp}{dt} = c^2 \frac{d\varrho}{dt}, \tag{A4}$$

where ϱ_* is the density of the moving object and A is the rate of accretion, counted positive if the object is a sink. If the object is a source of matter, then it would not be correct to use (A4) for both the ambient and injected matter since in general c^2 will differ for the two. If we treat the star as a point mass moving with velocity $\mathbf{V}_0$, we have

$$\varrho_* = M \, \delta(\mathbf{r} - \mathbf{V}_0 t) \tag{A5}$$

and

$$A = \lambda \pi R_A^2 V_0 \varrho \tag{A6}$$

where λ is not yet specified, but could depend on the solution. Let $\varrho = \varrho_0 + \delta\varrho$ with $|\delta\varrho| \ll \varrho_0$ and let $\mathbf{u}$ be small (as in usual acoustics). Because of the self-gravity, ϱ_0 is not a constant, but we make the usual Jeans swindle, and treat it so. Then, on dropping nonlinear terms, we obtain

$$\varrho_0 \frac{\partial \mathbf{u}}{\partial t} = - \nabla \, \delta p + \varrho_0 \nabla \, \delta\varphi, \tag{A7}$$

$$\frac{\partial}{\partial t} \delta\varrho + \varrho_0 \nabla \cdot \mathbf{u} = - \lambda \pi R_A^2 \varrho_0 V_0 \, \delta(\mathbf{r} - \mathbf{V}_0 t), \tag{A8}$$

$$\nabla^2 \, \delta\varphi = - 4\pi G (\varrho_* + \delta\varrho), \tag{A9}$$

$$\delta p = c^2 \, \delta\varrho, \tag{A10}$$

where λ is now a constant. These equations are readily combined into a wave equation for $\delta\varrho$ (with $k_J^2 = 4\pi G\varrho_0/c^2$):

$$\Box \Psi + k_J^2 c^2 \Psi = 4\pi GM \, \delta(\mathbf{r} - \mathbf{V}_0 t) + \lambda\pi R_A^2 \varrho_0 V_0^2 \, \delta(z - V_0 t) \, \delta(x) \, \delta(y).$$

(A11)

When $k_J = 0$ and $\lambda = 0$, this is Equation (11) of the text. The criterion for neglecting the k_J term is roughly that $k_J R_A \ll 1$, but even in that case certain long range effects are not correctly represented when k_J is set equal to zero. It also is clear that if we solve Equation (11) with $\lambda = 0$, the correction for $\lambda \neq 0$ is just to add a term $\lambda R_A \, \partial\Psi_0/\partial z$ to Ψ_0, where Ψ_0 is the solution with $\lambda = 0$.

We readily verify that Equations (14), (15), and (16) represent solutions of the linear equations.

Appendix B

In the frame of the star we assume steady fluid motions. If we can neglect gravity in the wake, we have the equations

$$\varrho\left(u\frac{\partial u}{\partial r} + \frac{v}{r}\frac{\partial u}{\partial \theta} - \frac{v^2}{r}\right) = -\frac{\partial p}{\partial r},$$

(B1)

$$\varrho\left(u\frac{\partial v}{\partial r} + \frac{v}{r}\frac{\partial v}{\partial \theta} + \frac{uv}{r}\right) = -\frac{1}{r}\frac{\partial p}{\partial \theta},$$

(B2)

$$\varrho\left(\frac{\partial u}{\partial r} + \frac{2u}{r} + \frac{1}{r}\frac{\partial v}{\partial \theta} + \frac{v}{r}\cot\theta\right) + \left(u\frac{\partial}{\partial r} + \frac{v}{r}\frac{\partial}{\partial \theta}\right)\varrho = 0,$$

(B3)

$$\left(u\frac{\partial}{\partial r} + \frac{v}{r}\frac{\partial}{\partial \theta}\right)p = \frac{\gamma p}{\varrho}\left(u\frac{\partial p}{\partial r} + \frac{v}{r}\frac{\partial}{\partial \theta}\right)\varrho.$$

(B4)

With the ansatz of Equations (24) we find

$$VU' - V^2 = g/f$$

$$VV' + UV = g'/f$$

$$U + V' + V\cot\theta + Vf'/f = 0$$

(B5)

$$(\gamma - 1)U = V\left(\frac{\gamma f'}{f} - \frac{g'}{g}\right)$$

where prime denotes differentiation with respect to θ. From (B5) we obtain two integrals:

$$g = Kf^\gamma(-fV\sin\theta)^{\gamma-1}$$

(B6)

and

$$\tfrac{1}{2}(U^2 + V^2) + \left(\frac{\gamma}{\gamma - 1}\right)\frac{g}{f} = \tfrac{1}{2}W_0^2,$$

(B7)

where W_0 and K are arbitrary constants. With these results the expansion of Equations (25) is readily obtained, and with this behavior near $\theta = 0$, we can find the solution of (B5) numerically.

Acknowledgements

I am grateful to a number of people for discussions of various aspects of these problems and I especially wish to thank among these S. I. Childress, M. Lampe, J. Theys, J. Toomre, and J.-P. Zahn who became sufficiently intrigued to do some relevant and enlightening calculations. Dr. Toomre also was kind enough to quickly produce the diagrams when the editor presented an ultimatum with his deadline. I also am indebted to Prof. Morgan for providing the plate of α Orionis and for pointing out the existence of the tail.

References

Benjamin, T. B. and Ellis, A. T.: 1966, *Phil. Trans. Roy. Soc. London* **A260**, 221.
* Bernstein, I. B. and Rabinowitz: 1959, *Phys. Fluids* **2**, 72.
Bondi, H.: 1952, *Monthly Notices Roy. Astron. Soc.* **112**, 195.
Bondi, H. and Hoyle, F.: 1944, *Monthly Notices Roy. Astron. Soc.* **104**, 273.
* Cohen, I. M.: 1963, *Phys. Fluids* **6**, 1492.
Danby, J. and Bray, T. A.: 1967, *Astron. J.* **12**, 219.
Danby, J. and Camm, G. L.: 1957, *Monthly Notices Roy. Astron. Soc.* **117**, 50.
* Davydov, B. and Zmanovskaya, L. J.: 1936, *Zh. Tekhn. Fiz.* **3**, 215.
Dokuchaev, V. P.: 1964, *Astron. Zh.* **41**, 33 (translation: 1964, *Soviet Astron.* **8**, 23).
Eddington, S. A.: 1959, *The Internal Constitution of the Stars*, Dover Publications, New York, p. 391.
* Hester, S. D. and Sonin, A. A.: Sixth Rarefied Gas Dynamics Conference (no further information available).
Kraus, L. and Watson, K. M.: 1958, *Phys. Fluids* **1**, 480.
* Lam, S. H.: 1965, *Phys. Fluids* **8**, 73.
Landau, L. D. and Lifshitz, E. M.: 1960, *Electrodynamics of Continuous Media*, Pergamon Press, London.
* Langmuir, I. and Mott-Smith, H. M.: 1926, *Phys. Rev.* **28**, 727.
Lyttleton, R. A.: 1953, *The Comets and Their Origin*, Cambridge University Press, p. 66.
McCrea, W. H.: 1955, *IAU Symposium No. 2, Gas Dynamics of Cosmic Clouds*, (ed. by H. C. van de Hulst and J. M. Burgers), North-Holland Publishing Company, Amsterdam, p. 186.
Mestel, L.: 1954, *Monthly Notices Roy. Astron. Soc.* **114**, 437.
Morgan, W. W., Strömgren, B. and Johnson, H. M.: 1955, *Astrophys. J.* **121**, 611.
Ozernoi, L. M. and Zasov, A. V. (in preparation).
Parker, E. N.: 1958, *Astrophys. J.* **128**, 664.
* Su, C. H. and Lam S H.: 1963, *Phys. Fluids* **6**, 1479.
Ward, G. N.: 1955, *Linearized Theory of High-Speed Flow*, Cambridge University Press.
Weymann, R.: 1962, *Astrophys. J.* **136**, 476.
Wickramasinghe, N. C., Donn, B. D. and Stecher, T. P.: 1966, *Astrophys. J.* **146**, 590.

13. DISCUSSION FOLLOWING SPIEGEL'S REPORT

(Thursday, September 18, 1969)

Chairman: I. S. SHKLOVSKII

[The text of this Chapter contains the first part of the Discussion on Thursday, September 18, 1969; the second part is given in Chapter 26. I added a long contribution on cosmogonical problems, presented by Dr. Ozernoi on September 10.]

Weymann: Dr. Spiegel, did you assume a (more or less) uniform distribution of matter throughout the Coma-cluster?

Spiegel: Yes. In the theory I have, one does the hydrodynamics as if the gas were adiabatic and then one calculates the radiation from the resulting conditions. So far I do not have a theory which includes radiation *ab initio*.

Weymann: What I worry about is whether or not the gas is condensed towards the center. Since the cooling rate is proportional to n^2 (n is the density) it is conceivable to me that an almost steady state is possible in which the gas in the center of the cluster is much cooler and much denser.

Spiegel: Quite so.

Ozernoi: In his exciting Report, Spiegel mentioned the interaction between galaxies and intergalactic gas. The importance of such interactions for the energy balance of radio galaxies was first suggested, as far as I know, by Shklovskii. I should like to discuss a recent paper (Ozernoi, 1969) on mass accretion of quasars and other systems. Consider Hubble's famous 'tuning-fork' diagram, which shows the morphological sequence of galaxies from E to Irr, including the branching into barred and non-barred spirals. I extended this fork by adding quasars and similar sources and got a sequence QSS-N-BCG-D-E-S0-Sa, etc. [BCG stands for blue compact galaxy. (Ed.).] In this sequence, from right to left, both the radio and the optical luminosities increase; the specific angular momentum decreases. I think that the source of energy [in the series QSS through D? (Ed.)] is an object in the nuclei of such galaxies, a magnetoid, i.e., a very massive magneto-dynamic plasma configuration rotating in a magnetic field, analogous to a giant pulsar. The magnetodynamic model can explain the physics of quasars in much detail. However, the question remains, what final reservoir supplies the central parts of galaxies with the required mass of gas? A first possible reservoir is the non-fragmented gas in the galaxies themselves. Between two active phases of the nucleus, cooling of the gas and condensation toward the center, may supply gas to the nucleus. A second possible reservoir is accretion of an intergalactic gas cloud by a galaxy. The rate of accretion depends on the relation between the velocity of a galaxy V and the parabolic velocity on its border $V_p \propto (M/R)^{1/2}$. For a galaxy of given morphological type the mean density $\langle \varrho_g \rangle =$ const, i.e., $R \propto M^{1/3}$, and the radius of accretion $R_A \propto M$. Consequently, for galaxies of small masses, $R > R_A$ and the rate of accretion

$dM/dt \propto M^{2/3}V$; meanwhile for most massive galaxies $R < R_A$, and $dM/dt \propto M^2 V^{-3}$. The parameters M, R, V, which characterize a galaxy, cannot be chosen freely but are determined by the cosmological past of a galaxy. The analysis shows that in all real situations $dM/dt \propto M^{\alpha}$ with $\alpha > 0$. Because the mean mass of a galaxy increases from Irr to E galaxies and, possibly, along all the extended sequence (from right to left), the rate of accretion would increase in the same direction. The accretion rate also depends on whether or not a galaxy is a member of a cluster. Field galaxies have velocity dispersions smaller by a factor of about 10 than galaxies in rich clusters. As a result the rate of accretion for galaxies of small masses would be greater inside clusters; meanwhile for most massive galaxies the rate of accretion would be greater outside clusters. This explains, possibly, the primary appearance of quasars and quasar-like phenomena outside clusters. After the gas has been captured by a galaxy, the rate at which it condenses into the center depends on the specific angular momentum of the galaxy. One expects that the rate of condensation would be greater, the smaller the specific angular momentum. Because the latter increases along the sequence from left to right, the luminosity of nonthermal emission (which is connected with transformed kinetic energy) would decrease in the same direction. The pressure of both radiation and relativistic particles as well as the outflow of the plasma from the nucleus restrict the inflow of metagalactic gas. Its condensation would repeat between active phases of the nucleus and, in principle, can explain the recurrent character of phenomena in quasars and the nuclei of galaxies. (Ozernoi, L. M.: 1969, *Astron. Tsirk. No. 469*.)

[The following part of the discussion took place on September 10.]

Ozernoi: *The Traces of 'Photon Whirls'*.

It is generally assumed that the observed inhomogeneities of the Universe (galaxies and galaxy clusters) originated as a result of gravitational instability of initially homogeneous matter with small, but finite (non-fluctuating) density perturbations. I should like, however, to discuss a different assumption, namely that the initial structure was not static, but dynamic: I assume that initially whirl (vortex) motions, both of the photon gas and the plasma, are superimposed on the cosmical expansion. I assume that vortex velocities are subsonic and (on sufficiently large scales) lead to a Kolmogorov spectrum, which is independent of the initial turbulence spectrum. The interval of k (the wave number) over which the Kolmogorov spectrum holds is determined at the lower side by the condition that the hydrodynamic time equals the cosmological time. When in the expansion the moment comes that the metagalactic plasma recombines, the sound velocity becomes significantly lower and the (originally subsonic) whirls are suddenly in the supersonic regime. This leads to motions and inhomogeneities in the distribution of matter, even if originally this distribution was uniform. Depending on whether the inhomogeneities are gravitationally bound or not, predictions can be made for (i) the future velocity spectrum, (ii) the possibility that internal motions will remain smaller than the differential velocity of cosmological expansion, (iii) the spectrum of the density inhomogeneities, and (iv) the density contrast between the inhomogeneities and the background. Comparing our predictions with observations of groups and clusters of galaxies (Karantchentsev, 1966), we conclude that there is a fair agreement between predictions and observations, provided that clusters of galaxies represent, on the average, gravitationally bound systems, except on the size scale of superclusters. The fact that there is agreement supports the basic assumption of the existence of primeval cosmological turbulence. More details may be found elsewhere (Ozernoi and Chernin, 1967, 1968). [Comment condensed. (Ed.).] [Karantchentsev, I. D.: 1966, *Astrofiz.* **2**, 81 (translation: 1966, *Astrophys.* **2**, 39); Ozernoi, L. M. and Chernin, A. D.: 1967, *Astron. Zh.* **44**, 1131 (translation: 1968, *Soviet Astron.* **11**, 907); 1968, *Astron. Zh.* **45**, 1137 (translation: 1969, *Soviet Astron.* **12**, 901).]

Spiegel: I do not understand why you would expect vortex motions. In an inviscid fluid the Hamiltonian density contains both longitudinal and rotational terms. Is it not true that, as in a superfluid, the lowest excitations are the longitudinal ones?

Zel'dovich: The longitudinal quanta will be smaller than the rotational quanta. So in thermodynamic equilibrium you will have phonons but not rotons. But in our primeval Universe there is no thermodynamic equilibrium. It is simply assumed that at the origin you have vortices and no strong sound wave motions. Ozernoi is not bound to a situation of thermal equilibrium at the beginning of the Universe. Therefore, your remark, Dr. Spiegel, seems to be irrelevant. There are other difficulties with Ozernoi's theory. I think that at the moment of chemical element formation it is incompatible with the Friedman model. Perhaps the theory could be made compatible by assuming many neutrinos.

Van de Hulst: Oort has recently also been interested in the formation of galaxies; and, like you, he traces it back to gas-dynamic motions in an earlier epoch. He told me that he was much impressed by the fact that in many cases the galaxies seem to form strings. Do you think that this structure is important? Does it fit into your theory?

Ozernoi: It seems to me that such clustering of galaxies is connected with the details of space correlation in the primeval whirl structure. According to the whirl model this correlation is the main reason for the formation of the clusters of galaxies. It explains some important features of the clusters, for instance, the relationship between the mean density of a cluster and the predominant morphological type of its galaxies.

Colgate: Dr. Ozernoi, can you suggest what could have caused the turbulence in the first place?

Zel'dovich: Dr. Colgate, I should like to ask you: from where came our Universe?

PART II

INTERACTION OF STARS AND INTERSTELLAR MEDIUM

14. MASS BALANCE OF INTERSTELLAR GAS AND STARS

Introductory Report

(Saturday, September 13, 1969)

E. E. SALPETER*

*Laboratory of Nuclear Studies, Physics Department, and Center for Radiophysics and Space Research,
Cornell University, Ithaca, N.Y., U.S.A.*

1. Introduction

In this paper I will deal almost exclusively with the interchange of matter between the interstellar gas and the stars. Other possible forms of interchange, such as mass-loss to and accretion from intergalactic matter *e.g.*, Oort's (1969) views on mass flowing into our Galaxy, will be discussed later I hope.

One topic of interest is the empirical study of the present-day rates of the mass interchange between gas and stars in our Galaxy. The present value of the birthrate of massive, short-lived stars can be obtained directly from the observed luminosity function and from the theoretically known lifetimes of stars on the main sequence. The rate at which mass is returned from ageing stars to the gas can also be obtained in principle, but this requires a theoretical knowledge of *late* stages of evolution of stars and is quantitatively less reliable. These rates are discussed in Section 2.

In Section 3 we review theoretical models of a semi-empirical kind for the evolution of our Galaxy, in particular for the variation with time of the stellar birthrate function and of the total mass of the interstellar gas $M_{gas}(t)$. We also discuss briefly the evolution of other types of galaxies from the point of view of such models. Such models are of interest for two reasons. First, they are required to infer the present birthrate of stars of low mass, whose main-sequence lifetimes are very long. Second, they should shed some light on the correlation between the birthrate function and some physical variables, such as the mean gas density (and possibly angular momentum, chemical composition, etc.). Unfortunately we shall find too many unknowns and too few observations at present for any definitive conclusions on such models. In Section 4 I shall review briefly physical theories which attempt to derive the mass interchange between gas and stars from first principles.

This paper should be considered as an introduction to the topics, not a review of them. Few references will be made to the recent literature, especially for the topics of Section 3, for the simple reason that I am not sufficiently familiar with it. Nevertheless, I hope that even a dated introduction can stimulate discussion.

2. Present Rates

Throughout this paper we shall be concerned with the birthrate function $\xi(M, t)$

* Dr. L. Mestel read this paper for Dr. Salpeter who, for personal reasons, could not be present.

Habing (ed.), Interstellar Gas Dynamics, 221–228. All Rights Reserved. Copyright © 1970 by IAU

which determines the rate at which stars of mass M are formed out of the interstellar gas at time t after the formation of the Galaxy. We define this function so that the number of stars dN formed in mass-interval dM and time-interval dt is given by

$$dN = \xi(M, t)\frac{dM}{M}\, dt\, M_{gas}(t),\tag{1}$$

where M_{gas} is the total mass of the gas in which the stars are formed. If

$$f(t) = \int_0^\infty \xi(M, t)\, dM\tag{2}$$

then f^{-1} is the exponential decaytime of the unprocessed gas. One question we shall ask is how strongly ξ and hence f depend on time t.

The present age t_0 of our Galaxy lies in the range of $(8 \text{ to } 20) \times 10^9$ yr, probably close to 10×10^9 yr (Rood and Iben, 1968). We shall express time t in units of t_0 and shall find that the precise value of t_0 is important only for a few considerations. For the total mass M_{Gal} of our Galaxy we adopt a value of $1.0 \times 10^{11}\, M_\odot$ (Inanen, 1966). The fractional mass in gaseous form, $M_{gas}(t)/M_{Gal}$, is also of interest and we assume the present value of this ratio to be about 0.2 in the vicinity of the Sun (integrated over a column perpendicular to the galactic disk) and about 0.04 for the whole Galaxy.

The most important observational datum for our present purposes is the present luminosity function φ for main-sequence stars near the Sun (also integrated over a column perpendicular to the galactic disk). At least for stellar masses M in the range $0.2\, M_\odot$ to $10\, M_\odot$, we know the visual magnitude, bolometric luminosity and lifetime t_{MS} on the main sequence as a function of M. It is convenient to discuss separately three mass-ranges of stars. For massive stars, $M \gtrsim 2\, M_\odot$, the main-sequence lifetime t_{MS} (which is roughly proportional to M^{-3}) is very much shorter than the present age t_0 of the Galaxy. Changes of the birthrate function $\xi(M, t)$ or of the gas-mass over the lifetime of one star can then be neglected. With the present luminosity function φ also re-expressed per logarithmic stellar mass interval and per gas mass, this gives

$$\varphi(M) = t_{MS}\xi(M, 1).\tag{3}$$

For stars of low mass on the other hand, $M \lesssim 0.5\, M_\odot$, we have $t_{MS} \gg t_0$. In this case stellar evolution can be neglected but galactic evolution enters the relation between φ and ξ,

$$\varphi(M) = \int_0^{t_0} dt\, \xi(M, t)\, [M_{gas}(t)/M_{gas}(1)].\tag{4}$$

For the intermediate mass range, $(0.5 \text{ to } 2)\, M_\odot$, the relationship is more complicated.

The relation between ξ and φ (each multiplied by M, on an arbitrary scale) is shown schematically in Figure 1. The original suggestion by Salpeter (1954) of Equation (3) is gratifying in two ways: (i) the combination $M\varphi/t_{MS}(M)$ is a slowly-varying

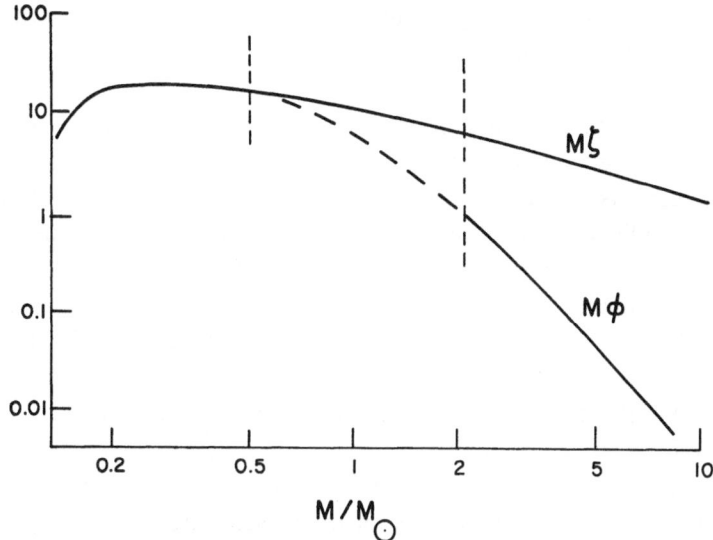

Fig. 1.   A schematic plot of the present luminosity function φ and of the birthrate function ξ (on a logarithmic scale) as a function of stellar mass M.

function of M for the massive stars, as is $M \varphi$ for the stars of low mass; (ii) the shape of the observed luminosity function for massive stars in very young clusters (Sandage, 1957; Van den Bergh, 1957) (where even massive stars have not yet evolved away from the main sequence) agrees with the shape of ξ (as given by Equation (3) for the vicinity of the Sun).

For the massive stars, $M \gtrsim 2 \, M_\odot$, observations plus Equation (3) give the present value of the birthrate function ξ (in the solar vicinity) with little ambiguity (but we have no direct information on values of ξ at earlier epochs). For the low-mass stars, on the other hand, Equation (4) only gives a weighted time-average of ξ and a model for the time-dependence is needed to infer the present value of ξ. Adopting such a model (discussed in the next section) one can then estimate $\xi \, (M, 1)$ for all M in the solar vicinity. If one uses the same value of $\xi \, (M, 1)$ throughout the Galaxy, the total rate of processing mass from interstellar gas into stars is about $1 \, M_\odot \, \text{yr}^{-1}$ for the whole Galaxy, i.e. about $2.5 \, M_{\text{gas}}/t_0$. This value is uncertain by factors of about 2 or 3, partly because of the uncertainty in the models for galactic evolution and partly because φ (and hence ξ) for stars of very low luminosity and mass $(M < 0.1 \, M_\odot)$ is poorly known (Luyten, 1968).

One difficulty we face is that we do not know how typical the solar vicinity is for the overall pattern of the interstellar gas and dust in other parts of the galactic disk. The *mean* density of neutral atomic hydrogen is not very different in regions closer to the galactic nucleus (where the density of stars is very much higher), but we have little direct information from these regions on dust density and on the nature of density fluctuations (regarding chemical composition, see Section 3). Since we do not even know in what direction these uncertainties affect the rate of star formation, we had to assume a uniform value for $\xi(M, 1)$ in the estimate above. Another unsolved and important question (especially regarding chemical composition) is whether thorough interchange (or net flow) of interstellar gas between various regions of the galactic disk can take place in timescales of the order of 10^{10} years.

We consider next the other aspect of the gas-star mass-interchange, the matter put back into the interstellar gas from highly evolved stars (see also in this volume the Introductory Reports by Pottasch, p. 272, and by Boyarchuk, p. 281). There are a number of possibilities: (i) continuous mass-loss (in the red-giant stage or in later evolutionary stages) leads to a white dwarf as the remnant star. Supernovae explosions could lead to a remnant core in the form of (ii) a white dwarf or (iii) a neutron star or (iv) a collapsed 'invisible' object. Observed masses of white dwarfs (Weidemann, 1968) lie mainly in the range of (0.4 to 1) $M_\odot$, with 0.6 $M_\odot$, or slightly larger (Greenstein and Trimble, 1967), a 'typical' value. Neutron star masses can only be estimated theoretically but are likely to be comparable to white dwarf masses. There is no upper limit to the mass of collapsed objects and in case (iv) it is possible that little mass returns to the interstellar gas. Estimates for the present rate of supernovae explosions (Minkowski, 1964; Katgert and Oort, 1968) vary from one per 300 years to one per 30 years, but even the higher rate would represent only about half of the present rate of star deaths. For the present Galaxy we shall therefore assume that for a star of initial mass M an amount $(M - 0.6 M_\odot)$ is eventually returned to the interstellar gas.

For the star deaths it is again convenient to discuss the three separate ranges of stellar mass. The low-mass stars, $M \lesssim 0.5 M_\odot$, have not evolved from the main sequence and mass loss from these stars can be neglected. For the massive stars, $M \gtrsim 2 M_\odot$, the uncertainty in the returned mass $(M - 0.6 M_\odot)$ due to uncertainty in the mass of the remnant star is small. Further, the time delay between birth and death for these massive stars is short, so that the rate of mass return from them is proportional to the present gas mass. This contributes about 0.5 M_{gas}/t_0 (about 0.2 $M_\odot$ yr^{-1} for the whole Galaxy) or about one fifth of the rate of mass transfer from the gas to stars.

The situation is more complicated and more uncertain for the stars of intermediate mass. Uncertainties in the present age of the Galaxy and inaccuracies in stellar evolution calculations lead to an uncertainty in the initial mass of Population II stars (of age near t_0) which have evolved past the red-giant stage. The value lies in the range (0.8 to 1.0) $M_\odot$ (Rood and Iben, 1968). Since white-dwarf masses are not much smaller and are also slightly uncertain, there is considerable uncertainty in the average mass returned to the gas from old stars of intermediate mass. For the solar vicinity this rate, and hence its uncertainty, is not very great and the total rate of mass conversion from stars back into gas is probably 0.3 to 0.4 times the rate from gas into stars. There is thus no doubt that the net result in the solar vicinity is mass drainage away from the gas. For the Galaxy as a whole the ratio of old stars to gas is very much greater, the mass interchange from gas to stars and back again is more nearly balanced (Partridge and Peebles, 1967). (The backward rate is probably 0.5 to 1 times the forward rate.) There is then the possibility that near the galactic nucleus the net result is actually a slight enhancement of the interstellar gas from the stars (with approximate balance most likely).

3. Evolutionary Models

We review next evolutionary models of a galaxy, i.e. the time-dependence of the birthrate function $\xi(M, t)$ and of the mass in gaseous form $M_{gas}(t)$. There clearly is not enough observational data to determine ξ uniquely, so one has to assume certain forms and study their consequences.

Salpeter (1959) investigated in detail the consequences of the simplest form, namely a time-independent 'universal' birthrate function $\xi(M)$. The quantity f in Equation (2) is then also time-independent and the fraction of the galactic mass in gaseous form would simply decay exponentially as e^{-ft} if there were no transfer of mass back from stars to gas. With this transfer included, the gas mass $M_{gas}(t)$ for the whole Galaxy drops rapidly at first as a function of time and then almost levels off. The next simplest assumption keeps a 'universal shape' for the birthrate function but allows the overall rate function $f(t)$ in Equation (2) to depend on time through the gas density. Schmidt (1959) discussed the specific form

$$\xi(M, t) = \xi(M)\left[M_{gas}(t)/M_{gas}(1)\right]^{n-1}. \tag{5}$$

These two papers suggest that $f(t)$ must have decreased with time. The data are probably compatible with $n \approx 2$ in Equation (5), although some arguments have been advanced (Reddish, 1962) against a unique dependence of the form of Equation (5) with $n > 2$.

The assumption of a 'universal shape' of $\xi(M, t) = \xi(M)f(t)$ has encountered difficulties (Limber, 1960). One manifestation is a difference in the shape of the observed luminosity function between star clusters (young and old) and the solar vicinity in the mass-range from $0.2 M_\odot$ to $0.8 M_\odot$: in the clusters the stars of lowest mass are less dominant, even though effects of stellar evolution are negligible in this mass-range. This discrepancy may in part be due to evaporation of stars from clusters, but at any rate clusters cannot furnish a 'universal function' and the time-dependence of $f(t)$ cannot be evaluated quantitatively at the moment.

Two important and relevant topics, which I will only mention but not discuss, are the helium abundance in the interior and surface of various stars and the spectroscopically-determined metal abundance Z in the surface of various stars (Cayrel and Cayrel-de Strobel, 1966). Very detailed correlations exist (Eggen et al., 1962) between Z and the kinematics and age of a star, as expected from the continual enrichment of the interstellar gas from star deaths. The data indicates that the Galaxy started without any metals and that an appreciable fraction of the Population II stars formed (Eggen et al., 1962) in much less than 10^9 years. Statistical data on Z also provides evidence against the universality of the birthrate function: an argument by Schmidt (1963) (independent of the quantitative time-development) shows that the earliest star formation favored stars which were more efficient at metal production, which probably means massive stars.

An interesting but unsolved question for these evolutionary models is whether they can give the helium abundance in the Sun and in the present-day gas in terms of He formation in stars (assuming pure H for the proto-Galaxy). The answer depends on details of mass-loss from various stars and on the age and state of mixing of the Galaxy. If t_0 is appreciably larger than 10×10^9 years and if the chemical composition is fairly uniform throughout the Galaxy, the answer is yes (Truran *et al.*, 1965). If $t_0 \lesssim 10 \times 10^9$ years and if the interstellar gas is unmixed, much of the helium in the solar vicinity would have to be primordial (but with appreciable He enrichment from stars near the galactic nucleus). There is conflicting evidence from the analysis of individual old stars, on the actual helium abundance of the primordial gas (Burbidge, 1969).

For other types of galaxies the present mass-ratio of gas to stars increases strongly from elliptical through spiral to irregular galaxies. The different galaxy types are not believed to be due to different ages (Tinsley, 1968) but must reflect differences in the birthrate function $\xi(M, t)$. With what physical properties of different galaxies are these differences correlated? It has been suggested that the generally higher total mass density in elliptical galaxies is the main cause (Holmberg, 1967; see also Fish, 1964; Salpeter, 1965). This requires a birthrate function positively correlated with mean gas density (such as Equation (5) with $n \gtrsim 2$) and predicts a present absolute gas density almost the same for different types of galaxies. This is more or less the case and, although there are some difficulties with Holmberg's hypothesis, there seem to be no attractive alternative suggestions for correlations with other physical variables (Reddish, 1968). Comparison between elliptical galaxies and the extreme stellar Population II in our Galaxy again indicates a lack in universality of the birthrate as a function of mass: although both represent old stars with little gas present, the mass-to-light ratio is appreciably greater for the elliptical galaxies. This indicates a preponderance of stars of low luminosity and presumably low mass. However, this could be due to either of two different features of $\xi(M)$ for ellipticals (i) a preponderance of faint main-sequence stars of very low mass ($M < 0.4 M_\odot$) or (ii) predominant production at early times of very massive stars ($M > 2 M_\odot$) which had a high luminosity then, but have become low-luminosity white dwarfs or neutron stars since.

To summarize the evidence on the birthrate function $\xi(M, t)$ and its integral $f(t)$, as defined in Equations (1) and (2): f has been decreasing with time in our Galaxy as the total gas mass has been decreasing. One cannot be quantitative yet, but f might well be proportional to the average gas density $\bar{\varrho}$ i.e. star formation rate per unit volume might be proportional to the square of $\bar{\varrho}$ (or possibly a slightly weaker dependence on $\bar{\varrho}$). The distribution of the birthrate among stars of different mass cannot be 'universal'; the more massive stars are probably favored at early times when $\bar{\varrho}$ was higher. The correlation of the birthrate function with chemical composition, turbulence, and angular momentum of the interstellar gas or with magnetic field intensity and cosmic-ray fluxes (which also may vary with time) is not yet known.

4. Physical Theories

I shall briefly review Oort's (1954) physical picture of small gas clouds coalescing, followed by star formation, followed by bright stars re-dispersing small clouds. Such physical theories are slightly modified in the light of more modern ideas (Pikel'ner, 1967; Field *et al.*, 1969) on static pressure equilibrium between interstellar clouds and a hotter, partially-ionized medium in H I regions [see the Report by Field, p. 51 and the remark by Pikel'ner, p. 359 (Ed.)].

When an interstellar gas cloud becomes large enough for gravitational instability, some stars are formed during the contraction and fragmentation. Some of these newly-formed stars are hot enough to produce copious ionization and dynamic effects. The dynamic effects are pictured as the dispersal of small clouds with velocities comparable with that of a hydrogen atom of kinetic energy equal to its ionization potential, $V \approx 51$ km sec^{-1}. If all the clouds coalesce after colliding (and remain at constant internal density) until they reach a critical mass for star formation, the mass-spectrum (Field and Saslaw, 1965) and velocity distribution (Penston *et al.*, 1969) of the clouds can be predicted. Such estimates are in reasonably good agreement with observation, although the observed velocity dispersion decreases more slowly with increasing cloud mass then predicted.

The overall rate of star formation as a function of average gas density can be calculated with such a picture only if an assumption is made on how the internal density ϱ_i in a cloud depends on $\bar{\varrho}$, the overall gas density. If one assumes with Field and Saslaw (1965) that ϱ_i is independent of $\bar{\varrho}$ (assuming $\varrho_i \gg \bar{\varrho}$, of course), then the rate per unit volume of cloud-cloud collisions and hence of star formation is proportional to $\bar{\varrho}^2$. If the thickness of the galactic disk did not change with time, then $\bar{\varrho}$ decreased proportionally to M_{gas} and the picture would lead to $n=2$ in Equation (5). However, to dissipate enough energy by radiation after a cloud-cloud collision, the temperature (and possibly magnetic field strengths) would have to be correlated with $\bar{\varrho}$. The thickness of the galactic disk might also have evolved slightly with time.

Attempts at deriving the birthrate ξ as a function of stellar mass M from first principles have so far been made only for one rather specific model (Reddish and Wickramasinghe, 1969) − fragmentation in clouds which have been cooled to about 3 K by efficient grain radiation. This model predicts $\xi \propto M^{-(1 \text{ to } 1.5)}$ within a certain range of masses, in rough agreement with observation. On this model, and probably more generally, the dominant masses of the forming stars are inversely correlated with the internal density in the condensed cloud at the onset of fragmentation. Unfortunately, it is not clear how this density is related to the density ϱ_i of a cloud before its gravitational contraction, or to the mean gas density $\bar{\varrho}$ of the galactic disk.

Acknowledgements

This work was supported in part by the Office of Naval Research and by a National Science Foundation grant (GP 9621).

References

Burbidge, G. R.: 1969, *Comments Astrophys. Space Res.* **1**, 101.
Cayrel, R. and Cayrel-de Strobel, G.: 1966, *Ann. Rev. Astron. Astrophys.* **4**, 1.

Eggen, O., Lynden-Bell, D., and Sandage, A.: 1962, *Astrophys. J.* **136**, 748.

Field, G. B. and Saslaw, W. C.: 1965, *Astrophys. J.* **142**, 568.

Field, G., Goldsmith, D., and Habing, H.: 1969, *Astrophys. J. Lett.* **155**, L149.

Fish, R. A.: 1964, *Astrophys. J.* **139**, 284.

Greenstein, J. L. and Trimble, V. L.: 1967, *Astrophys. J.* **149**, 283.

Holmberg, E.: 1967, *Ark. Astron.* **3**, 387.

Inanen, K. A.: 1966, *Astrophys. J.* **143**, 153.

Katgert, P. and Oort, J. H.: 1968, *Bull. Astron. Inst. Netherl.* **19**, 239.

Limber, N.: 1960, *Astrophys. J.* **131**, 168.

Luyten, W. J.: 1968, *Monthly Notices Roy. Astron. Soc.* **139**, 221.

Minkowski, R.: 1964, *Ann. Rev. Astron. Astrophys.* **2**, 247.

Oort, J. H.: 1954, *Bull. Astron. Inst. Netherl.* **12**, 177.

Oort, J. H.: 1969, *Astron. Astrophys.*, in press.

Partridge, R. and Peebles, P. J.: 1967, *Astrophys. J.* **148**, 377.

Penston, M. V., Munday, A., Strickland, D., and Penston, M. J.: 1969, *Monthly Notices Roy. Astron. Soc.* **142**, 355.

Pikel'ner, S.: 1967, *Astron. Zh.* **44**, 1915.

Reddish, V. C.: 1962, *Observatory* **82**, 14.

Reddish, V. C.: 1968, *Quart. J. Roy. Astron. Soc.* **9**, 409.

Reddish, V. C. and Wickramasinghe, N. C.: 1969, *Monthly Notices Roy. Astron. Soc.* **143**, 139, 189.

Rood, R. and Iben, I.: 1968, *Astrophys. J.* **154**, 215.

Salpeter, E. E.: 1954, *Astrophys. J.* **121**, 161.

Salpeter, E. E.: 1959, *Astrophys. J.* **129**, 608.

Salpeter, E. E.: 1965, in *The Structure and Evolution of Galaxies* (13th Conference of the Solvay Institution of Physics, Brussels), Interscience, New York, p. 71.

Sandage, A.: 1957, *Astrophys. J.* **125**, 422.

Schmidt, M.: 1959, *Astrophys. J.* **129**, 243.

Schmidt, M.: 1963, *Astrophys. J.* **137**, 758.

Tinsley, B. M.: 1968, *Astrophys. J.* **151**, 547.

Truran, J., Hansen, C., and Cameron, A. G. 1965, *Can. J. Phys.* **43**, 1616.

Van den Bergh, S.: 1957, *Astrophys. J.* **125**, 445.

Weidemann, V.: 1968, *Ann. Rev. Astron. Astrophys.* **6**, 351.

15. SUPERNOVAE AND THE INTERSTELLAR MEDIUM

Introductory Report
(Saturday, September 13, 1969)

L. WOLTJER

Dept. of Astronomy, Columbia University, New York, N.Y., U.S.A.

1. Introduction

The nature of the supernova event is still poorly understood. A variety of models has been proposed and the interpretation of the observations remains ambiguous. About all that is certain is that in the supernova event much of the matter is ejected at speeds of typically $10\,000$ km sec^{-1} and that about 10^{49} erg of visible light is emitted. The amount of matter ejected and the bolometric correction to be applied to the optical radiation are very uncertain. The discovery of a pulsar in two supernova remnants (Crab Nebula and Vela X) suggests that frequently a neutron star or other condensed object results following the outburst.

In most models it is assumed that some instability near the end of the evolution of a star causes the interior to collapse; and that, following this, the envelope is driven out, possibly by the pressure from neutrinos generated in the core. Colgate and White (1966) have made a detailed study of the effect of this pressure pulse on the envelope. A strong, essentially adiabatic shock moves out and gains speed because of the steep density gradient. The outermost parts of the star come off at relativistic speeds, so that the whole event may be important as an acceleration mechanism for cosmic rays. Pacini (1967) and several authors following him have drawn attention to the large rotational energies expected in neutron stars and the possibility that electromagnetic waves generated by rotating magnetic neutron stars could energize supernova remnants even after the initial explosion.

2. Simple Hydrodynamical Expansion

To isolate some of the more important aspects of the expansion of the supernova remnant we shall assume that in the supernova explosion at some initial moment an amount of energy ε_0 is released in a small volume and imparted to a mass of gas M_0 which acquires an initial expansion velocity V_0. We treat the system as spherically symmetric and consider its interaction with the surrounding interstellar medium (hydrogen density per cm^3 n_{is}, mass of hydrogen atom m). We suppose for the moment that the interaction can be described in macroscopic hydrodynamical terms and that magnetic field and relativistic particle effects can be neglected. Four phases of the expansion can be distinguished.

Phase I — As long as $(4\pi/3)n_{is}mR^3 \ll M_0$, the expansion of the remnant will be essentially free and much of the developments will depend on the initial conditions.

Habing (ed.), Interstellar Gas Dynamics, 229–235. All Rights Reserved.

Phase II — As the remnant expands, the mass of the interstellar matter that has been swept up will gradually increase and exceed M_0. As first pointed out by Shklovskii (1962), we then enter the phase where the behavior of the expanding gas can be described by the simple similarity solution discussed by Sedov (1959). In this solution an amount of energy ε_0 (but no mass) is injected into a uniform medium at time $t=0$. A shock propagates into the undisturbed medium behind which the gas is compressed. As the amount of matter swept up becomes very large compared to M_0, the assumption of no initial mass ejection becomes less serious. In the similarity solution the ratio of thermal to kinetic energy is constant in time and we may write for the total energy

$$\varepsilon = \varepsilon_0 = \eta \tfrac{2}{3}\pi n_{is} m R^3 V^2, \tag{1}$$

with R and V the radius and velocity of the shock. For a gas with $\gamma=\tfrac{5}{3}$ we have $\eta=1.37$ from Sedov's solutions. Of course the solution is only valid if $(4\pi/3)n_{is}mR^3 \gg M_0$. Integrating Equation (1) with $R=0$ at $t=0$ we have for $\gamma=\tfrac{5}{3}$

$$R = \left(\frac{75}{8\pi}\right)^{1/5} \left(\frac{\varepsilon_0}{n_{is}m\eta}\right)^{1/5} t^{2/5} = \tfrac{5}{2}Vt. \tag{2}$$

The shock separating the hot gas from the undisturbed medium is strong. Consequently the density behind the shock front n^* is equal to $4n_{is}$, while for the temperature T^* we have

$$T^* = \frac{3}{32}\frac{m}{k}V^2. \tag{3}$$

The hot gas will radiate and by the time that the integrated radiation becomes comparable with ε_0 the adiabatic solution can no longer be valid. At high temperatures ($T^* \gg 10^6$ K) free-free radiation of hydrogen dominates for material of interstellar composition. The emission per unit mass is proportional to $nT^{1/2}$. With $T \propto V^2 \propto t^{-6/5}$ and the total mass $M \propto R^3 \propto t^{6/5}$, the total radiative loss rate is proportional to $t^{3/5}$ and the time integrated loss to $t^{8/5}$ (cf. Shklovskii, 1968). Hence the loss is small at early times (see also Heiles, 1964) and it is easily verified that for a wide range of energies the nonadiabatic effects are negligible until temperatures near or below 10^6 K are reached. But at those temperatures radiation by heavy elements becomes much more important than that by hydrogen. Pottasch (1965) has computed the total radiative loss from a plasma of standard composition which is ionized and excited collisionally. Over the interval $10^5 < T < 4 \times 10^6$ the loss rate per unit volume can be represented reasonably well (to within a factor of 2) by

$$\begin{aligned} j &= Qn_e^2 T^{-1} \\ Q &= 8 \times 10^{-17} \text{ cgs}. \end{aligned} \tag{4}$$

From what we said above about the unimportance of the free-free emission in the early phases, it is clear that negligible error is incurred in our calculations if we make use of Equation (4) also at higher temperatures. We shall estimate the radiative loss rate assuming that all matter is concentrated in a shell with volume $(\pi/3)R^3$, electron density $n_e = 4n_{is}$ and temperature T^*. This results in an overestimate, because behind the shock the density actually decreases while the temperature goes up. The radiative loss rate of the remnant now becomes

$$\frac{d\varepsilon}{dt} = -\frac{\pi}{3} R^3 Q \, 16n_{is}^2 \frac{32}{3} \frac{k}{m} \frac{1}{V^2}$$

or making use of Equation (2) and introducing a suitable constant $\mathfrak{A}$ (equal to 9.3×10^{-2})

$$\frac{d\varepsilon}{dt} = -\frac{3200}{9} \left(\frac{75}{8\pi}\right)^{1/5} \frac{\pi k Q}{m^{6/5} \eta^{1/5}} n_{is}^{9/5} \varepsilon_0^{1/5} t^{12/5} = -\frac{17}{10} \mathfrak{A} n_{is}^{9/5} \varepsilon_0^{1/5} t^{12/5}, \qquad (5)$$

corresponding to a time integrated loss of

$$\int_0^t \frac{d\varepsilon}{dt} dt = -\frac{1}{2} \mathfrak{A} n_{is}^{9/5} \varepsilon_0^{1/5} t^{17/5}. \qquad (6)$$

The validity of the adiabatic solution requires

$$\int_0^t \left|\frac{d\varepsilon}{dt}\right| dt \ll \varepsilon_0. \qquad (7)$$

Denoting the variables at the time when the radiative loss is equal to $\frac{1}{2}\varepsilon_0$ by t_{rad}, R_{rad}, and V_{rad} we have

$$t_{rad} = \mathfrak{A}^{-5/17} \varepsilon_0^{4/17} n_{is}^{-9/17}$$
$$R_{rad} = (15/4\pi)^{2/5} (\eta m)^{-1/5} \mathfrak{A}^{-2/17} \varepsilon_0^{5/17} n_{is}^{-7/17} \qquad (8)$$
$$V_{rad} = (2/5)(15/4\pi)^{2/5} (\eta m)^{-1/5} \mathfrak{A}^{3/17} \varepsilon_0^{1/17} n_{is}^{2/17}.$$

Poveda and Woltjer (1968) derived analogous equations. They used a different criterion to estimate the validity of the adiabatic solution, namely that per unit volume of the shocked gas the thermal energy exceeds R/V times the radiative loss. This criterion leads to the same results for t_{rad}, if the constant Q (and therefore also $\mathfrak{A}$) is multiplied by a factor of 3, a value of V_{rad} that differs by about 25 per cent, and a value of R_{rad} that differs even less.

Inspecting Equation (8) we note that V_{rad} in particular is very insensitive to the physical parameters involved.

Phase III — When the radiative losses become dominant, the structure of the object changes. The matter that passes through the shock front cools rapidly and the density becomes high. In this phase the remnant can be described as a rather thin shell ploughing through the interstellar medium. The thermal energy in the object now is small and we may consider the shell to move with constant linear momentum. Consequently, as first discussed by Oort (1951), the equation of motion simply becomes

$$R^3 V = C_1 \qquad (9)$$

or upon integration

$$R = (4C_1 t + C_2)^{1/4} \tag{10}$$

with C_1 and C_2 constants.

Phase IV — After the shell has slowed down for some time the expansion velocity becomes comparable to the thermal or random motions in the surrounding interstellar gas. The object gradually loses its identity and becomes part of the interstellar medium.

3. Effects of Relativistic Particles

If in Phase I or II relativistic particles are present and coupled effectively to the gas, the situation will not change too much. During the expansion the relativistic particles will quickly lose their energy, transferring it to the gas and the final situation will not be very different from that described before, provided that the initial cosmic-ray energy is added to ε_0. In Phase III the situation is different. According to Equation (9) the kinetic energy of the remnant (without cosmic rays) is proportional to V or to R^{-3}. The cosmic-ray energy on the other hand is proportional to R^{-1}. Hence as the object expands, the cosmic-ray energy tends to increase in relative importance and its dynamical effect on the shell cannot be neglected.

Following Kahn and Woltjer (1967) we consider a simple model of a shell of swept-up interstellar matter (mass per unit solid angle $\frac{1}{3}n_{is}mR^3$) in which at one time the radius was R_0 and the total cosmic ray energy E_0. Assuming all cosmic rays to be confined we have at later times $E = E_0 R_0/R$. The force exerted by the cosmic rays per unit solid angle of the shell is $\mathbf{P}R^2$ where the pressure $\mathbf{P}$ is equal to one-third of the energy density $(3E/4\pi R^3)$. The equation of motion becomes

$$\tfrac{1}{3}\varrho_0 R^3 \frac{d^2 R}{dt^2} + \varrho_0 R^2 \left(\frac{dR}{dt}\right)^2 = \frac{E_0 R_0}{4\pi R^2}, \tag{11}$$

where the first term represents the mass of the shell multiplied by the acceleration, the second the rate of change of momentum of the matter swept up by the shell, and the third the cosmic-ray force; all terms refer to a unit solid angle. Upon integration we have

$$\left(\frac{dR}{dt}\right)^2 = \frac{3}{4\pi} \frac{E_0 R_0}{n_{is} m R^4} + \frac{C}{R^6} \tag{12}$$

with C a constant of integration. If $E_0 = 0$, Equation (12) leads to Equation (9). Multiplying with the mass M we can write Equation (12) as

$$MV^2 = E\left[1 + \frac{R_0^2}{R^2}\left(\frac{M_0 V_0^2}{E_0} - 1\right)\right] \tag{13}$$

with M_0 now the mass of the shell at the time that the radius was R_0. Hence in sufficiently late phases, the kinetic energy tends to be one half of the cosmic ray energy.

4. Other Effects

Several important factors are not yet included in our analysis. Interstellar magnetic fields will be compressed along with the interstellar gas. Current evidence indicates a systematic magnetic field in the solar neighborhood of about $3\mu G$; possibly the random component of the field is somewhat stronger. In a medium with a density of 0.1 cm^{-3} the corresponding Alfvén velocity is about 20 km sec^{-1}. The field is likely to have two effects: (i) When the expansion velocity drops much below 100 km sec^{-1}, the compression at the shock will be decreased. (ii) More importantly, already at higher expansion velocities the field will affect the compression further behind the shock, for the main pressure will ultimately be provided by the compressed fields. (In the absence of the field this pressure will be provided by the matter compressed after cooling.) As a consequence rather strong fields will occur behind the shock and this is particularly important in the analysis of the radio emission in supernova remnants. The dynamics of a supernova shell propagating through a medium with a uniform magnetic field have been further discussed by Bernstein and Kulsrud (1965).

Inhomogeneities in the interstellar gas will affect the propagation of the shock and the appearance of the supernova shell. At the same time, Rayleigh-Taylor type instabilities may well lead to a rather inhomogeneous shell. Especially in the early phases, when the velocity of expansion is high, it would be likely that dense concentrations of matter would not be accelerated too much by the passing shell and could be left behind. In fact, the stationary filaments in Cas A may show this effect.

5. Supernova Energetics and the Interstellar Medium

From Equations (8) and (9) we obtain for a remnant in Phase III

$$R^3 V = R_{rad}^3 V_{rad} = (\tfrac{2}{5})(15/4\pi)^{8/5}(\eta m)^{-4/5}\mathfrak{A}^{-3/17}\varepsilon_0^{16/17}n_{is}^{-19/17}. \tag{14}$$

For an object like the Cygnus Loop we can measure R and V and if n_{is} can be estimated ε_0 is obtained from Equation (14). If cosmic rays drive the shell in Phase III, as discussed in Section 3, then Equation (9) has to be replaced by Equation (13). If $R \gg R_0$ this leads to $R^2 V = R_{rad}^2 V_{rad}$ (since $M/E \propto R^4$) and subsequently to a value of ε_0 which is about a factor R_{rad}/R times the value derived from Equation (14). This factor is less than 1; it is of the order of $\tfrac{1}{2}$ to $\tfrac{1}{3}$. If the incoming gas swept up by the Cygnus Loop is ionized, then every proton that passes through the shock ultimately recombines and yields one Balmer photon. It can be shown quantitatively that the importance of direct collisional excitation of Balmer lines is quite minor in this case. If the incoming gas is neutral the collisional excitation might be somewhat more important, because in the process of collisional ionization some excitation would also take place. The photoelectric measurements of Parker (1967) show that the Balmer decrement in the Cygnus Loop is of the recombination type. Consequently we can find the rate at which photons are swept up by observing the rate at which Balmer photons are emitted. The available data for the Cygnus Loop suggests $n_{is} = 0.2$ or 0.3

cm^{-3}. With $R=20$ pc and $V=100$ km sec^{-1}, we obtain from Equation (14) $\varepsilon_0 = 4 \times 10^{49}$ erg. For IC 443 the value of ε_0 may be even smaller. Most of the other large supernova remnants are likely to be also in Phase III, but precise observational data on velocities and intensities are not available.

With regard to the younger supernova remnants, the Crab Nebula is undoubtedly still in Phase I. Its present kinetic energy is about 2×10^{49} erg. Cas A may be at the transition from Phase I to Phase II, while following Minkowski we assign Tycho's supernova remnant to Phase II. The radio isophotes indicate emission from a rather smooth shell; optically very little is seen. Minkowski (1964) has estimated $\varepsilon_0 = 3 \times 10^{51}$ erg from Equation (1) obtaining V from the radius and the known age of the object and taking $n_{is} = 1$ cm^{-3}. The estimate is very uncertain because of the uncertainty in n_{is} and that in R and V. The latter two are proportional to the distance D which has been estimated from 21-cm absorption data; note that $\varepsilon_0 \propto D^5$. Tycho's supernova was of type I and therefore the chance that it exploded in the tenuous intercloud medium is much greater than that the event took place in a cloud. The regular shape of the object also would be unexpected if an interstellar cloud were involved. Consequently an interstellar density of 0.1 cm^{-3} may be more appropriate.

If we consider that a supernova remnant effectively transfers its kinetic energy to the interstellar medium when it has been slowed down to 10 km sec^{-1}, an object like the Cygnus Loop, which now has an energy of 2×10^{49} erg left, would give an effective input of 2×10^{48} erg; the remainder is radiated away. With one Cygnus Loop-type supernova per 60 years, the input rate becomes 1×10^{39} erg sec^{-1} in the Galaxy. The total dissipation rate in the interstellar medium was estimated by Kahn and Woltjer (1967) to be 8×10^{39} erg sec^{-1}. Of course an additional energy input comes from H II regions and perhaps from rarer, more energetic supernovae like those whose huge expanding rings are seen in the Magellanic Clouds.

With regard to cosmic rays, it is easily shown that the total energy input into the Galaxy amounts to

$$dE/dt = ucM/q \qquad (15)$$

irrespective of the confinement volume of the Galaxy (Woltjer, 1968)*. With the energy density $u = 1 \times 10^{-12}$ erg cm^{-3}, c the velocity of light, the total mass of interstellar gas in the confinement volume $M = 4 \times 10^9$ $M_\odot$, and the amount of matter traversed by a typical cosmic-ray particle $q = 4$ g cm^{-2} we have $dE/dt = 6 \times 10^{40}$ erg sec^{-1} or 10^{50} erg per supernova. Supernovae may inject cosmic rays in two phases. During the first explosion phase the outer envelope may come off at relativistic speeds (Colgate and White, 1966) and, in the cases where a rotating magnetic neutron star is left, effective acceleration is possible over a more prolonged period (Gold, 1969; Gunn and Ostriker, 1969). If the shell energy is as small as in the case of the Cygnus Loop, the contribution of the first process is unlikely to be large. In the rotating neutron stars, enough energy could be present; but in the one case about which we have information — the Crab

---

* See also the remark by Syrovat-skii in the Discussion following this Report. (Ed.)

Nebula — it is unlikely that a large cosmic-ray energy is present. We know that the shell has a very low kinetic energy and those cosmic rays that are present will accelerate the shell effectively, since coupling between cosmic rays and shell motion is to be expected. Perhaps the main producers of cosmic rays are, again, rarer, more powerful supernovae; cosmic-ray acceleration in supernovae is still far more speculative than is frequently assumed.

References

Bernstein, I. B. and Kulsrud, R. M.: 1965, *Astrophys. J.* **142**, 479.
Colgate, S. A. and White, R. H.: 1966, *Astrophys. J.* **143**, 626.
Gold, T.: 1969, *Nature* **221**, 25.
Gunn, J. E. and Ostriker, J. P.: 1969, *Phys. Rev. Lett.* **22**, 728.
Heiles, C.: 1964, *Astrophys. J.* **140**, 470.
Kahn, F. D. and Woltjer, L.: 1967, *IAU Symposium No. 31, Radio Astronomy and the Galactic System* (ed. by H. van Woerden), Academic Press, London, p. 117.
Minkowski, R.: 1964, *Ann. Rev. Astron. Astrophys.* **2**, 247.
Oort, J. H.: 1951, *Problems of Cosmical Aerodynamics*, Central Air Documents Office, Dayton, Ohio, p. 118.
Pacini, F.: 1967, *Nature* **216**, 567.
Parker, R. A. R.: 1967, *Astrophys. J.* **149**, 363.
Pottasch, S. R.: 1965, *Bull. Astron. Inst. Netherl.* **18**, 7.
Poveda, A. and Woltjer, L.: 1968, *Astron. J.* **73**, 65.
Sedov, L. I.: 1959, *Similarity and Dimensional Methods in Mechanics*, Academic Press, New York.
Shklovskii, I. S.: 1962, *Astron. Zh.* **39**, 209 (translation: *Soviet Astron.* **6**, 162).
Shklovskii, I. S.: 1968, *Supernovae*, Wiley, Interscience, New York.
Woltjer, L.: 1968, 2nd Stony Brook Summer Institute for Astronomy and Astrophysics, in press.

16. DISCUSSION FOLLOWING THE REPORTS BY SALPETER AND WOLTJER

(Saturday, September 13, 1969)

Chairman: L. DAVIS

Editor's remarks: During the second half of this Discussion van de Hulst commented on the overall flow of energy in the interstellar medium. That part of the Discussion has been transferred to the Final Discussion (p. 362). The rest has been rearranged into the form presented here.

1. Discussion of Salpeter's Report

Mestel: I would like to repeat a point I made this morning when presenting Salpeter's paper. Pagel has emphasized that there is no evidence in support of a steady increase of the heavy-element content of stars continuing from the galactic birth. This picture is a theorist's idealization. If you consider M67, the prototype of the oldest galactic clusters, you will find that it does not contain significantly fewer heavy elements than the Sun. But the oldest globular clusters are certainly deficient in heavy elements. Salpeter concludes that there was an early phase of rapid heavy-element formation, perhaps during the collapse of the Galaxy, and subsequently only a very modest increase. This corresponds to very different star formation laws in the two phases, which is reasonable enough, as the gas dynamical conditions were certainly different.

Pikel'ner: For some problems it is important to know what fraction of the interstellar gas is still primordial, and what fraction has gone through a stellar state. One such problem, for example, is the question of how many quarks there are in our vicinity. Together with Okun' and Zel'dovich, I estimated that between 10 and 30 per cent of the gas is still primordial. But I think that a more reliable result can be obtained from Salpeter's fractions.

Mestel: I do not know if any of the primordial gas is left over.

2. Discussion of Woltjer's Report

Bisnovatyi-Kogan: I should like to give a short description of a mechanism for supernova explosions that is not based on neutrino emission and detonation of a nuclear explosion. The model I have in mind is the explosion of a rotating star (Bisnovatyi-Kogan *et al.*, 1967). Let us consider a star at the edge of rotational stability. The stability is lost and the star starts to collapse at the moment disintegration of iron ($Fe \rightarrow n, p, \alpha$) begins. In the core the collapse will stop after the formation of a neutron star; in the outer part centrifugal forces will stop the matter and form a flattened disk with a mass of the order of or greater than the mass of the neutron star. The disk will

Habing (ed.), Interstellar Gas Dynamics, 236–248. *All Rights Reserved. Copyright* © *1970 by IAU*

be in differential rotation with the rotational velocity decreasing outwards. If there is some momentum transfer from the core to the disk-like envelope, then compression of matter can occur, a shock wave can be produced, and an explosion may follow. The momentum transfer can be accomplished by a magnetic stress following from a field of at least 3×10^9 G. In this way we can understand why supernova remnants have rotational velocities much less than pulsars.

Finally I want to draw attention to calculations by Imshennik and Nadezhin from the Institute of Applied Mathematics in Moscow who studied supernova models of $10 M_\odot$ and $30 M_\odot$ and found a mass loss of 5 per cent and 3×10^{50} erg of kinetic energy in outflowing matter. [Bisnovatyi-Kogan, G. S., Zel'dovich, Ya. B., and Novikov, I. D.: 1967, *Astron. Zh.* **44**, 525 (Translation: 1967, *Soviet Astron.* **11**, 428).]

Krat: The transfer of angular momentum from the center of a star to the atmosphere can be one of the causes of stellar explosion. As far as I remember the effect was first noted by Hoyle. The efficiency of the mechanism is very strongly dependent on form and strength of the magnetic field. It may remove much angular momentum from stars to the interstellar medium, so it is a very important effect to study.

Colgate: I have a couple of comments. First, I want to discuss the iron production in a supernova. There are arguments that per supernova explosion $\frac{1}{4} M_\odot$ of iron is produced. (a) This amount is suggested by the present theories of nucleo-synthesis by Truran *et al.* (1967) and by Bodansky *et al.* (1968), which predict that all iron is produced through $Ni^{56} \to Co^{57} \to Fe^{56}$. (b) A recent paper by Clayton and Silk (1969) indicates that the gamma rays produced in this process may explain the observed overabundance of gamma rays in the Universe. They require $\frac{1}{4} M_\odot$ of iron per supernova. (c) Recent work by McKee and myself (Colgate and McKee, 1969) ascribes the optical emission of supernovae to the sequence of beta-decays mentioned above. $\frac{1}{4} M_\odot$ of iron synthesized to Ni^{56} gives a beta-decay energy of 10^{49} erg – which is the total amount of light put out by a supernova.

To explain all the iron in the Galaxy by supernova explosion, you need one supernova per ten years. That seems pretty high. But according to Salpeter's Report there was a rapid formation of metals in the early history of the Galaxy, and that would agree with a high supernova rate in the young Galaxy, but a lower rate now. Fortunately there is a large probability that within the next two years measurements will select the correct theory. The measurements follow from a suggestion by Clayton *et al.* (1969), that we look in a supernova for gamma rays from decaying Ni, Co, and Fe. If a supernova occurs in one of the 50 to 100 nearest galaxies, one should be able to see the gamma-ray lines and the intensity as a function of time. This function of time gives you the thickness of the matter expanding around the star; and from this you can get both the total mass and a mean velocity distribution.

In the second place, I should like to raise an objection against the Morrison-Sartori (1969) theory of the optical emission of supernovae, which is based upon fluorescence of an envelope around the supernova by UV radiation from the supernova star. Since the fluorescing medium is He, we consider only UV photons with $h\nu > 54$ eV. No more than 10^{-1} of the energy of all photons will be in this part of the spectrum;

kinetic energy cannot be converted into photons with an efficiency higher than 10^{-1}; the efficiency of the fluorescence mechanism is at best 10^{-2}. This means that with a light output of 10^{49} erg we need a supernova energy of 10^{53} erg. This large an energy seems unlikely.

In the third place, the present theories of supernovae with neutron star formation, including my own work (Colgate and White, 1966) and that of Arnett and Schwartz, are, according to a recent paper by Arnett and Truran (1969), all necessarily wrong. They are wrong because they result in the ejection of too much heavy-element matter. Instead Arnett and Truran would claim that the vast majority of supernovae were thermonuclear ($C^{12} + C^{12}$) in origin and that neutron stars are formed without significant mass ejection. (Truran, J. W., Arnett, W. D., and Cameron, A. G. W.: 1967, *Can. J. Phys.* **45**, 2315; Bodansky, D., Clayton, D. C., and Fowler, W. A.: 1968, *Phys. Rev. Lett.* **20**, 161; *Astrophys. J. Suppl.* **16**, 299; Clayton, D. C. and Silk, J.: 1969, *Astrophys. J. Lett.* **158**, L43; Colgate, S. A. and McKee, C.: 1969, *Astrophys. J.* **157**, 623; Clayton, D. C., Colgate, S. A., and Fishman, G. J.: 1969, *Astrophys. J.* **155**, 75; Morrison, P. and Sartori, L.: 1969, *Astrophys. J.* **158**, 541; Colgate, S. A. and White, R. H.: 1966, *Astrophys. J.* **143**, 626; Arnett, W. D. and Truran, J. W.: 1969, *Astrophys. J.* **157**, 339.)

Sunyaev: I agree with Colgate that 10^{52} erg in UV radiation per supernova is too much. Such a flux would give a background radiation from the Metagalaxy which is 10 to 100 times more than is observed. Also, if the UV flux is in 100 eV photons the effects on the interstellar medium become important. For the heating of all interstellar matter in the Galaxy we need 10^{40} to 10^{41} erg sec^{-1}; this allows per supernova only 10^{49} to 10^{50} erg in UV photons.

Colgate: To bring up another point, there is also the question of the energy spectrum of ejected matter. The existing theories predict about 10^{52} erg in the form of particles with a kinetic energy between 2 to 5 MeV per nucleon. About 10^{50} erg is in relativistic matter. Apparently the large mass of low-energy matter is not observed, I think for the following reason: the slowing-down time by collision is 10^6 yr. But in 3×10^4 yr the matter is so far expanded (radius 3×10^{21} cm), that the magnetic energy inside this volume is 10^{52} erg, and the matter can easily escape along the field lines without forming a shock. Instead, a small fraction of the mass, whose velocity is only 3×10^8 cm sec^{-1} will slow down in 100 yr by collisions (the collisional cross-section $\sigma \propto V^{-4}$) and form an easily visible (and sharp) shock front at a radius of 10^{19} cm.

Syrovat-skii: From the total mass of the interstellar gas in the Galaxy and the total amount of matter traversed by cosmic rays we can estimate that all the sources in the Galaxy produce (1 to 3) $\times 10^{40}$ erg sec^{-1} in cosmic rays. If there is one supernova per 30 yr, you have (1 to 3) $\times 10^{49}$ erg of cosmic rays per supernova. This is less than Colgate mentioned, although not by a large amount.

Tsytovich: I am not convinced that the 2 to 5 MeV particles escape without forming a shock front. If fast particles penetrate into a plasma they excite waves, for example Langmuir waves. The group velocity of these waves is much smaller than the particle velocities, and the waves accumulate near the surface where the particles enter the plasma. The layer of accumulated waves can act as a very sharp shock wave and reflect cosmic rays. Theoretical work on this subject has been done by Feinberg and Shapiro.

Colgate: I have argued many times for such an electrostatic shock. The problem is that the energy distribution of the ejected matter is monotonically decreasing with time; it has no peak. Therefore, I am not convinced that it will excite electrostatic waves. Are the cosmic rays electrostatically unstable? I am not sure.

Tsytovich: I think they are. If you have a monotonic distribution inside the super-nova, the faster particles enter the surroundings more quickly and some kind of instability must exist. In addition the outgoing cosmic rays have an anisotropic velocity distribution locally, where they penetrate into the undisturbed plasma.

Colgate: But the density ratio between the high energy particles and the ambient plasma is probably very small. When you put in the Landau damping, the density ratio, the velocity gradient, and in addition, the magnetic field diffusion for the outside field, the problem is very difficult to solve.

Tsytovich: I agree, but still I think that possibilities for instabilities exist. May I now ask Dr. Woltjer what evidence he has that the high-energy particles in the Crab Nebula have been accelerated recently?

Woltjer: In the case of the electrons there is evidence from the spectrum of the synchrotron radiation. The radio emission can be explained as coming from electrons injected shortly after the supernova explosion. But the optical synchrotron radiation, emitted by electrons with energies of several hundreds of GeV, cannot be explained in this way. The combined effect of expansion and synchrotron radiation gives these electrons a lifetime shorter than the age of the Nebula. The argument holds even more strongly for the X-rays, but here we are not certain that the X-rays are really synchro-tron radiation.

Tsytovich: I think that there is a possibility for compensation of the energy losses of fast particles. Usually the energy losses are proportional to the square of the particle energy and to the square of the magnetic field strength. But in a turbulent region the acceleration rate $d\bar{\varepsilon}/dt$ is also proportional to the square of the particle energy. It there-fore exactly compensates the losses. However, the acceleration depends also very strongly on the density and the temperature of the turbulent plasma. In the Report by Kadomtsev and me, we give a table (Table I, p. 129) that predicts the acceleration time. It shows that in a plasma with a temperature of 10 eV the acceleration time is very short for a density of 10^{20} cm^{-3} and very long for 10^2 cm^{-3}. So the question is, in what density does the acceleration take place? Another point is that one usu-ally assumes that the emission is synchrotron radiation. But plasma turbulence may emit light with the same spectrum. And induced effects can give polarization without a magnetic field.

Woltjer: First, the energetic electrons do not radiate only near the star, but they radiate all through the nebulosity. The surrounding density must be quite low, cer-tainly less than 100 cm^{-3}, and probably much less. I think, therefore, that you can not compensate for the radiated losses of the particles by plasma turbulence. The clue to the energy input in the Nebula may well be given by the wave-like phenomena that are observed near the center of the Nebula, near the location of the pulsar. As Baade first observed, near the location of the pulsar you see from time to time small waves moving out with velocities on the order of one tenth the velocity of light. (Of course, the fact that these waves propagate so fast tells you that the density there must be very low.) Second, I agree, of course, that you can obtain some polarization with a suitable anisotropy. But the linear polarization observed in the Crab Nebula goes, at some places, up to 60 and 70 per cent. On whether you can get that much polarization just from anisotropy in the turbulence, I would still be a bit doubtful.

Tsytovich: I think that it is possible to have a very high degree of polarization. But the plasma turbulence should give a rapid change in polarization angle.

Woltjer: The effect of polarization is extremely regular in the Nebula, and this would suggest that magnetic field is the cause.

Syrovat-skii: I have some remarks to Tsytovich concerning the acceleration of particles by high-frequency plasma waves. There are reasons to believe that for most of the cosmic rays high-frequency plasma turbulence is not the acceleration mechanism. The first reason is that the acceleration by high-frequency plasma turbulence is not effective at high energies. We have seen in the Kadomtsev-Tsytovich Report (p. 108) that at high energies the acceleration is inversely proportional to the energy. It means simply that the acceleration by plasma turbulence heats the particle to some effective temperature of the order of the temperature of the plasma disturbances. But in cosmic rays we have particles up to 10^{20} eV and this seems too high a temperature for a turbulent plasma in interstellar space. The second reason lies in the chemical composition of the accelerated particles. All the statistical mechanisms of acceleration are based essentially on thermal injection. It means that the particles are accelerated from the tail of the Maxwell distribution. In this case the acceleration is very sensitive to the mass and the charge of the particles. It was shown about ten years ago, by Korchak and myself (Korchak and Syrovat-skii, 1958), that, if you have thermal injection, then the heavy ions are accelerated faster than the protons or electrons. The same argument may be used for plasma turbulence. But we observe that the cosmic-ray particles have almost the cosmic abundance, and in addition, there is a large number of electrons. So it seems to me that a more realistic mechanism is cumulative acceleration. The first example of such a mechanism has been given by Colgate (Colgate and White, 1966). In this mechanism all particles are accelerated together in the thin outer envelope of the exploding supernova star, because an ultrarelativistic shock wave blows up the outer layer of the star. However, this theory has difficulties with the structure of the shock wave and may not work. Another possibility of cumulative acceleration comes from studies of solar flares and the magnetosphere. An example of such a mechanism is the current sheet which develops near magnetic zero (neutral) lines. When the current sheet is disrupted, then all the plasma particles are accelerated to very high energy. Perhaps a mechanism of this sort can account for the interstellar cosmic rays. [Korchak, A. A. and Syrovat-skii, S. I.: 1958, *Dokl. Akad. Nauk SSSR* **122**, 792 (1958, *Soviet Phys. Dokl.* **3**, 983); Colgate, S. A. and White, R. H.: 1966, *Astrophys. J.* **143**, 626.]

Tsytovich: I want to say that the efficiency of the acceleration by plasma turbulence increases with particle energy up to very high values. For example, if a supernova of approximately one solar mass explodes, the density is very high in the first stage, and particles with an energy of 10^{16} eV can be produced in the time in which the supernova doubles its radius. Near the neutron star one can produce particles up to 10^{18} to 10^{19} eV, and therefore it seems possible to produce all cosmic rays by plasma turbulence in supernova explosions. As regards the problem of injection: Melrose has shown that Alfvén-turbulence leads to preferential acceleration of heavy ions. But Langmuir-turbulence is very effective for acceleration of protons and electrons. As an example, Shklovskii has stated that supernovae seem to produce more relativistic electrons than relativistic protons.

Zel'dovich: Dr. Woltjer, you described the shock wave as a front between undisturbed gas and gas set in motion. But is it not possible that the radiation should ionize the gas even in advance of the shock wave? One must therefore distinguish a region where gas is ionized but not moving, and then the shock wave going into ionized gas. And only in the later stages, perhaps, will the shock wave overtake the ionization front.

Woltjer: In the very early stages, the energy per particle behind the shock is so large compared to the ionization energy, that it makes little difference for the velocity of the shock whether the gas in front of the shock is ionized or not.

Zel'dovich: No, perhaps that is true for the velocity, but the physical state of the gas is not the same, and there may be a large difference in the UV and X-ray emission.

Woltjer: In the case of the Cygnus Loop, one can show that depending upon the amount of UV radiation emitted in the original supernova explosion, the gas which is swept up at this moment could easily be either ionozed or not ionized. In principle, this gives small differences in the observable spectra. At this moment it is not possible to distinguish observationally between the two possibilities.

Zel'dovich: The shock wave is a very stable phenomenon. This can be demonstrated for shock waves moving in water. The shock waves reflect light. If one perturbs the shock front the image wiggles and after a short time it is normal again. But there is another front, namely that which divides the compressed interstellar gas and the expanding shell. There is pressure equilibrium, but the front is unstable. My question now is: does one observe this instability?

Woltjer: The possible instability of such a front is very interesting. We have been considering the possibility that the filamentary structure of the envelope in the Crab Nebula may be due to the instabilities that you mentioned.

Lüst: I would like to discuss an experiment which is of interest for the question of instabilities occurring in a spherical expansion. This year we have carried out an artificial plasma cloud experiment at 12.5 Earth radii in the distant magnetosphere. The density in this region is so low that collisions among particles are of no importance; furthermore in the initial phase the kinetic energy density in the radially expanding cloud was large as compared to the magnetic energy density of the Earth's magnetic field. The initial spherical cloud broke up very soon into filaments elongated along the magnetic lines of forces, and we believe that this disruption was caused by flute instabilities. The theoretical estimates of the characteristic times are in good agreement. In this case the observed structures are not related to the injection mechanism but are due to the acceleration in the expanding cloud.

Silk: Dr. Woltjer, your explanation of the energy in the Cygnus Loop was based on Hα emission. Is it not possible that there is a substantial region of high-temperature gas which would be emitting lines observable primarily in the soft X-ray region?

Woltjer: The cooling time of the high-temperature gas is shorter than the lifetime of the Cygnus Loop. That means that at the present time there is a steady state behind the shock wave. For every incoming proton, there is one in the gas behind the shock that has to recombine.

Silk: My comment relates supernova explosions to observations of HI regions. The discussion by Spitzer and Tomasko of the heating of HI regions by low-energy cosmic rays indicated that $\approx 10^{51}$ erg per Type I supernova must be injected in low-energy cosmic rays, with a frequency of one supernova per hundred years. The alternative explanation of the observed temperatures and electron densities in HI regions makes use of soft X-rays. This mechanism has been proposed by Sunyaev in the Soviet Union and by Werner and myself in the United Kingdom. Supernova remnants may be strong sources of soft X-rays. We require something like 3×10^{30} erg of soft X-rays per supernova. Heiles (1964) and Shklovskii (1968) have suggested that a substantial amount of the initial kinetic energy of the supernova shell is radiated in bound-bound emission during the deceleration phase. These authors considered only O VII and O VIII emission at ≈ 20 Å, where the interstellar medium is essentially transparent. I would like to point out that appreciable emission will also occur in the lines of C V at 34 Å and C VI at 40 Å and in two photon decays from the $2s$ states of these oxygen and carbon ions. Most of the photons will not escape from the Galaxy, and so a substantial fraction of the supernova energy may provide a soft X-ray input to the interstellar medium. For example, if we take $\approx 3 \times 10^{51}$ erg for the initial kinetic energy of the supernova envelope, and assume that 10 per cent of this energy is converted into soft X-rays of energy below 0.3 keV, then a supernova rate of one per century is sufficient to give a heat input to the interstellar medium of 3×10^{-26} erg cm^{-3} sec^{-1} and a hydrogen ionization rate ζ of 5×10^{-16} sec^{-1}. (Heiles, C.: 1964, *Astrophys. J.* **140**, 470; Shklovskii, I. S.: 1968, *Supernovae*, Wiley, London.)

Sunyaev: I should like to continue this and discuss a specific mechanism for the production of UV and soft X-ray quanta. The mechanism is based on the supernova model (unpublished) by Morozov and Imshennik from the Institute of Applied Mathematics in Moscow, which shows a shock in a low-density stellar atmosphere with an electron temperature of about 10^6 K behind the shock. In the dense regions of the atmosphere, which are optically thick for bremsstrahlung, optical radiation is produced with a temperature of about 10^4 K. When this low-temperature radiation diffuses outward through the 10^6 K electron gas, its temperature rises due to inverse Compton scattering, a process discussed first by Kompaneyets in the U.S.S.R. and later by Weymann in the U.S.A. In this process the number of quanta does not change. The resulting spectrum is sketched, qualitatively, in Figure 1. The spectral form depends only on one parameter ($y = \int [kT_e/(m_e c^2)] d\tau$, where τ is the optical depth for Thompson scattering). If $y > 1$, the maximum of the spectrum is around $h\nu \approx kT_e$, which means, for $T_e = 10^6$ K, in the soft X-ray region.

Pikel'ner: In the first place, I want to discuss the origin of filamentary structure in supernova shells. I studied this problem many years ago (Pikel'ner, 1954) and found that there are two types of mechanism which can form filaments. When the supernova shell is very young, it is accelerated by cosmic rays and magnetic pressure. Gravity acts on the shell and a Rayleigh-Taylor instability can develop in a short time,

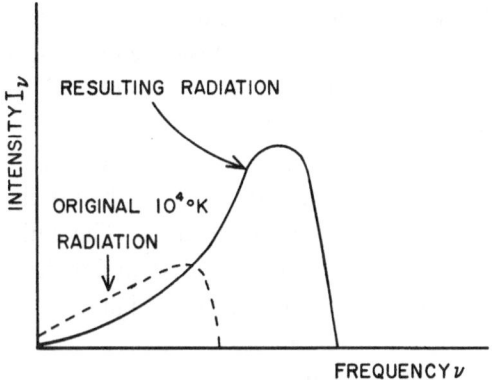

Fig. 1. (See the remark by Sunyaev.) Radiation resulting from interaction between photons emitted at 10^4 K and an electron shell of 10^6 K. The interaction process is the inverse Compton effect.

forming filaments. In an old shell like the Cygnus Loop, this mechanism does not work, because the shell is not accelerated. (Actually it is retarded.) In this case, some other formation mechanism may work, perhaps crossing of shock waves. The velocity of a shock wave is dependent on the density of gas in front of the wave. If there is some density fluctuation in the gas, the shock wave behaves as if it were focused by a lens, where the part of the lens is played by the density fluctuation. The focusing is not into a point but into a filament. Every density fluctuation should lead to a filament.

A second comment. Woltjer calculated the Hα emission in supernova shells assuming that one Hα quantum is produced per incoming proton. In my calculations of the emission of supernova shells like the Cygnus Loop (Pikel'ner, 1954), I found that collisional excitation of hydrogen can occur in a very thin, hot layer behind the shock front. This leads to an increase in total emission by a factor of about 2 to 3, so the layer is not very important. It would be interesting to know the new data on old supernova shells by Losinskaya from the Sternberg Institute in Moscow. She studied two old supernovae, IC 443 (velocity of expansion 65 km sec^{-1}), and Simeis 22 (velocity of expansion 35 km sec^{-1}). (Pikel'ner, S. B.: 1954, *Izv. Krymsk. Astrofiz. Observ.* **12**, 94. A discussion in English of this work may be found in Sections 9 and 10 of *Interstellar Gas Dynamics* by S. A. Kaplan, Pergamon Press, Oxford 1966).

Woltjer: I have a comment on the Hα emission. First, the assumption was that each atom that comes in makes one recombination photon, not necessarily Hα. It may also be a photon of one of the other Balmer lines. But the point here is that the gas flowing into the shock front may already be in the ionized state due to the original UV radiation from the supernova. If the gas enters in that state, it remains ionized. Behind the shock there is no neutral hydrogen to be collisionally excited. And by the time that recombination becomes effective, the temperature is too low for direct excitation.

Pikel'ner: The emission in the filaments can be observed only if the density ahead of the shock front is not less than 10 particles cm^{-3}, and so the recombination time is less than the age of old nebulae, which is 20 000 years.

Woltjer: If the density is that high, that would be true.

Field: Dr. Pikel'ner, if the supernova explosion emits X-rays, the ionized gas in front of the shock may have a very high temperature, 10^5 or 10^6 K, which would reduce the recombination coefficient.

Pikel'ner: The temperature of the gas behind the shock front is dependent on the density in front of the shock. If this density is that of the rarefied intercloud medium, we cannot observe emission behind the shock front. The emission is only observed when the shock penetrates into a rather dense cloud, and in dense clouds the temperature of the gas due to X-ray emission cannot be very high.

Sunyaev: But near the supernova it can be.

Pikel'ner: It is more than 30 pc away.

Field: I have been doing some calculations of the X-ray ionization around a supernova. If I remember correctly, I obtained an envelope of 100 pc completely ionized gas for a density of one atom per cm^3. I did not calculate the thermal equilibrium, but I guess that the temperature will exceed 10^5 K. Of course, for a surrounding medium of higher density the envelope will be smaller.

Pikel'ner: I do not know these calculations and cannot discuss them directly, but I think there is some observational answer. If the gas in front of the shock wave were ionized and strongly heated, it should expand and not leave behind condensations of 10 to 20 particles cm^{-3}. If, therefore, emission is actually observed from condensed gas, the gas in front of the shock wave is not ionized and not strongly heated.

Field: You are probably correct, but I think that one ought to look into the expansion of a dense region with a temperature of 10^5 K.

Kardashev: I want to suggest that per supernova explosion we may observe two flares: an optical flare and a radio flare. The optical flare corresponds to the moment of explosion. The radio flare corresponds to the moment when in the supernova remnant magnetic fields and cosmic rays have maximum energy. The time difference between two flares should be large, possibly many tens of years, because we would obtain too large energy losses for the cosmic-ray particles if we assumed that the radioflare occurs during the initial explosion. Also the radio emission at this moment would be too large. Obviously, it is very important to observe the radio emission of the initial stage after the supernova explosion and also the initial stage of the pulsar. I mentioned before (p. 193) that in the remnants of supernovae magnetic fields can be generated by regular twisting of the magnetic flux between the pulsar and the expanding and rotating envelope. Such a process gives an increase of the magnetic energy; subsequently the magnetic pressure accelerates the envelope. The total magnetic energy at maximum is of the order of 10^{49} erg. The structure of the twisted magnetic field is perhaps related to that of the ripples observed near the pulsar in the Crab Nebula. There are three possibilities: (1) If the coupling between pulsar and envelope continues to exist up to the moment of observation, we get the picture of a central concentration of brightness, like that in the Crab Nebula. (2) If the coupling disappears after a short time, we obtain an envelope with a radial magnetic field like that in Cassiopeia A (and probably in the Tycho and Keppler remnants). This type of coupling may occur if the conductivity is small during the collapse of the star to the gravitational radius. This is probably the case in supernovae of Type II which have large masses. (3) If the time needed for generation of magnetic fields is too small, there will be no remnants left with non-thermal radio emission, and we obtain an isolated pulsar. So in summary, the three types of objects observed after a supernova explosion (thin envelopes, nebulae with central condensation, and isolated pulsars) may be explained by different initial conditions before explosion relating to the mass, the angular momentum, and the magnetic field of the star.

Verschuur: A question to the theoreticians: If the supernovae inject kinetic energy into the interstellar medium, how could one observe this? I have made 21-cm observations in the vicinity of nine pulsars and in two cases (CP 1133 and CP 0950) the hydrogen structure looks pretty interesting. It looks as if I see two shells, either expanding or contracting. Velocities involved are of the order of 5 km sec^{-1}.

Woltjer: When the Cygnus Loop becomes a few times 10^6 yr old, the expanding shell will have a velocity of the order of 10 km sec^{-1}. So you may be correct.

Colgate: To answer Verschuur, when a supernova injects 10^{52} erg in low-energy cosmic rays (2 to 5 MeV per nucleon), then certainly one would expect to see the bubbles that have been discussed by Parker.

Syrovat-skii: In discussing the content of cosmic rays in supernovae, Woltjer assumed that the cosmic rays are confined and accelerate the shell. But what about the observational data? Do the radio isophotes coincide with the optical boundary? Or is there some radio radiation outside the sharp boundary which indicates diffusion of cosmic rays out of the shell?

Woltjer: I think that the photometric evidence available at the moment suggests that the outermost isophotes coincide with the filamentary shell. I don't think, however, that this is very strong proof that the particles are contained by the shell. For the surrounding interstellar medium the magnetic field strength is likely to be down by a factor of the order of 100, and the emission from escaped particles would be down by an even larger factor and, therefore, would be very weak. I think, however, that, unless you say that the particles escape completely freely, the estimate of the upper limit of the cosmic-ray content cannot be changed very much.

Menon: One of the characteristics of the radio contours of most of the supernova remnants is the extremely steep intensity gradient at the edge of the shell. But in addition, it is also a common characteristic that the shell is not complete. That is, in

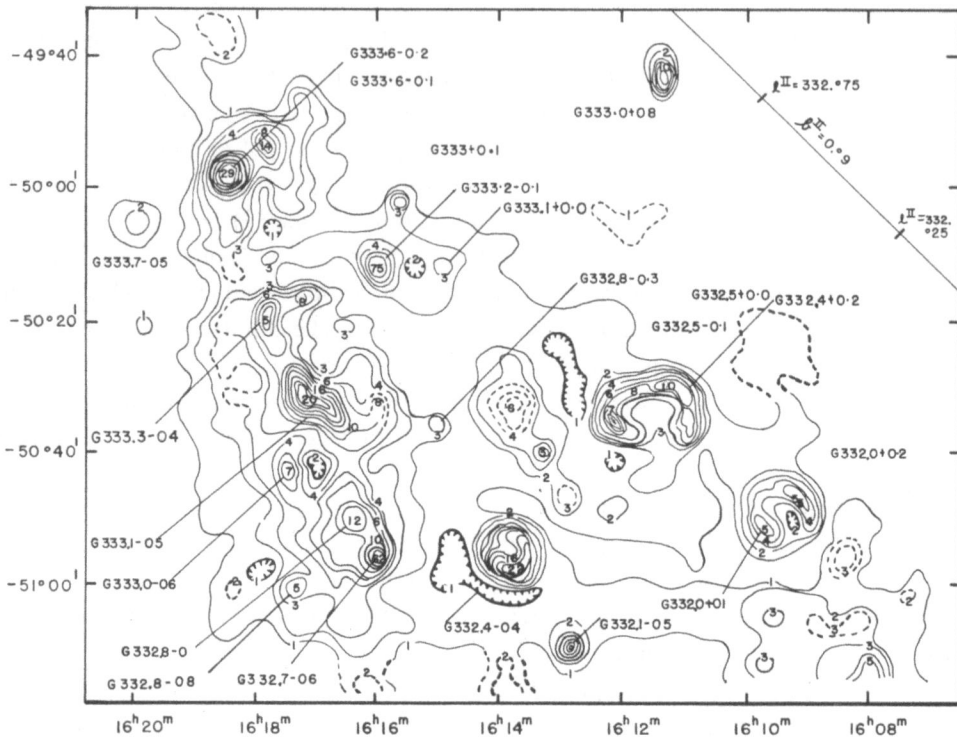

Fig. 2. (See the remark by Mills.) Radioisophotes in the constellation Norma at 408 MHz observed with the Molonglo radiotelescope. It shows several suspected supernova remnants. [This figure has been redrafted in the editorial office in order to improve the presentation. This was done, however, at the loss of some accuracy.]

almost all well-resolved cases, there appears to be a tail, perhaps a real blowout of particles. An example is the hole that you see in the Cygnus Loop. Often the structure shows symmetry, though not spherical symmetry. Usually the shell consists of two opposite arcs which are of fairly uniform brightness with some microstructuring inside them. In addition there is a very bright circular top part and a fairly well-defined bottom part. There are at least seven supernovae which show this structure. Symmetry of this type of structure is seen in objects, which are as different in age as the Cassiopeia remnant and the Cygnus Loop. I do not have an explanation, but I want to stress that this is a common characteristic and on probability arguments seven is a high number for formation on a random basis.

Mills: I have a slide (Figure 2) which shows that spherical supernova shells are an exception. The slide shows an area in the constellation Norma, observed with the Molonglo radio telescope; the resolution is 3'. The slide is from a paper by P. A. Shower and W. M. Goss, which is in press in the *Supplement Series* to the *Australian Journal of Physics*. The sources G332.0+0.2, G332.4−0.4, and G332.4+0.2 are non-thermal and undoubtedly supernova remnants. However the sources G332.1−0.5 and

G333.3−0.4 are both very dense H II regions in a ring-shaped structure. The remaining sources are also thermal. This is the sort of picture we find all along the Milky Way; and very, very rarely do the non-thermal sources have a nice symmetrical shape.

Woltjer: If you take a sample of supernova remnants, very likely it contains mainly old supernova remnants that have a long lifetime, and those are very thick. The first stages of a supernova remnant last for a much shorter time; therefore you have to select very special cases in order to discuss shells in the early stages.

Mills: Would you say then that the remnants in Figure 2 are all in Phase III?

Woltjer: Yes, if they are supernova remnants at all.

Menon: In your discussion of the various phases of the supernova phenomena, you put all observed remnants in the later stages except the Crab. On what parameters will the duration of Stage I depend?

Woltjer: The duration of Stage I will probably depend largely on the amount of matter injected, compared to the amount of interstellar matter in the same volume.

Menon: But then, is it not a coincidence that of all the remnants known to us and identified, all the others have velocities of 7000 km sec^{-1} and more, whereas the Crab has a velocity of only about 1000 km sec^{-1}? If you classify supernovae on the basis of light curves, even Minkowski himself would agree that the classification of the Crab is most uncertain, because the light curve is not very reliable. Describing the Crab as a Type I supernova is, I think, rather dangerous, because the Crab has no parameters in common with all the other remnants.

Woltjer: It is true that one cannot classify the Crab Nebula as a supernova of Type I, Type II, or any other type. However, if you look at the current energy content, both in the pulsar and in the form of relativistic particles, it is clear that in the future the Crab Nebula will never reach an expansion velocity of the order of 7000 km sec^{-1}. There simply is not enough energy to accelerate the present shell to that kind of velocity. On the other hand, the total energy of the Crab at this moment is not very different from what you infer for the supernova that gave rise the Cygnus Loop.

Shklovskii: I want to point out a difficulty in comparing the Kepler supernova with the Tycho supernova. The properties of both supernovae, i.e., dimensions, brightness, etc., are practically equal. This would mean that the density of the matter surrounding the Kepler nova should be similar to that of the Tycho nova. But the Kepler nova is far from the plane (between 600 and 1000 pc) and the Tycho nova is not (100 pc). So here is a problem. Is the outside density the same?*

Woltjer: I would like to quote a counter example, namely the supernova of AD 1006. If this object has been properly identified, it is at about $b = +15°$, and it seems to have a radius that is far larger than that of the Tycho object. Perhaps one can interpret this in terms of a low interstellar density at that level, $z = 1000$ pc.

Ozernoi: In supernova explosions gravitational waves are also emitted. A few years ago I made some calculations (Ozernoi, 1965) which indicated that one can observe

---

* Dr. R. Minkowski (private information) suggests that the difference in distances to the Kepler and the Tycho supernovae, respectively, may have been overestimated in the past, the ratio between the distances being closer to 1.3 than to 2 as supposed by Shklovskii. (Ed.)

the gravitational radiation of an anisotropic supernova at a distance of about 10 Mpc. But the precise value of gravitational radiation depends on the degree of anisotropy in the explosion. I think that in the Crab Nebula there is an indication that the initial process of explosion was anisotropic. Does Dr. Woltjer agree with this? [Ozernoi, L. M.: 1965, *Zh. Eksp. Teor. Fiz. Pis. v. Red.* **2**, 83 (*Soviet Phys., JETP Lett.* **2**, 52).]

Woltjer: Yes, I do.

Weymann: May I ask Dr. Mestel why there used to be a discrepancy between the rate of star deaths and the supernova rate? Do the theorists think it is possible that stars greater than, say, $1.5 M_\odot$ can die without leaving a trace? Can a collapse occur but not be observable as a supernova? And do the statistics still support the discrepancy which I mentioned? In the past we could not account for this discrepancy by appealing to slow continuous mass loss.

Mestel: I recall that when the figure of one supernova per 500 years was accepted, it then looked difficult to account for all the white dwarfs, assuming that a star evolving into a white dwarf must go through a nova or supernova phase. The new upper limit of one supernova per 30 years clearly releases the conditions, but more and more white dwarfs tend to be discovered. However, there are other ways in which stars can lose mass, e.g., by stellar wind-type mass loss in the red-giant phase, which may enable stars a little above the (modified) Chandrasekhar limit to evolve into white dwarfs. More to the point is the fairly violent phenomenon that enables an evolved star to transmute into a planetary nebula surrounding a hot central star which subsequently cools into a white dwarf. However, I am not well up on the statistics.

Weymann: I think that the planetary nebula phenomenon occurs only on those stars that are already near the white dwarf limit, stars of about 1.2 to $1.4 M_\odot$. Could someone answer the question whether or not it is possible for a collapse to occur without producing supernova luminosity?

Field: I cannot answer that question. However, if Webber is correct in his estimate of the intensity of gravitational radiation by the Galaxy, the production of gravitational radiation is 10^6 times that of electromagnetic radiation over all observed wavelengths (visible, infrared, and radio wavelengths).

Colgate: Lingenfelter (1970) has calculated the frequency of pulsar formation in our neighborhood, taking the distance from the electron dispersion and the time from the slowing down of the pulse rate. He found a frequency of pulsar creation of one per 100 years for the whole Galaxy. If one assumes that not all pulsars are observed, because they emit their radiation in beams, then the formation rate would be higher than this. In other words, 1 per 100 years is probably a lower limit. (Lingenfelter, R. E.: 1970, VIth International Conference on Cosmic Rays, Budapest, 1969.)

Woltjer: I want to comment on what Colgate said. Many of us have tried to make the connection between pulsars and supernovae and to discuss the frequency, and I think the general experience has been that you can get whatever you want. For example, an error in the distance scale of a factor of two gives you immediately a factor of ten in the frequency.

17. THE SOLAR WIND – AN EXAMPLE OF A COSMICAL PLASMA AND A STELLAR WIND

Introductory Report

(Sunday, September 14, 1969)

R. LÜST

Max-Planck-Institut für Physik und Astrophysik, Institut für extraterrestrische Physik,
Garching/München, Germany

1. Introduction

When Parker developed the theory of the solar wind, he suggested that other stars might also be surrounded by an expanding atmosphere. This suggestion has been worked out by Deutsch (1958, 1960, 1961, 1969) and by Weymann (1960, 1963), who discussed coronal evaporation as a possible mechanism for mass loss in red giants and used hydrodynamical equations for describing the phenomena. During the last few years detailed observations of the solar wind have been made, and the theoretical approach has also been further developed. Both might help us in our understanding of problems of a quasi-stationary mass loss of stars. But besides its relevance for the study of other stellar winds, the solar wind is important, as it is the best observed cosmical plasma available for guidance in the study of other cosmical, collisionless plasmas. What makes the interplanetary plasma so important is that beyond several solar radii from the Sun there are no collisions among the gas particles (one solar radius $[R_\odot]$ is 7.0×10^{10} cm; 1 AU is 215 $R_\odot$). Only near the Sun the collision rate is sufficiently high to maintain isotropy of the thermal particle motions and to produce ordinary fluid behavior. At larger distances from the Sun the fluid behavior is maintained by irregularities in weak magnetic fields and by micro-instabilities in the plasma.

It will be impossible to discuss here all aspects of the solar wind. I shall try to summarize those points which might be of particular interest. In Section 2 a summary of observational results will be given. In Section 3 theoretical models and in Section 4 transport of angular momentum will be discussed. We shall deal with waves and fluctuations in Section 5. Finally, in Section 6, the problem of the interaction of the solar wind with the surrounding interstellar medium will be studied.

For more details and also for the omitted subjects (such as interaction with cosmic rays, comets, planets, moon, and magnetosphere) I mention a number of review articles: Lüst (1962); Mustel' (1964); Parker (1963, 1965); Dessler (1967); Axford (1968); Ness (1968); Wilcox (1968); and in particular the two recent reviews by Hundhausen (1968b) and by Parker (1969) which helped greatly in preparing this Report.

2. Observational Data

A. QUIET STATE OF THE SOLAR WIND

Observational data about the quiet state of the solar wind were obtained mainly by

Habing (ed.), Interstellar Gas Dynamics, 249–262. All Rights Reserved. Copyright © 1970 by IAU

Mariner 2 during its 1962 flight; by Vela 2, Vela 3, and Vela 4 in 1964 to 1967; and by IMP I in 1963 to 1964. The data were obtained near the orbit of the Earth (1 AU). According to the measurements by the Vela satellites, the flow or bulk velocity is near 320 km sec^{-1} with a general inclination of 1.5° from the radial direction in the sense of a co-rotation with the Sun. However, results from IMP I indicate an average inclination of $-15°$ (anti-corotation) (Egidi *et al.*, 1969). The proton density is about 5 cm^{-3}; α-particles amount to about 4.5% compared to protons (new observations on Explorer 34 by Ogilvie and Wilkenson, 1969, give about 5% α-particles) and other positive ions number less than 0.5% of the protons. The total positive-ion flux in the quiet solar wind is thus near 1.75×10^8 electronic charges cm^{-2} sec^{-1}. He$^+$ has been measured on certain occasions with an abundance of $n_{He}/n_p \leqslant 10^{-3}$. O^{+6}, O^{+5}, and O^{+7} ions have been detected with variable relative abundances.

The solar-wind protons have a most probable temperature of about 4×10^4 K under quiet conditions. But even then the temperature varies considerably, occasionally dropping below 10^4 K. The He^{++} ions have temperatures about four times higher than the protons. The electron temperature is near 10^5 K, which is about three times the proton temperature. The ion distribution functions are normally anisotropic, with the proton temperature along the interplanetary magnetic field lines about twice as high as normal to the lines. The electron distribution functions are less anisotropic than those of the protons. Both the electrons and protons are conducting heat outward from the Sun along the magnetic field lines, at rates of about 10^{-2} and 10^{-5} erg cm^{-2} sec^{-1}, respectively.

The average magnetic field strength is about 5×10^{-5} G. The magnetic energy density is much smaller than the kinetic energy density of the solar wind, and the magnetic field is carried along by the solar wind. The field either points away from the Sun or towards it in the form of an Archimedian spiral. The field lines have their origin in the Sun.

B. TIME AND SPATIAL VARIATIONS OF THE SOLAR WIND

The wind is highly variable in its velocity (which can go up to about 850 km sec^{-1}), in its direction ($\pm 15°$), in its temperature (up to 9×10^5 K), and in its composition (up to 20% α-particles). Variations over the solar cycle have been observed. Also there often exist streams of hot, high-velocity plasma which recur at 27-day intervals. The magnetic field frequently shows a sector structure, which is correlated with the persistent high-velocity plasma streams. In addition a filamentary structure in the plasma and the magnetic field has been observed.

Short-time fluctuations – mainly observed by magnetometers since plasma detectors require larger sampling times – indicate plasma waves of various types, shock waves, and other discontinuities. They will be discussed in Section 5.

3. Theoretical Description of the Large-Scale Properties of the Solar Wind

The basic theory of the hydrodynamic expansion of extended stellar atmospheres has been developed by Parker (1963). The observations have confirmed his general ideas and are helping us now to improve the details of the different possible models. In his first models Parker demonstrated that from the Bernoulli equation alone – assuming an isothermal corona – a supersonic expansion at large distances would be implied if a small expansion velocity exists near the base of the corona and if the pressure at large distances from the Sun is small.

Assuming stationary conditions and spherical symmetry, we obtain from the equation of continuity

$$r^2 \varrho V = \text{const} \tag{1}$$

(r=distance from the origin, ϱ=density, V=radial velocity), and from the momentum balance, neglecting viscosity and magnetic field

$$\varrho V \frac{dV}{dr} = - \frac{dp}{dr} - \varrho \frac{M_\odot G}{r^2} \tag{2}$$

(p=pressure, $M_\odot$ = solar mass, G=gravitational constant) the relation

$$\left[2 - \frac{M_\odot G}{c_s^2 r} \right] \frac{dr}{r} = \left[\frac{V^2}{c_s^2} - 1 \right] \frac{dV}{V}. \tag{3}$$

Here c_s=sound velocity; assuming adiabatic processes $dp/d\varrho = c_s^2$. Now $|dp/dr| > \varrho M_\odot G/r^2$, because the corona is very hot and there exists no pressure far away from the Sun. Therefore, it follows from Equation (2) that $dV/dr > 0$, i.e., the corona is expanding at an ever increasing speed. Equation (3) demonstrates that the expansion velocity will increase outward from subsonic to supersonic if $M_\odot G/(c_s^2 r) > 2$ for $V < c_s$ and if $M_\odot G/(c_s^2 r) < 2$ for $V > c_s$. Sonic velocity is reached at a critical point at $r = r_c$, where

$$r_c = M_\odot G/2c_s^2. \tag{4}$$

For an isothermal, solar corona with $T = 1.0 \times 10^6$ K and $c_s = 1.7 \times 10^7$ cm sec^{-1} we obtain from Equation (4) $r_c = 3.5\ R_\odot$. If the corona is not isothermal it will expand finally with supersonic velocity provided that the radial temperature dependence keeps the critical point at a finite distance. If we assume a polytropic relation $p \propto \varrho^\alpha$ this condition is satisfied as long as α remains smaller than $3/2$ (for an isothermal atmosphere $\alpha = 1$). The solar gravitational force chokes or constricts the flow as does the converging section of a Laval nozzle (Clauser, 1960), and in this way it permits the development of sonic and supersonic flow. Also, the coronal temperature has to be within certain limits. It must be high enough so that the gravitational field cannot hold the corona in static equilibrium (i.e., $|dp/dr|$ has to be sufficiently large), but if the temperature were too high, the corona would be expanding only subsonically, since the influence of the gravitational field would be rather small [in Equation (3), c_s would be very large]. The exact lower and upper bounds for the average temperature depend on radial variation in the temperature, which are determined by the thermal processes involved. The maximum temperature T_m at which supersonic expansion can occur is determined by Equation (4) with $r_c = R_\odot$ and $c_s^2 = 2\gamma kT/m$ (γ = ratio of specific heats, k = Boltzmann constant, m = mass of proton). However, since the temperature decreases towards the chromosphere, a lower temperature in the chromosphere may yield $r_c > 1\ R_\odot$ and the coronal expansion will always be supersonic.

To obtain the values of the density, the velocity, and the temperature near the orbit of the Earth, it is necessary to take into account also the energy equation, in which thermal conduction plays a significant role. However, a number of calculations with a two-fluid model (Sturrock and Hartle, 1966; Hartle and Sturrock, 1968) have shown that thermal conduction alone is not sufficient for supplying the necessary energy. One obtains a too-high electron temperature and a too-low ion temperature. Furthermore, as in one-fluid hydrodynamic models, the density of the wind at 1 AU tends to be higher and the velocity lower than observed. As a consequence, besides conduction and convection, other forms of energy transport must be operative. During quiet conditions these other forms should be effective out to at least $r = 3\ R_\odot$, but during active periods (indicated by higher temperatures, high densities, and high wind velocities) the energy input should be enhanced out to the orbit of the Earth. The additional energy is presumably delivered by waves or by turbulence. The mechanism responsible for heating near the Sun might also be effective in the outer corona, since not all energy in the shock waves will be absorbed in the inner corona. In addition other waves might be heating, for in a collisionless plasma hydromagnetic waves behave different from those in a collision-dominated plasma. During quiet times the fast-mode of magnetoacoustic waves could heat effectively close to the Sun, as proposed by Barnes (1968, 1969). These waves would have frequencies small compared to local ion frequencies, and wavelengths long compared to local Larmor radii. Collisionless damping of hydromagnetic waves is probably important also in a number of other astrophysical phenomena, first of all for heating astrophysical plasmas. Although other kinds of plasma waves (e.g., ion cyclotron waves) are more efficient plasma heaters, in astrophysical situations most wave energy seems to exist in the hydromagnetic modes associated with large-scale plasma motions. Second, wave damping is intimately related to scattering of charged particles by turbulent plasma.

Barnes estimates that a flux of about 5×10^{26} erg sec^{-1} propagates outwards into the outer corona in the form of fast-mode magnetoacoustic waves. Most of this energy is transformed into electron and ion thermal energy within a few $R_\odot$, and it is expected to increase the velocity of the wind at the orbit of the Earth. Beyond a few $R_\odot$, the surviving fast-wave energy (about 5×10^{24} erg sec^{-1}) is propagating parallel to the magnetic field, and hence it is not dissipated. This situation persists out to $r \approx 10\ R_\odot$, where the solar field ceases to be radial. At this point the remaining waves will damp out, supplying additional energy which goes mainly to the protons because of the direction of wave propagation. Since the waves increase only the kinetic temperature parallel to the main magnetic field, the transverse temperature remains unchanged. In this way we can understand the observed anisotropies in the temperature. The higher electron temperature is most probably due to the larger electron heat conduction.

The fast-mode magnetoacoustic waves are generated in the solar corona as remnants of processes occurring in the solar photosphere, such as the granulation or spicules. The waves will have damped out before reaching the orbit of the Earth. Therefore the waves observed at the orbit of the Earth – in particular those with shorter wavelengths (less than 10^7 km) – are generated in the solar wind and do not directly reflect processes occurring at the surface of the Sun. The wavelength spectrum at the orbit of the Earth may reflect some sort of equilibrium between generation and decay of hydromagnetic waves. The strong damping of waves provides a significant heat source for the solar wind.

However, waves generated by internal plasma instabilities cannot heat the wind, since they contribute only to the equalization of the anisotropies, and probably do not produce the long-wavelength waves of interest. Jokipii and Davis (1969) proposed that the interaction between streams, or sectors distributed in solar longitude, of different wind velocity would provide a dominant source for long-wavelength waves that heat the plasma. The velocity differences between such streams represent an energy source corresponding to a thermal velocity of a few hundred km sec^{-1} which is certainly adequate to maintain the solar wind temperature around some 10^5 K. The energy source has also the required property of being variable. In addition it would heat the particles proportionally to their mass, since the interaction produces equal velocity spreads. Therefore we would expect that during perturbed times the ion temperatures are proportional to their mass. This would indeed explain the fact that during active times the temperature of α-particles is about four times that of protons while during cold and quiet times the temperatures are more nearly equal. Another interesting feature of this model is that the sharpest fluctuations and strongest heating occurs within 2 to 3 AU from the Sun, since the fluctuations should be generated as long as the stream-velocity differences persist and interact strongly. If there are, for instance, four fast streams distributed around the Sun with $\Delta V = 100$ km sec^{-1} over the intervening 300 km sec^{-1} slow streams, and each fast stream lags the preceding slow stream by about 45°, then the fast streams, if undeflected by the slow streams, would pass across the slow streams at about 3 AU. More gradual, very-long-wavelength fluctuations are confined to a region whose heliocentric radius has an upper limit of 15 to 30 AU. Beyond this point the wind should cool adiabatically, and both the magnetic structure and the velocity field seem likely to be quite uniform except perhaps for the short-wavelength fluctuations generated by the anisotropy of the temperature. This result is in agreement with recent observations of galactic cosmic rays which indicate that the boundary of sensible modulation by the solar wind is roughly at 5 to 10 AU at solar minimum (Meyer *et al.*, 1956; O'Gallagher and Simpson, 1967; Simpson and Wang, 1967). The propagation or diffusion of cosmic-ray particles in the solar system may be determined from the observed power spectrum of magnetic irregularities with wavelengths roughly equal to the particle cyclotron radius. These would be of the order of 10^5 to 10^7 km for cosmic-ray particles with energies between 10 MeV and 10 GeV per nucleon in the 5×10^{-5} G average interplanetary magnetic field.

As just outlined the model calculations showed that thermal conduction and convection alone is not sufficient for producing the observed temperature, but that the damping of certain waves can explain the observed parameters in the solar wind. Direct evidence of the coronal gas before it enters the wind is contained in the observed state of ionization of oxygen. Calculations by Hundhausen *et al.* (1968), show that the state of ionization remains essentially fixed beyond about two solar radii since no more recombinations take place. In the solar wind O^{+6} is more abundant than O^{+7}, implying that the temperature in the low corona is usually below about 2.2×10^6 K. But sometimes the observations show O^{+7} several times more abundant than O^{+6}, indicating the temperature of the coronal source region near 3×10^6 K. The observations with the Vela space vehicle show, furthermore, that coronal sources at widely varying temperatures can give rise to a solar wind with the same low temperature ($\approx 10^4$ K) at 1 AU. This is another indication that the ion temperature in the interplanetary space is largely determined by processes affecting the particles well out in the interplanetary region.

Finally it should be mentioned that the observed He^+ is not consistent with the establishment of the ionization state of helium deep in the corona and thus must indicate some local interplanetary modifications of that state.

4. The Transport of Angular Momentum

A. TRANSPORT OF ANGULAR MOMENTUM BY THE MAGNETIC FIELD

So far only radial motions have been considered. Now we discuss also the azimuthal motions and their relation to the magnetic field. While the magnetic-energy density is small compared to the kinetic energy density at 1 AU, the situation is reversed near the Sun. There, assuming a 1 G field, the magnetic energy density is larger than the thermal energy density even in quiet regions. Therefore the magnetic field will certainly influence the motions of the gas.

The large-scale picture of the magnetic field in the interplanetary space is determined by the outward motion of the solar wind, and the solar fields are extended out into the solar system. Due to the rotation of the Sun the field should have a spiral structure which is indeed observed. For the simple case of a constant radial wind velocity V the field beyond r_0 is given by (Parker, 1958):

$$B_r(r, \theta, \varphi) = B(\theta, \varphi^*)\left(\frac{r_0}{r}\right)^2 \tag{5}$$

$$B_\varphi(r, \theta, \varphi) = B(\theta, \varphi^*)\frac{r_0}{r}\frac{r_0\Omega}{V}\sin\theta \tag{6}$$

$$B_\theta(r, \theta, \varphi) = 0, \tag{7}$$

when $B(\theta, \varphi)$ is the field at $r = r_0$, when θ and φ represent the polar and azimuthal angles, when $\Omega (= 2.5 \times 10^{-6}$ radian sec^{-1}) is the angular velocity of the Sun and when $\varphi^* \equiv \varphi + r\Omega/V$. Assuming a field of 1 G at the surface of the Sun one obtains 3×10^{-5} G at the orbit of the Earth, giving an Alfvén speed $V_A = 30$ km sec^{-1} for a particle density of 4 proton cm^{-3}. Thus at 1 AU the Alfvén speed is of the same order as the thermal velocity at a temperature of about 5×10^4 K.

Near the Sun the magnetic field is sufficiently strong, and it will force the gas to co-rotate. In this way the gas will take away some angular momentum from the Sun. But the magnetic field, which has an azimuthal component B_φ, will also remove solar angular momentum. Before we discuss this problem in more detail, we shall estimate the torque exerted on the Sun by the Maxwell stress of the magnetic field (Parker, 1958). The total torque exerted on the Sun is given by

$$Z = r^3 \int_0^\pi d\theta \sin^2 \theta \int_0^{2\pi} d\varphi \, \frac{B_r B_\varphi}{4\pi}. \tag{8}$$

If we assume that $B(\theta, \varphi)$ is uniform over the entire Sun with a field strength B_0, it follows that the total torque (and therefore the rate of outward transport of angular momentum) is given by

$$Z = \frac{1}{3} r_0^4 \frac{\Omega}{V} B_0^2 = \frac{1}{3} \frac{\Omega}{V} (r^2 B_r)^2. \tag{9}$$

This rate is independent of r. Assuming $V = 400$ km sec^{-1} and $B_0 = 2$ G (corresponding to 5×10^{-5} G at 1 AU in the radial direction), the torque is 4×10^{30} dyne cm. If the Sun rotates rigidly its angular momentum is about 2×10^{48} g cm^2 sec^{-1} and the outer convection zone might have about one tenth of this. Hence the characteristic deceleration time is 0.5×10^{18} sec or 2×10^{10} yr for the entire Sun and 2×10^9 yr for the convection zone alone, as compared with the present age of 5×10^9 yr of the Sun.

B. TRANSPORT OF ANGULAR MOMENTUM BY THE OUTFLOWING GAS

So far we have neglected any azimuthal motion and thereby any transport of angular momentum by the gas. While for the Sun this approximation is sufficient (as we will see), the situation is different for stars with higher rotational velocities and stronger magnetic fields. We will now discuss the study of this problem by Weber and Davis (1967a). Again stationary conditions, spherical symmetry and infinite electrical conductivity are assumed and the dynamical equations are considered only in the vicinity of the equatorial plane. The additional equations which have to be taken into account are, first,

$$\text{curl}[\mathbf{V} \times \mathbf{B}] = 0 \tag{10}$$

and, second, the equation of motion in the φ-direction. The last one can be integrated yielding the sum of the angular momentum of the gas and the torque by the magnetic field. This sum, called L, must be a constant. Then,

$$rV_\varphi - \frac{rB_r B_\varphi}{4\pi\varrho V_r} = L = \Omega r_A^2. \tag{11}$$

Here V_φ = azimuthal velocity and r_A the distance where $V_r = B_r/\sqrt{(4\pi\rho)}$. Thus L, the total angular momentum per unit mass, is determined by the position of the 'Alfvén critical distance' r_A. The azimuthal velocity at this distance is of the order of the velocity of rigid rotation. At larger distances V_φ has the form

$$V_\varphi \approx \frac{\Omega r_A^2}{r}\left(1 - \frac{V_{rA}}{V_{r\infty}}\right) \tag{12}$$

where V_{rA} and $V_{r\infty}$ are the radial velocities at $r = r_A$ and for large r respectively. Weber and Davis used the following parameter values: $V_r = 400$ km sec^{-1}, $n = 7$ proton cm^{-3}, $B = 5 \times 10^{-5}$ G at 1 AU and $B = 2.4$ G at $r = r_\odot$. Adopting a polytropic index $\alpha = 1.229$, they find $V_\varphi = 1$ km sec^{-1} at 1 AU. Three quarters of the angular momentum is carried away by the magnetic field and one quarter by the gas ($r_A = 24.6\, r_\odot$ and V_φ reaches a maximum with $V_\varphi = 2\, r_\odot\Omega$ at $r = 11.5\, r_\odot$).

C. PROBLEMS FOR THE SOLAR CASE; APPLICATION TO OTHER STARS

The Weber and Davis model predicts an azimuthal velocity of the solar wind near the Earth's orbit of about 1 km sec^{-1} (conservation of angular momentum would yield only about 10^{-2} km sec^{-1}). The major part of the angular momentum is transported by the magnetic field. Therefore the simple picture outlined in Section 4a is not substantially changed. However, the theoretically predicted azimuthal velocity does not agree with the observed value, for the observed angle of about 1.5° co-rotation between the direction of the solar wind and the radial direction gives about 10 km sec^{-1}. From the directions of comet tails Brandt and Heise (1970) had found a similar value. An azimuthal velocity as high as 10 km sec^{-1} at the orbit of the Earth would give a considerably higher transport of angular momentum. The exerted torque would be of the order of 7×10^{30} dyne cm^{-1}, assuming $V_\varphi \propto \sin\theta$. Such a large torque would decelerate the rotation of the entire Sun in 10^{10} yr and the outer part of the convection zone in 10^9 yr.

There still remains the difficulty of understanding how the solar wind can attain this high azimuthal velocity. Weber and Davis (1967b) as well as Meyer and Pfirsch (1969) have re-examined the problem with inclusion of viscosity and an anisotropic pressure tensor. Taking into account the latter, Meyer and Pfirsch obtain for the azimuthal velocity at large distances from the Sun:

$$V_\varphi = \frac{\Omega r_A^2}{r}\left[1 - \frac{r^2}{r_A^2}\frac{B_r^2}{4\pi\varrho V_r^2}\left(1 - \frac{p_\parallel - p_\perp}{B^2/4\pi}\right)\right] \tag{13}$$

where $p_\parallel$ and $p_\perp$ are the pressure components parallel and perpendicular to the magnetic field respectively.

This shows that an anisotropy of the pressure can increase the tendency towards co-rotation beyond 5 $R_\odot$ and can give a much higher azimuthal velocity near the orbit of the Earth. For instance if the solar wind is near the limit of the 'firehose' instability with

$$p_\parallel - p_\perp \approx B^2/4\pi \tag{14}$$

then the azimuthal velocity would be of the order of

$$V_{\varphi} \approx r\Omega r_A^2/r^2 = 7.5 \text{ km sec}^{-1}. \tag{15}$$

Schubert and Coleman (1968) have examined the transport of angular momentum from the Sun by hydromagnetic waves and showed that this mechanism may also be important in producing co-rotation of the plasma. Finally, it should be mentioned that the interaction of fast and slow streams can give a higher rotational velocity directed towards co-rotation, as Siscoe *et al.* (1968) have shown.

The problem of transport of angular momentum from a star to the surrounding interstellar gas had been investigated some time ago by Lüst and Schlüter (1955). They did not take into account the corpuscular radiation from the star, but assumed that force-free (or nearly force-free) fields connected to the star would be strong enough to enforce co-rotation out to some distance from the star. The angular momentum would be transported to distances from which turbulent friction would be effective.

Schatzman (1959) later connected this with the existence of a stellar wind, and he pointed out that if gas emitted by a star is magnetically constrained to co-rotation with the star out to distances large compared to the stellar radius, then a small amount of mass-loss would yield a significant loss of angular momentum, because of the effective increase of the moment of inertia of the gas during the outflow.

Mestel (1968) has linked this general picture with the theory of a quasi-stationary stellar wind, introducing important modifications to include the presence of strong centrifugal and magnetic fields. Mestel assumes a poloidal stellar field, which near the star should be roughly dipolar and strong enough to force the flow to follow the field and to keep the gas approximately co-rotating with the star. Farther out the gas flow drags the field, and each element approximately conserves its angular momentum. The two regions will be separated by a limiting field-line, the one where at the equator the wind speed just equals the local Alfvén speed. The theory is not applicable to the Sun and similar stars since the lines of force in the interplanetary space are not part of the Sun's polar field but emerge from near-equatorial regions with somewhat stronger fields than the polar regions. But Mestel's theory should describe the phenomena for a star contracting towards the main sequence, either down the Hayashi track, with the bulk of the star convective, or on the subsequent Kelvin-Helmholtz-type path, with the outer layers still strongly convective, and so still generating a corona.

Mestel showed that such a star can lose a substantial part of its angular momentum without losing a large percentage of its mass if the stellar magnetic field is just strong enough to keep the gas beneath the coronal base co-rotating with the star.

5. Fluctuations, Discontinuities and Waves

The observations of the solar wind and particularly of the imbedded magnetic fields show that there exist fluctuations in the plasma and magnetic fields which reflect the 'turbulent' structure of the motion. Furthermore one can distinguish different forms of discontinuities – sometimes shock waves have been identified and sometimes other waves. Very probably these phenomena are important in understanding the nature of the solar wind, in particular for the energy transport and the energy distribution with respect to the direction of the magnetic field as well as among the different constituents of the solar wind.

In analyzing the observed data one has to take into account that the Alfvén velocity in the wind at 1 AU is typically 60 km sec^{-1} (5×10^{-5} G and 4 H-atom cm^{-3}) and the thermal velocity is 15 to 100 km sec^{-1} (10^4 to 5×10^5 K). Both these velocities are small compared to the 300 to 700 km sec^{-1} observed for the velocity of the wind. Thus the observed temporal variations can be transformed directly into spatial variations along the solar wind direction. The frequency f of a fluctuation corresponds to the wavelength $\lambda = V/f$ in the radial direction where V is the wind velocity.

Finally it should be mentioned that most observations of fluctuations refer to magnetic fields since the time resolution of the magnetometers is at present much higher than that of particle detectors. However, for a real understanding of the observed phenomena measurements of the other quantities are also needed.

The power spectra of the three magnetic field components and of the total field show about the same dependence on f, and normally the lowest power is in the magnitude of the field, indicating that the fluctuations are transverse waves. The power in the radial component is usually less, by a factor of almost two, than the power in the two transverse components. Therefore the fluctuations below about 1 Hz could well be transverse hydromagnetic waves. Waves with a frequency of 10^{-1} Hz have a wavelength of the order of 500 km (assuming a propagation velocity of 50 km sec^{-1}) and appear as fluctuations with a frequency $f \geqslant 1$ Hz in the fixed frame of reference.

Below 10^{-5} Hz the power spectrum is flat. The spectrum at these very low frequencies originates mainly in time variations on the Sun and in features which co-rotate with the Sun. At higher frequencies the power spectrum declines with increasing f. There seems to be a change in the power spectrum over the years; the results so far are: 1962: $p \propto f^{-1}$ for $10^{-5} \leqslant f \leqslant 10^{-2}$; 1965: $p \propto f^{-3/2}$ for $10^{-4} < f < 10^0$ and, 1966: $p \propto f^{-1}$ to $p \propto f^{-3/2}$ for $10^{-6} < f \leqslant 10^{-4}$ and $p \propto f^{-2}$ for $10^{-4} < f < 10^{-1}$. The spectrum is the same on active, intermediate, and quiet days. On active days the r.m.s. fluctuations are in excess of the large-scale field, and even in quiet days the fluctuations have occasional peaks comparable to B.

Scintillation of radio point sources is due to fluctuations in the interplanetary electron density. Observations indicate that the fluctuations have a correlation length of about 100 to 200 km corresponding to frequencies of about 1 to 10^{-1} Hz. However, fluctuations in the magnetic field have a correlation length of about 2×10^6 km according to a recent investigation by Jokipii and Hollweg (1970). This length is defined as the characteristic scale beyond which the two-point correlation function falls rapidly towards zero. It corresponds to the point below which the frequency spectrum is nearly flat. Jokipii and Hollweg show that the large value of the correlation length is nevertheless in agreement with the observed interplanetary scintillation and that the 100 to 200 km scale refers to the 'inner scale' of the fluctuations of that wavelength below which there is little power. Most probably the observed fluctuations supply significant energy to the solar wind. In Section 3 I mentioned that the hydromagnetic waves are strongly damped. Therefore it is unlikely that waves with wavelengths less than about 2×10^6 km ($f \geqslant 2 \times 10^{-4}$ Hz) will reach the Earth. Thus the longer wavelengths (lower frequencies) may be of solar origin but the rest of the spectrum, apparently, must originate in space. The mechanism proposed by Jokipii and Davis (see Section 3) may be responsible for the occurrence of these fluctuations.

The other mechanism which may explain the fluctuations at higher frequencies ($f \geqslant 10^{-3}$ Hz), is based on thermal anisotropies in the wind. These have been observed, but it is not yet clear whether they are able to drive the instabilities, since the theoretical interpretation of the observation is rather complicated. In summary only the following should be stated here. If the temperature parallel to the magnetic field $T_{\parallel}$ exceeds the temperature perpendicular to the magnetic field $T_{\perp}$, we may expect the so-called 'firehose' instability. If one takes into account the resonance effect at the ion cyclotron frequency and the effect of the finite ion cyclotron radius, the instability can occur only when the total pressure of the gas exceeds the magnetic pressure. Under normal conditions ($n = 5$ cm^{-3} and $T_p \approx 5 \times 10^4$ K) the ion pressure is too small, and the ion anisotropy cannot drive the firehose instability, except during very active times when the temperature is very high. But the electrons are very hot and they may be able to produce both instabilities, namely the resonance instability occurring near the ion cyclotron frequency, and the 'firehose' instability. In this way the observation of fluctuations with high frequencies ($f > 10^{-3}$ Hz) might be understood.

Besides these general fluctuations, individual fluctuations have been observed such as shock waves, shear planes, and oscillatory waves. But we shall not discuss them here.

6. The Transition into the Interstellar Space

A. THE TRANSITION SHOCK FRONT

The solar wind will finally mix with the interstellar gas. The related transition has been studied by several investigators (Axford *et al.*, 1963; Parker, 1961; Hundhausen, 1968a; Brandt, 1964). Davis (1955) pointed out that the dynamical pressure of the solar wind, which falls off as r^{-2}, will be balanced at some heliocentric distance r_s by the galactic magnetic field. Since the solar wind has supersonic velocities, it has to pass through a collisionless shock transition at r_s (Parker, 1962, 1966). The dynamical pressure of the solar wind is about

$$p_s \approx 2 \times 10^{-8} \frac{1}{r^2} \quad \text{dyne cm}^{-2} \tag{16}$$

whereas the interstellar pressure is of the order $p_i \approx 10^{-12}$ dyne cm^{-2} (this is the sum of magnetic pressure, gas pressure, and cosmic-ray pressure). These values of p_i and p_s give $r_s = 100$ AU. The main uncertainty in this estimate is due to our imperfect knowledge of p_i in the neighborhood of the Sun. But other effects – to be discussed below – enter too. At 100 AU the solar wind density before the shock front would be about 10^{-4} proton cm^{-3}. The shock raises the temperature to about 10^7 K by thermalization of kinetic energy of mass motion; the density increases by a factor 4 if the ratio of specific heats is $\frac{5}{3}$. If other processes take place in this transition, the density can be higher.

B. THE SOLAR HII REGION

It certainly would make a difference whether the solar wind would be embedded in a HI or in a HII region. The Sun is a rather cool star so that its Strömgren sphere will have a radius much less than a parsec. Williams (1965) has calculated the size of the solar HII region taking into account the observed far-ultraviolet solar spectrum and assuming that the interstellar gas is at rest with respect to the Sun. It turns out that the outer boundary of the solar HII region is not very sharp, but drops off gradually, a consequence of the presence of penetrating radiation with wavelengths around 200 Å. The ionization falls to 50% at 1500 AU if $n_i = 1$ atom cm^{-3} (n_i = density of the interstellar gas), at 400 AU if $n_i = 10$ atom cm^{-3} and at 120 AU if $n_i = 100$ atom cm^{-3}. According to this calculation the shock transition would lie inside the HII region. The solar wind passing through the shock front would remain ionized due to the large recombination times.

C. THE EFFECT OF CHARGE EXCHANGE WITH NEUTRAL INTERSTELLAR ATOMS

The surrounding interstellar gas is moving with respect to the Sun with a velocity of 20 km sec^{-1} towards the solar apex near $\alpha = 18^h$ and $\delta = +30°$. In addition it may fall towards the Sun due to the solar gravitation. At 10^3 AU the gravitational escape velocity is only of the order of 1 km sec^{-1} where the radiation pressure in the solar Ly-α line has been taken into account. If we assume that the gas is neutral at large distances from the Sun ($r > 10^4$ AU) the neutral hydrogen atoms will still come quite close to the Sun before they are ionized either by photoionization or by charge exchange with the solar wind. The photoionization time scale τ_{ph} is about

$$\tau_{ph} \approx r^2/\alpha \quad [\text{sec}] \tag{17}$$

where r = distance from the Sun in AU and α = photoionization rate at 1 AU = 10^{20} cm^2 sec^{-1}. The neutral atoms will be ionized by photoionization between 5 to 10 AU from the Sun depending on the direction with respect to the apex. The time scale for charge exchange τ_{ch} is given by the lifetime of a neutral atom in a plasma of N atom cm^{-3}

$$\tau_{ch} \approx 10^7/N \quad [\text{sec}] \tag{18}$$

for velocities of 10^7 to 10^8 cm sec^{-1}. With numbers chosen by Parker one obtains a radius r_{ch} of about 30 AU which an inward-moving neutral hydrogen atom will reach before it is ionized by charge exchange with the solar wind.

We see that charge exchange can influence the solar wind quite appreciably and the wind might be degraded [*i.e.*, neutralized, Ed.] before it reaches the shock front. This is possible when the flux of neutral interstellar atoms is as large or larger than the flux of ions in the solar wind. Thus serious degradation cannot occur unless there is at least one interstellar atom supplied for each wind ion. The wind flux VN_0 falls to the level of the interstellar flux of neutral atoms $N_H U$ at

$$r_d = r_0 \left(\frac{VN_0}{UN_H} \right)^{1/2}. \tag{19}$$

With $V = 400$ km, $N_0 = 5$ atom cm^{-3}, $U = 3$ km sec^{-1} (400 K) and $N_H = 1$ cm^{-3} we get $r_d = 15$ AU. The values of r_d and r_{ch} suggest that the solar wind could be seriously degraded by interstellar hydrogen at 30 AU. This effect can bring the shock front closer to the Sun than is estimated above. Therefore, taking into account all the uncertainties mentioned, we conclude that the shock front may be found anywhere between 30 and 300 AU.

The charge exchange will convert solar wind protons into fast neutral hydrogen atoms and interstellar hydrogen atoms into relatively slow ions. If this occurs mainly outside the shock front, the fast or 'hot' neutral hydrogen atoms will have almost random direction. Ultimately the ions and electrons will disappear by radiative recombination but this is a very slow process and it requires a large volume. The lifetime for radiative recombination τ_r for a hydrogen plasma is given by

$$\tau_r \approx 3 \times 10^{21} \, (T^{3/4}/n^2) \quad [\text{sec}]. \tag{20}$$

If the recombination is in equilibrium with the production one obtains for $T = 10^4$ K and $n = 1$ cm^{-3} a radius of the order of 10^3 AU.

In the region between this outer boundary and the shock front, the lines of force of the solar magnetic field will mix with those of the interstellar field. Plasma-instabilities, such as the Rayleigh-Taylor instability, could be of importance.

With respect to stars other than the Sun I should finally mention the discussion by Newman and Axford (1968) of an isothermal corona expanding to such great distances that recombination occurs. This study may be of interest for stars with high surface temperature, such as planetary nebulae, with an outflow of 50 km sec^{-1} and 10^4 atom cm^{-3}. In such a case a recombination front occurs before one passes through a shock transition.

References

Axford. W. J.: 1968, *Space Sci. Rev.* **8**, 331.
Axford, W. J., Dessler, A., and Gottlieb, B.: 1963, *Astrophys. J.* **137**, 1268.
Barnes, A.: 1968, *Astrophys. J.* **154**, 751.
Barnes, A.: 1969, *Astrophys. J.* **155**, 311.
Brandt, J. C.: 1964, *Icarus* **3**, 253.
Brandt, J. C. and Heise, J.: 1970, *Astrophys. J.* **159**, 1057.
Clauser, F. H.: 1960, Contribution from the Johns Hopkins University AFOSR TN 60-1386.
Davis, L.: 1955, *Phys. Rev.* **100**, 1440.
Dessler, A. J.: 1967, *Rev. Geophys.* **5**, 1.
Deutsch, A.: 1958, *Astron. J.* **63**, 49.
Deutsch, A.: 1960, in *Stars and Stellar Systems*, Vol. VI (ed. by J. Greenstein), University of Chicago Press, Chicago, p. 543.

Deutsch, A.: 1961, in *Aerodynamic Phenomena in Stellar Atmospheres* (ed. by R. N. Thomas), Zanichelli, Bologna, p. 238.

Deutsch, A.: 1969, in *Mass Loss from Stars* (ed. by M. Hack), Reidel, Dordrecht, The Netherlands, p. 1.

Deutsch, A.: 1969, in *Mass Loss from Stars* (ed. by M. Hack), Reidel, Dordrecht, The Netherlands, p. 1.

Egidi, A., Pizella, G., and Signorini, C.: 1969, *J. Geophys. Res.* **74**, 2807.

Hartle, R. E. and Sturrock, P. A.: 1968, *Astrophys. J.* **151**, 1155.

Hundhausen, A. J.: 1968a, *Planetary Space Sci.* **16**, 783.

Hundhausen, A. J.: 1968b, *Space Sci. Rev.* **8**, 690.

Hundhausen, A. J., Gilbert, H. E., and Bame, S. J.: 1968, *Astrophys. J. Lett.* **152**, L3.

Jokipii, J. R. and Davis, L.: 1969, *Astrophys. J.* **156**, 1101.

Jokipii, J. R. and Hollweg, H. V.: 1970, *Astrophys. J.* **160**, 745.

Lüst, R.: 1962, *Space Sci. Rev.* **1**, 522.

Lüst, R. and Schlüter, A.: 1955, *Z. Astrophys.* **38**, 190.

Mestel, L.: 1968, *Monthly Notices Roy. Astron. Soc.* **138**, 359.

Meyer, P., Parker, E. N., and Simpson, J. A.: 1956, *Phys. Rev.* **104**, 768.

Meyer, F. and Pfirsch, D.: 1969, *Kleinheubacher Berichte 243*.

Mustel', E.: 1964, *Space Sci. Rev.* **3**, 319.

Ness, N. F.: 1968, *Ann. Rev. Astron. Astrophys.* **6**, 79.

Newman, R. C. and Axford, W. J.: 1968, *Astrophys. J.* **151**, 1145.

O'Gallagher, J. J. and Simpson, J. A.: 1967, *Astrophys. J.* **147**, 819.

Ogilvie, K. W. and Wilkenson, T. D.: 1969, *Solar Phys.* **8**, 435.

Parker, E. N.: 1958, *Astrophys. J.* **128**, 664.

Parker, E. N.: 1961, *Astrophys. J.* **134**, 20.

Parker, E. N.: 1962, *Planetary Space Sci.* **9**, 461.

Parker, E. N.: 1963, *Interplanetary Dynamical Processes*, Interscience, New York.

Parker, E. N.: 1965, *Space Sci. Rev.* **4**, 666.

Parker, E. N.: 1966, *Planetary Space Sci.* **14**, 371.

Parker, E. N.: 1969, *Space Sci. Rev.* **9**, 325.

Schatzman, E.: 1959, *IAU Symposium No. 10*, Cambridge University Press, Cambridge, p. 129.

Schubert, G. and Coleman, P. J.: 1968, *Astrophys. J.* **153**, 943.

Simpson, A. J. and Wang, J.: 1967, *Astrophys. J. Lett.* **149**, L73.

Siscoe, G. L., Davis, L., Coleman, P. J., Smith, E. J., and Jones, D. E.: 1968, *J. Geophys. Res.* **73**, 61.

Sturrock, P. A. and Hartle, R. E.: 1966, *Phys. Rev. Lett.* **16**, 628.

Weber, E. J. and Davis, L.: 1967a, *Astrophys. J.* **148**, 217.

Weber, E. J. and Davis, L.: 1967b, *Trans. Am. Geophys. Union* **48**, 171.

Weymann, R.: 1960, *Astrophys. J.* **132**, 380.

Weymann, R.: 1963, *Ann. Rev. Astron. Astrophys.* **1**, 97.

Wilcox, J. M.: 1968, *Space Sci. Rev.* **8**, 258.

Williams, R. E.: 1965, *Astrophys. J.* **142**, 314.

18. DISCUSSION FOLLOWING LÜST'S REPORT

(Sunday, September 14, 1969)

Chairman: R. WEYMANN

Editor's remark: This discussion actually took place after the Reports by Lüst and by Pottasch. However, the remarks pertaining to Pottasch's Report have been combined with the discussion following Boyarchuk's Report, since the latter two Reports appeared to have more in common than the first two. I have condensed several contributions, notably a very long remark by Gordon.

Severnyi: I would like to discuss a comparison of Crimean measurements of the average solar magnetic field with direct measurements of interplanetary magnetic fields made with the aid of Explorer 33 and 35 (Wilcox *et al.*, 1969). Let me mention first that we have been measuring the average solar magnetic field since 1967. We use the parallel beam of solar light passing through the entrance slit of the solar magnetograph. So we are recording the average magnetic field of the Sun, as if it were a star, where the weighting function is the distribution of brightness over the disk. Figure 1 shows the results. In the upper half we plot as a function of time the average solar field and the results on the interplanetary fields by the Explorers (a small time delay is taken into account). You see that there is an almost complete coincidence of sign between the general magnetic field and the interplanetary field. Even very small peaks in the general field, which I did not think had a real meaning, correspond closely to the reversals obtained by the Explorers. I want to stress that the general magnetic field is essentially a background field and not the resultant field of all the sunspots together. Indeed, in the lower half of Figure 1 the total sunspot field is shown. It clearly is in antiphase with the general magnetic field. This probably means that there is some tendency to compensate the excess of magnetism of background field. Now I wish to stress that such a close correspondence between the solar magnetic field and the interplanetary field can arise only if the field on the disk is ordered on a large scale. This supposition is supported by results obtained a couple of years ago by Wilcox for the solar wind. (Wilcox, J. M., Severnyi, A., and Colburn, D. S.: 1969, *Nature* **224**, 353.)

Meyer: I would like to add to this picture some recent results obtained by Howard at Mt. Wilson Observatory, concerning large-scale velocity structures on the solar surface. With high Doppler resolution Howard observed apparently systematic upward and downward velocities appearing in alternating areas covering the Sun in about the same way that these magnetic sectors do. It will be very interesting to learn about a possible correlation between the two phenomena and their relation to the hypothetical 'giant granulation cells'. Structures of a similar kind are also indicated by recent hydrodynamic investigations by Busse dealing with the convection zone patterns in rotating stars and in the Sun.

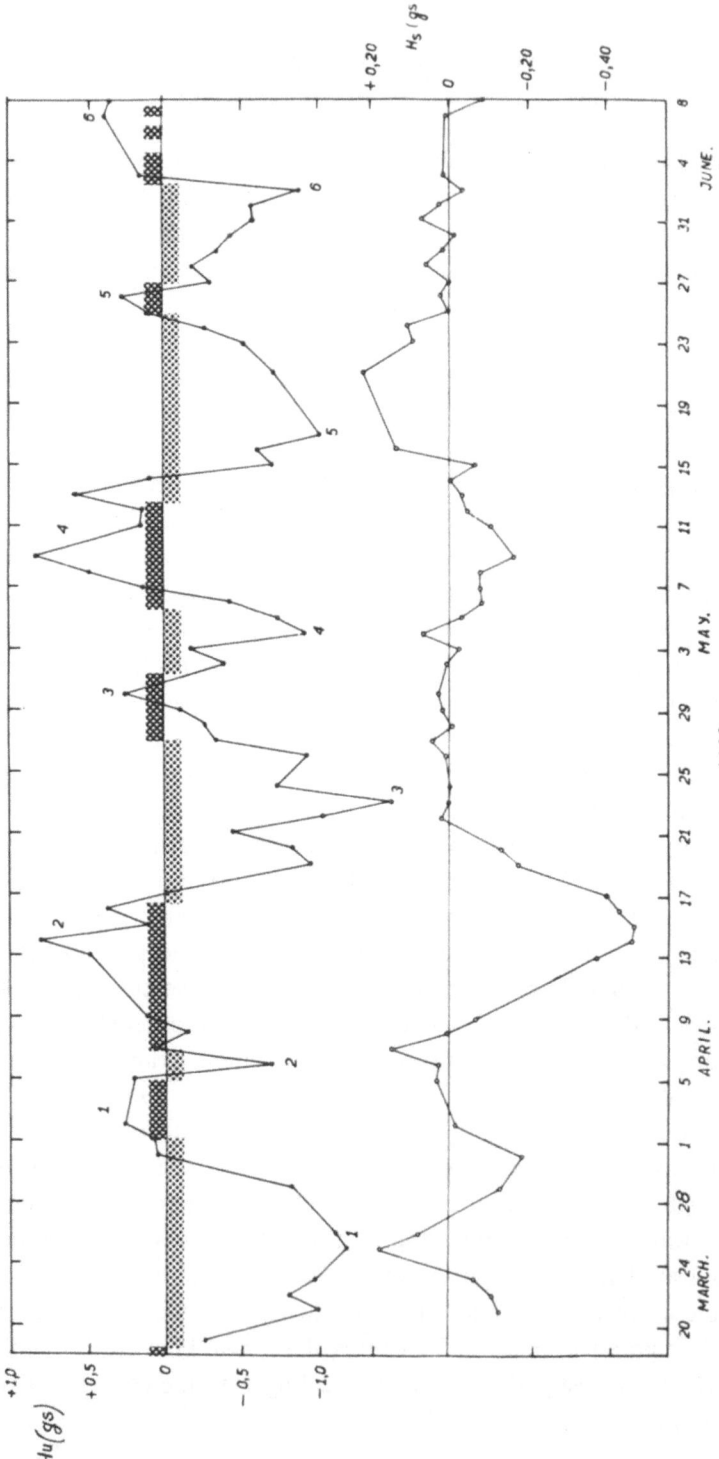

Fig. 1. (See the remark by Severnyi.) Top: mean value of the solar magnetic field (dots) and polarity of the interplanetary magnetic field (bars). The interplanetary field is displaced by five days to allow for the transit time of solar wind plasma from the Sun to the Earth. Bottom: contribution of sunspot magnetic fields to the mean solar field shown above. (Figure taken from Wilcox, J. M., Severnyi, A., and Colburn, D. S.: 1969, *Nature* **224**, 353.)

Burke: Dr. Severnyi, to get a good agreement what time delay was put in between the solar measurements and the satellite data?

Severnyi: A time delay of 4.2 days.

Mestel: A question to Dr. Lüst. I presume that you imagine a 2 G solar field being extended radially all the way to the Earth, yielding 50 μG by the inverse square law of flux conservation. But I thought Wilcox's work had shown that the field observed near the Earth is correlated with regions at the solar equator where the field is rather stronger than the 1 G or 2 G field at the solar poles. In that case the field near the Sun would have to fall off somewhat more rapidly than $1/r^2$ – say, rather like $1/r^3$ – going over into a $1/r^2$ law further out. The point is important for the braking problem (especially for the stars with strong general magnetic fields). The torque on the star is quite sensitive to the detailed field structure.

Lüst: The 2 G field is obtained when one uses the simple expression given in my Report [Equation (5), p. 254] that the radial component force is equal to $1/r^2$ and takes 50 μG near the Earth. The azimuthal angle at the distance of the Earth transcriber is then 135°.

Parker: You can add, that in general the gas pressure in the corona is capable of extending a field that is weaker than 2 G. Stronger fields resist expansion and you know that the magnetic field in space is from the weak field regions of the Sun.

Mestel: I agree that of the flux emanating from a local bipolar region with a strong field, only a part should be able to escape from the Sun and reach the neighborhood of the Earth; and I believe that Wilcox's observations confirm this. I would expect near the Sun both a dead zone, in which the field is strong enough to resist distortion, and a wind zone, with the flow near the Sun channelled along the quasi-dipolar field-lines, but with the same lines further out too weak to resist being pulled into a roughly radial form by the wind. The relationship between the field strength seen at the Earth and that at the base of the same line as it emerges from the Sun will be more complicated than a simple $1/r^2$ law.

Meyer: In one other respect the 2 G extrapolation must be quite good. Though the flux tubes leading into the solar wind might have come from several restricted regions on the solar surface, as Mestel just mentioned, the flux must have smoothed out any irregularities by the time it has reached distances of the order of 10 to about 20 $r_\odot$, and must have become very nearly radial everywhere. The reason is that at this distance all the multipole contributions of the inhomogeneous distribution on the surface have dropped off to quite small values. On the other hand the magnetic tensions are still the dominant dynamical feature. This requires that the magnetic field is in equilibrium with itself, that is, that the field is radial (except for the small 'garden hose' angle) and of equal strength on spheres around the Sun. Thus, though one only observes the magnetic flux at the orbit of the Earth, the 2 G should also stand for all other directions.

Konyukov: I want to discuss the solar wind solution found by Parker and, in particular, the problem of heat flow. Essential to Parker's solution is the existence of a singular point, where the flow becomes supersonic. Under the assumption that the gas flow is stationary and spherically symmetrical I have expanded the heat flow function W in a series with the inverse Reynolds number, R_{ey}^{-1}, as an expansion parameter. The expansion is valid only if energy dissipation is unimportant. In the zero-order approximation the heat flow is the same as in Parker's solution. However, for the coronal and interplanetary plasma $R_{ey}^{-1} \approx 1$ and the expansion is invalid, implying that Parker's solution is also invalid. [Condensed. (Ed.)] [Konyukov, M. V.: 1967, *Geomagn. Aeronom.* (Akad. Nauk SSSR) **7**, 577.]

Lüst: In a recent review, Parker (1969) has discussed different models of the solar wind under different assumptions for the energy equation. (Parker, E. N.: 1969, *Space Sci. Rev.* **9**, 325.)

Konyukov: I only wanted to mention that, if $R_{ey}^{-1} \approx 1$, we have no singular (sonic) point. I understand that magnetic fields and plasma turbulence decrease R_{ey}^{-1}, but these processes introduce other difficulties.

Thomas: Dr. Lüst, the energy input into the fast modes that you require is more than 10^4 erg cm^{-2} sec^{-1}. What was the figure you and Biermann derived?

Lüst: I remember a figure of about 10^6 erg cm^{-2} sec^{-1} for the chromosphere and the corona. We derived this number in two ways: (a) By estimating the production of acoustic noise at the upper part of the hydrogen convection zone. (This estimate is uncertain since it is difficult to calculate the efficiency of turbulence in generating acoustic noise); (b) By estimating the energy losses at the various levels. Both figures were in reasonable agreement and indicated about 10^6 erg cm^{-2} sec^{-1}.

Thomas: You remember that at the preceding Gas Dynamics Symposium in Nice there was a big uncertainty about this figure. And now you are saying that 1 per cent of this flux will go out to $10r_\odot$ and that you really need this! Well, all I will do is caution. The aerodynamic/hydrodynamic theory is very uncertain. Perhaps if we fix what is required to heat the chromosphere, actually the low corona, and if we fix the requirement at large distances from the Sun, we can perhaps tie down all that is needed. I would prefer such an estimate, because I do not believe any theoretical predictions on heat production and energy transport.

Tsytovich: Dr. Lüst, you say that the region in which the solar wind is observed is collisionless and that there are something like colliding beams. If these collisions are collective, what efficiency should you have for heating?

Lüst: I would like to pass the buck* to Davis.

Davis: The temptation is to assume essentially 100 per cent efficiency over a sufficient distance. Near the Earth streams are observed with velocities ranging from 300 km sec^{-1} to 700 km sec^{-1}. These streams will interact and give all kinds of waves; the waves will damp, according to Barnes, and eventually, by the time you get to Jupiter, there are no streams left.

Baranov: Dr. Lüst, you mentioned that the temperature anisotropy of electrons is less than the temperature anisotropy of ions in the solar wind plasma. This is an experimental result obtained by Vela 4. How do you explain this effect?

---

* To pass the buck: Colloquial (American) expression meaning: to throw the responsibility to another person. Buck also means a gymnastic apparatus for vaulting over. Here both interpretations seem possible. (Ed.)

Lüst: It can be explained by the mechanism proposed by Barnes. According to him the fast hydromagnetic waves would transfer energy to the protons, preferentially along the magnetic lines of force.

Parker: There is another effect, too, that the electrons have a large enough thermal conductivity to keep themselves warm at large distances from the Sun. But the protons have a very small thermal conductivity, and cannot keep themselves warm in the face of all the expansion. Lüst mentioned the opposite effect, i.e., warming the protons, rather than the electrons, by dissipation of hydromagnetic waves.

Davis: I would like to continue the discussion about the anisotropy of the electron and proton distribution functions in the solar wind. If you use only simple two-body collisions, they all should be much less isotropic than they are observed to be. It is clear that there are instabilities and collective modes which are maintaining the isotropy, to some extent, of both the electrons and the protons. The real issue is: Why are these collective modes more efficient for electrons? I suspect that any plasma physicist can give many reasons why the electrons cannot get a very anisotropic distribution.

Nikolaiev: I have a comment on Lüst's Report. Following earlier work (e.g., that of Parker) Korobeinikov and I studied the propagation in the solar wind of disturbances after a solar flare. Our model consisted of a strong point explosion in a gas, in which the density varies as $1/r^2$ and the solar wind velocity is constant. We obtained an exact analytical solution. This non-similarity solution was compared with experimental data of the interplanetary plasma measured by two space probes simultaneously in different points of space. Résumé: (1) Only strong shock waves are rather well described by the similarity solutions. (2) If we include the motion of the solar wind, the calculated energy in a solar flare is decreased by a factor between 3 and 5. (3) The effective adiabatic exponent γ can take both values $\frac{5}{3}$ (a perfect gas) and $\frac{2}{3}$ (one-dimensional gas flow). Gamma depends on the density and the velocity of the solar wind. (4) The propagation of a shock wave in a plasma with weak magnetic fields leads to formation of a magnetic trap for the solar protons.

I have also a question to Lüst. For the determination of the radius of the shock front between the solar wind and the interstellar medium we should include the subcosmic-ray pressure. The interstellar pressure, including the magnetic pressure ($\approx 10^{-11}$ erg cm^{-3}) and the subcosmic-ray pressure ($\approx (3$ to $4) \times 10^{-12}$ erg cm^{-3}), gives a radius of 20 or 30 AU but not 200 AU.

Lüst: The shock waves should be located at roughly 100 AU. But what pressure did you take? I used 10^{-12} dyne cm^{-2} for the pressure of the interstellar medium. If you use your numbers, I agree with you that you have the shock wave closer to the Sun. As I said at the end of my talk, one might find the shock between 30 and 300 AU. But I think the radius of about 5 AU quoted earlier from Brandt seems to me too small.

Van Woerden: I think we do not know very well the local interstellar gas density. We might be in a cloud of 10 atom cm^{-3}; we might be between clouds, so that the density is 0.2 atom cm^{-3}.

Stecher: I disagree with Van Woerden. I think that the Ly-α measurements make it

fairly clear that around the Sun the density within 100 pc is quite low, between 0.1 and
0.2 cm^{-3}.

Lüst: But, if you take such a low gas density, the gas pressure is unimportant compared to the other pressure components.

Baranov: I have a question for Dr. Lüst, which is connected to the problem of the
interaction of the solar wind and the interstellar medium. The interstellar medium
flows past the Sun with a velocity of 20 km sec^{-1}, which is a supersonic velocity. But
the solar wind plasma is flowing with supersonic velocity too, and the velocity of the
solar wind plasma is decreasing. Therefore we must have two shock waves; one in the
interstellar plasma and one in the solar wind plasma. Could you comment on the
distance between these shock fronts and the distance of these shock fronts from the
Sun?

Lüst: You are certainly correct that one must take into account the motion of the
solar system as a whole against the surrounding interstellar medium. This would
cause a standing shock wave in front of the solar system (with respect to its apex
motion). The distance of this shock wave from the boundary of the region where the
magnetic field lines are still connected with the Sun may be estimated in a similar way
as for the standing shock waves in front of the Earth's magnetosphere.

Davis: In the discussion this morning, Lüst said the solar system was surrounded
by a sphere with a radius of the order of 1000 AU, filled up with old solar wind which
had gone through the shock and was sitting there waiting to recombine. Is not this
going to have some effect on the charge exchange with the possible incoming neutral
particles he also mentioned? The neutrals in the interstellar gas have a chance to
exchange charge in this region. The density is very low but the extent is very large.
This may have a significant effect on the calculations.

Baranov: Dr. Lüst, Williams (1965) calculated the radius of the sphere ionized by
solar radiation. He found a radius of about 1000 AU. You have a much lower value.
Why? (Williams, R. E.: 1965, *Astrophys. J.* **142**, 314.)

Lüst: Williams assumed in his calculation for the extended H II region around the
Sun that the solar system is at rest. In this way he obtained a radius of about 500 to
1000 AU. However, since the solar system is moving with a velocity of about 20 km
sec^{-1}, the neutral hydrogen atoms coming from a distance larger than 1000 AU will
penetrate into the H II region and will be ionized by charge exchange or by photoionization only at about 30 to 100 AU.

Parker: There is a basic question discussed earlier as to whether the interstellar
medium some distance outside the solar system is in fact neutral or ionized. If it is
neutral, one has the situation Lüst has described. But if it were ionized, there would be
no possibility for charge exchange, and one has the interplay of two ionized gases.

Field: A general comment concerning the nature of the interstellar medium in the
vicinity of the Sun. We heard from Stecher that Ly-α absorption lines in spectra of
nearby stars indicate a density of about 0.1 to 0.2 cm^{-3}. Is it possible that this medium
is in fact the intercloud medium that has been discussed earlier in this Symposium? If
so, its temperature is expected theoretically to be about 7000 K. And if the medium

were unaffected by the Sun, its pressure would be approximately 0.3×10^{-12} erg cm^{-3}. If, on the other hand, it is significantly ionized, either by the Sun or possibly by another nearby star, its pressure would be about twice that value. Some people have suggested that the intercloud medium is in fact fully ionized by perhaps quite distant early-type stars. In one paper it was suggested that in fact we live inside the H II region of the Gum Nebula, which is believed to be ionized by ζ Puppis and γ Velorum, two stars that Stecher mentioned (see p. 295).* In that case we expect that the medium is in fact completely ionized out at least to several tens of parsecs from the Sun. That may be of some importance in calculating the interaction with the interstellar medium. Finally, a comment on the pressure of the surrounding magnetic field, which Lüst emphasized. Presumably, such a pressure is highly anisotropic, and consequently the shock itself may be nonspherical. Is there any possibility of testing that? The observations of Kurt and Sunyaev might throw light on this question.

Lüst: I don't know of any observations, but I would agree that the shock would not be spherical.

Spiegel: Concerning anisotropy produced by a surrounding field, I worried about the same question *re* supernovae remnant observations. There was a paper specifically on this by Bernstein and Kulsrud (1965). Woltjer mentioned that they found a very small anisotropy of the shell for a supernova explosion in an ambient magnetic field. (Bernstein, I. B. and Kulsrud, R. M.: 1965, *Astrophys. J.* **142**, 479.)

Parker: You can also do a stationary solution, and again the shock is very nearly spherical. Not quite, but deviations are small.

Meyer: There is another source of anisotropy in the solar wind post-shock region that might produce even larger effects. If the magnetic field remains frozen in to some degree, it will be substantially wound up in the equatorial plane but not in the polar direction. The resulting large 'magnetic pressure' gradients in latitude must have important dynamical effects on the flow direction.

I. M. Gordon: *On the Interpretation of Radar Studies of the Sun*

The origin of the solar wind is one of the most difficult problems of solar physics. The main obstacle for the solution of this problem is the absence of observational evidence on the physical conditions and the motions of the coronal plasma at distances r from the center of the sun between 1.4 and 10 $R_\odot$. It is between these distances that acceleration of the solar wind probably takes place. In principle, direct measurements of the velocities of coronal plasma can be carried out with the aid of the radar explorations of the Sun. This method appears promising because of the existence of an optimum frequency range extending from 40 MHz down to lower frequencies. The range corresponds to reflection layers with $r \geqslant 1.4\,R_\odot$ (r represents distance from the center of the Sun). However, the interpretation of the results of long standing radar

* Carruthers (1968, *Astrophys. J.* **151**, 269) reports the presence of interstellar Ly-α absorption in both γ Vel and ζ Pup, with hydrogen surface densities of (3 and 6) $\times 10^{19}$ cm^{-2}, respectively. This excludes the possibility that the Sun is inside the Gum Nebula. (Ed.)

experiments at 38 MHz (James, 1964, 1966) is rather problematic because the almost symmetrical distribution of frequency shifts and the changes of effective cross sections observed are incompatible with the conventional theories of formation of reflected signals (Gordon, 1967, 1968a, b). Recently, it has been argued that the reflected signal may be formed by scattering of electromagnetic waves on turbulent pulsations in the coronal plasma (for this process see the Report by Kadomtsev and Tsytovich, p. 108). In the light of this new concept the radar investigations of the Sun form diagnostics of the coronal plasma on a large scale. The observed frequency displacements are no longer Doppler shifts, but result from combination scattering on ion acoustic waves (Gordon, 1968b, c) or from four-plasmon scattering on high frequency plasma waves as was supposed by Liperovski, Tsytovich, and myself. The major portion of the reflected signals is formed in the corona above the plages, while reflections from the quiet corona are weak or completely absent (Gordon, 1968c, 1969). Identification of the place of origin of the reflected signal enables us to define the velocity of the scattering layer by the average 'violet' shift. An analysis of the spectra of radar echoes from the Sun (James, 1964) reveals that a sometimes rapid rise takes place in the velocity of the solar wind in the corona above the plages. In different experiments the gradient of velocity changes over a quite wide range. From the scanty data published I reached the preliminary conclusion that higher effective cross sections (Gordon, 1969) are accompanied by steeper velocity gradients. In a further development of the theory of the solar wind (Bisnovatyi-Kogan and Gordon, 1969) we assumed that the rapid increase of velocity is caused by strong heating which leads to the acceleration of subsonic flow in a relatively thin layer. The correlation of both strong radar echoes and steep velocity gradients with the increase of level and fluctuations of the natural radio noise from the Sun suggests that the heating derives from coronal plasma turbulence excited by beams of accelerated electrons (Gordon, 1968a). From this model, we predict a rise of temperature up to 4×10^6 to 10^7 K. Velocities up to 120 km sec^{-1} are reached at a relatively low level $r \approx 1.5\ R_\odot$.

The next desirable step is the investigation of the structure of the layer, where the acceleration takes place, including transport phenomena. In this connection some important questions arise: How frequent are these accelerations of plasma above the plages? What is the duration of these events? Are they stationary? In addition to these a problem of great importance is to determine the correlation of these events with other characteristics of active regions and also with geophysical indices.

Two of the most important observational problems connected with the radar explorations of the Sun are the origin of weak scattered components and the variation of the structure of the reflected signal on a time scale of about 1 sec. The necessity of continuous radar patrolling of the Sun at different frequencies with adequate resolving power both in time and space can hardly be overestimated. Simultaneous observations of natural radio emission by the Sun at different frequencies may be of great value (Gordon, 1968a, b). The precise measurements of the coordinates of sources of various kinds of outburst, their small scale structure, motions, and frequency drifts (Wild, 1967; Philips, 1968) are very desirable. [Condensed. (Ed.)]

References [to Gordon's remark]

Bisnovaty-Kogan, G. S. and Gordon, I. M.: 1969, *Astron. Zh.*, in press.
Gordon, I. M.: 1967, Astron. Circ. Akad. Nauk SSSR, N447.
Gordon, I. M.: 1968a, *Astron. Zh.* **45**, 1002.
Gordon, I. M.: 1968b, *Astrophys. Lett.* **2**, 49.
Gordon, I. M.: 1968c, Astron. Circ. Akad. Nauk SSSR, N487.
Gordon, I. M.: 1969, *Astrophys. Lett.* **3**, 181.
James, J. C.: 1964, *IEEE Trans. Ant. Propag.* **AP-12**, 876.
James, J. C.: 1966, *Astrophys. J.* **146**, 356.
Philip, K. W.: 1968, *Astron. J.* **73**, S197.
Wild, J. P.: 1967, *Proc. Astron. Soc. Austr.* **1**, 59.

Van de Hulst: I think that it is quite clear that these radar measurements contain a lot of information. My main point of doubt is: do they contain relevant physical information or irrelevant geometrical information? The point being that if you have in the radar beam a perfectly reflecting sphere, large compared to the wavelength, then you have one reflection point in the middle, and the radar cross-section is equal to πr_s^2, where r_s is the radius of the sphere, which in this case is equal to the r_c radius of curvature at the center point. If, however, the sphere is buckled somewhat, $r_c \neq r_s$ at the reflection point and you will get many values of the cross section πr_c^2 which may be both larger or smaller than πr_s^2. Radial velocities may be explained in this way, if by a slight motion the main reflection shifts from a nearer to a farther point. My question is: Do you have any good criterion to discriminate against these geometrical effects?

Gordon: (1) The quiet corona is too poor a reflector to explain the very strong signals that occur and (2) there is a strong correlation between plage areas and reflective cross sections. [Answer condensed. (Ed.)]

Severnyi: Observations at the Crimea Observatory give predominantly downward motions, whereas you find an average upward motion.

Gordon: The downward motions mentioned by Severnyi are in the chromosphere. The radar signal cannot penetrate in such a low part of the solar atmosphere, but is reflected high in the corona. Somewhere in the corona the acceleration has to take place of plasma which initiates the solar wind.

19. MASS LOSS FROM STARS

Introductory Report

(Sunday, September 14, 1969)

S. R. POTTASCH

Sterrekundig Laboratorium Kapteijn, Groningen, The Netherlands

1. Introduction

In this summary we shall attempt to evaluate the mass loss from several kinds of high luminosity stars, especially planetary nebulae, OB supergiants and M giants and supergiants. The purpose is to give an observational basis for the discussion of the mechanism of mass loss and of the consequences of stellar mass loss for the interstellar medium and for stellar evolution. For reasons which will presently be discussed, we are now certain that mass loss is occurring in all the objects mentioned, and probably to a similar extent in all high luminosity stars as well. The precise values of the mass loss rate are uncertain at present; for some objects the uncertainty will be large (two orders of magnitude) and have important influence on the consequences of the mass loss. Therefore we shall discuss in some detail how the different loss rates quoted in the literature have been obtained and what assumptions have been made (see also the Report by Boyarchuk, p. 281). On the basis of this discussion we will indicate the most probable loss rates and their consequences, always remembering the possible influence of the uncertainties.

When comparing space densities and rates of mass loss over the whole Galaxy, we assume that the space density given is an average over a volume of 10^{12} pc^3 (a radius of 15 kpc and an extent in the z direction of 600 pc on either side of the plane). For most of the objects discussed this is a reasonable assumption. Only in the case of the OB supergiants is the z extent substantially smaller, about 200 pc. We have therefore lowered the space density artificially by a factor of 3, so that the total mass loss will be approximately correct.

2. The Sun

For comparison we will briefly discuss the Sun (see also the Report by Lüst, p. 249). Biermann (1951) was the first to demonstrate that particles flow from the Sun. Since then much work on this subject culminated in the actual measurement of a 'solar wind' by space probes removed from the influence of the Earth. Measurements for quiet periods on the Sun (Ness, 1968) are shown in Table I. The velocity observed, about 350 km sec^{-1}, is considerably in excess of the solar escape velocity at the Earth orbit, 42 km sec^{-1}, and there can be no doubt that this material is being lost from the Sun. The mass-loss rate, given in column 6, is typical of the quiet Sun. At times of solar activity the mass-loss rate increases, sometimes considerably, but it is difficult to determine the precise average increase. There is also a variation with the solar cycle.

In column 7 the space density of F, G, and K dwarf stars is given. If it is assumed

Habing (ed.), Interstellar Gas Dynamics, 272–280. All Rights Reserved. Copyright © 1970 by IAU

TABLE I

Mass loss rates for various kinds of objects

1	2 Observed velocity (km sec⁻¹)	3 Distance R from star at which velocity observed (cm)	4 V_{escape} at R (km sec⁻¹)	5 V_{escape} at surface (km sec⁻¹)	6 Mass loss rate ($M_\odot$ yr⁻¹)	7 Space density (stars pc⁻³)	8 Mass loss to interstellar medium ($M_\odot$ pc⁻³ yr⁻¹)
Sun	400	1.5×10^{13}	42	618	2×10^{-14}	2×10^{-2}	4×10^{-16}
Planetary nebulae	20	4×10^{17}	0.4	800	[a] 1.1×10^{-5}	1.5×10^{-8}	1.5×10^{-13}
OB	1400	3×10^{12}	500	600	1.5×10^{-6}	[b] 10^{-7}	1.5×10^{-13}
MI	10	3×10^{15}	2.5	125	4×10^{-6}	2×10^{-8}	8×10^{-14}
MII	10	3×10^{15}	2.5	125	4×10^{-7}	4×10^{-7}	1.5×10^{-13}
M5-8 III	8	10^{15}	2	120	2×10^{-8}	3×10^{-7}	6×10^{-15}
M3-4 III	12	10^{15}	2	140	4×10^{-9}	2×10^{-6}	8×10^{-15}
Novae	1000				$10^{-4} \, M_\odot$	(40 yr^{-1})	4×10^{-15}
Supernovae	5000				$1 \, M_\odot$	$(10^{-2} \text{ yr}^{-1})$	10^{-14}

[a] A lifetime is assumed of 35 000 yr.
[b] This number is lowered by a factor of 3 (see text, p. 272).

that these stars have a mass loss similar to the Sun (why should the Sun be an exception?) then the total mass loss to interstellar medium is given in column 8.

3. Planetary Nebulae

Our knowledge of mass loss from planetary nebulae is more certain than from the other objects we shall discuss. This is because we observe the mass lost in the form of a nebula which is optically thin in many important lines. The velocity of expansion of the nebulae can be directly observed: it is at least an order of magnitude higher than the escape velocity at the distance of the nebula (assuming 1 $M_\odot$ for the mass of the central star).

To obtain the mass of the nebula, one needs to know the density, and the distance of the nebula. The density of the ionized material follows directly from measurements of the forbidden lines, or of the Balmer lines, but in the latter case the square root of the distance also enters. The distance of the nebula is a more difficult question which I do not wish to discuss here (see Aller and Liller, 1968, for a summary). Suffice it to say that distances appear to be known for individual objects with about 50 per cent uncertainty. Statistically the distances are considerably better.

The mass thus derived is the mass of ionized hydrogen in the nebula. In some nebulae a correction must be made for the fact that the entire nebula is not ionized and in all nebulae the total mass must be increased by about 40 per cent to take helium into account. While all these considerations introduce uncertainties in the mass determination, it seems generally accepted that for an average nebula a value of 0.4 $M_\odot$ will be correct within a factor of 2.

The average rate of formation of planetary nebulae is probably somewhat more difficult to assess. In addition to the usual problems associated with space density, we are also required to know the lifetime of a planetary nebula and assume that only one shell has been emitted in this lifetime. Knowledge of the size of the largest nebulae, together with the velocity of expansion which is probably constant during the existence of the nebula, leads to a lifetime of about 35 000 yr. From this lifetime the rate of formation and the average mass-loss rate follow.

The space density listed in Table I, column 7, is an average of two very similar estimates by Cahn (1968) and O'Dell (1968). That these estimates are so similar leads us to regard the value as reasonably well determined. It is likely that the mass lost to the interstellar medium by planetary nebulae, given in column 8, is reliable to within a factor of 3.

4. M Giants and Supergiants

It has been known for more than 30 years that in all high-luminosity M-type stars, the strong absorption lines arising from the ground state of atoms and ions are composed of two components. In addition to a broad, relatively shallow line probably formed in the 'photosphere', there is a deep, narrow component with near-zero central intensity. The component is displaced toward the violet by a small amount, perhaps 10 km sec^{-1}.

This is evidence that material is moving away from the star; since the escape velocity of the surface of these stars is at least 100 km sec^{-1}, it is not direct evidence for mass loss.

That mass loss occurs was first demonstrated by Deutsch (1956). He analyzed the spectrum of the visual companion of the binary star α Hercules, the primary being an M giant. In the spectrum of the companion the sharp lines are still seen, indicating that the absorbing material is present at a distance of at least 10^{16} cm ($=700$ AU) from the M giant and is still expanding at essentially the same velocity, 10 km sec^{-1}. At a distance of 10^{16} cm, the escape velocity is between 1 and 2 km sec^{-1}, thus indicating quite definitely that the matter is being lost from the star. This picture is confirmed by observations in spectroscopic binaries, where these narrow, circumstellar lines remain at a fixed velocity, while the photospheric lines shift with the changing position and velocity in the orbit.

The determination of a quantitative value for the mass loss is more uncertain. This is generally done in the following steps:

(1) the surface density of Ca$^+$ ions is measured from the H and K lines using the curve of growth;

(2) the density is related to the total number of Ca atoms in the line of sight and thus to the total number of H atoms, provided an abundance of Ca is assumed;

(3) the size of the emitting region is determined; if in addition spherical symmetry and constant velocity are assumed the value of the mass loss can be derived.

There are uncertainties in each of these steps. Consider the first step. Since the H and K lines lie on the flat portion of the curve of growth, small errors in the equivalent width determination will lead to large uncertainties in the number of ions integrated over the line of sight. For example the values given by Deutsch (1956) and Weymann (1962) for α Hercules differ by a factor of 20, which is mostly due to this effect. In the second step the computation of the ionization equilibrium introduces an uncertainty. Deutsch assumed that the Ca is predominantly singly ionized throughout the expanding envelope, while Weymann has suggested that no Ca$^+$ is found within about 10 stellar radii of the M star (the Ca probably being in the form of Ca^{++}). If Weymann is correct, it would be necessary to increase Deutsch's estimate of the mass loss by an order of magnitude again. This demonstrates that the present estimate of the mass loss rate for M stars is uncertain by at least a factor of 10.

There are two additional effects which are also shown in the table. First, the more luminous the star, the higher its mass loss. There is about an order of magnitude difference between each of the luminosity classes listed (Deutsch, 1960). Second, the giants may show an increase in mass loss for decreasing surface temperature. The supergiants do not seem to show this effect and in all likelihood the mass loss is roughly the same for all spectral types. In the giants however there is a pronounced relation between the spectral type and the strength of the narrow H and K, in the sense that the lower temperature stars show stronger lines. We have here assumed that this effect means more mass loss for the later spectral types. But it might simply mean an increase in ionization of Ca$^+$ in earlier spectral types and in that case the mass loss might actually be as high in the earlier M giants as in the late M giants. Notice that many of the uncertainties point in the direction of a higher mass loss than is listed in Table I.

It appears that the kinetic temperatures in the outflowing mass in M type stars are significantly lower than in the solar wind. First, lines of several neutral elements, e.g., Fe, Ca, and Cr, are observed. Second, the level of excitation is quite low: absorption is observed principally from the ground level, although weak absorption occurs from lines as high as 1 eV above the ground state. Far from the star only the ground level absorptions are seen.

The space density of the M stars is given in column 7. It is taken from Blanco (1965) for the giants and from the compilation of Allen (1963) for the supergiants. The values of Blanco agree quite well with the earlier results of Neckel (1958); they are probably quite reliable. On the other hand the supergiant densities are less reliable. I have considered MII stars to have an M_v between -1.5 and -3.5, and all brighter objects to be MI. The values of total average mass loss to the interstellar medium is given in column 8. The estimates are rather conservative. It can be seen that similar amounts of material are lost from the M supergiants as from the planetary nebulae. But the uncertainties are such that the mass loss from M supergiants may be an order of magnitude higher.

5. OB Supergiants

The qualitative evidence for mass loss from the early type supergiants is derived essentially from the following argument: The spectrum shows the presence of emission lines with violet displaced absorption edges, the so-called P Cygni profiles. These are thought to be formed in an outer expanding atmosphere with an excitation temperature lower than the stellar photosphere. That is a reasonable explanation of the P Cygni profiles can be shown by analogy to the novae where similar profiles are observed near maximum light and where we later see the shell expanding at about the same velocity as was indicated by the absorption line.

When we restrict ourselves to observations in the visible part of the spectrum, the velocities of expansion indicated by the P Cygni profiles are rather on the low side, between 100 km sec^{-1} to 500 km sec^{-1}. P Cygni itself shows 200 km sec^{-1}. These velocities are rather smaller than the escape velocity from the surface of the star, about 600 km sec^{-1}. Since there is no indication in the spectra of material returning to the star, one has assumed that the material is accelerated to still higher velocities further out in the atmosphere. Actually there are indications for a velocity gradient in the atmosphere, e.g., those Balmer lines which have the highest absorption coefficient and are thus formed further out in the atmosphere, show systematically higher velocities. Hα always has the highest velocity.

Morton (1967) has recently observed the spectra of three OB supergiants between 1000 Å and 2000 Å. He found P Cygni profiles for all strong lines, but now the velocity shift of the absorption component was about 1400 km sec^{-1}, indicating definite mass loss. Interestingly, all the lines observed showed roughly the same absorption velocity shift, not only in a given star, but in all three stars!

The lines observed in the UV are resonance lines of abundant elements, and originate therefore further out in the atmosphere. This verifies the hypothesis that the velocity of expansion increases outward in the atmosphere until it exceeds the escape velocity.

Interesting also is the comparison of the UV spectra with the visible spectra of the same star. For one of the stars (δ Ori, O9.5 II) there is no indication for mass loss in the visible spectrum. For the other two stars (ζ Ori, O9.5 Ib and ε Ori, B0 Ia) only a very weak Hα P Cygni profile is seen, giving a velocity shift of 100 km sec^{-1}, and no other emission lines. A possible conclusion from this fact is that stars, which show stronger mass loss indications in their visible spectrum, are actually losing mass at a higher rate than δ, ε, and ζ Ori.

The actual values of the mass loss are more difficult to ascertain. There are several estimates in the literature (e.g., Morton, 1967; Hutchings, 1968, 1969a, b). It is my opinion that Morton's analysis of the three Orion stars is the most trustworthy since he makes the minimum of assumptions. I do not wish to go through his analysis here, but I will point out his principal assumptions:

(1) the weak N v absorption line is unsaturated;

(2) the absorption all takes place in a region where the velocity is uniform;

(3) the ionization equilibrium is determined by photo-ionization (due to the stellar radiation) and radiative recombination. Together with assumption 2 and the continuity equation, this leads to an ionization equilibrium independent of distance.

Morton finds similar values for the mass loss for the three stars. The average value is given in Table I, which may be an underestimate since, as I have said above, other OB supergiants may be losing substantially more material. The analysis by Hutchings seems to be less solidly based. He attempts (and succeeds to a limited extent) to reproduce the observed P Cygni profile in different stars. To do this he must assume velocity, density, excitation temperature and ionization as a function of distance in the atmosphere. It is my opinion that the single observational fact, the line profile, is not sufficient to decide whether he has chosen a correct model. It may be that he has sufficient insight into OB atmospheres that his results are reasonable. But these model quantities are still so uncertain in the outer solar atmosphere with so much better observational material that one is sceptical. In addition his models do not satisfy the equation of continuity, so that his computed mass loss varies with height in the atmosphere. For example in the star HD 152408, for which Hutchings quotes a mass loss of 10^{-4} $M_\odot$ yr^{-1}, his actual values are 10^{-5} $M_\odot$ yr^{-1} at the surface of the star, increasing to almost 10^{-2} $M_\odot$ yr^{-1} at 7×10^{12} cm (or 0.5 AU) above the surface.

Hutchings finds similar mass loss values to Morton for the Orion stars. For other stars he finds higher values. For P Cygni itself he finds a value of 5×10^{-4} $M_\odot$ yr^{-1}. De Groot (1969) derived a similar value for the mass loss in this star, but here again the analysis (specifically, the density determination) makes one very sceptical of the resultant value.

Turning to the space density of OB supergiants, there is again a good deal of uncertainty. Allen (1963) gives a space density of OB stars brighter than $M_v = -6$ as

4×10^{-9} pc^{-3}. Sharpless (1965), however, in a less definite statement, says that several thousand OB supergiants have been found in surveys and that most of these are probably within a distance of 2–3 kpc. If we have 10^3 stars within a cylinder of radius 2.5 kpc and 600 pc thickness*, the space density is at least 10^{-7} star pc^{-3}, more than an order of magnitude higher than the previous estimate. We shall adopt this rough estimate of Sharpless, however, noting the uncertainty.

6. Discussion

A. MASS LOSS TO THE INTERSTELLAR MEDIUM

The mass loss rates are given in the last column of Table I. (We have also included the novae and supernovae in this table for comparison). The main-sequence stars of about the solar type probably do not contribute an important amount of material to the interstellar medium, nor do the novae. It appears that the planetary nebulae, the OB supergiants, the M supergiants, and to a lesser extent, the supernovae contribute roughly equal amounts to the interstellar medium. Is this amount significant? The total listed in column 8 is $6 \times 10^{-13} M_\odot$ pc^{-3} yr$^{-1} = 2.5 \times 10^{-11}$ particle cm^{-3} yr^{-1}, which in 10^{10} yr amounts to 0.25 particle cm^{-3}. This is of the same order as the gas density observed at present, averaged over a region 500 to 600 pc on either side of the galactic plane – roughly the region occupied by these various objects (with the exception of the OB supergiants). It is doubtful that all four mass-loss rates have been substantially overestimated and it is likely that this total mass-loss rate is on the low side; the mass loss from the OB supergiants and from the M supergiants may both be substantially higher. It may also be that substantial mass loss occurs in an as yet unrecognized form, perhaps in the K giants or perhaps in the M dwarfs. If the mass loss were a factor of 10 higher it would mean that stellar mass loss is replenishing the interstellar medium in 10 per cent of the lifetime of the Galaxy.

B. IMPORTANCE FOR STELLAR EVOLUTION

There are 0.13 star pc^{-3} near the Sun, yielding a mass density in stars of 5×10^{-2} $M_\odot$ pc^{-3}. With a mass-loss rate of $6 \times 10^{-13} M_\odot$ pc^{-3} yr^{-1}, this means that in 10^{10} yr 10 per cent of the observed stellar mass is being lost from the stars. This is a strong indication that mass loss will be important in stellar evolution. The consequences of a mass-loss rate 10 times higher would have to be very carefully considered.

C. ENERGY TRANSFER TO THE INTERSTELLAR MEDIUM

In the second column of Table II the kinetic energy which can be transferred to the interstellar medium is given. The energy is appreciable only for the OB stars and the supernovae. Even for these objects it does not reach the value 6×10^{-26} erg sec^{-1} cm^{-3} given by Spitzer (1968) as the energy which the hot stars radiate below the Lyman limit. And it may be that cosmic rays or X-rays are a more important source

---

* This thickness has been adopted to facilitate comparison with the other stars, instead of 200 pc, which is probably a limit to the actual spatial extent of these objects.

TABLE II

Energy considerations

1	2 Kinetic energy of flow at R (erg sec^{-1} cm^{-3})	3 Potential energy required to bring the material from the surface to infinity (erg sec^{-1})	4 Stellar luminosity (erg sec^{-1})	5 Ratio of potential energy loss and stellar luminosity
Sun	4×10^{-31}	2×10^{27}	3.9×10^{33}	5×10^{-7}
Planetary nebulae	6×10^{-31}			
OB stars	3×10^{-27}	2×10^{35} [a]10^{36}	10^{39}	10^{-3}
MI	10^{-31}	2×10^{34}	3×10^{38}	7×10^{-5}
MII	2×10^{-31}	2×10^{33}	4×10^{37}	5×10^{-5}
M5–8 III	5×10^{-33}	10^{32}	7×10^{36}	1.5×10^{-5}
M3–4 III	10^{-32}	3×10^{31}	4×10^{36}	8×10^{-6}
Novae	5×10^{-29}			
Supernovae	3×10^{-27}			

[a] The kinetic energy is given because in this case it is greater than the potential energy. In all the other cases it is substantially less.

of energy than the hot stars. We can therefore conclude that, while stellar mass loss may supply the material to the interstellar medium, it does not at the same time supply the energy.

D. ENERGY SOURCE FOR THE MASS LOSS

The loss of potential energy required to carry the material from the surface of the star to infinity is given in column 3, Table II. It is generally greater than the kinetic energy in the mass flow, except in the case of the OB stars, where the kinetic energy is also listed in the table. We may compare this energy with the energy which the star radiates, its luminosity shown in column 4. This ratio is given in column 5. The ratio is seen to vary by several orders of magnitude for the different objects; it may be an important number in considering the origin of the mass loss. For example if the radiation pressure drove the mass flow in OB supergiants it would have to be quite an efficient process [see also the relevant remarks made in the following Discussion (Ed.)].

Acknowledgement

I would like to thank D. Beintema for an interesting discussion of mass loss from OB stars.

References

Allen, C. W.: 1963, *Astrophysical Quantities*, Athlone Press, London.
Aller, L. H. and Liller, W.: 1968, in *Nebulae and Interstellar Matter* (ed. by L. H. Aller and B. Middlehurst), University of Chicago Press, Chicago, p. 483.

Biermann, L.: 1951, *Z. Astrophys.* **29**, 274.

Blanco, V. M.: 1965, in *Galactic Structure* (ed. by A. Blaauw and M. Schmidt), University of Chicago Press, Chicago, p. 241.

Cahn, J. H.: 1968, in *IAU Symposium No. 34, Planetary Nebulae* (ed. by D. E. Osterbrock and C. R. O'Dell), Reidel, Dordrecht, The Netherlands, p. 44.

Deutsch, A. J.: 1956, *Astrophys. J.* **123**, 210.

Deutsch, A. J.: 1960, in *Stellar Atmospheres* (ed. by J. L. Greenstein), University of Chicago Press, Chicago, p. 543.

De Groot, M.: 1969, *Bull. Astron. Inst. Netherl.* **20**, 225.

Hutchings, J. B.: 1968, *Monthly Notices Roy. Astron. Soc.* **141**, 219; **141**, 329.

Hutchings, J. B.: 1969a, *Monthly Notices Roy Astron. Soc.* **144**, 235.

Hutchings, J. B.: 1969b, preprints.

Morton, D. C.: 1967, *Astrophys. J.* **150**, 535.

Neckel, H.: 1958, *Astrophys. J.* **128**, 510.

Ness, N. F.: 1968, *Ann. Rev. Astron. Astrophys.* **6**, 79.

O'Dell, C. R.: 1968, in *IAU Symposium No. 34, Planetary Nebulae* (ed. by D. E. Osterbrock and C. R. O'Dell), Reidel, Dordrecht, The Netherlands, p. 361.

Sharpless, S.: 1965, in *Galactic Structure* (ed. by A. Blaauw and M. Schmidt), University of Chicago Press, Chicago.

Spitzer, L., Jr.: 1968, in *Nebulae and Interstellar Matter* (ed. by L. H. Aller and B. Middlehurst) University of Chicago Press, Chicago, p. 1.

Weymann, R.: 1962, *Astrophys. J.* **136**, 844.

20. MASS LOSS FROM ERUPTIVE STARS

Introductory Report

(Tuesday, September 16, 1969)

A. A. BOYARCHUK

Krymskaya Astrofizicheskaya Observatoriya, U.S.S.R.

1. Methods of Determination of the Mass Loss Rate

The previous reports by Lüst and by Pottasch concerned quasi-stationary mass loss. The loss is called quasi-stationary because the basic characteristics of a star, its temperature and its radius, do not change significantly during the time that a particle needs to flow from the stellar surface out to a point where the stellar influence is negligible. However there are many stars which show significant changes of temperature and radius within a short time. Very often these changes are accompanied by instantaneous ejection of a great amount of material. For example, during the outburst of a nova the ejected mass can reach $10^{-4} M_\odot$. Such effects are very appreciable in the so-called 'eruptive variables', in which the brightness changes spasmodically. It is very likely that the rate and the mechanism of mass loss of eruptive variables are different from that of normal stars.

I would like to say a few words on the determination of the mass loss rate. The mass loss rate is found mainly from three kinds of observations:

(i) Direct observation of nebulae, which are formed by the ejected material.

(ii) Investigation of line profiles in stellar spectra.

(iii) Observations of changes in the periods of binary systems.

The first kind gives the most reliable data about the mass-loss rate. One can derive the mass M of an optically thin nebula from the total energy $E_{H\beta}$ radiated at $H\beta$. The mass of the nebula is given by

$$M = V n_e m_H, \tag{1}$$

where V is the volume of the radiating material. Similarly we have for the total emission

$$E_{H\beta} = V n_e^2 \varepsilon_{H\beta}(T_e), \tag{2}$$

where $\varepsilon_{H\beta}$ is the volume emissivity (erg cm^{-3}) calculated for $n_e = 1$ cm^{-3}. Eliminating n_e between Equations (1) and (2) we obtain

$$M = m_H \sqrt{\left(\frac{E_{H\beta} V}{\varepsilon_{H\beta}} \right)}. \tag{3}$$

From observations we obtain both the $H\beta$ flux at the Earth (which yields $E_{H\beta}$) and the angular size of the nebula (which gives V). Since $E_{H\beta}$ is proportional to R^2 and V to R^3, it is clear that the mass of the nebula is proportional to $R^{5/2}$, where R is the distance to the nebula. Thus the main error in the value of the mass of the nebula

Habing (ed.), Interstellar Gas Dynamics, 281–290. All Rights Reserved. Copyright © 1970 by IAU

arises from the uncertainty in the distance to the nebula. In addition the assumptions of optical thinness and of homogeneity of the nebula may not always be correct and may introduce errors.

The second method of determining the mass of the ejected material is based on line profiles. It is well known that emission lines with an absorption component at the short wavelength side are formed in an expanding gaseous envelope surrounding the star. The absorption component is formed in that part of the envelope which is projected on the stellar disc while the rest of the envelope gives the emission component. The mass loss rate is equal to

$$\mathrm{d}M/\mathrm{d}t = \varrho V_{ej} 4\pi R_{\mathrm{env}}^2, \tag{4}$$

where ϱ and R_{env} are the density and the radius of the envelope and V_{ej} is the expansion velocity. All three values in the right-hand part of Equation (4) can be obtained from the observations. Kuhi (1964) developed a detailed theory of the line formation in an expanding envelope, which allows one to get ϱ, V_{ej}, and R_{env} from the observed profile. Rough estimates of these values can be obtained from a visual inspection of spectrograms. The displacement of the absorption component relative to the emission component gives the expansion velocity. The relative intensities of the lines that are sensitive to dilution allow the determination of the radius of the envelope. A set of quality estimates as to the presence or absence of forbidden lines, the assumption that the density varies as R^{-2} and so on, can give some information about the value of the density. Therefore the error in the density may reach several orders of magnitude if one does not undertake a detailed investigation. The stratification of radiation may bring in an additional difficulty. Finally, one needs to be certain that the material of the envelope leaves the star completely and does not return to the stellar surface in an invisible form. As a whole this second method gives less accurate results than the first.

The third method is based on the consideration of a change of the period of binary systems. Huang (1956) has shown that mass loss leads to an increase of the period, while mass exchange between the components leads to a decrease. Under simplifying assumptions Huang derived the following equation, which connects the rate of the mass loss from the primary component Δm_1 with a period change Δp

$$\Delta p/p = 2\Delta m_1/(m_1 + m_2). \tag{5}$$

It is difficult to estimate the errors introduced by the simplifying assumptions. Also it is necessary to observe the period of binary systems with high precision in order to discover its change.

Now we will consider separately different types of eruptive stars: novae, symbiotic stars, and stars of the T Tauri type, the U Geminorum type and the UV Ceti type.

2. Novae

Novae are the uncontestable examples of objects which have significant rates of mass

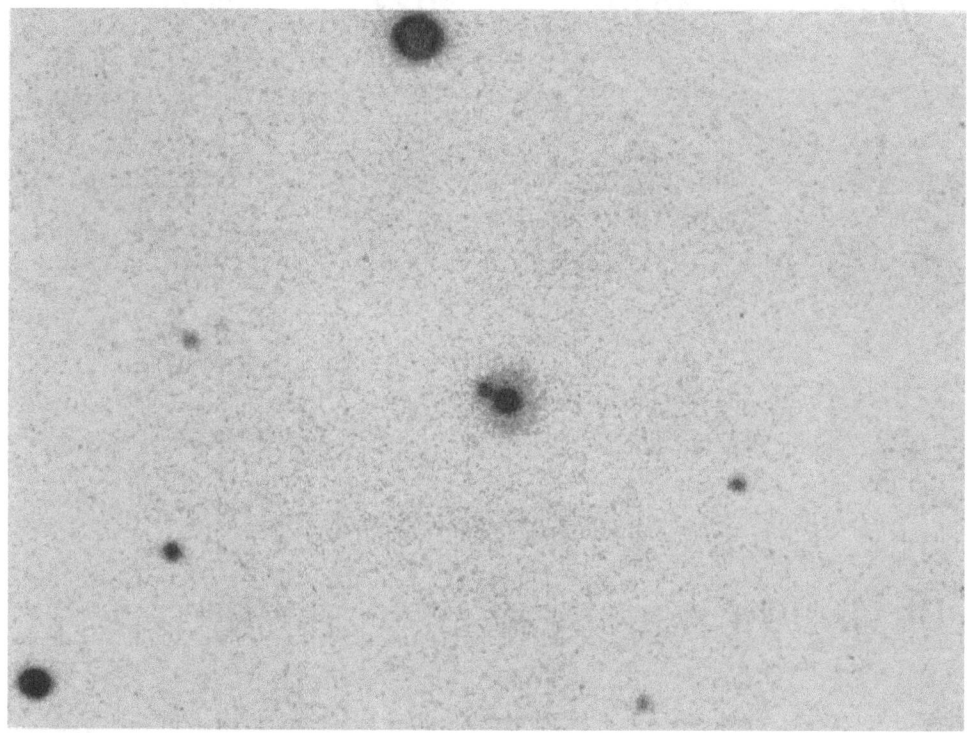

(a)

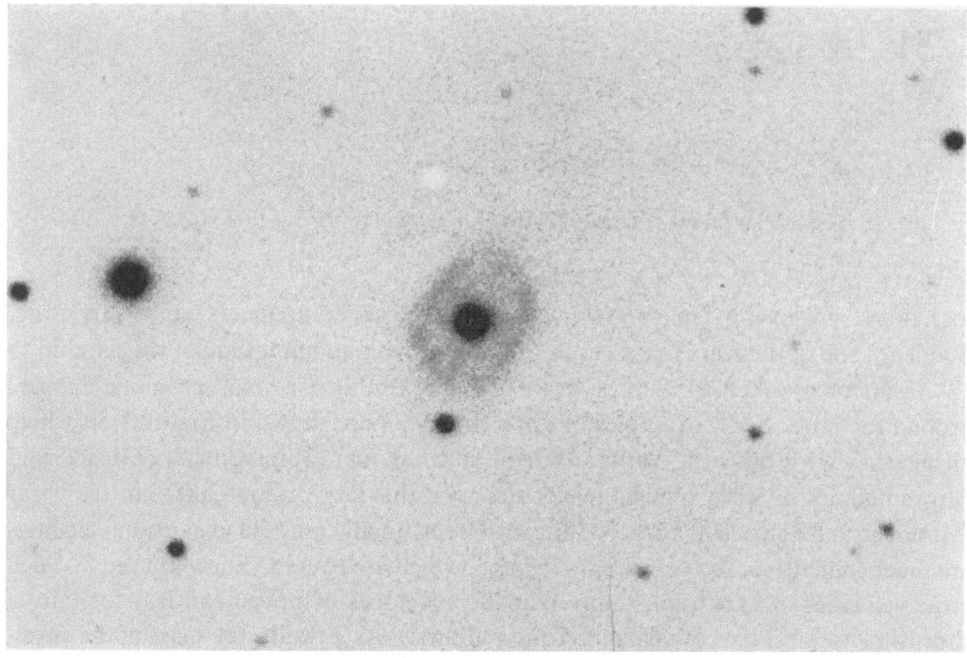

(b)

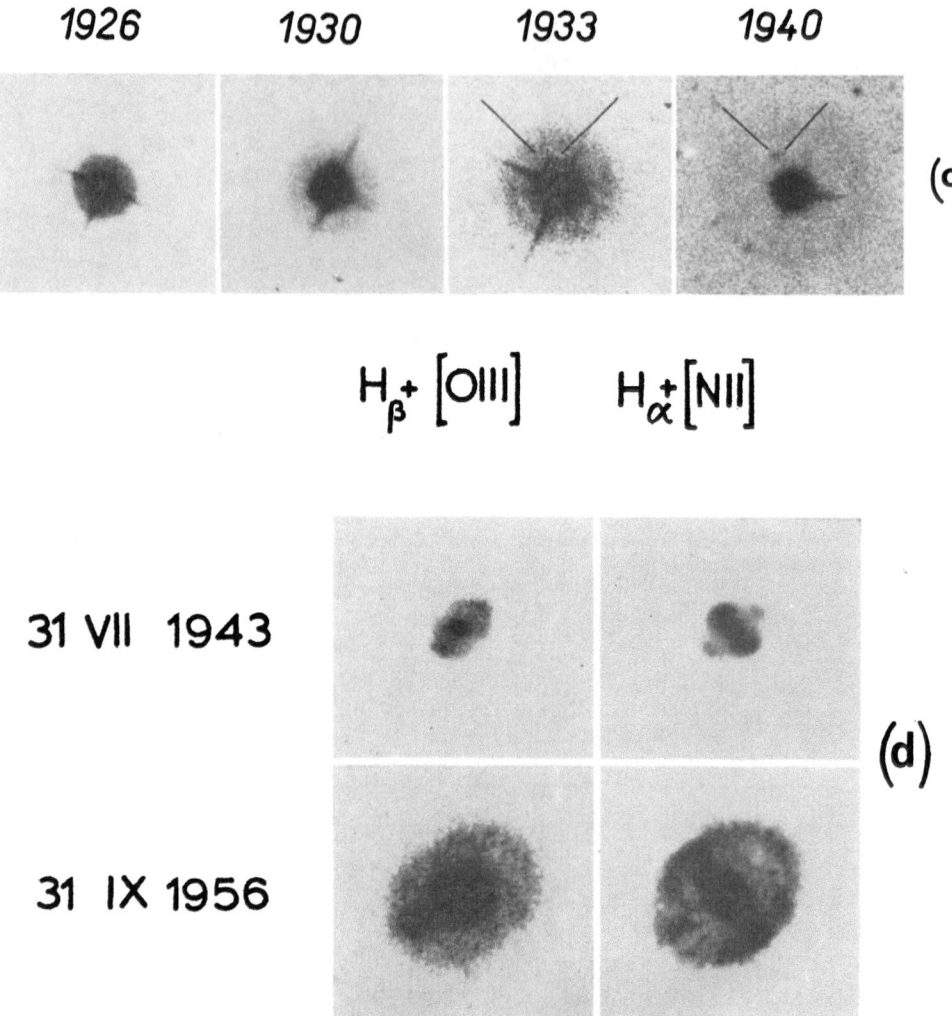

Fig. 1.   Several novae: (a) N Cyg 1920; (b) N Aur 1891; (c) N Aql 1918; (d) N Her 1934.

loss. Most novae are binary systems in which both components are dwarf stars. Suddenly one of the stars increases in brightness, with an amplitude of the flare up to 12^m (a factor of 100 000). Several years after the outburst a small envelope appears around the nova. Some examples of nova envelopes are shown in Figure 1, in which all pictures were taken at Mount Wilson and Palomar Observatories (Mustel' and Boyarchuk, 1970). The expansion velocities are rather large: about 2000 km sec^{-1} for fast novae and about 300 km sec^{-1} for slow ones. In all cases the expansion velocities are much larger than the velocities of escape, which are equal to a few km sec^{-1}. Such large velocities of expansion imply that the envelopes of novae can last for only a short time (several tens of years after the outburst). As a result, the mass of the envelope has been determined for only very few novae, namely for those with the outburst

TABLE I

Mass loss from novae

Name	V_{exp} (km sec^{-1})	Mass of shell ($M_\odot$)	References
N Lac 1936	2200	3.6×10^{-5}	1
N Aql 1918	1700	1.4×10^{-4}	1
N Per 1901	1340	3.8×10^{-5}	1
N Pic 1925	800	10^{-4}	1
N Her 1934	290	4.6×10^{-5}	2
Recurrent Nova			
RS Oph 1933		$5 \ \times 10^{-6}$	3
1958	1000	$2 \ \times 10^{-7}$	4

(1) Pottasch (1959); (2) Mustel' and Boyarchuk (1970); (3) Sayer (1937); (4) Folkart *et al.* (1964).

in the last century. Table I shows the expansion velocities and the masses of the envelope for several novae. The average mass equals $7 \times 10^{-5} \ M_\odot$. According to Kukarkin (1954) and Allen (1955) about 100 novae explode in the Galaxy per year, so that novae put back into the interstellar medium a mass of about $7 \times 10^{-3} \ M_\odot \ yr^{-1}$.

It appears that a nova may lose mass for a long time after the dramatic explosion. For example, recently Nather and Warner (1969) have found that the period of the binary system DQ Her ($=$ N Her 1934) increases with time. This means that DQ Her continues to lose mass 35 years after the outburst. The value of the period change corresponds to a mass loss of $1.1 \times 10^{-7} \ M_\odot \ yr^{-1}$. One can assume that other old novae also have mass loss still going on at about the same rate, because all of them show emission lines in their spectra indicating gaseous streams. In order to estimate a total rate of mass loss for all novae after the outburst we need to know how long old novae continue their activity. DQ Her had its outburst in 1934; T Aur had its outburst in 1891, and it still seems active. There are many objects in the Galaxy which have spectral and photometric characteristics similar to that of old novae. Therefore, one can assume that the activity of old novae continues at least 10^2 yr. If we assume that each old nova has the same rate of mass loss as that of DQ Her then the total rate will be more than $1.1 \times 10^{-3} \ M_\odot \ yr^{-1}$, less than the rate of mass loss during the outbursts.

Finally consider the 'recurrent novae', a small group of novae, for which more than one outburst has been observed. Only for one recurrent nova – RS Oph – has the rate of mass loss during the outburst been determined. Table I shows the expansion velocities and the mass of the envelope. One can see that in this case the mass of the ejected material is a hundred times less than that of the ordinary novae. The total number of recurrent novae in the Galaxy is at least ten times less than that of the novae. We can conclude that the mass ejected by all recurrent novae during many outbursts is probably less than that ejected by all novae during single outbursts.

3. Symbiotic Stars

The term 'symbiotic stars' is widely used for objects whose spectra represent a combination of absorption features of low temperature stars (TiO bands; CaI, CaII absorption lines, etc.) with emission lines of high excitation ([FeVII], [NeV], etc.). The brightness of these stars varies with amplitudes up to 3^m and with a period of several years. The current hypothesis states that the object is a binary. One of the components is a late-type giant and the other one is a hot, small star which is the source of excitation of a nebula surrounding both components. One can assume that the nebula has been ejected from one of the two stars. Unfortunately the angular sizes of the nebulae are much less than $1''$ and nobody has seen them on direct pictures. Several astronomers (Merrill, 1959; Boyarchuk, 1967) have measured the widths of emission lines, which characterize the expansion velocity of the nebula. Table II shows the expansion velocities and the masses of nebulae for some symbiotic stars. The total mass of both stars is about $10 \, M_\odot$, the radius of the nebula is about $10^3 \, R_\odot$ (Boyarchuk, 1969) and so the escape velocity is about 60 km sec^{-1}, which is less than the observed expansion velocities for most of the stars. From these data it follows that the average mass in the nebula of a symbiotic star is about $5 \times 10^{-4} \, M_\odot$. Unfortunately we do not know the lifetime of symbiotic stars. If we assume that the lifetime of symbiotic stars is close to

TABLE II
Mass loss of symbiotic stars

Name	$V_{\exp}$ (km sec^{-1})	Mass of shell ($M_\odot$)
AG Dra		1.5×10^{-4}
Z And	70	10^{-4}
AG Peg	100–400	10^{-4}
AX Per		4×10^{-4}
CI Cyg		$< 5 \times 10^{-5}$
BF Cyg		4.2×10^{-3}

that of planetary nebulae then the rate of mass loss is about $5 \times 10^{-8} \, M_\odot$ yr^{-1}. The total number of symbiotic stars in the Galaxy is about 10^3 (Boyarchuk, 1970a) and the total rate of mass loss of symbiotic stars is about $5 \times 10^{-5} \, M_\odot$ yr^{-1}.

One can obtain some information about the mass loss in large flares. It is known that during a large flare the spectra of symbiotic stars show many lines with P Cygni-type profiles (Swings and Struve, 1940; Gauzit, 1955). The relative displacement of absorption and emission components corresponds to an expansion velocity of about 120 km sec^{-1}. Boyarchuk (1970b) studied the large flare of Z And of 1961. The hot component had a radius of $70 \, R_\odot$. If the mass equals $1 \, M_\odot$, the escape velocity is 80 km sec^{-1}, less than the observed velocity. The density of the region where the absorption lines were formed was found to be 5×10^{-14} g cm^{-3} and the continuous loss rate $2 \times 10^{-6} \, M_\odot$ yr^{-1}. Since the P Cygni-type profile has been observed during two or three months and the frequency of flares is approximately one per two years, the rate of mass loss is about $10^{-7}\ M_\odot\ \mathrm{yr}^{-1}$. The total mass loss from all symbiotic stars is about $10^{-4}\ M_\odot\ \mathrm{yr}^{-1}$.

4. T Tauri Stars

T Tauri-type stars are characterized by irregular light variations which may reach a few magnitudes and by the presence of emission lines, especially fluorescent lines of FeI. The stars have spectral classes G through M and they are located in the main-sequence region and in the subgiant region of the H-R diagram. The majority of these stars are evidently connected with dark and with bright diffuse nebulae. There is considerable evidence that T Tauri stars are very young and still in the process of

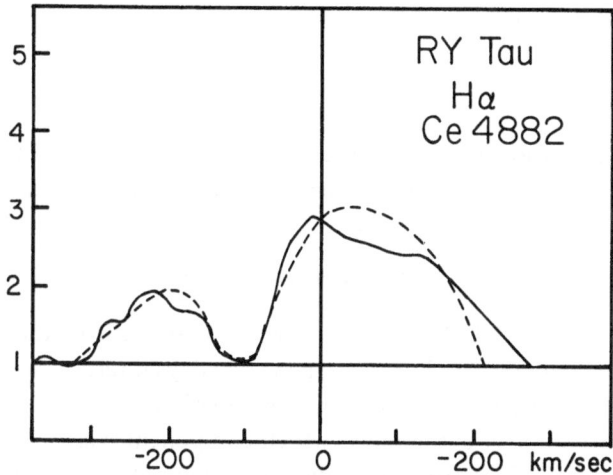

Fig. 2.   Comparison of the observed profile of Hα (dashed line) and the predicted profile (solid line) in the spectrum of RY Tau (Kuhi, 1964).

gravitational contraction. The strongest emission lines – HI and CaII (H and K) – usually have P Cygni-type profiles: emission components 5 to 6 Å wide which have violet-displaced absorption features. Such profiles indicate the presence of material moving away from the star. As there is no sign of the returning of material, we can conclude that real mass loss takes place. Several determinations of the mass loss rate have been made. From a rough investigation of spectrograms of the very luminous T Tau star LK H_α 120 Herbig (1962) estimated a mass loss rate of about $10^{-5}\ M_\odot\ \mathrm{yr}^{-1}$. Varsavsky (1962) used the density determined by Osterbrock (1958) from the relative intensities of the [OII] $\lambda\lambda 3727$, 3729 doublet, to obtain the result that the mass loss rate from T Tau itself is about $10^{-7}\ M_\odot\ \mathrm{yr}^{-1}$. The most detailed analysis of the mass loss rate has been carried out by Kuhi (1964). He assumed that the stars have a spherically symmetric envelope moving radially outward and subjected only to the deceleration of gravitational forces. The inner part of the envelope provides the emission lines while the outer part of an envelope provides the absorption feature. Kuhi calculated a set of model profiles and by fitting these to the observed profile he determined the density at the stellar surface and the expansion velocity (see Figure 2). Then he obtained the mass-loss rates. The results for six T Tauri stars are shown in Table III. The weighted mean rate is $3.7 \times 10^{-8} \, M_\odot \, yr^{-1}$.

TABLE III

Mass-loss rates of T Tauri stars (after Kuhi, 1966)

Star	V_{exp} (km sec^{-1})	$dM/dt \, (M_\odot \, yr^{-1})$
RU Lup	200	$1.4 \ \times 10^{-7}$
Lk H$_\alpha$120	180	$5.9 \ \times 10^{-7}$
AS 209	~ 0	0.65×10^{-7}
GW Ori	80	0.35×10^{-7}
T Tau	140	0.35×10^{-7}
RY Tau	100	0.31×10^{-7}

If we assume that the total number of T Tauri-type stars in the Galaxy is 10^6 (Kuhi, 1966) they are returning about $0.04 \, M_\odot \, yr^{-1}$ to the interstellar medium.

5. U Geminorum Stars

The main characteristic of U Geminorum stars is that the brightness increases now and then by several magnitudes, but after one to two days it returns to its former value. It has been established that U Geminorum stars are binary systems, both components of which are dwarf stars. It is unknown which component has an outburst. Nobody has yet observed nebulosities around the stars. Grant and Abt (1959) and Zuckermann (1962) have found that the color of SS Cyg is affected by reddening. They decided that the reddening is due to the absorption in a circumstellar envelope which was formed by the ejected material. Some doubt about their conclusion arises from the uncertainty of the intrinsic color of SS Cyg. All this leads to difficulty in the determination of the mass-loss rate. Walker and Chincarini (1968) have investigated the radial velocity variation of SS Cyg and have found that the length of the period increases with time. The change of the period corresponds to a mass-loss rate of $1.8 \times 10^{-7} \, M_\odot \, yr^{-1}$. The total number of U Geminorum stars in the Galaxy is 10^7 (Gorbatskii 1970). If we suppose that SS Cyg is typical, then the total mass loss rate from U Geminorum stars will be about $2 \, M_\odot \, yr^{-1}$.

6. UV Ceti Stars

The UV Ceti variables are dwarf stars of spectral classes dM3e – dM6e. They are characterized by rare and very short (several minutes) flares with amplitudes up to 6^m. So far no nebulae have been seen around these stars. The faintness of UV Ceti stars and the short duration of the flares does not permit as yet the careful investigation of the line profiles. The nature of flares is still unknown. The best hypothesis is the assumption that the observed flares arise in chromospheres and coronae surrounding UV Ceti stars. On the basis of this hypothesis Kahn (1969), using the observations by Lovell (1969) and by Kunkel (1969), has found that the star YZ CMi loses $3 \times 10^{-12} M_\odot \ \mathrm{yr}^{-1}$. Gershberg (1968) has evaluated an upper limit to the mass loss from an analysis of the light changes during the flares. He has found that the rate of the mass loss is less than $10^{-11} M_\odot \ \mathrm{yr}^{-1}$. According to Gershberg (1968) the Galaxy contains 10^9 UV Ceti stars. Then the total loss rate will be about $10^{-2} M_\odot \ \mathrm{yr}^{-1}$.

7. Summary

Table IV is a summary. The first column gives the type of variable stars, the second the mass loss rate per star, the third the number n of such stars in the Galaxy, the fourth the total mass loss rate of all stars of the type considered, the fifth the ejection velocities.

TABLE IV

Summary of mass-loss rates

Name	$dM/dt \ (M_\odot \, \mathrm{yr}^{-1})$	n	$n(dM/dt) \ (M_\odot \, \mathrm{yr}^{-1})$	V_{exp} (km sec^{-1})
Novae				
outburst		10^2	$7 \ \times 10^{-3}$	1000
cont. eject.	1.1×10^{-7}	10^4	10^{-3}	
Symbiotic stars	$5 \ \times 10^{-8}$	10^3	$5 \ \times 10^{-5}$	100
T Tau stars	3.7×10^{-8}	10^6	$4 \ \times 10^{-2}$	100
U Gem stars	1.8×10^{-7}	10^7	1.8	
UV Cet stars	10^{-11}	10^9	10^{-2}	

It would follow from Table 4 that U Geminorum stars are the main contributor of the material into space. However this conclusion is not convincing because the value of the mass loss rate of U Geminorum stars is based on only one star and the total number of these stars in the Galaxy may be in error by a factor of 10. The last statement may be true for other kinds of eruptive stars as well.

References

Allen, C. W.: 1955, *Monthly Notices Roy. Astron. Soc.* **114**, 387.
Boyarchuk, A. A.: 1967, *Astron. Zh.* **44**, 12 (1967, *Soviet Astron.* **11**, 8).
Boyarchuk, A. A.: 1969, in *Proc. Fourth IAU Colloquium on Variable Stars*, Budapest, p. 395.
Boyarchuk, A. A.: 1970a, in *Eruptive Stars*, Nauka, Moscow, p. 113.
Boyarchuk, A. A.: 1970b, *Astrofiz.*, in press.
Folkart, W., Pecker, J.-C., and Pottasch, S. R.: 1964, *Ann. Astrophys.* **27**, 252.
Gauzit, J.: 1955, *Ann. Astrophys.* **18**, 354.
Gershberg, R. E.: 1968, in *Problems of the Stellar Evolution and Variable Stars*, Nauka, Moscow, p. 50.
Gorbatskii, V. G.: 1970, in *Eruptive Stars*, Moscow, in press.
Grant, G. and Abt, H. A.: 1959, *Astrophys. J.* **129**, 323.
Herbig, G. H.: 1962, in *Symposium on Stellar Evolution* (ed. by J. Sahade), Observatorio Astronómico, La Plata, Argentina, p. 23.

Huang, S. S.: 1956, *Astron. J.* **61**, 49.

Kahn, F. D.: 1969, *Nature* **222**, 1130.

Kuhi, L. V.: 1964, *Astrophys. J.* **140**, 1409.

Kuhi, L. V.: 1966, in *Stellar Evolution* (ed. by R. F. Stein and A. G. W. Cameron), Plenum Press, New York, p. 373.

Kukarkin, B. V.: 1954, *Variable Stars and Stellar Evolution*, Moscow.

Kunkel, W. E.: 1969, *Nature* **222**, 1129.

Lovell, B.: 1969, *Nature* **222**, 1127.

Merrill, P. W.: 1959, *Astrophys. J.* **129**, 44.

Mustel', E. R. and Boyarchuk, A. A.: 1970, *Astrophys. Space Sci.* **6**, 183.

Nather, R. E. and Warner, B.: 1969, *Monthly Notices Roy. Astron. Soc.* **143**, 145.

Osterbrock, D. E.: 1958, *Publ. Astron. Soc. Pacific* **70**, 399.

Pottasch, S. R.: 1959, *Ann. Astrophys.* **22**, 394.

Sayer, A. R.: 1937, *Ann. Harv. Coll. Observ.* **105**, 21.

Swings, P. and Struve, O.: 1940, *Astrophys. J.* **91**, 546.

Varsavsky, C. M.: 1962, in *Symposium on Stellar Evolution* (ed. by J. Sahade), Observatorio Astronómico, La Plata, Argentina, p. 33.

Walker, M. F. and Chincarini, G.: 1968, *Astrophys. J.* **154**, 157.

Zuckermann, M.-C.: 1962, *Ann. Astrophys.* **24**, 431.

21. DISCUSSION FOLLOWING THE REPORTS
BY POTTASCH AND BOYARCHUK

(Sunday, September 14, 1969 and Tuesday, September 16, 1969)

Chairmen: R. WEYMANN and S. A. KAPLAN

Editor's remark: This Discussion contains the remarks made during the discussion after Boyarchuk's Report on Tuesday, September 16, and about half of the remarks made in the discussion after the Reports on Sunday, September 14. The other half of the latter discussion is presented in Chapter 18. At the end of the Session on Tuesday Thomas asked Busemann to comment on the interrelation between aerodynamics and interstellar gas dynamics. Busemann's response has been transferred to the Final Discussion (Chapter 27). Some general comments by Goldsworthy have been moved to Chapter 5, Section 6.

1. Dynamical Interaction of Stellar Winds and Surrounding Matter

Weymann: Inasmuch as we are dealing with aspects of the interstellar medium, Pikel'ner would like to try to tie today's deliberations into this general topic.

Pikel'ner: I would like to discuss the interaction of stellar winds with matter in diffuse nebulae. Shcheglov at the Sternberg Institute in Moscow has observed radial motions of little blobs of gas in the Omega Nebula and some other nebulae. The velocities of these motions go up to 50 to 70 km sec^{-1}. At the Prague meeting of the IAU Menon suggested that such velocities may be connected with stellar winds blowing from stars inside the Nebula. I have made calculations for such a process (Pikel'ner and Shcheglov, 1968) which I want to discuss now. Suppose that we have a star, which produces a stellar wind, and a nebula around it. Usually one makes the assumption that the wind produces a thin shell of compressed interstellar gas having a radius R from the star and moving away at a velocity V. One derives easily from momentum-conservation and energy-conservation equations (see Pikel'ner and Shcheglov, 1968) that $VR =$ constant. The constant depends on a number of known parameters and can be estimated. If this model is applied to the Omega Nebula, one obtains $V \approx 50$ to 100 km sec^{-1} at $R \approx 0.2$ pc; the observed value of R is, however, 1 to 2 pc. I therefore considered a different model, which is shown in Figure 1. In addition to the layer of compressed interstellar gas (layer 3 in Figure 1) we have a layer of compressed wind (layer 2), that is not very thin, as was assumed previously. I assumed a stellar wind velocity of 1500 km sec^{-1}. Since in shock front I the kinetic energy of the stellar wind is converted almost completely into thermal energy, the temperature in layer 2 is about 10^8 K. The density in layer 2, n_2, is about 0.1 cm^{-3}, so the gas cools very slowly by radiation and adiabatic cooling dominates. Since the sound speed in layer 2 is very large, the pressure p_2 will be uniform. It is easily shown that $p_2 \propto R_1^{-2}$. It is also seen that

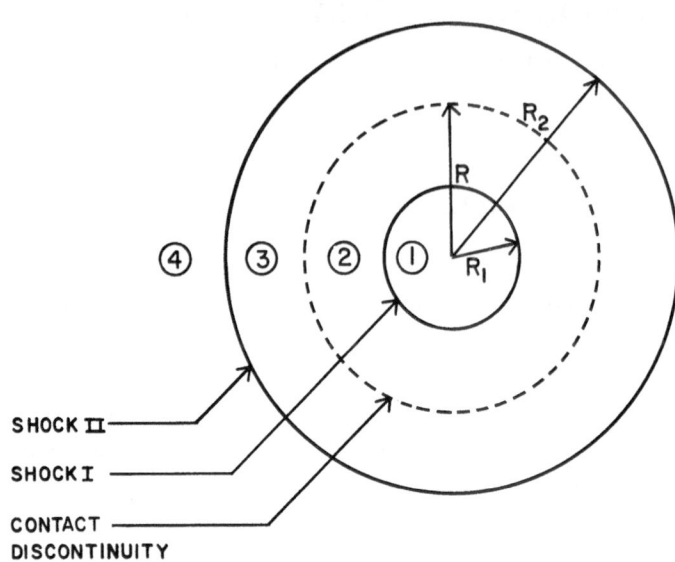

Fig. 1. (See the remark by Pikel'ner.) Schematic picture of the different zones connected with outflow of matter from a star, which is put at the center. 1 and 2 are the stellar wind region; 3 and 4 are the interstellar gas region.

shock front II is driven by p_2. For if the velocity v_{II} of shock front II were too low, R_1 would decrease and p_2 would increase, which would increase v_{II}. This situation resembles a hydraulic press. Calculations show that layer 3 is very compressed ($n_3 = 10^3$ cm^{-3}; layer thickness 10^{-2} pc). It is the layer that shows up in the observations; it can have a velocity of 50 to 70 km sec^{-1}, even for $R \approx R_2 \approx 1$ to 2 pc. The emission measure of layer 3 is about 10^4 cm^{-6} pc, which is large enough to explain the observations. The velocity of this layer is determined essentially by the equation $p_2 \approx \varrho_4 v_{II}^2$, its temperature by the UV flux from the central star. If the density in layer 4 is not uniform, shock front II will not maintain its spherical symmetrical shape. Let us suppose that there are some globules in layer 4. The globule will not share in the motion and, after some time, it will be in layer 1, where the surrounding pressure is ten or more times higher than in layer 4. The globule will therefore be compressed to a density of perhaps 10^5 cm^{-3}. Fluctuations in the interstellar gas density can increase in the same way. Perhaps the high emission measure of the center of the Orion Nebula may be explained by a compression of the kind described.

Finally I tried to apply these ideas to Wolf-Rayet stars. There are several Wolf-Rayet (WR) stars with a nebula around them; the best known of such nebulae is NGC 6888, which has a filamentary structure. This nebula, together with several others, has been studied recently by T. Losinskaia. Together with V. Avedisova and T. Losinskaia I made calculations to explain these nebulae as a consequence of stellar winds coming from the WR stars. We took a mass loss of 10^{-5} $M_\odot$ per year and a wind velocity of 1400 km sec^{-1}. Calculations give an expansion velocity for the nebulae up to 100 km sec^{-1}, which is in agreement with observations. The elongated form of the nebulae cannot be explained by the influence of magnetic fields unless the field strength is at least 100 μG. So it seems more reasonable to suppose the stellar wind to be asymmetric. [Pikel'ner, S. B. and Shcheglov, P. V.: 1968, *Astron. Zh.* **45**, 953 (translation: 1969, *Soviet Astron.* **12**, 757).]

Thomas: Why would there be an asymmetry of the stellar wind?

Pikel'ner: I do not know. Perhaps active surface regions give a stronger wind than quiet regions, as in the Sun.

Dyson: Two comments. First, Dr. Meaburn of the University of Manchester has made observations with a Fabry-Pérot interferometer on several nebulae observed by Shcheglov and has not confirmed the 50 to 70 km sec^{-1} velocities. Second, I have a comment on the suggestion that ionized globules are compressed by the expanding stellar wind. In the case of the Orion Nebula, there is some disagreement as to where the *Trapezium* stars should be located relative to the ionized gas. There is a possibility that the Trapezium is not spatially associated with the larger part of the ionized gas comprising the Nebula. The observations of Münch and Wilson seem to indicate splitting of lines, probably due to shock waves, associated with the densest regions of the Nebula. Therefore it is possible that the displacement of the exciting stars may cause this mechanism to be ineffective.

Pikel'ner: Shcheglov thinks that his very sensitive photoelectric equipment permits the use of his Fabry-Pérot étalons with higher accuracy. I cannot discuss Meaburn's observations since I am not an observer. I try only to explain observations made by others. But, in any case, if stars have winds of the indicated magnitude the winds affect the surrounding nebulae in about the way I indicated. As for the Orion Nebula, Menon pointed out, several years ago, the difficulty that the very steep gradient of surface brightness in the center of the Nebula indicates a density concentration which will expand in a few thousand years. This difficulty has not been solved before, so far as I know. It has been suggested (see e.g., the Report by Mezger, p. 336) that there are strong density fluctuations in the center of the Orion Nebula, since locally high emission measures are observed. Such fluctuations can persist for a considerable time only if there is some external pressure which keeps them in quasi-equilibrium. So if the model I described is not valid, it is necessary to find some other explanation for the high external pressure.

Dyson: Is there a possibility of confusion with the forbidden N II lines?

Pikel'ner: I do not think so, that mistake seems too elementary to make.

Dyson: I have proposed a somewhat different mechanism involving the existence of neutral condensations in the central regions of the Orion Nebula. If the ionized blobs in the Orion Nebula are simply fluctuations in the ionized gas, they will smooth out rapidly due to expansion, as Pikel'ner has mentioned. Consider however, the situation where the density fluctuation is of neutral hydrogen. On the surface of this condensation is an ionization front which eats its way into the neutral gas. Ionized gas streams off the surface with a supersonic velocity. The interaction of these supersonic gas streams produces a system of shock waves. Computed line profiles for the simplest case of interaction of two streams from adjacent condensations are in good agreement with the results given by Münch and Wilson. Admittedly one should take into account other emission in the line-of-sight which would affect the detailed profiles. This mechanism would not produce velocities much greater than perhaps 30 to 40 km sec^{-1}.

Thomas: These problems were also discussed by Lindsey Smith in the Symposium on Wolf-Rayet stars held in Boulder, Colorado, in 1968 (Gebbie and Thomas, 1968). Expanding shells carrying material out from WR stars were discussed classically a long time before the concept of a stellar wind was introduced. A stellar wind implies the existence of a high temperature corona as discussed first by Parker. The old idea was simply one of an initial expansion, without any idea of the source of the expansion; and no one thought of the WR star as having a corona. Lindsey Smith posed the problem of why only a few, and not all WR stars have such nebulae. If I remember correctly, she concluded that an observable nebula could only be swept up in regions of space having a higher space density than normal. In other regions of space, there is too little material. (Gebbie, K. B. and Thomas, R. N. (eds.): 1968, *Wolf-Rayet Stars*, U.S. Government Printing Office, Washington, D.C.)

Stecher: In this connection, Bless, Fischel, and I suggested we might get X-ray emission from the expanding shells, when they run into the interstellar medium. We also used a figure of $10^{-5} M_\odot$ yr^{-1} for the mass loss. We tentatively suggested as a candidate the WR star, γ Velorum, which is now observed to be emitting low-energy X-rays quite strongly. So, Pikel'ner's model seems quite reasonable to me.

Silk: If I remember well, in that area of the sky there is quite a large margin of error as to which of the objects actually emits X-rays.

Stecher: Yes, the observations have only about 1° resolution.

Pottasch: Dr. Pikel'ner, a question on your mass loss of $10^{-5} M_\odot$ yr^{-1} for WR stars. Is that a result of your calculations or an assumption?

Pikel'ner: An assumption, taken from what is in the literature.

Pottasch: Then I caution against the use of this figure. It's derived by assuming the mass-loss proportional to $R^2 V \varrho$, and the density values at the point where we know the velocity are so uncertain that a great deal of reserve is necessary when using this number.

Pikel'ner: I agree. But we relied also on Morton's statement that the usual OB stars have a mass loss of some $10^{-6} M_\odot$ yr^{-1}. The envelopes of WR stars are brighter and therefore should have a higher mass loss.

Thomas: I went through these arguments before (Thomas, 1949), using the data on the density distributions in 444 Cygni as determined by Shapley and Kopal. I reached a figure of $10^{-5} M_\odot$ yr^{-1} by using the observed velocity. I pointed out that this was a very dubious result, because, for example, γ Velorum has not changed its magnitude in about 3000 years; and with such a high rate of mass loss the star would lose several per cent of its mass over this period. If you find such a high mass loss rate, you must argue that a several per cent change in the mass does not appreciably affect the luminosity. I therefore agree completely with Pottasch's caution. (Thomas, R. N.: 1949, *Astrophys. J.* **109**, 500.)

Weymann: The photometric errors were quite large 3000 years ago.

Thomas: That does not matter. Relative photometry was fairly good; and in

3000 years γ Velorum has not changed its relative magnitude by more than a half magnitude. How much change in luminosity do you expect from a few per cent mass loss?

Pikel'ner: Perhaps the WR state is a very short one in the life of a star. If a star loses a significant portion of its mass, its luminosity should change. But I am not discussing these stars; I am using published figures to compute the effects on the interstellar medium.

Mestel: There exists an upper limit on the mass loss rate derived by Lucy and Solomon (1970). Their argument does not deal with the classical stellar wind model, but with ejection under radiation pressure in resonance lines. The upper limit is just L/c^2 where L is the stellar luminosity. This means that the mass loss is never greater than the mass loss from conversion of hydrogen into helium. (Lucy, L. B. and Solomon, P. M.: 1970, *Astrophys. J.* **159**, 879.)

Field: Can one convert that upper limit to solar masses per year? [For a star with $M_{bol} = -8$ one finds $L/c^2 = 7 \times 10^{-9}\ M_\odot\ \text{yr}^{-1}$. According to Table I in Pottasch's Report, p. 273, Morton gives a mass loss rate of $1.5 \times 10^{-6}\ M_\odot\ \text{yr}^{-1}$ for OB stars. (Ed.)]

Silk: Jokipii (1969) has suggested that the low-energy cosmic rays between 1 and 2 MeV, that have been observed near the Earth, have been produced in the region of interaction between the solar wind and the interstellar gas. My question is a more general one. Could someone comment on the general contribution of stellar winds and their interactions to an interstellar low-energy cosmic-ray flux? (Jokipii, J. R.: 1968, *Astrophys. J.* **152**, 799.)

Parker: The contribution to interstellar space would be extremely small. The suggestion by Jokipii is that the interaction of the wind with the interstellar medium would generate a high density of low-energy cosmic rays, within the region occupied by the solar wind. That is a small region. I do not think that it would significantly affect the abundance of the low-energy cosmic rays in interstellar space. It is too small a contribution to fill that vast a region.

Silk: Would this also apply to the OB stars possessing more energetic stellar winds?

Parker: I would have to do some arithmetic before I can answer.

2. Evidence of Stellar Winds

Stecher: I would like to show two slides with direct observations of stellar winds from OB stars (Figures 2 and 3). Figure 2 shows the UV spectrum of the star ζ Puppis, an O5 star, that — with γ Velorum – illuminates the Gum Nebula. The wavelength range is from 1000 to 2500 Å. Note the CIV lines around 1500 Å both in emission and in absorption within a 100 Å interval. The separation of the peaks corresponds to 15 Å, or 3000 km sec^{-1}. The same combination of emission and absorption in one line shows up in all the resonance lines of the star. One would assume then a model, in which the velocity is low close to the star, but increases as one goes out because of the density decrease. The emission is produced by a shell around the star. The absorption

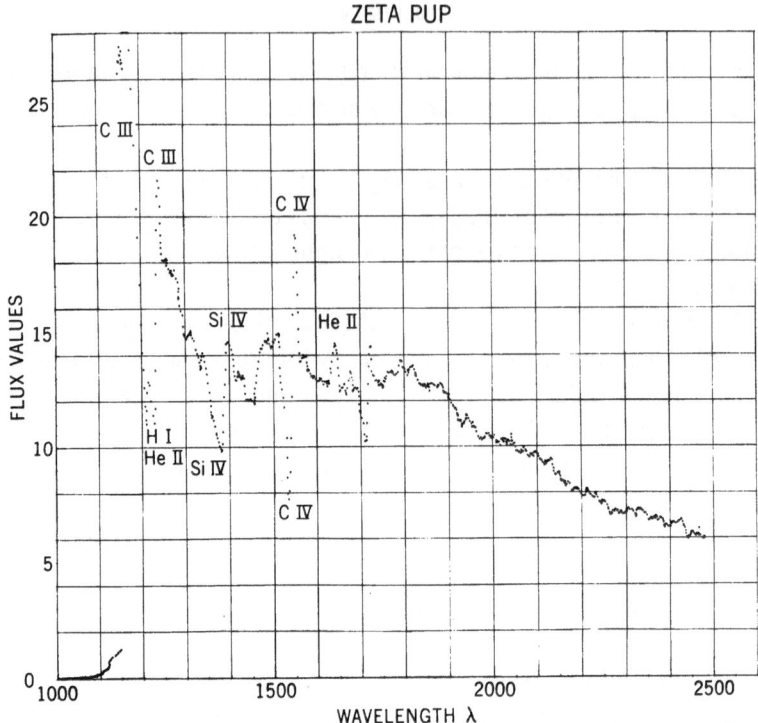

Fig. 2. (See the remark by Stecher.) The spectrum of ζ Pup between 1000 and 2500 Å. The flux is in units of 10^9 erg cm^{-2} sec^{-1} Å^{-1} at the top of the Earth's atmosphere.

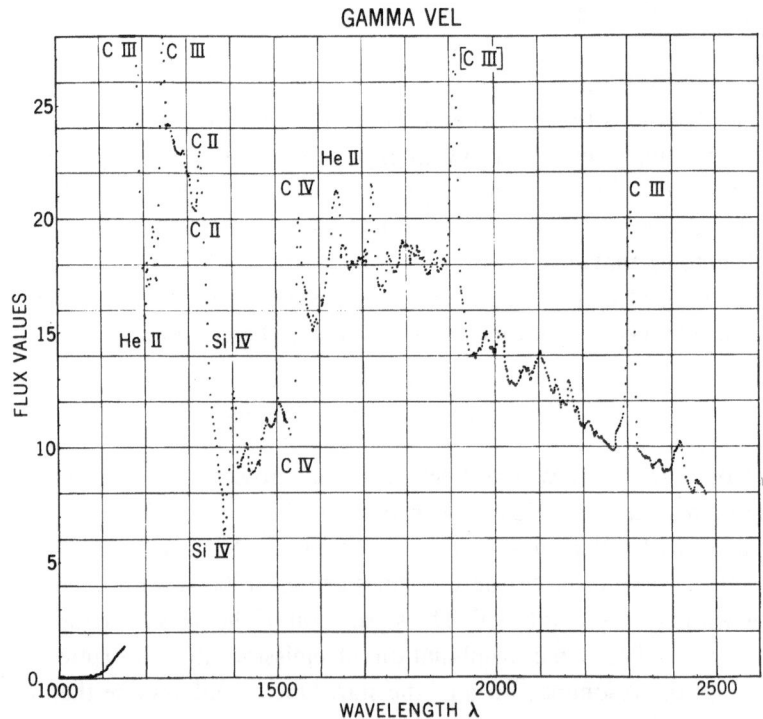

Fig. 3. (See the remark by Stecher.) The spectrum of γ Vel between 1000 and 2500 Å. The flux is in units of 10^9 erg cm^{-2} sec^{-1} Å^{-1} at the top of the Earth's atmosphere.

line originates in shell matter moving away from the star and seen in projection against the stellar disk. The inferred velocity of the gas is model-dependent because some of the absorption may occur against the emitting shell, instead of against the stellar disk. The total amount of gas flowing out is model-dependent because we do not know whether the line broadening is radiation damping or Doppler broadening. Figures 2 and 3 contain photoelectric observations obtained with a 32-cm telescope, flown in a rocket. Figure 3 shows a WR star, γ Velorum which, as I mentioned, also participates in the excitation of the Gum Nebula. Again CIV shows in emission and absorption with about half the Doppler shift of ζ Puppis, but the density is higher. The line at $\lambda 1909$ is an intercombination line of CIII with an A-value of 190 sec^{-1}, while the $\lambda 2296$ line is a permitted line in the singlets of CIII with an A-value of 10^8 sec^{-1}. D. K. West and I are now engaged in producing a set of models and attempting to derive density and mass loss from this star. We have no final conclusions yet.

Thomas: I would remark on the reliability of a direct, naive interpretation of such results as Stecher summarized; his statement that the results are model-dependent should be in flashing letters. Stecher's interpretation is the historical one, applied originally to WR and P Cyg spectra. It assumed a classical model atmosphere, with excitation and electron temperature decreasing outward, surrounded by an optically thin, tenuous atmosphere expanding outward. Possibly there are dilution effects; but aside from these all excitation functions are derived from LTE. In this model the emission lines are observed *only* because the line-producing region has a larger volume than the continuum-producing region. But this kind of model is probably wrong. I argued many years ago that the WR atmosphere was one big chromosphere-corona and that the interpretation of the line-profiles must be based on the same kind of physics we use for the solar outer atmosphere. The evidence is now very strong that this suggestion was correct; you may consult the Proceedings of the Wolf-Rayet Symposium (Gebbie and Thomas, 1968). Recent work by Castor at the Joint Institute for Laboratory Astrophysics in Boulder gives strong evidence for optical depth effects; hence radiative transfer must be included. All of these considerations must be applied to interpret these new rocket UV spectra of the supergiants. On this basis, please note that the observation of the emission line profiles gives us the possibility of inferring density distribution, electron temperature distribution, and velocity gradients. The simplest illustrations of the techniques arise in the Sun, where the velocities and the velocity gradients are not so high, but where the effects are present. These are the reasons we return always to the Sun as a guide – in the solar-stellar wind, in the solar-stellar line profiles. So, be cautious, and demand physical consistency of models. (Gebbie, K. B. and Thomas, R. N. *op. cit.*)

Field: Dr. Thomas, is there any escape from the qualitative conclusion that there is an outflow of about 1000 km sec^{-1} from the star, even though the amount of such flow is very uncertain?

Thomas: No, there is no uncertainty that such high velocities exist. The uncertainty lies in the exact value of the outflow density, and in where it occurs in the atmosphere.

One wants to base a final interpretation on a completely realistic model of the atmosphere, viewing it as one in which the electron temperature increases outward, due to the mechanical heating, which in turn arises from the aerodynamical motions whose geometry we try to infer. Let me remind you that, before these observations, most astronomers disbelieved that such stars even *had* chromospheres-coronas and the required underlying aerodynamic motions. The preobservational predictions were based wholly on the classical, non-aerodynamical, non-chromospheric models. Let us please be physically consistent, now, in making any more models and interpretations.

Stecher: To amplify Thomas's remark: In one case we have tried to compute a model. Now take the intercombination line of CIII, at $\lambda 1909$. It is barely thick in the thermal case. That is, by solving the transfer problem in the moving atmosphere we can reproduce the equivalent width of the emission line. The line has no absorption core and we have an opacity due to electron scattering of about 10^{-2}. For the resonance line $\lambda 977$, from which we are trying to get the mass-flow, this model implies a many-times-greater opacity in electron scattering. I agree completely with Thomas that the problem is very difficult.

Field: Would Dr. Pottasch comment on his statement that some stars having observable mass outflow in the UV had not been observed to show this in the visual. What about the reverse? And is there any star where we observe mass loss in both spectral regions?

Pottasch: In the three stars observed by Morton, which I discussed, there is essentially little or no evidence for mass loss in the visible part of the spectrum (no P Cyg-type profiles). In one star you do not see anything at all which you could associate with mass loss. In the other two stars, you do see Hα in emission. But before you saw the UV spectra nothing would have led you to suppose that there was a great deal of mass loss.

Weymann: In one of these objects Abt looked very hard for the metastable He line at 3389 Å. There was no sign of it.

Field: Are there many stars where you do see evidence for mass outflow in the visible spectrum? Are they the same type of stars?

Pottasch: They are exactly the same type of stars. Perhaps the difference is that they are somewhat more luminous. If you were to make a semi-qualitative plot of evidence of mass loss – say, P Cyg profiles, helium absorption, and Hα emission – there would be a correlation between the luminosity of the star and these features.

Field: What about the stars Stecher described?

Pottasch: Both have been observed in the visible. One is a WR star. There, the evidence for mass loss is the conventional one inferred from broad profiles. It is hard to see how you can avoid some sort of mass loss from WR stars.

Stecher: The star ζ Puppis has emission lines in the visible, and thus some indication of mass loss. Ten years ago, Robert Wilson in Edinburgh found some absorption lines that were blue-shifted, suggesting mass loss. But the lines were very weak.

Bisnovatyi-Kogan: There has been some discussion on the effects of mass loss on the evolution of a star. I want to mention some computer results on mass loss of massive stars during their evolution. Madejin and I are computing the evolution of stars with large masses, not just the static phase, but also in the phase when there is a static core and outflow of matter in the atmosphere. The important point is that the rate of mass loss is not arbitrary, but follows from the theory and is a unique function of the stage of evolution. The preliminary results for $30\,M_\odot$ stars are as follows: after the production of a helium core and the beginning of helium burning the surface temperature reaches a value of about $4600\,K$, the luminosity about $2.3 \times 10^5\,L_\odot$, and the radius about $610\,R_\odot$. At this moment the star begins to lose mass rapidly. The time of mass loss is smaller than the time of nuclear burning. We expect that mass will continue to flow out until the star has lost about $15\,M_\odot$ and the hydrogen layer source will be exhausted. The computation for a $9\,M_\odot$ star did not give a large mass loss during the phase of helium burning.

Weymann: Can you describe briefly the mechanism responsible for the mass out-flow?

Bisnovatyi-Kogan: The physical mechanism is radiation pressure. The mechanism works because of high opacity in the zone of incomplete hydrogen ionization. [Bisnovatyi-Kogan, G. S. and Zel'dovich, Ya. B.: 1968, *Astron. Zh.* **45**, 241 (translation: 1968, *Soviet Astron.* **12**, 192).]

Stecher: Dr. Boyarchuk, I noticed that in your list the T Tau stars have a mass loss of $10^{-7}\,M_\odot\,\mathrm{yr}^{-1}$. That seems more than the upper limit set by Lucy and Solomon on account of radiation pressure. Are there other stars in your list that are not luminous enough to have radiation pressure as a driving mechanism?

Boyarchuk: I do not know about the correct mechanism for the mass loss from T Tau stars. Many people think that it may be something like prominence activity. But it seems to me that for none of the eruptive stars do we have a good mechanism that can produce mass loss.

Mestel: Is it ruled out that in the T Tau phase we are seeing essentially a violent Parker-style stellar wind? It should be remembered that the wind theory does not in itself determine the mass loss; the density at the coronal base (or alternatively at the sonic point) is known only when we have a model for the chromospheric-coronal structure for the particular star considered. But it seems to me that the T Tau phenomenon will be more usefully studied as a quasi-stationary rather than as a non-stationary phenomenon.

Thomas: Again I should like to make the point that in discussing mass loss one has to make a distinction. First there is the sort of model that Boyarchuk used; it consists only of an expanding envelope. But second, we have the solar-wind-type models. And there we do not know what produces the corona. There must be some sort of mechanical energy transport going up in the atmosphere. I think we should discuss here the fundamental thing, the production of the chromosphere-corona. If, as some of us have argued, all stars probably have chromosphere-coronae (not just those stars for which there is a convective zone that can become unstable), then this subject is a whole region of interest which we must consider.

Mestel: I am not opposed to this view; however, it is worth remembering that a magnetically-controlled centrifugal wind can cause substantial mass loss from a rapidly rotating star, even if the surrounding corona is cool. The mechanism is clear: the magnetic field tries to preserve co-rotation, and the centrifugal forces drive the gas out. The Parker critical point is now given by the Bernoulli equation, modified to include the rotational terms. If the star rotates rapidly, then on most of the field-stream lines this point is at a level where the density is high. Of course, the process involves loss of angular momentum. If the star is not contracting then its rotation rate slows down, the sonic point moves out to a position of lower density, and the process cuts off. But T Tau stars are contracting towards the main sequence, and so can maintain a powerful centrifugal wind.

3. Mass Loss from Binary Systems

Krat: I want to comment on mass interchange in close binaries. If the time scale of mass interchange is of the order of 10^5 or 10^6 yr, it can be comparable to the time scale of propagation of energy from the inner parts of a star to its boundary. If the mass of a star changes significantly on a time scale less than 10^5 yr due to a super-ficial process of mass capturing or to mass ejection, and if the luminosity remains constant, because the changes in internal energy production take longer to reach the boundary, then certain kinds of instabilities will develop. The expanding star will accelerate its expansion, and the star receiving the flow of extra mass will contract. The first process will form rarefied gaseous shells around the expanding star; the second will lead to a burst, and the star can become variable. In some specific cases, supernova and nova explosions can occur. The short time scale of mass interchange changes a continuous evolution process to a disruptive one and radically changes the composition of stellar populations and the chemical composition of interstellar matter. *Vice versa* the interstellar material can be processed in a comparatively short time. The time scale of the interchange processes plays a very important role in the cosmogenic picture of stellar evolution.

Gorbatskii: I want to comment on the input of matter from close binaries to the interstellar medium. In close binaries there are gaseous streams forming disk-like envelopes and jets. For binary systems of very small absolute luminosity ($M_v \approx +9^m$) such as old novae, U Gem stars and similar ones, the gaseous streams are very important for the explanation of the observed features. Recently gaseous streams in such systems were studied theoretically at the Leningrad University (Gorbatskii, 1967). The motion of gaseous jets was calculated, including expansion of the jet due to gaseous pressure as well as the effects of centrifugal forces and Coriolis forces. For a wide range of initial conditions a fraction of the gas escapes from the system due to gaseous pressure at the first Lagrangian point; the escaping fraction of the jet can be as much as 0.5. By comparing the theory with the observational data it was found that the gas flow from the jet to the surface of the other star is of the order of $10^{-10} M_\odot$ yr^{-1}. The total number of the binaries under discussion in the Galaxy may be about 10^8. There-fore the mass loss from all these systems may be as high as $10^{-2} M_\odot$ yr^{-1}. There are many more close binaries of other types in the Galaxy that contain gaseous streams and may lose mass effectively. The mass loss from such systems has not been considered quantitatively yet, but many reasons let us believe that the total mass loss from close binaries may amount to $1 M_\odot \text{ yr}^{-1}$. This is of much importance for the mass balance of the interstellar medium. I have a second comment referring to the heating of interstellar medium by UV and X-ray radiation emitted by close dwarf binaries. Calculations show that the velocity of the jet, when it collides with the disk-like envelope, is of the order of 500 km sec^{-1}. The collision leads to a strong shock wave which corresponds to what the observers call a 'hot spot'. The gas is heated in the shock to a temperature of several million degrees, and emits mainly in the far UV and soft X-ray region. The total energy radiated in this range for the Galaxy may be 10^{39} to 10^{40} erg sec^{-1}. This is a considerable quantity compared with the output of other UV and X-ray sources in the Galaxy. [Gorbatskii, V. G.: 1967, *Astrofiz.* **3**, 245 (translation: 1968, *Astrophys.* **3**, 116).]

Field: Did you skip a decimal? Did you say $1 M_\odot \text{ yr}^{-1}$?

Gorbatskii: Correct, $1 M_\odot \text{ yr}^{-1}$ from close binaries. Boyarchuk mentioned an even larger figure from U Gem stars alone. Commenting upon Boyarchuk's figure of $1.8 \times 10^{-7} M_\odot \text{ yr}^{-1}$ for SS Cyg, I would like to say that, if we have a mass loss of that order of magnitude, a very dense envelope will develop around the binary system and will make the system invisible. However, U Gem, SS Cyg and others are very clearly observed with sharply defined lines at the minimum. This fact indicates that the mass loss is at most $10^{-8} M_\odot \text{ yr}^{-1}$.

Mestel: Since I presented Salpeter's paper, I feel obliged to comment on one point. Boyarchuk gave a very high upper limit of $1 M_\odot \text{ yr}^{-1}$ for mass loss from U Gem stars alone, as compared with Salpeter's figure of a fraction of $1 M_\odot \text{ yr}^{-1}$ from stars of all types. The figures are reached by quite different methods: Boyarchuk's, quasi-empirically; Salpeter's, from a general theoretical picture of the evolution of stars into white dwarfs. Salpeter also estimated current star formation at a rate of $1 M_\odot \text{ yr}^{-1}$. If we accept this star formation figure together with Boyarchuk's upper limit on mass injection in interstellar space, we must conclude that the interstellar medium may actually be gaining mass at the moment, and primarily from members of U Gem systems, which will for the same reason end up as white dwarfs of very low mass. (My own feeling is that future work will greatly reduce the contribution of U Gem stars.)

We should also note that if Kuhi's results on T Tau stars are accepted (see the Report by Boyarchuk, p. 281), then during this pre-main sequence stage, mass loss can sometimes be so great as to affect very markedly stellar evolution computations, as noted in particular by Iben. [See also Kuhi, L. V. and Forbes, J. E.: 1970, *Astrophys. J.* **159**, 871. (Ed).]

Drobishevskii: I should like to show that a classification of magnetic stars supports the evidence for mass ejection by some types of stars. (1) I want to point out two difficulties with the hypothesis that magnetic stars are to be found among those with large convection zones and rapid rotation. The first difficulty is that most magnetic stars are of spectral class A1 and earlier $(M_* \geqslant 3\,M_\odot)$ where the convective flux is unimportant (the maximum flux is attained between spectral types A5 and F0). The second difficulty is that the magnetic field, generated in the convective envelope, diffuses too slowly into the interior, radiative equilibrium part of the star. Diffusion by ohmic dissipation takes a time equal to or longer than the stellar age. (2) Both difficulties disappear if one postulates that mass accretion on the stars forces the convective zone to move outward, leaving the frozen-in magnetic field behind in the radiative equilibrium zone. One can put lower limits to the mass accretion rate in two ways. One: the accretion rate has to be so large that the rate at which gas flows into the interior of the star is longer than the rate at which magnetic energy diffuses out of the radiative equilibrium zone. Two: the rate of generation of magnetic flux in the convective zone has to be at maximum. Both lower limits turn out to be $10^{-12}\,M_\odot$ yr^{-1} to $10^{-13}\,M_\odot\,\mathrm{yr}^{-1}$ for stars of types A through F. (3) It is certain that accretion of matter from interstellar space is smaller than the lower limits presented. Therefore we consider the possibilities of mass-exchange in close binaries under the supposition that essentially there are two types of mass outflow from stars: one, a stellar wind-type outflow, to be found in stars with $1\,M_\odot \leqslant M_* \leqslant 1.5\,M_\odot$. Two, a large outflow of unspecified nature in stars with $M_* \geqslant 5\,M_\odot$. On this supposition we predict the existence of three groups of magnetic stars. Group 1, where the stellar mass M_{*1} is between 1.5 and $3\,M_\odot$; group 2, where $M_{*2} = M_{*1} + 1.5\,M_\odot$; and group 3 where $M_{*3} = M_{*1} + 5\,M_\odot$. (4) The prediction of the existence of three groups of magnetic stars is in agreement with the results obtained by Babcock, Preston, and others. There is also quantitative agreement between the magnetic field strength observed and that calculated by using an EMF produced by Coriolis forces and given in Drobishevskii, 1968. [Condensed. [Ed.).] [Drobishevskii, V. G.: 1968, *Astrofiz.* **4**, 537 (under translation for *Astrophys.*).]

4. Various Questions to Boyarchuk

Verschuur: I have made 21-cm line observations in directions of four recurrent novae. I have found that a peculiar cloud structure exists in the direction of two of these novae (for the other two the data are not yet reduced). To interpret these data, I need answers to the following questions: How long do recurrent novae operate? What is the age of a recurrent nova? What is the total mass ejected during the lifetime? Can we expect to see this mass directly? Can we observe this mass in a neutral state if it is ejected as hot matter and then cools at great distances from the star? Can we see the effect that the nova had on the surrounding medium? Can the energy ejected by the nova effectively be converted to kinetic energy of the surrounding medium? I do not think answers to all these questions exist, but I would like to hear a discussion of them, either by Boyarchuk or by Pottasch.

  Boyarchuk: I am afraid that we have no good answers to all these questions, but I will discuss some of them. We know eight recurrent novae and only two have been known for a long time – T Cor Bor and RS Oph. T Cor Bor was recorded at the end of the last century, therefore we can say only that the activity of recurrent novae has been continuous for 100 years at least. Now about the mass. It is very difficult to give an accurate value for the mass per outburst for the following reasons. Suppose a recurrent nova has consecutive outbursts separated by 30 years. During the outburst, we see many emission lines, but we cannot be sure where the emission lines are formed. They may be formed in an envelope consisting of material ejected during the ongoing outburst, or in an envelope consisting of material from all previous outbursts. Because material is ejected with high speed, a shock wave can be generated exciting all the material collected in the envelope. Therefore, we do not know whether the figure of $10^{-7} M_\odot$ in the shell refers to one outburst or to many. Another difficulty is that the velocity of expansion is large for recurrent novae (perhaps 4000 km sec^{-1}) and therefore the material disappears very quickly. I do not know enough about the other questions.

Pottasch: I could discuss one question, perhaps. There is evidence in RS Oph that during the last outburst there was already a substantial amount of material surrounding the star (Pottasch, 1968). How it got there is an open question. It might have arrived there between outbursts, or it might have been the product of a previous outburst. The evidence for this mass is the existence of very narrow absorption and emission lines that are observed during the outburst and the fact that the material in the outburst and the shell seems to be decelerated. It looks as if there is a mass out there in approximate hydrostatic equilibrium that extends quite a distance from the star. Presumably, between outbursts this cools and might be neutral and observable. (Pottasch, S. R.: 1968, *Bull. Astron. Inst. Netherl.* **19**, 227.)

Thomas: It seems to me that Pottasch is saying that material is being pumped out which does not go far enough to escape.

Pottasch: It looks as if there were a lot of material surrounding the nova. The evidence for it is the Hα line you observe. Superimposed on the usual, broad P Cyg-type profile you see a very narrow P Cyg-type profile (with a width of the order of 10 km sec^{-1}). When the outburst takes place, it sweeps up the material that surrounds the star, and this narrow line decreases in intensity with time, since less material remains. At the same time the absorption velocity of the main profile shows that the material involved in the outburst is being decelerated. The deceleration probably occurs because of the effect of the sweeping up of the existing medium far out. This all leads to a consistent picture with rough hydrostatic equilibrium in a large amount of material surrounding the nova before its outburst. It may be moving with small speeds, of the order of 5 or 10 km sec^{-1}.

Thomas: Is it moving away or back?

Pottasch: It is moving away from the nova.

Habing: Dr. Pottasch, could you give an angular size for this shell?

Pottasch: You can give it an absolute size of about 10^{15} cm before the density gets so low that it is no longer observable.

Habing: Such sizes exclude observation of associated 21-cm emission-line profiles.

Shklovskii: Can a regular nova be in fact a recurrent nova with a very long period between outbursts, say of the order of 2000 yr?

Boyarchuk: There are differences. In the first place, a regular nova consists of a binary system in which the two components are both dwarf stars, and the rotation period is only a few hours. But the only recurrent nova that has been studied closely, T Cor Bor, presents a different picture. Again, it is a binary system; but one component is a red giant and the other component is a hot, small star. The rotation period is something like 250 days. In the second place there are spectroscopic differences. The recurrent novae have a higher degree of excitation (coronal lines from Fex, etc.). In regular novae we never have such strong coronal lines. Third, the mass of the envelope of regular novae is about $10^{-4} M_{\odot}$ but for recurrent novae, we have $10^{-7} M_{\odot}$ in the envelope.

Gordon: Is there some difference in the space distribution between the ordinary and the recurrent novae?

Boyarchuk: We do not know the space distribution of the recurrent novae because only eight recurrent novae are known.

Van de Hulst: Dr. Boyarchuk, I got the impression that each mass loss determination was based on several spectrograms taken at one time. Did you repeat the analysis subsequently with a number of spectra taken at several different times after the nova outburst, and were the consecutive results consistent?

Boyarchuk: I studied nova envelopes only a long time after the outburst and so it does not matter if you investigate today or next year, because the ejected material is the same and the excitation conditions do not change strongly. But during the outburst, or just after the outburst, of course, these conditions change rapidly.

Shklovskii: Dr. Boyarchuk, if you determine the mass of emission nebulae, the result will be strongly influenced by inhomogeneities in the density. But if you use absorption lines, the inhomogeneities will not be so important. So the question is: Is there some difference between the mass determined from absorption lines and that from emission lines?

Boyarchuk: There is no single object, for which we have reliable mass determinations in the two ways you described. In the case of nebulae and novae where we have large envelopes, we have no absorption lines. But for T Tau stars, where we use absorption lines for mass-loss determinations, we do not observe an envelope. Only for the symbiotic stars is there in principle the possibility for such a comparison. But here the comparison is not very meaningful, because there are many uncertainties, such as the frequency of large flares, how long they last, and the amount of material ejected per flare. In the one case available in the literature both methods give the same order of magnitude for the mass-loss rate.

Shklovskii: But if you have the same results from the two cases then it follows that there are no inhomogeneities in the envelope.

Thomas: I have a simple pictorial comment on your question. In the literature one often finds the optical thickness of flares, hence the line-of-sight mass in a flare, determined from the equation

$$I_F = I_p e^{-\tau} + \int_0^\tau B e^{-\tau} \, d\tau,$$

where I_F is the intensity in the flare. To determine τ one assumes that I_p, the intensity at the base of the flare, can be obtained from observations of non-flare regions; and by comparing I_F and I_p, one gets τ by making some assumption on B in the flare. The problem is that this can lead to orders of magnitude error in the value of τ, hence in the value of the mass of the flare. The error lies in the fact that I_p depends very much on whether or not there is overlying material. This procedure is therefore completely invalid. One requires an analysis that takes into account the influence of the material in the flare upon the value of I_p. This is the same kind of problem, intrinsically, as that of determining the amount of material in the chromosphere, or in the mass of a shell. As an example, early estimates gave $\tau \approx 1$ for the chromosphere in Hα; modern determinations give $\tau \approx 100$. The same kind of error can enter the determination in the cited cases of flares and stellar shells. The absorption part of the P Cyg profile would be treated as above. The emission part would be treated differently. Thus, depending upon whether or not we use a correct theory for each, we can find differences in results which can reflect an error in the theory of analysis rather than in the physical situation or the presence of inhomogeneities.

Gershberg: I should like to say a few words about UV Ceti flare stars and their possible relevance for the interstellar medium. This type is the most numerous of all eruptive stars. We have one UV Ceti-type star per 12 pc^3 and out of 45 stars nearer than 5 pc, 8 belong to this type. In quiet periods the stars show strong H and Ca emission lines. A preliminary analysis indicates that the stars have important chromospheres and that the number of radiating H-atoms under 1 cm^2 of their photosphere is one or two orders of magnitude higher than in the Sun. The radii are in the range of 0.1 to 0.3 $R_\odot$ and the masses in the range of 0.1 to 0.3 $M_\odot$. Because of the presence of an important chromosphere and the similarity of the gravitational potential ($g \propto M/R$) for the Sun and the stars we expect a big corona and stellar wind. Kahn (1969) arrived at similar conclusions from the analysis of the radio flare of YZ CMi. Evaluation of the soft X-ray emission from UV Ceti-type stars gives a few per cent of the total observed soft X-ray flux. But if this flux is mainly from an extragalactic origin, the stellar source will be dominant in the direction of highly obscured regions like the Ophiuchus and the Taurus areas. The soft X-ray flux from UV Ceti-type stars is not yet observed during the flare, however. Two years ago we compared simultaneous photoelectric observations of several flare stars and observations of radio emission at 13 and 26 MHz. During stellar flares no radio emission was observed and an upper limit of $F_x/F_{opt} < 10^5$ was obtained. This observed upper limit is very much larger than the expected ratio. But one has to keep in mind that in stellar flares the ratio F_{radio}/F_{opt} is sometimes 100 times higher than in solar flares. (Kahn, F. D.: 1969, *Nature* **222**, 1130.)

22. INTERSTELLAR GRAINS AND SPIRAL STRUCTURE*

Introductory Report

(Wednesday, September 17, 1969)

J. MAYO GREENBERG

State University of New York and Dudley Observatory, Albany, N.Y., U.S.A.

1. Introduction

The problem of the interstellar grains is in many ways still unsettled. The theories of the formation, growth, and destruction of the grains have undergone a number of modifications because of the very significant increase in our understanding of the dynamical processes in our Galaxy. I will emphasize here how the density wave theory of the origin of the spiral structure may lead to a new broad concept in dust formation and its generic relationship with star formation. The basic physical processes are not changed but rather the framework in which we consider them.

We should bear in mind that on all theories of dust there are key limitations based on rather 'solid' observations. There are definite optical boundary conditions imposed on the grain properties by well-established observations of the amount and the wavelength dependence of extinction and polarization. However, where possible in the following presention we will try to avoid grain characteristics whose values depend on detailed models and we will focus our attention on the most broadly-founded features as they relate to galactic structure.

The principal theme will be the question of time scales. In the earliest days of the discussion of interstellar grains, the interstellar medium was pictured as being rather amorphous and having little structure that might have significance to the physics of the grains. In such a context, the only time scale of importance was the age of the Galaxy, which was believed to be of the order of 10^9 to 10^{10} yr (thirty years ago the Galaxy was about ten times younger than it is now). When Oort and van de Hulst (1946) approached the problem, they were concerned with the fact that in such a time a grain would grow so large that its optical characteristics would be inconsistent with the observed wavelength dependence of extinction. The notion of cloud collisions was used to introduce the possibility of destroying the grains within the clouds. In this way the size of a grain had an upper limit, which depended upon the frequency of cloud collisions. Another, and important, result of this type of theory was that it gave rise to an average or steady state distribution of particle sizes with a time scale for averaging of the order of 10^7 yr. It may be seriously questioned whether the notion of clouds and of cloud collisions is any longer crucial in relation to grain destruction. Their role in limiting grain growth may be negligible, because of the existence of other grain destroying processes, such as collisions of grains and cosmic rays, sputtering by high-energy atom collisions, evaporation in the neighborhood of hot stars.

* Work supported in part by NASA Grant NsG 113 and by NSF Grant GP 11925.

However we will consider these processes only from a limited point of view, namely in relation to the broader problems connected with the spiral structure.

2. Basic Observations and Interpretations

Some representative observations of the wavelength dependence of extinction and polarization are shown in Figure 1. The extinction is roughly linear up to $\lambda^{-1} \approx 9\ \mu^{-1}$. The polarization is caused almost certainly by the same agency as the extinction and is due to a difference in the extinction for radiation polarized along two orthogonal directions. This implies that the particles are oriented and nonspherical. The wavelength dependence of polarization for stars whose polarization relative to extinction is

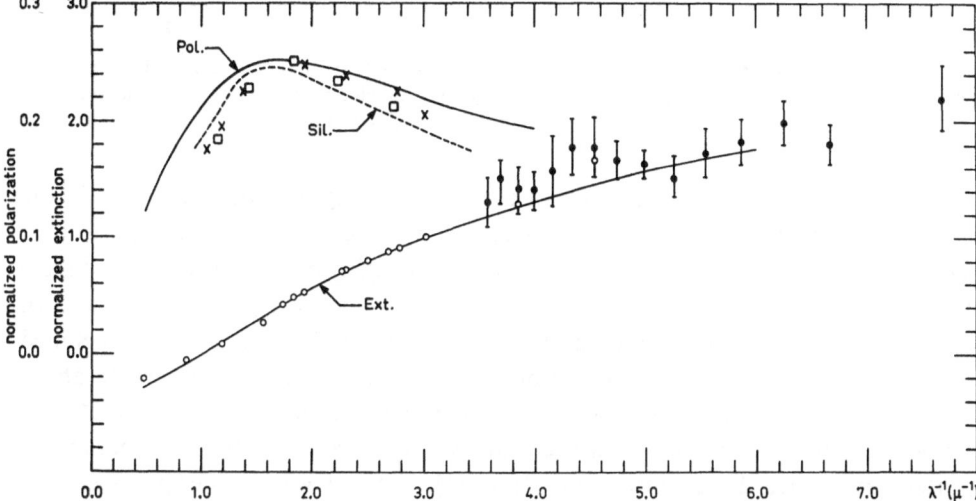

Fig. 1. Extinction observations are given by open and closed circles as specified by Stecher (1965). Polarization observations (crosses and squares) are as averaged in Greenberg (1968). Lower curve and upper solid curve labeled 'pol' are extinction and polarization by a size distribution $n(a) = 49 \times \exp[-5(a/0.2)^3] + \exp[-5(a/0.6)^3]$ of particles with index of refraction $m = 1.33 - 0.05i$. Dashed curve is polarization by a size distribution $n = \exp[-5(a/0.1)^3]$ of particles with $m = 1.66$. The theoretical polarization curves are normalized to one magnitude of extinction between $\lambda^{-1} = 1\mu^{-1}$ and $\lambda^{-1} = 3\mu^{-1}$.

greater than the median observed value ($\Delta m_p/\Delta m \gtrsim 0.025$) generally exhibits a broad maximum in the vicinity of $\lambda^{-1} \approx 2\ \mu^{-1}$. Averaged over a number of stars the values shown in the upper portion of Figure 1 are found. We can readily establish a reliable estimate of the characteristic particle size, if we assume that the particles are made of some solid material. The guideline is given by calculations of extinction by infinite dielectric cylinders. Predicted extinction cross sections are shown in Figure 2 for particles with an index of refraction $m = 1.33 - 0.05i$; this index reproduces in the visible the optical properties of 'dirty ice'. Although more refined results based on theory and measurement of extinction by other kinds of particles are available they do not modify the size estimate significantly.

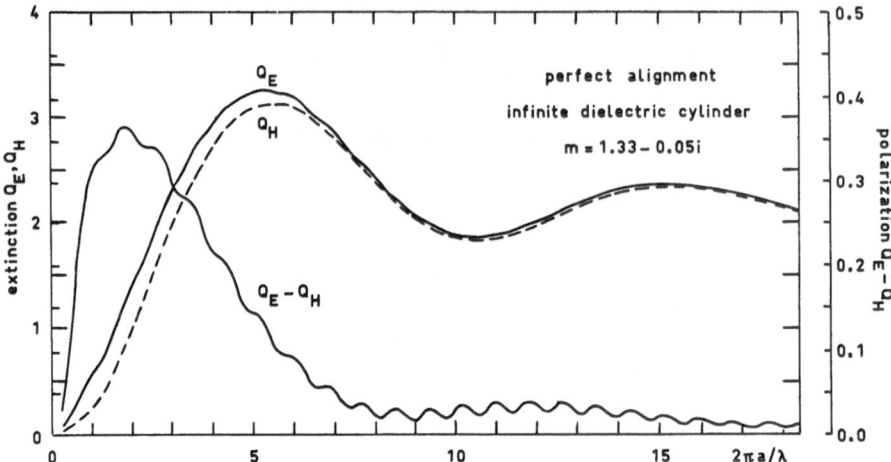

Fig. 2. Extinction efficiencies for an infinite circular cylinder with radiation perpendicular to the cylinder axis. Subscripts E and H specify respectively that the electric vector or magnetic vector of the radiation is parallel to the cylinder axis.

As the criterion for grain size we can use the wavelength of the maximum in the curve of polarization vs. wavelength (Figure 1). We then see from a comparison of Figures 1 and 2 that the $2\pi a/\lambda$ scale of the latter is roughly equivalent to the λ^{-1} scale of the former. In other words, elongated particles with a real part for the index of refraction $m' = 1.33$ should have a thickness given by $2a = 0.3\,\mu$. More detailed considerations lead to a size distribution with a range of radii from zero to about $0.2\,\mu$. For the purpose of our discussion a characteristic radius $a = 0.1\,\mu = 10^{-5}$ cm will be used. From extinction measurements we obtain as a fundamental parameter for the particle size the quantity $4\pi a\lambda^{-1}(m'-1)$. Thus, if we consider particles with different indices of refraction, the typical size varies in such a way that $a(m'-1)$ remains invariant. The only solid, non-dielectric particle seriously considered consists of graphite or carbon; its size spans a limited range with a maximum perhaps half that of the dirty-ice particles. Large molecules of the type suggested by Donn (1968) involve an entirely different type of scattering and are not considered here. Finally, we should point out that the interpretation of the scattered light from reflection nebulae (Greenberg and Roark, 1967) tends to reinforce our notion of solid dielectric particles whose size is of the order of $0.1\,\mu$.

The total mass of the interstellar dust is obtained from average values of the extinction. A reasonable average value would appear to be about 1 mag kpc^{-1}. The extinction in magnitudes is (ignoring polarization) given by the simple formula

$$\Delta m = 1.086 \sum \text{extinction cross section/unit area.} \qquad (1)$$

Thus, for one magnitude of extinction the total extinction cross-section is about 1 cm^2 per 1 cm^2. For the particle sizes we use, the extinction cross-section is approximately given by the projected area of the particle. Using spheres, for simplicity, the number of particles N in a path length D is obtained from

$$\Delta m = 1.086 \, N\pi a^2 / D. \tag{2}$$

The mass density ϱ_g of grains of specific density s is

$$\varrho_g = N\tfrac{4}{3}\pi a^3 s \, D = \tfrac{4}{3} as (\Delta m/D) \times \tfrac{1}{1.086}.$$

For $D=1$ kpc, $\Delta m=1$ we obtain

$$\varrho_g \approx 0.5 \times 10^{-26} \text{ g cm}^{-3} \tag{3}$$

where we use $s \approx 1$, valid for dirty ice.

It turns out that ϱ_g is rather insensitive to the type of grain material. This can be understood qualitatively if $\varepsilon - 1 \propto s$, where ε is the dielectric constant of the medium. This relation, of course, holds best if for a particular material one varies the degree of porosity. The correct formula is $(\varepsilon-1)/(\varepsilon+2) \propto s$. In that case we can say that approximately

$$\varrho_g \propto as \propto a \, \frac{m'^2 - 1}{m'^2 + 2} = a(m' - 1)\left(\frac{m' + 1}{m'^2 + 2}\right). \tag{4}$$

Since we take $a\,(m'-1)=$ invariant, we find that $\varrho_g \approx$ constant for $m' \approx 1$. The above derivation is not valid for metallic particles.

We see from Equation (3) that the total space density of the obscuring matter is of the order of 1 percent of the density of observed interstellar atoms and molecules. This statement holds for all grain models (including the large molecule model which we have not considered here). In the solar neighborhood the hydrogen density is of the order of 20 per cent of the total of gas plus stars. Therefore, we expect that, regardless of the grain composition, of the order of 10^{-3} to 10^{-4} of the mass of our Galaxy is in the form of obscuring matter. It is interesting to note that if all planetary systems in our Galaxy have the same mass ratio to their respective stars as ours, then even if a fraction 10^{-2} of the stars (by mass) have such planetary systems, there is 10 times more mass in the form of dust than in the form of planets.

3. Distribution and Dynamics of Gas, Dust and Stars

I summarize here some simplified observational and theoretical features of spiral structure which pertain directly and critically to the interstellar grain story. The principal sources of data are the Schmidt (1965) model, the density wave theory of spiral structure (Lin and Shu, 1964; Lin et al., 1969), the work of Roberts (1969) on gas motion and shock phenomena (the two-armed-spiral shock picture), and the recent studies of Lynds (1970) on the distribution of dust in Sc galaxies.

BASIC DATA AND ASSUMPTIONS

(1) The rotation period of the Milky Way is about 2.5×10^8 yr $= 2\pi \times 10$ kpc/(250 km sec^{-1}).

(2) The grand design is a two-armed spiral.

(3) The density wave rotates (at the solar position) with $\frac{1}{2}$ the speed of the gas and stars.

(4) Based on photos of external spiral galaxies, dust lanes increase in width with distance from the galactic center and are usually found in the inner edges of the spiral arm. Dust lanes have a width of the order of, or greater than, $\frac{1}{4}$ of the spiral arm width. In the solar neighborhood (if we are in a principal arm) we would expect the dust lane to be about 300 to 500 pc wide.

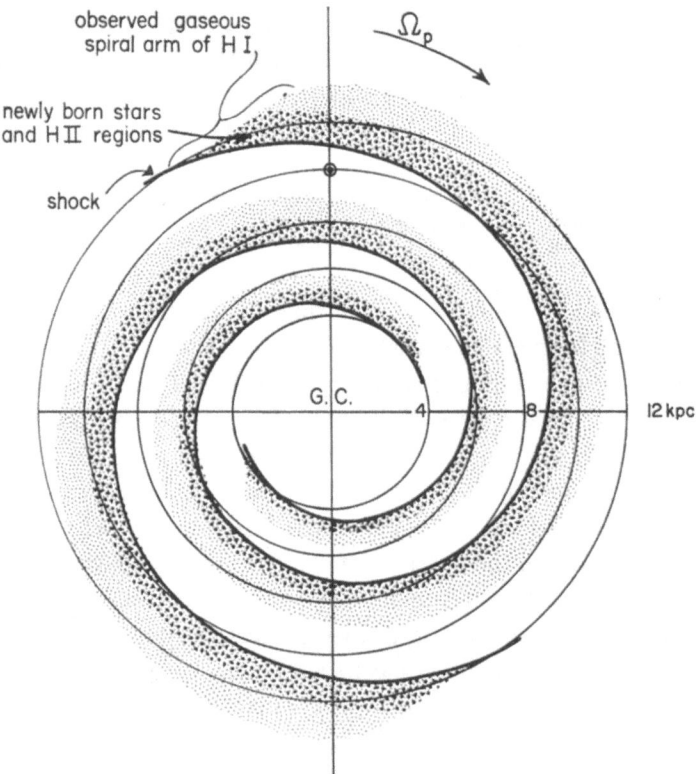

Fig. 3.   Shock and background spiral pattern of the Galaxy. The shock occurs slightly on the inner edge of the background spiral arm. Dust condensation initiated just ahead of shock. (Figure from Roberts, 1969).

(5) The H II regions (presumably regions of star formation) appear to lie within and at the edge of the dust lanes.

(6) The gas follows a path as shown in Figure 3. Its velocity component normal to the spiral arm is approximately given by the projected difference in velocity between the density wave and the material circular motion. This is roughly true because of the combination of two effects: first, going through the shock, the gas speeds up; second,

behind the shock, the stream lines are closer to the tangent to the arm. The result is that the galactic material passes through the spiral arm in about the same time as it passes between the spiral arms.

4. Formation of Dust

The first question we ask is: Can grains grow out of the gas within the time available? We assume here that there exists a sufficient number of nucleation points for grain growth and that we can ignore the problem of where the nuclei come from. Several suggestions for the origin of such small nucleation particles have been made, ranging from (i) formation in stars followed by ejection to (ii) the residue of dust particles destroyed during star formation. Conceivably sufficient nuclei will be produced. We further assume that as a result of radiative equilibrium with the ambient radiation field the grains are at a sufficiently low temperature that atoms like O, C, and N readily stick to the surface if they collide with the grain. For grain temperatures of about 10 K and gas temperatures of about 100 K it is most likely that the fraction of such atoms which stick is close to unity. Although for high gas temperatures $(T_{gas} \approx 1000 \, \text{K})$ experimental verification on the sticking fraction is lacking, it appears that some atoms (for example oxygen atoms) will stick rather well; namely those, that are bound more strongly than can be done by van der Waals forces only. This is a problem which should be studied in the laboratory. Hot H I gas, with temperatures in the range of 1000 to 10000 K may prevail in the interarm regions and also perhaps as a medium surrounding normal H I-clouds (Spitzer and Scott, 1969; Goldsmith et al., 1969). We will call this hot hydrogen gas H I'.

For spherical particles the rate of growth by attachment of atoms may be shown to be

$$\frac{da}{dt} = \frac{\gamma n_A}{4s} (3m_A kT)^{1/2}, \tag{5}$$

where γ = sticking fraction, n_A and m_A are number density and mass of the condensing atoms. If we let the inner edge of a spiral arm be a 'standard' H I region with hydrogen density $n_H = 10 \, \text{cm}^{-3}$ and gas temperature $T = 100 \, \text{K}$ we find (Greenberg, 1968) that

$$\dot{a} = f\gamma \sqrt{(TA)} \, n_H \times 2.6 \times 10^{-20} \, \text{cm sec}^{-1} \tag{6}$$

where $f \approx 10^{-3}$ is the standard cosmic abundance ratio of condensible atoms to H atoms, and A is the atomic weight. The high number density is achieved as a result of the compression at the inner edge of the arm. We find, under these conditions, that it takes about 3×10^7 yr for a particle to grow from a negligible radius to $a = 10^{-5}$ cm if γ is close to unity. In H I' regions, however, the growth rate may be smaller than in H I regions. If pressure equality exists between H I and H I' regions, i.e., $n_H T$ = constant, then Equation (6) predicts that the growthrate in H I' regions is smaller by a factor $(T_{HI}/T_{HI'})^{1/2}$.

If we use Lynds' estimate of dust-lane width of $\frac{1}{3}$ or greater of the arm width we find that in this dust lane the gas spends about $(\frac{1}{6}) \times 2.5 \times 10^8 \approx 4 \times 10^7$ yr. It thus appears that the time and conditions are not inconsistent with the idea of dust grains growing out of the gas. To complete the picture we would then have the dust participating with the gas in the formation of new stars, thus supplying the band of H$_{II}$ regions following the dust lanes on the inside of the spiral arms, as noted by Lynds and others.

A recent suggestion by Herbig (1969) on the formation of stars and dust appears to give, at least initially, a very similar distribution of gas, dust and stars in the spiral arms when combined with the shock picture. As a matter of fact, Herbig's suggestion may not differ in basic processes as much from the above picture as it appears at first although a quite different chemical composition of the dust results. Herbig suggests, that during the prestellar condensation the condensation of dust and the formation of complex molecules takes place under conditions of a very high density, a moderate radiation field and a moderate temperature ($T \approx 1000$ K). The resulting dust would be in the form of non-volatile solids of a type similar to the carbonaceous chondrites in our solar system. The excess volatile constituents would come off in the form of molecules consisting of the more abundant elements. Herbig estimates that the total dust output by such a process carried on over the age of the Galaxy would be sufficient to account for the total observed mass of dust. This raises the question: why isn't a sizeable fraction of the dust used up again in the subsequent formation (next generation) of new stars? If this were the case, then the total *net* production of dust would be lowered and might perhaps be insufficient to account for the present mass. But even if for this reason the 'solar nebula' hypothesis for dust formation were inadequate, it may be an important source of nucleation particles. The subsequent accretion in average interstellar conditions at the leading edge of the spiral arms will then lead to a more consistent picture with respect to the observed distribution of dust lanes and H$_{II}$ regions. The conclusion is that one should expect to find particles with core materials consisting of the heavier elements, Si, Fe, Al, etc., and with mantles consisting of the 'dirty ice' mixture of frozen water, methane and ammonia.

There has been some evidence in recent years for the formation of solid particles in the atmospheres of cool stars. The only direct clue is the spectroscopic identification in some stars of (possibly) olivene in emission (Knacke *et al.*, 1969). [Olivene = (Mg, Fe) SiO$_3$. (Ed.)]. Observations of a more indirect nature and theoretical calculations have led to the inference not only of such silicate particles but also of carbon or graphite particles. Can a sufficient amount of such materials be ejected from the stellar atmospheres to provide the observed dust mass? To answer this question let us estimate an upper limit on all material (gas as well as dust) lost by M stars.

According to Pottasch (see p. 272) all M stars together lose mass to the interstellar medium at a rate of less than $2.5 \times 10^{-13} M_\odot$ pc^{-3} yr^{-1}; stars of all kinds together lose certainly less than $10 \times 10^{13} M_\odot$ pc^{-3} yr^{-1}. For the M stars the amount of material supplied in 10^8 yr is $2.5 \times 10^{-5} M_\odot$ pc^{-3}, compared to a value of ϱ_g of 10^{-3} to $10^{-4} M_\odot$ pc^{-3}. The mass loss by M stars is therefore inadequate by at least an order of magnitude. Even the mass loss from all stars appears inadequate unless the ejected mass is completely in the form of dust. I should reiterate that this conclusion is based on the hypothesis of a regeneration every 10^8 yr because I have assumed the dust to be largely used up in star formation. However, even if this condition is relaxed completely, the M stars must produce dust for at least 10^9 yr with all ejected mass in the form of dust. If only one tenth of the mass loss is in the form of solid particles the time required equals the age of the Universe and would appear to be prohibitively large.

5. Magnetic Orientation and Interstellar Polarization

In this meeting, as in previous meetings in this series, one has inferred (in part at least) the existence of magnetic fields from the observation of optical interstellar polarization. Although at the moment we do not believe that these fields play an important role in the formation of the gross spiral structure, they certainly are still important in more local considerations involving gas dynamics. Magnetic field observations of a more direct nature have led to lower and lower field estimates and the entire question of associating optical polarization with magnetic orientation has been revived.

There is a rather good theoretical basis for the statement that even magnetic fields as small as 5 μG are adequate to orient 'standard' dielectric (ice or silicate) grains sufficiently for the required degree of interstellar polarization. This conclusion is based on rather detailed calculations (Greenberg, 1969) which take into account not only the scattering properties of nonspherical particles, but also the kinetics of the Davis-Greenstein (1951) orientation mechanism with physically reasonable assumptions of the gas collision processes. With a reasonable modification of the magnetic properties (Jones and Spitzer, 1967) it is possible to bring this limit down by an order of magnitude. Furthermore, the excellent correlation (Brouw, 1962) between the directions of optical and radio polarization (orthogonality) still is, to me, a convincing argument in favor of a theory in which a magnetic field is the primary agency orienting grains. Nevertheless, there are time limits in the Davis-Greenstein orienting mechanism which should be kept in mind when we consider other dynamical processes.

The relaxation time for magnetic orientation of a sphere is given by

$$\tau_0 = \frac{\omega I}{\chi'' V B^2} = \left(\frac{\omega}{\chi''}\right) \frac{2a^2 s}{5B^2}, \tag{7}$$

where ω is the angular velocity, I the moment of inertia, χ'' the imaginary part of the magnetic susceptibility, V the volume, and s the density of the grain. B is the magnetic field strength. For paramagnetic substances the value of χ'' is given by $\chi'' \approx 2.5 \times 10^{-12}$ ω/T_g where T_g is the grain temperature. For $a = 10^{-5}$ cm, $s = 1$, $B = 5$ μG and $T_g = 10$ K we get $\tau_0 = 2 \times 10^5$ yr. For other dielectric grain models the result is about the same. We see that the optical polarization can only follow gas processes with time scales of at least 10^5 yr.

6. Grain Temperature

The grain temperature is a critical parameter not only with respect to the magnetic orientation but also to such physical processes as molecule formation and the sticking of atoms on grain surfaces. For example, it has been suggested that hydrogen in solid form can accrete on grains at sufficiently low temperatures. Here it is impossible to present details of the calculations of grain temperatures. These calculations depend sensitively on the chemical constituents and the physical state of the grains since these properties determine the wavelength dependence of absorptivity and emissivity of electromagnetic radiation from the ultraviolet to the microwave regions. However, it is not difficult to present a qualitative determination of grain temperatures.

The interstellar radiation field has been estimated to have an energy density in the range 0.5 to 1.0×10^{-12} erg cm^{-3}. Since a considerable portion of this radiation comes from rather hot stars, a representation (useful for our purposes) of the interstellar radiation is by means of a radiation temperature $T_R = 10\,000$ K and a dilution factor $W = 10^{-14}$. Together these give an energy density of 0.8×10^{-12} erg cm^{-3}. If a body is placed in this radiation field it will reach an equilibrium temperature when its emission is equal to its absorption. If the body has wavelength independent absorptivity (and emissivity) the radiation law gives (disregarding proportionality constants) a black body temperature T_B according to

$$\text{Absorption} = \text{Emission}$$

or

$$WT_R{}^4 = T_B{}^4$$

with the result that $T_B = (10^{-14} \times 10^{16})^{1/4} = 3.2$ K. Very large bodies in space would reach this temperature (disregarding other processes). However, the interstellar grains are quite small and are increasingly poor emitters as the wavelength increases. For such small particles the absorptivity is proportional to the volume rather than to the area and therefore is proportional to λ^{-1}. We may then show that the equilibrium temperature is determined by $WT_R^5 = T_g^5$ which gives $T_g = 10^{6/5} = 15.8$ K. It turns out that, under normal interstellar radiation conditions, dirty ice grains and silicate grains have temperatures between these two values (generally about 10 K), whereas graphite or metallic particles tend to have temperatures between 20 to 40 K. Even in the center of very dense clouds with heavy attenuation of the ultraviolet radiation, the grain temperatures are unlikely to be less than 5 K. In the center of a cloud with 20 magnitudes of extinction dielectric grains of especially high infrared emissivity and low visible absorptivity may attain temperatures as low as 4 K. With regard to condensation of solid hydrogen (Greenberg and De Jong, 1969) this temperature is prohibitively high. However, if such very-low-temperature grains exist, they could help to bring down the gas temperature in dense clouds. For example, Heiles (1969) found OH spin temperatures as low as 4.5 K. Could such temperatures be produced by cooling of gas on grains? It may be shown that if each atom loses all of its energy on colliding with a grain the fractional change in the gas energy per unit time is given by the simple formula

$$\frac{dE}{dt}/E = \frac{dT}{dt}/T \approx \frac{\Delta m V}{R} \tag{8}$$

or $T \approx T_0 \exp(-t/\tau_T)$, where $\tau_T = R/(\Delta m V)$. Δm is the extinction through the cloud,

R is the cloud radius, V is the gas atom velocity ($\propto T^{1/2}$). For a 1 pc cloud with extinction of 5 magnitudes and $V = 1.5$ km sec^{-1} (initial gas) we get $\tau_T \approx 0.15 \times 10^6$ yr. (Here we assumed that V is constant. Actually $V \propto T^{1/2}$, but, as can be shown easily, this fact does not alter our conclusion.) In other words, to go from 50 K to 5 K would take $t \approx 10^6$ yr. This seems a bit long to be generally effective but not so long that a number of cool clouds could not be observed.

References

Brouw, W.: 1962, private communication.
Davis, Jr., L. and Greenstein, J. L.: 1951, *Astrophys. J.* **114**, 206.
Donn, B.: 1968, *Astrophys. J. Lett.* **152**, L129.
Goldsmith, D. W., Habing, H. J., and Field, G. B.: 1969, *Astrophys. J.* **158**, 173.
Greenberg, J. M.: 1968, in *Nebulae and Interstellar Matter* (ed. by B. M. Middlehurst and L. H. Aller), University of Chicago Press, Chicago, p. 221.
Greenberg, J. M.: 1969, *Physica* **41**, 67.
Greenberg, J. M.: 1970, *Astron. Astrophys.*, in press.
Greenberg, J. M. and De Jong, T.: 1969, *Nature* **224**, 251.
Greenberg, J. M. and Roark, T.: 1967, *Astrophys. J.* **147**, 917.
Heiles, C.: 1969, *Astrophys. J.* **157**, 123.
Herbig, G. H.: 1969, paper presented at the VIth International Astrophysical Symposium, Liège.
Jones, R. V. and Spitzer, Jr., L.: 1967, *Astrophys. J.* **147**, 943.
Knacke, R. F., Gaustad, J. E., Gillett, F. C., and Stein, W. A.: 1969, *Astrophys. J. Lett.* **155**, L189.
Lin, C. C. and Shu, F. H.: 1964, *Astrophys. J.* **140**, 646.
Lin, C. C., Yuan, C., and Shu, F. H.: 1969, *Astrophys. J.* **155**, 721.
Lynds, B. T.: 1970, in *IAU Symposium No. 38, The Spiral Structure of Our Galaxy* (ed. by W. Becker and G. Contopoulos), Reidel, Dordrecht, p. 26.
Oort, J. H. and Van de Hulst, H. C.: 1946, *Bull. Astron. Inst. Netherl.* **10**, 187.
Roberts, W. W.: 1969, *Astrophys. J.* **158**, 123.
Schmidt, M.: 1965, in *Galactic Structure* (ed. by A. Blaauw and M. Schmidt), University of Chicago Press, Chicago, p. 513.
Spitzer, Jr., L. and Scott, E. H.: 1969, *Astrophys. J.* **158**, 161.
Stecher, T. P.: 1965, *Astrophys. J.* **142**, 1683.

23. INTERSTELLAR MOLECULES

Introductory Report

(Wednesday, September 17, 1969)

THEODORE P. STECHER

Goddard Spaceflight Center, Code 670, Greenbelt, Md., U.S.A.

1. Introduction

The observation of molecules in the interstellar medium offers an excellent oppor- tunity to specify the physical conditions present there. This must be done through the combined use of theory and observation. In the last few years the advancement in this area has accelerated spectacularly. McNally (1968) reviewed the subject and Spitzer (1968) included molecules in his more general review; these are both excellent references and the reader is referred to them for the development and state of the subject in 1966.*

The subject may be divided into several topics: observations, formation processes, destruction mechanisms, and excitation processes.

2. Observations

Optical observations have been made of (i) a number of electronic transition lines of CH, CH$^+$, and CN and, (ii) the as yet unidentified diffuse interstellar lines. There are negative optical searches for the electronic transitions of OH (Herbig, 1968) using conventional instruments and of H$_2$ (Carruthers, 1967; Smith, 1969) using rockets, but a positive measurement of H$_2$ in emission has been made in the quadrupole lines using a scanning interferometer (Werner and Harwit, 1968). Perhaps the most sophisticated application of classical techniques is the work by Bortolot and Thaddeus (1969) who added up large numbers of high-dispersion spectrograms and who were able to detect C^{13}H molecules. The most complete study of the interstellar line spectrum of a single star is that of ζ Oph by Herbig (1968). Polyatomic molecules were looked for and not found in Herbig's study.

Radio techniques have produced some very exciting results in recent years. The OH radical has been observed in absorption and emission in a variety of excitation conditions. OH in emission has been observed from sources with angular diameters $< .01''$ (assuming that the sources are not coherent so that a Michelson interferometer will resolve them). The resulting brightness temperatures are of the order of 10^{13} K. These sources are believed by some to be protostars where maser action is producing the high brightness temperature (see the Introductory Report by Mezger, p. 336). Litvak (1969) has discussed the models proposed to date.

In the last year three polyatomic molecules have been discovered and they also

---

* A status report on H₂ by E. E. Salpeter appears as an Appendix to this Report.

Habing (ed.), Interstellar Gas Dynamics, 316–319. All Rights Reserved. Copyright © 1970 by IAU

exhibit a variety of physical conditions. The molecules are NH_3 (Cheung et al., 1968), H_2O (Cheung et al., 1969), and H_2CO (Snyder et al., 1969). The water molecule is observed in a higher level transition and is apparently also a maser. NH_3 is observed through the inversion transition.

3. The Destruction Mechanisms

To explain the observed molecular abundances one has to consider both the formation and the destruction mechanisms. In this section, we shall discuss destructive mechanisms. Bates and Spitzer (1951) showed that photodissociation in a typical radiation field would give a CH molecule a lifetime of about 1000 yr. A similar calculation by Stecher and Williams (1966) gave about 10^4 yr for OH and NH and about 10^6 yr for CO and CN. Within times of these orders, molecules can be transported only over small distances, so that the existence of formation mechanisms is required at the site where the molecules are observed. For molecules less strongly bound than H_2 destruction will occur through chemical exchange reaction with atomic hydrogen, provided the temperature is above 500 K and the density is normal. With respect to H_2 itself, it was thought for many years that photodissociation had to be done by photons with $\lambda < 850$ Å. These photons are absent in HI regions and the H_2 content of HI regions was therefore considered very high. However, Stecher and Williams (1967) found another kind of destruction mechanism: excitation into electronic states followed by a decay into the vibrational continuum of the ground states. The lifetime of an H_2 molecule is only 10^3 yr in the average radiation field of an HI region. Predissociation has also shortened the expected lifetime of CH (Dressler, 1969; Herzberg and Johns, 1969).

4. Formation Processes

Bates and Spitzer (1951) showed that two-body radiative association of C and H would fail by a factor of 10^4 to produce the observed number of CH molecules. Herbig (1968) found that due to a revised cross section the two-body process produces CH at a sufficiently large rate (provided one has a high density), but more rapid destruction processes discussed by Dressler (1969) nullified this conclusion. Molecular hydrogen cannot form by the two-body process because the stabilizing transition is forbidden. Evaporation from grains has been suggested as a production source, but this leaves the explanation of the origin of the grains in a rather unsatisfactory state.

Formation of molecules on grains, with the grain acting as a third body and taking up the energy of formation in lattice vibration, has been one of the most promising suggestions. However, Knaap et al. (1967) showed that the first atom impinging on the grain would evaporate before the second atom arrived, unless the grain were at a much lower temperature than typical conditions indicate. This work assumed that the grains are perfect crystals. A solution to the evaporation problem (Hollenbach, 1969) is to have imperfections in the surface that would trap atoms and bind them more tightly to the grain. In shielded regions, where the temperature is low and the density is high, this may well be the fastest way to form molecules.

When it looked as if physical adsorption would fail under all conditions, Stecher and Williams (1966) proposed chemical exchange reactions between an incoming atom and one chemically bound to the grain. The process requires an activation energy which the particles can obtain only during cloud-cloud collision or when the grains are driven through the gas at high velocity. There is some evidence that the right molecules are produced in the observed amounts around hot stars where radiation pressure can drive the grains. For example, it seems possible to produce the CN molecule in this way, whereas other mechanisms do not seem capable of doing so.

5. Excitation of Interstellar Molecules

The purpose of this Symposium is to specify and discuss the kinematics and dynamics of the interstellar material. The state of excitation of atoms and molecules may yield information on the physical conditions present and may help in deducing the kinematics. Litvak (1969) has reviewed OH and H_2O masers and discussed in detail the problems associated with their excitation for which chemical, IR, UV, and collisional pumping are possible mechanisms. OH is also observed in thermal emission and absorption as well as in anomalous absorption. When the line formation of the OH radical is understood, it will help us considerably in specifying the physical environments in which this molecule is seen.

The CN radical has rotational levels whose excitation has been used to determine the universal 3 K black-body radiation (Field and Hitchcock, 1966). Other molecules have been used to place upper limits on this radiation at higher frequencies. One of the most interesting observations is that of the H_2CO molecule by Palmer *et al.* (1969), who found it in absorption against the universal black-body radiation, which implies that the molecule must be refrigerated. Townes and Cheung (1969) have suggested a collisional mechanism to accomplish this. The C^{12}/C^{13} ratio has been determined as a by-product of the CN work (Bortolot and Thaddeus, 1969) and of the H_2CO work (Zuckerman *et al.*, 1969). This ratio is of fundamental importance for the study of stellar mass loss.

6. Conclusion

The pace of progress is currently so rapid in this field of science that any complete review would have a lifetime measured in months. I hope I have stated some of the reasons for studying interstellar molecules in connection with the study of cosmical gas dynamics.

Note added in proof, June 1970. Since the symposium, H_2 has been found in large amounts in the direction of χ Per by Carruthers. CN has been found at radio wavelengths. CO has been discovered with an interesting isotopic abundance ratio. HCN and HC_3N have been found as well as more states of OH.

References

Bates, D. R. and Spitzer, L.: 1951, *Astrophys. J.* **113**, 441.

Bortolot, V. J., Jr. and Thaddeus, P.: 1969, *Astrophys. J. Lett.* **155**, L17.

Carruthers, G. R.: 1967, *Astrophys. J. Lett.* **148**, L141.

Cheung, A. C., Rank, D. M., Thornton, D. D., Townes, C. H., and Welch, W. J.: 1968, *Phys. Rev. Lett.* **21**, 1701.

Cheung, A. C., Rank, D. M., Thornton, D. D., Townes, C. H., and Welch, W. J.: 1969, *Nature* **221**, 626.

Dressler, K.: 1969, *Physica* **140**, No. 1.

Field, G. B. and Hitchcock, J. L.: 1966, *Astrophys. J.* **146**, 1.

Herbig, G. H.: 1968, *Z. Astrophys.* **68**, 243.

Herzberg, G. and Johns, J. W. C.: 1969, *Astrophys. J.* **158**, 399.

Hollenbach, D. J.: 1969, Ph.D. thesis, Cornell University, Ithaca.

Knaap, H. F. P., van den Meydenberg, C. J. N., Beenakker, J. J. M., and van de Hulst, H. C.: 1967, *Interstellar Grains*, U.S. Government Printing Office, Washington, D.C., NASA SP-140, p. 253.

Litvak, M. M.: 1969, *Science* **165**, 855.

McNally, D.: 1968, *Adv. Astron. Astrophys.* **6**, 173.

Palmer, P., Zuckerman, B., Buhl, D., and Snyder, L. E.: 1969, *Astrophys. J. Lett.* **156**, L147.

Smith, A. M.: 1969, *Astrophys. J.* **156**, 93.

Snyder, L. E., Buhl, D., Zuckerman, B., and Palmer, P.: 1969, *Phys. Rev. Lett.* **22**, 679.

Spitzer, Jr., L.: 1968, *Diffuse Matter in Space*, Interscience, New York.

Stecher, T. P. and Williams, D. A.: 1966, *Astrophys. J.* **146**, 88.

Stecher, T. P. and Williams, D. A.: 1967, *Astrophys. J. Lett.* **149**, L29.

Townes, C. H. and Cheung, A. C.: 1969, *Astrophys. J. Lett.* **157**, L103.

Werner, M. W. and Harwit, M.: 1968, *Astrophys. J.* **154**, 881.

Zuckerman, B., Palmer, P., Snyder, L. E., and Buhl, D.: 1969, *Astrophys. J. Lett.* **157**, L167.

Appendix. Molecular Hydrogen in the Interstellar Gas

E. E. SALPETER

Laboratory of Nuclear Studies, Physics Department, and Center for Radiophysics and Space Research, Cornell University, Ithaca, N.Y., U.S.A.

There are two levels of interest in the abundance of molecular hydrogen in the interstellar gas. (i) Even a low abundance can enhance the cooling rate at a few hundred K. (ii) Densities comparable with or larger than those of atomic hydrogen could affect the dynamics in the galactic disk, if widespread, or at least affect star formation, if restricted to isolated large clouds.

Extensive calculations of growth rates were carried out some time ago by Gould and Salpeter (1963) and by Gould *et al.* (1963), and the situation as of a few years ago can be summarized as follows (Field *et al.*, 1966). At densities normal for ordinary interstellar clouds ($n_H \approx 10$ cm^{-3}) a hydrogen atom strikes the surface of a dust grain about every 10^7 yr. For grain temperatures T_{grain} below about 20 K the recombination efficiency γ (probability of an H-atom to end up in a molecule after hitting the grain surface) was thought to be quite high, $\gamma \approx \frac{1}{3}$. The threshold for photodissociation of H_2 was thought to be beyond the Lyman continuum, in which case molecular hydrogen would break up only when a cloud came near an OB star. The net result was the belief that molecular hydrogen might generally be about as abundant as atomic hydrogen and much more abundant in dense clouds.

In recent years, ideas of high H_2-abundances received two separate blows. (i) The inclusion of quantum mechanical zero-point energy lowers the adsorption binding energy for a hydrogen atom on a perfect crystal surface (Knaap *et al.*, 1966). It was concluded that the atoms would evaporate instead of recombine unless $T_{grain} < 8$ K. Pure graphite grains would have temperatures well above 20 K and the recombination efficiency γ would be very low. (ii) Stecher and Williams (1967) found that about 10 per cent of excitations of H_2 to a set of discrete levels, brought about by photons with $\lambda > 912$ Å, are followed by dissociation. The lifetime of a hydrogen molecule in normal interstellar space (H I-region) against this radiative destruction is only about 10^3 yr. One might then have expected the H_2-abundance to be quite negligible throughout interstellar space.

More recently still, the first difficulty was overcome in two ways: (a) Grains are now thought to contain enough impurities and/or dielectric mantles to radiate fairly efficiently in the infrared, so that grain temperatures well below 15 K are now expected. (b) Surface defects of various types on realistic grains are expected to provide some sites with increased adsorption binding energy. Estimates indicate good recombination efficiency as long as $T_{grain} \lesssim 25$ K (Hollenbach, 1969; Hollenbach and Salpeter, 1970).

The second difficulty persists under normal interstellar conditions: in an 'ordinary' interstellar cloud ($n_H \approx 10$ to 30 cm^{-3}, optical depth in the visible of $\tau_{vis} \lesssim 0.2$), the

Habing (ed.), Interstellar Gas Dynamics, 320–321. All Rights Reserved. Copyright © 1970 by IAU

molecular abundance of hydrogen is only about 10^{-5} times the atomic abundance. H_2 is therefore unlikely to play a dominant role in the dynamics of the galactic disk as a whole. However, the situation is radically different in the denser clouds which presumably are the precursors of star formation. Since the relevant photodisintegration is produced by line-radiation, the interior of a dense enough cloud can be shielded from this line-radiation by self-absorption in H_2. The H_2 is replenished by continuous recombination in the outer layers of the cloud and the attenuation of the radiation is also helped by the usual grain absorption. Estimates show that hydrogen should be mainly molecular in any cloud with $n_H \gtrsim 100$ cm^{-3} and $\tau_{vis} \gtrsim 0.5$ (Hollenbach 1969; Solomon, to be published). One direct observation (Werner and Harwit, 1968) and some observed anticorrelation (Garzoli and Varsavsky, 1966; Heiles, 1969) between atomic hydrogen and dust-density in dense clouds seem to corroborate these estimates. *

References

Field, G. B., Somerville, W., and Dressler, K.: 1966, *Ann. Rev. Astron. Astrophys.* **4**, 207.

Garzoli, S. L. and Varsavsky, C.: 1966, *Astrophys. J.* **145**, 79.

Gould, R. J., Gold, T., and Salpeter, E.: 1963, *Astrophys. J.* **138**, 408.

Gould, R. J. and Salpeter, E. E.: 1963, *Astrophys. J.* **138**, 393.

Heiles, C.: 1969, *Astrophys. J.* **156**, 493.

Hollenbach, D. J.: 1969, Ph.D. thesis, Cornell University, Ithaca.

Hollenbach, D. J. and Salpeter, E. E.: 1970, *Astrophys. J.* (in press).

Knaap, H., v. d. Meijdenberg, C. J. N., Beenakker, J., and Van de Hulst, H. C.: 1966, *Bull. Astron. Inst. Netherl.* **18**, 256.

Solomon, P. M. (to be published).

Stecher, T. P. and Williams, D. A.: 1967, *Astrophys. J. Lett.* **149**, L29.

Werner, M. W. and Harwit, M.: 1968, *Astrophys. J.* **154**, 881.

* (*Note added in proof.*) Carruthers (1970, *Astrophys. J. Lett.* **161**, L 81) announced the detection of interstellar H₂ in the spectrum of χ Per. (Ed.)

24. DISCUSSION FOLLOWING THE REPORTS
BY GREENBERG AND STECHER

(Wednesday, September 17, 1969)

Chairman: H. C. van de Hulst

Editor's remark: The material of this Discussion has been rearranged in five sections: 1. Grain Orientation; 2. The Diffuse Galactic Light; 3. The Extinction Curve in the Far UV; 4. Formation and Destruction of Grains; 5. The OH Molecule. Except for rearranging, very few changes have been made.

1. Grain Orientation

Verschuur: Dr. Greenberg, what happened to the argument that alignment of grains required fields of the order of 10 μG?

Greenberg: During the past several years an enormous amount of new information has become available on the polarizing properties of elongated particles, and that, combined with my recent calculations on the particle orientation mechanism based on Jones and Spitzer's work (1967) gave a value of 8 μG without stretching any one of the parameters. (Jones, R. V. and Spitzer, Jr. L.: 1967, *Astrophys. J.* **147**, 943.)

Verschuur: Would you have to stretch the parameters to go down to 2 μG?

Greenberg: Yes, if I still demand the maximum ratio of polarization to extinction. By reducing n_H from 10 cm^{-3} to, say, 2 cm^{-3} the required field is down by a factor of 2. If I further allow the ratio of polarization to extinction to be 0.025 rather than 0.06, then a magnetic field of 2 μG would be sufficient. I would like to add that, as stressed earlier by Woltjer (p. 195), the optical polarization gives essentially the squared field. Perhaps variations in the direction of the magnetic field reduce the linear average, which you see in the Zeeman and in the Faraday measurements, while a sufficiently large squared average remains to produce the optical polarization. Also, a small contribution by super-paramagnetism could decrease the field requirement further.

Kurt: The grain charge depends on the nature of the dust particles (metallic or dielectric) on the UV radiation and on the interstellar electron density. For the specific conditions in interstellar space we may have both positive or negative charges. What is your opinion, Dr. Greenberg?

Greenberg: I think that the dust is mainly negatively charged. If there is an electric or (because of rotation) a magnetic dipole, there are other forces, but I do not expect that they are large enough to affect the orientation.

Van de Hulst: The charge is also important in connection with the drag between gas and dust.

Dolginov: If the grains are charged, but the center of charge distribution is not the

Habing (ed.), Interstellar Gas Dynamics, 322–335. *All Rights Reserved. Copyright* © 1970 *by IAU*

center of gravity because the grain has a complex shape, then an electric dipole results and electrical forces occur when the grain moves with respect to the interstellar magnetic field. If the dust and the magnetic field move with respect to each other with differential velocities of at least 1 km sec^{-1}, the electrical forces are large enough to destroy the orientation of the particles. There is also another possibility to orient the grains: place the dust particles in a gas flow; if the relative flow velocity is about 10 km sec^{-1}, then one can calculate (Dolginov, 1968) that within 10^3 yr the grains are oriented perpendicular to the flow velocity. It takes about 10^6 yr for the grains to attain the flow velocity, after which the orientation will be random. This process of grain orientation (originally proposed by Gold) will work in the solar wind and in stellar winds. A somewhat similar, but quantitatively (usually) somewhat less important process is that of particles placed in an anisotropic radiation field. [Dolginov, A. Z.: 1968, *Dokl. Akad. Nauk SSSR* **179**, 1070 (translation: 1968, *Soviet Phys. Dokl.* **13**, 281).]

Greenberg: The grains are slowed down in about a third of a parsec for a gas of any density. They do not maintain a sufficient relative velocity unless there is a driving mechanism. Recently Purcell (1969) has looked into the polarization produced by grains driven through a gas by anisotropic stellar radiation. The result was not enough to account for the polarization. One should combine all proposed mechanisms and see whether they combine to give the observed amount of polarization relative to extinction, or whether they detract from each other. I think that the magnetic field by itself is sufficient. [Purcell, E. M.: 1969, *Physica* **41**, 100 (Proceedings International Conference on Laboratory Astrophysics, ed. by J. Rosenberg, Lunteren, 1968).]

2. The Diffuse Galactic Light

Rozhkovskii: At the Astrophysical Institute of the Kazakian Academy of Sciences, we are making photometric studies of the diffuse galactic radiation. The methods of our observations and other details of the investigation are described elsewhere (Rozhkovskii 1969a, b). My remarks concern only the albedo γ of the interstellar grains. Our observations show $0.3 < \gamma < 0.6$, with a probable error of 0.3 accounting for uncertainties in the observational data. However, even with so large an uncertainty it seems that the optical properties of grains are sufficiently different from those of perfect dielectric particles. Our result differs significantly from that of Witt (1968), who found that γ must be extremely close to 1 (about 0.98).

The discrepancy between the two results has a simple explanation. For the interpretation of his observation, Witt used Henyey and Greenstein's (1941) theoretical treatment. Their formulae are only approximate. I found, however, a rigorous analytic solution of the equation of transfer in Eddington's approximation in the special case when $\gamma = 1$ and the scattering is isotropic. It appears that the predicted intensity is higher in our solution than in that of Henyey and Greenstein. Therefore Witt found too high an albedo for the grains compared to what actually is required. (Rozhkovskii, D. A.: 1969a, *Vest. Akad. Nauk Kazakhskoi SSR* **9**, Rozhkovskii, D. A.: 1969b,

Vest. Akad. Nauk Kazakhskoi SSR **7**, 63; Witt, A. N.: 1968, *Astrophys. J.* **152**, 59; Henyey, L. G. and Greenstein, J. L.: 1941, *Astrophys. J.* **93**, 70.)

Van de Hulst: You have been speaking of the differences between two computed curves. In Leiden de Jong and I made a similar study by numerical means rather than by analytic solution of the equation of transfer (van de Hulst and de Jong, 1969) and we disagree in the same sense as you do with Witt's conclusion regarding γ.

Let me mention that at the same observatory Glozhkov and others have done much work on reflection nebulae (e.g., Glozhkov, 1968) but nobody has been asked to talk about those here. [van de Hulst, H. C. and de Jong, T.: 1969, *Physica* **41**, 151 (Proceedings International Conference on Laboratory Astrophysics, ed. by J. Rosenberg, Lunteren, 1968); Glozhkov, Yu. I.: 1968, *Trudy Astrofiz. Inst. Akad. Nauk Kazakhskoi SSR* **11**, 57.]

3. Extinction Curve in the Far UV

Kurt: I should like to have a discussion on Stecher's newest measurements of the extinction curve in the UV.

Stecher: The curve you mentioned has been published recently (Stecher, 1969). Let me summarize this paper. With an Aerobee rocket I obtained spectra between 1150 Å and 4000 Å of ζ Persei and ε Persei. Figure 1 presents the difference in magnitude (ζ Persei $-\varepsilon$ Persei), for each 2 Å interval. The differences have been multiplied by a

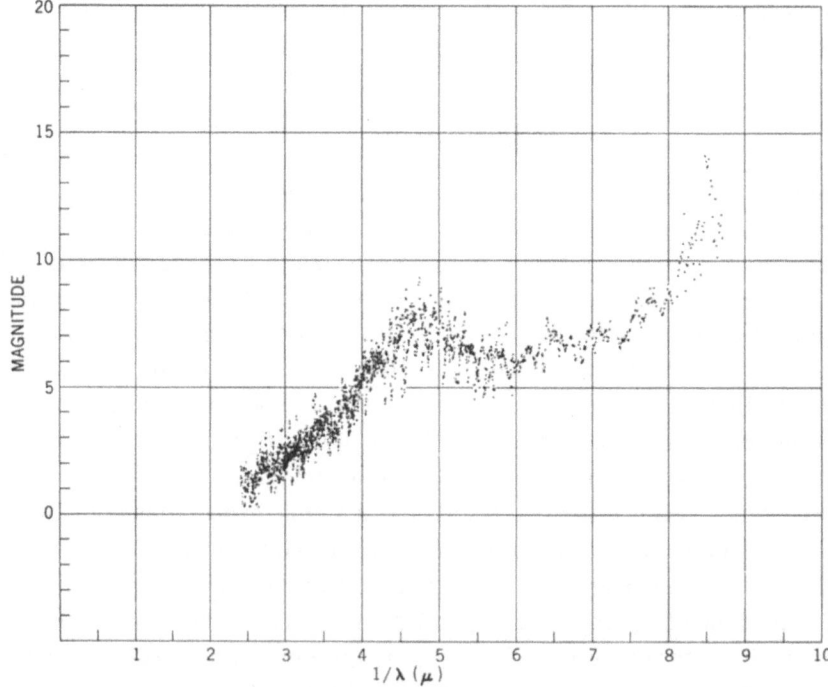

Fig. 1.   (See the remark by Stecher.) Interstellar extinction in magnitudes as a function of inverse wavelength determined from ζ and ε Persei. The curve is normalized to $B-V=1$ mag and $V=0$.

factor of 3.57 to normalize the observations to $B-V=1$. The two stars are quite close in visual magnitude and therefore the curve goes through $V=0$ without any additional normalization. When a small correction for the geocorona Ly-α radiation is applied, the curve will show a very large extinction near 8.22 μ^{-1} due to the interstellar Ly-α absorption line. Only the region within 100 Å of Ly-α is affected by this. The difference in spectral type is of minor importance (ζ Persei is BIIb and ε Persei is B0.5V).

The present extinction curve may be considered a confirmation of earlier work (Stecher, 1965). Its main characteristic is the hump at about 4.7 μ^{-1}, which is quite clearly present. Stecher and Donn (1965) and Wickramasinghe and Guillaume (1965) have pointed out that a transition in graphite would give rise to a similar feature. At the shorter wavelength another material is necessary if graphite is assumed. Atomic and molecular gases (Stecher and Williams, 1969) as well as coated particles could serve as such. Some minor features are atomic and are probably circumstellar. (Stecher, T. P.: 1965, *Astrophys. J.* **142**, 1683; Stecher, T. P.: 1969, *Astrophys. J. Lett.* **157**, L125; Stecher, T. P. and Donn, B.: 1965, *Astrophys. J.* **142**, 1681; Stecher, T. P. and Williams, D. A.: 1969, in preparation; Wickramasinghe, N. C. and Guillaume, C.: 1965, *Nature* **207**, 366.)

Kurt: Dr. Stecher, in the slide [Figure 1. (Ed.)] we can see an absorption line at about 1538 Å, 1408 Å and 1178 Å. Is this real or is it an instrumental effect?

Stecher: There are several lines, but I believe them to be circumstellar. For instance C IV at 1549 Å is present.

Greenberg: Figure 2 shows some model calculations I made. The dash-dot curve is pure H_2O-ice. Characteristically all dielectric-grain models show some absorption

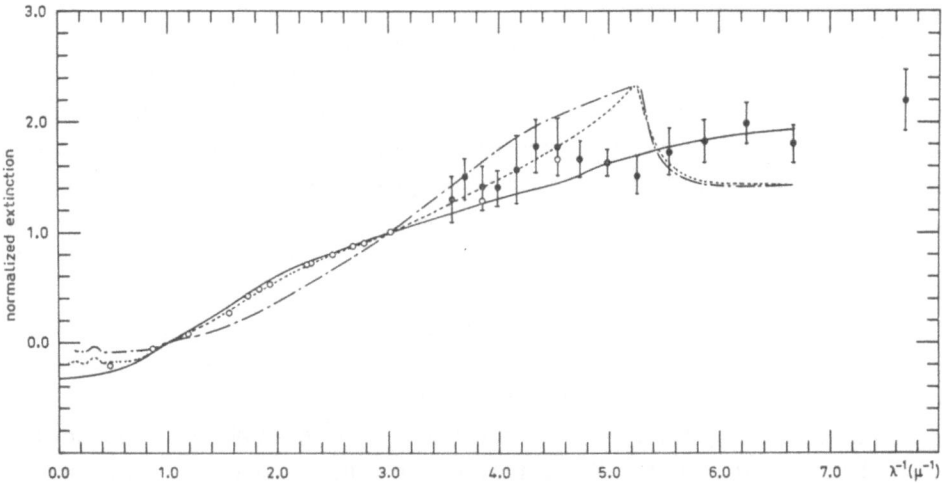

Fig. 2. (See the remark by Greenberg.) The solid curve shows the extinction by a graphite core (radius 0.054 μ) with a dielectric ($m = 1.33 - 0.05i$) mantle (radius 0.16 μ); the dashed curve is for a 'dirty-ice' mantle, whose index of refraction varies according to the best available measurements; dash-dot curve is for extinction by a solid 'dirty-ice' grain (no graphite core) with a radius of 0.16 μ. The rise and drop in the latter two curves is caused by an absorption edge for the ice.

at about $5\,\mu^{-1}$, so that the drop in the curve is easy to make with dielectric particles. The drop does not reproduce the absorption band exactly, but one could use other molecules to smooth the effect a little. Regardless of whether we have the graphite core or some dielectric core (like silicates) it is always a problem to find a material which causes a continuous rise in extinction after the drop. This situation has to do with the fact that if an absorbing particle has the size of the wavelength, its extinction continues to drop. But what happens when the surfaces of particles are not smooth? I have shown that irregular particles will smooth out the kinds of details shown in Figure 2 because they would give rise to an effective absorption in the wavelength region before the onset of real absorption. The result is that real absorptivity would not show up strongly. If the particles are rough, then all spectral characteristics would have to be based on more elaborate discussions of the scattering and absorbing properties of particles of that nature. Stecher pointed out that graphite particles could give the observed hump, but the graphite model alone does not produce the continued rise in the extinction observed beyond $5\,\mu^{-1}$, and Stecher needs to put in another material as well. Further any model with enough graphite to produce the hump does not satisfy the observed wavelength dependence of polarization. We also note that graphite particles cannot be oriented by the magnetic field; therefore the polarization has to be produced by that extra ingredient alone. If you add enough of that material (but not enough to obscure the hump), then you cannot get the observed ratio of polarization to extinction, regardless of what the mechanism of orientation is.

Shulman: Can the rise for $\lambda^{-1} > 5\,\mu^{-1}$ be explained by the presence of large organic molecules?

Stecher: A large molecule with benzene rings in it would probably have the sort of extinction required. The amount of extinction yields at once the amount of H_2^+ in the Galaxy; one obtains $n(H_2^+) = 2 \times 10^{-4}\ \text{cm}^{-3}$.

Shulman: I should like to suggest that large molecules can exist in grains. Let us consider a grain covered by the simplest radicals mentioned here (CH, OH, etc.). Suppose that the grain is exposed to cosmic rays, then it is easy to show that the energy input is of the order of 10^{-14} eV sec^{-1} per radical. It is a well-known chemical fact that synthesis of heavy molecules can take place in solids induced by radiation. The rate of radiative synthesis depends on many conditions such as temperature, structure, chemical composition and others. Gains are found ranging from 10^{-2} to 10 reactions per eV. The exact value in the case considered here is unknown. If we estimate one reaction per eV, almost any radical will take part in a chemical reaction in a time of the order of 10^7 yr. Actually not only radiative synthesis occurs, but also radiative destruction of complex molecules. However, chemical experiments have shown that radiative synthesis is more probable. It is, for example, an interesting fact that solid methane (CH_4) at 77 K when exposed to gamma radiation, transforms into oil with a large molecule of the form $C_{20}H_n$ where n may have various values. So we can expect more complex molecules than CH (and similar ones) on interstellar grains. Can anybody tell me whether it is possible to detect such molecules?

Van de Hulst: I might answer by saying no, because, first of all even if you had these molecules nicely chemically isolated in a gas, then you would have difficulty recognizing their spectra among those of similar molecules. But here, where they are imbedded in a particle, the absorption of the bulk material might show up as a reduction (thereby an apparent emission) in the star spectrum. I would think that a chance of a positive identification for such molecules is very small.

Field: I disagree with van de Hulst on the question of whether or not one might identify organic compounds. In particular, several people who discussed the relationship between the unidentified lines and the grains suggested that the width of the lines can be interpreted as due to molecules imbedded in the grains and interacting strongly with the grain matrix. F. M. Johnson, in the U.S.A., investigated possible identification of these lines with heavy organic molecules (Johnson, 1967). Also in response to Shulman's remark, we should consider the effects of X-ray emission by discrete sources, such as Sco XR-1. If Shulman is correct, one might expect differences in the grain properties near X-ray sources, because the X-ray quanta have similar effects to those of cosmic rays in changing the chemical compounds in the grain. One might, for example suggest observing the unidentified feature at λ 4430, in the neighborhood of such X-ray sources. I have been considering a model for explaining the intense interstellar calcium lines in the spectrum of Sco XR-1 by the removal of material from grains by X-ray interaction. I can explain the calcium in the gas phase as due to ejection from the grain by 1-keV photons which generate very energetic electrons within the grains. It is interesting that in Scorpius the calcium in the gas phase is generally 1000 times less than would have been predicted from solar abundances. We therefore need to remove only about 1 per cent of the calcium locked up in the grains (if that is, in fact, where the missing calcium is), in order to explain the observations in the Scorpius region. [Johnson, F. M.: 1967, in *Interstellar Grains* (ed. by J. M. Greenberg and T. P. Roark), Office of Technology Utilization, National Aeronautics and Space Administration, Washington, D. C.]

Shulman: I agree with Field, because only the total energy input has significance in such a process, but not the nature of the radiation. X-radiation and cosmic protons give the same effect.

Greenberg: I agree that the unidentified lines may be one of the keys to unlocking at least one of the problems of the interstellar grains. I have done a calculation on the polarization of an absorption band in the grain material. Historically the consideration of producing diffuse lines by atoms or molecules in grains goes back to our Chairman (H. C. van de Hulst). Many years ago I myself tried calcium as a source of λ 4430. The problem is, however, that any kind of absorbing material whose general optical characteristics give rise to the observed extinction and polarization, does not give an absorption line at 4430 Å but rather a dispersion curve (i.e., a profile with an emission wing at shorter wavelengths and absorption at longer wavelengths). This is a general and unavoidable result. If you put the material in a metal grain which is also already absorbing, the absorption line is relatively reduced or you do not find the line at all. From an observational point of view it is very difficult to measure the exact shape of the λ 4430 line because of all the background hash. However, several people have recently claimed that the λ 4430 line is indeed an asymmetric feature with a strong emission wing at the shorter wavelength. I refer specifically to Seddon at Edinburgh, but Walker at Victoria also proposed a rather significant emission wing at the shorter wavelength. Some time ago I made a calculation in which I tried to eliminate the emission wing by varying particle size distribution, since the observations of that time said that λ 4430 was symmetric. But, whatever I tried, the absorption band always turned out to be of the dispersion type. In my most recent calculations for cylinders, I make a prediction of the polarization of λ 4430. From an observational point of view the polarization is independent of the star and is therefore a more suitable source of unambiguous information than the shape of the absorption band. The key to unlock the puzzle of the grains may well be to look at polarization in the diffuse lines. The shape of the polarization can distinguish whether the particles are dielectric or metallic.

Field: I want to emphasize that NH_3 has a photodissociation edge at exactly 4.7 μ^{-1}. Is it possible to explain the bump as a rise with wavelength in the extinction coefficient at the long wavelength side of the bump, followed by a drop when one goes over the absorption edge of NH_3?

Greenberg: On the curve that I presented (Figure 2) I included only ice. Including other molecules would shift the hump. I discovered, however, that I still had difficulty getting the extinction to rise in the far UV beyond the drop. Possibly surface roughness will modify the sizes of the particles somewhat and still maintain a continued rise in the extinction curve. A purely theoretical but not very quantitative calculation indicates the possibility of such a rise by using dielectric material.

I should also point out that, thus far, this bump has only been confirmed in rather nearby, very hot stars. I would like to see it confirmed more generally. Is it possible that this hump is a selection effect? After all, these are rather small extinctions, therefore relatively local. I have still another point. There is an old anomaly in the near UV extinction of θ' Orionis. The observations in the far UV by Carruthers indicate that the extinction curve levels off instead of continuing to rise, a behavior which agrees with the classical result. I would like to see the shape of the 4.7 μ^{-1} hump for this star. It would help us to find out what characteristics, such as grain size, can do to absorption features.

Lynds: Your criticism of the UV portion of the extinction curve holds for the entire extinction curve, which might well represent only the nearby region of the Galaxy.

Field: I did some calculations in 1963 that showed that in the presence of NH_3, the curve will rise smoothly, then drop suddenly at 4.7 μ^{-1}. After this it either rises or stays constant, I do not remember which. I therefore wonder whether Dr. Greenberg does or does not assign a specific meaning to the 4.7 μ^{-1}?

Greenberg: No, I don't.

Field: I do not understand that reply because the dissociation potentials of the molecules are determined very well in the laboratory. The limit for H_2O is 1500 Å; for CH_4 1350 Å; and NH_3 gives the only potential that fits. On the other hand, we hear from Stecher that the hump is caused by graphite. Therefore, I am lost.

Van de Hulst: In order to identify the unknown constituent the theorists should settle first the question of whether to look at the frequency of 4.7 μ^{-1}, or whether to look at the combined feature of hump and valley, saying that it all belongs to the same resonance. I have always had the feeling that once you have the right frequency then a second uncertainty may show up because many chemicals will fit into a particular feature. Does Dr. Greenberg think that is true?

Greenberg: Yes. To begin with we must study the effects of spectroscopic candidates as they show up in small, solid particles. First, the observed feature is very broad. Second, if it is due to something in the interior composition of the grain, the shape of the feature would be size-dependent. Therefore, to attribute a specific feature to something at a precise frequency is a bit premature. There are many dielectric materials which have absorption edges around this frequency. For this reason, I prefer to make a model that fits the whole curve, including the right shape of the hump. Experimental studies of absorptivities of various mixtures of the simple (H_2O, NH_3, CH_4) molecules should be made.

Field: I suggest that the original van de Hulst model, which has a well-defined composition, including some NH_3, should be tested by calculations.

Van de Hulst: The model was only a guess.

Stecher: I think that with a reasonable size distribution, a graphite particle will fit the data, out to $\lambda^{-1} \approx 6 \mu^{-1}$; beyond that, graphite does not provide enough extinction.

It is my impression that the uniformity of the extinction curve in different directions is quite remarkable. This may be partly because of observational selection of close-by clouds; but on the other hand we have variations in gas density of a factor of 10 to 100. The accretion should be proportional to the square of the density and therefore, the grains should grow quite rapidly to large sizes and give only geometric extinction. But we still observe the $1/\lambda$ variation in every direction.

Van de Hulst: Like Stecher, I was always impressed by the rather amazing uniformity which still holds in spite of the differences in the extinction curve you see from place to place. I am somewhat relieved by the new UV observations in the following respect. In the old observations of the extinction curve, where $\lambda^{-1} < 3 \mu^{-1}$, the most obvious way to extrapolate was to have a maximum near $\lambda^{-1} = 4 \mu^{-1}$, which puts the main importance in fitting not on the material of the grains but on their sizes. Now we have new observations which go to $\lambda^{-1} = 9 \mu^{-1}$ and the gross features of the extinction curve should correspond more to a smooth curve with a maximum near $\lambda^{-1} = 8$ or $10 \mu^{-1}$, leveling off beyond that. This automatically means smaller grains, relatively more absorption in the visual range, and a less sensitive dependence on the assumed size. So this relieves the difficulty, but of course it does not solve the complete problem before you have solved all details of the wiggles, which are definitely dependent on composition.

Shklovskii: Many times in the past I have drawn attention to one remarkable fact. We know that the density of interstellar grains is about one per cent of the density of interstellar gas. On the other hand, the density of heavy elements in interstellar gas relative to hydrogen is also close to one per cent. Since the grain particles consist of heavy elements I have the impression that perhaps 50 per cent of the heavy atoms in the interstellar medium is condensed in dust and 50 per cent is in the gas. It is a very unexpected situation and may not be explainable by chance. I will not attempt to give an answer; but perhaps here is an argument in favor of the origin of grain particles in the atmospheres of red giants; maybe so, maybe not.

4. Formation and Destruction of Grains

Van de Hulst: I had one further question on the dust from stars. Like Greenberg I find it difficult to believe that most of the dust comes from stars. Linking up with the data on the mass loss of stars (p. 273 and p. 289) I made this calculation.

Dust present in the Galaxy $10^{11} M_{\odot} \times 10^{-4}$ $\qquad = 10^7 M_{\odot}$
Rate of supply from stars $\quad 1(M_{\odot} \, yr^{-1}) \times 10^{-1} \times 10^{-2} = 10^{-3} M_{\odot} \, yr^{-1}$
Supply time $\qquad\qquad\qquad\qquad\qquad\qquad\qquad\qquad = 10^{10} \, yr.$

The factor in the first line is the estimate that about 10^{-2} of the total mass of the Galaxy is gas, about 10^{-2} of the gas is condensed in dust. The factors in the second line signify that perhaps one kind of star out of ten supplies dust and that by the abundances only 10^{-2} get into the grains. The 10^{10} yr seems to me too long to pretend that all dust is supplied in this manner. I know other people have made such calculations and have reached a different conclusion, but I do not know if anybody here wants to defend that position.

   Greenberg: I agree with such a calculation. However, ejection from stars may be one of the mechanisms for creating growth nuclei (grain seeds) in interstellar space. There is one other mechanism for dust production that I should mention, although I believe it is less effective by orders of magnitude. In this process, proposed by Dorscher (1967), particles are ground in a planetary system around a star and are ultimately ejected, as happens to the particles in the solar system.

   In general, I think that grain seeds do not form directly in interstellar space, but that little particles are injected into space onto which molecules or gas atoms can come and stick. (Dorscher, J.: 1967, *Astron. Nachr.* **290**, 171.)

   Stecher: I am an author of a paper which purports to get enough grains from stars. We used Schwarzschild's value of 10^9 yr for turnover time in the interstellar material and considered the grains to be refractory. With the new sputtering rate the grains are not destroyed until they actually go into the stars. Under these conditions 10^9 yr seems to be long enough.

   Weymann: Dr. Greenberg, there are many places now where there is a suspicion of coexistence of dust and H II gas. Can you bring us up to date on what is known about destruction of dust by O stars? I would like to know particularly how the destruction rate depends on the energy of the photons. Is it just a matter of total energy density upon the surface, or do we need to know more details about the spectral distribution of the radiation? Also, can you give a simple criterion for deciding when the grains are more rapidly destroyed by hot gas than by the radiation field?

Greenberg: I think it is well established that in H II regions there is dust; there is certainly reflected continuum radiation in the Orion Nebula (O'Dell *et al.*, 1966). Atomic collisions at 10 000 K are exceedingly inefficient in knocking anything off, for the threshold for sputtering from experimental results is, I believe, something like 10 eV. Energies of a few keV are required for an efficient sputtering process. As far as the UV radiation is concerned, it may raise the temperature of the grains. You need grain temperatures of 90 K or so in normal stellar surroundings to evaporate ice off the grain. If the material is somewhat more refractory than ice, you would not evaporate it at all. But to get 90 K temperatures you have to be very close to the star; otherwise T is considerably lower. (O'Dell, C. R., Hubbard, W. B., and Peimbert, M.: 1966, *Astrophys. J.* **143**, 743.)

Ozernoi: Are there any differences between the properties of interstellar grains, in particular with respect to their chemical composition, in the direction of the galactic center and in the opposite direction? And, similarly, what are the properties of the dust grains in radio galaxies?

Van de Hulst: When you investigate clouds in the direction of the galactic center, you still talk about clouds which are very nearby, generally, and not at the galactic center.

5. The Gas/Dust Ratio

Habing: There have been several investigations in which one compares dust clouds with 21-cm observations. Implicit in these discussions has been the assumption that everywhere the ratio of the number of grains to the number of hydrogen atoms in the line-of-sight is constant. To me, such an assumption seems questionable. Could Dr. Greenberg comment?

Greenberg: I think that there is evidence for an average or general correlation of hydrogen and dust but I am not aware that the local correlation needs to follow. If the dust grows out of the gas *and* there are no effective mechanisms to bring about a subsequent separation, then we would find an excellent correlation between hydrogen and dust. Now we ask: what happens in very dense clouds where there is apparently an anticorrelation? Does it occur because of a separation of gas and dust after dust formation or perhaps because the gas is in an at present unobservable form – say molecular hydrogen? But if the dust does not exclusively grow from the interstellar gas we must ask instead: Why should there be a correlation? As a matter of fact, if Herbig's ideas have some significance, it is possible that one could have molecules and dust grains thrown into clouds from a star formation process without the existence of a great amount of hydrogen. We really must therefore find out what is in those dust clouds. There may be a correlation of dust and gas even if there is no correlation of dust and hydrogen.*

Habing: So the conclusion that if one does not observe any excess hydrogen at the

---

* (*Note added in proof.*) There are now several dust clouds known that show up in the 21 cm line profiles. The interpretation is uncertain – either the atomic hydrogen is deficient or it is cool. (Heiles, C. E.: 1969, *Astrophys. J.* **156**, 493; Sancisi, R. and Wesselius, P. R.: 1970, *Astron. Astrophys.* **7**, 341.)

position of a dust cloud, therefore all the extra hydrogen is molecular is not necessarily true?

Greenberg: There is no firm theoretical basis for this assumption but it is a possibility.

Van de Hulst: From direct photographs the dust clouds occasionally have very sharp edges. I have never heard of a mechanism working on the dust that can create such sharp edges. Such mechanisms, however, do exist for the gas, and this has always been a convincing argument for me, that dust and gas go together.

Zimmermann: I have calculated dust motions during a cloud collision (Zimmermann, 1968). During the collision the heavy particles move into the inner parts of the new-built cloud and take the form of a dense sheet. They remain there during the expansion period. So at the end, when pressure equilibrium with the intercloud medium is reached, we have the heavy dust particles in the middle of the configuration while the light dust particles are distributed more or less uniformly among the gas of the cloud. Therefore we have a dynamical separation process of particles and dust. (Zimmermann, H.: 1968, *Astron. Nachr.* **290**, 193.)

Field: If a separation occurs between gas and dust then clouds exist without the cosmical abundance, i.e., with large amounts of heavy elements and no hydrogen. How does one then explain that there exist no stars with such a composition?

Greenberg: Clouds without hydrogen would have smaller masses than are obtained by assuming a standard hydrogen-to-dust ratio. Maybe the conditions for star formation are then not favorable.

Field: I should have thought that clouds of dust would collapse very easily under their own self-gravitation and external radiation pressure, as the internal pressure is negligible.

Van Woerden: I have a small comment about the gas/dust ratio. In considering the lack of correlation between gas and dust, one should not only take into account the possibilities for molecular hydrogen or even for abnormal abundances, but one should also consider the effect of possible variations of the gas temperature. In the milder cases of non-correlation, such as that in the Taurus region discussed by Garzoli and Varsavsky (1966), assumption of a gas temperature below 125 K restores the normal gas-to-dust ratio, as has been shown by Sancisi and Wesselius (*op. cit.*). (Garzoli, S. L. and Varsavsky, C. M.: 1966, *Astrophys. J.* **145**, 79.)

Field: I have a question for Dr. Lynds. What is the evidence about the existence of dust in regions where the gas density is very low, for example what do we know about extinction in stars, say 100 pc away, and with very weak interstellar lines?

Lynds: All I know is the old work on the general correlation of the strength of the calcium lines with extinction. There is a much wider scatter in this relation than can be interpreted as errors in observations. I do not know what the upper limit you refer to would be.

6. The OH Molecule

Van de Hulst: Dr. Weaver has been challenged to give his view on the future of all the OH observations.

Weaver: I certainly do not know what the future of all the OH observations will be, but there are some aspects of the OH problem that perhaps should be discussed here. One thing that has impressed me very much in looking at the OH absorption observations in conjunction with neutral H absorption observations is that if you observe an OH absorption line and a neutral H absorption line they have the same velocity, within observational error. I would suggest that this evidence indicates that there must be a very close association of the OH molecule formation mechanism and the concentration of the neutral H seen as a 'cloud' in a particular direction. The aspect of the OH that is rather less well-correlated with the H is the line width. One would predict on any reasonable basis that the OH line is much narrower than the H line. In fact, in many instances the half-width of the OH line is very much greater than the half-width of the H line. There is a very peculiar situation in regard to the velocity distribution of the OH molecule in the cloud seen in absorption. These properties, it seems to me, are directly related to some of the aspects of the Symposium topics.

Greenberg: If, in absorption, the OH line is wider than the H I, does this not indicate that there is no thermal equilibrium?

Shklovskii: The velocities are not thermal, but result from macroscopic motion.

Menon: The most conspicuous case is the absorption toward the galactic center. The OH absorption is the strongest in those features which are the widest in 21 cm. I do not know how that observation fits with your anti-correlation.

Weaver: These very broad features Menon mentions are examples of another special class of OH sources. OH is a molecule that seems to disobey all standard rules and to appear spectroscopically in almost any form that one can imagine. I think the OH features in the direction of the galactic center occur in very special structures in the interstellar medium in which molecules are very numerous. In these regions towards the galactic center one finds NH_3, H_2CO, and so on. The areas are small, diameters less than about 10'. They must represent regions where molecule formation goes on very strongly. They are the best place in the sky to look for new molecules. But I really did not have these strange spots in mind when I spoke about the relationship of line-width between OH and H. The abnormality in OH widths occurs in the more common sources seen in various positions in the galactic plane. I was trying to confine my remarks to the general distribution of OH rather than to specialized regions.

Shklovskii: What is the situation now for the relative intensity of the different absorption components of the OH line? Do they satisfy the Boltzmann ratios?

Weaver: They tend to show non-LTE ratios. In some instances they may show almost-LTE ratios; whether this appearance is real or accidental, I do not know.

Varshalovich: Usually the description of the state of atoms and molecules in space takes into account particle density distribution, velocity distribution, and the degree of ionization and excitation. But spin states of particles have never been considered. Apparently spins were assumed to be oriented at random and to play a negligible role in the problems considered. This assumption is correct for the conditions here on Earth. But it does not necessarily hold in the interplanetary and in the interstellar medium, where there is no thermodynamic equilibrium. In an analysis of physical conditions in various astrophysical objects I have shown (Varshalovich, 1965) that the alignment of spins should be a rather common phenomenon in the Universe. I considered two possible mechanisms of orientation of spins: (1) equilibrium orientation by a magnetic field; and (2) non-equilibrium dynamical orientation by directed fluxes of photons or particles. Magnetic orientation is not important in outer space, because $\mu B \ll kT$. But the second mechanism is important. Resonance scattering of directed fluxes of nonpolarized radiation can be an effective mechanism when a magnetic field also is present. Necessary for the process to operate are the following conditions: (a) the interaction of particles with radiation is greater than the interaction with other particles; (b) the radiation is anisotropic. Spin alignment due to resonance scattering of unpolarized radiation has been observed in the laboratory by A. Kästler and others. In the laboratory the system quickly relaxes to the isotropic equilibrium state. At the same time, there are vast regions in outer space where all the necessary conditions are fulfilled and spin alignment is maintained permanently by nature itself: nebulae in the interstellar medium, upper layers of stellar atmospheres, comets, and so forth. These objects have in common that the radiation fluxes are rather large and essentially anisotropic (because the optical thickness is small $\tau \lesssim 1$), and interparticle collision relaxation is negligible, because the density is small.

Optical properties (extinction and refraction) of the medium which contains spin aligned particles depend strongly on the direction of observation and on the polarization. Both decrease and increase of the optical thickness of resonance radiation are possible. If the ground state has a fine or superfine structure, spin alignment may give inverse populations of the upper state magnetic sublevels relative to the corresponding lower state ones and can lead to amplification of resonance radiation. Initially unpolarized radiation, while propagating through such a medium, acquires linear polarization. Spin alignment may substantially affect resonance radiation transfer. One should take it into account in plotting the curve of growth, by introducing the corresponding changes in the optical thickness. In some cases it may give an appreciable correction for the derived numbers of atoms and molecules; i.e., it may change the derived chemical composition, degree of ionization, and degree of excitation. Moreover, analysis in terms of spin alignment gives quite new information on the anisotropy of the prevalent physical conditions and, particularly, on the anisotropy of the radiation field. Finally, the fact that spin alignment produced by optical pumping depends on the direction of the field makes it possible to determine this direction in a given region of space. [Varshalovich, D. A.: 1965, *Astron. Zh.* **42**, 557 (translation: *Soviet Astron.* **9**, 442).]

Van de Hulst: It is obvious that if alignment could be observed, it would indeed give very good clues. Do you have a particular molecule in mind where it would be observable?

Varshalovich: As an example, I have considered in detail the spin alignment of NaI due to the pumping of solar radiation in the head of a comet.

Shklovskii: It is a long way from the physical considerations to astrophysical applications. The physical side is evident, but in astrophysical conditions the mean free path of the resonance quanta is short, and it is very difficult to obtain more or less strong anisotropies in the radiation field. Perhaps sometimes the situation is favorable for spin orientation, but in the general astronomical situation it is difficult to fulfill the necessary conditions.

Van de Hulst: At the end of the discussion I should like to add a few comments, because I often find that the history starts too late, and that earlier contributions are not mentioned at all. To my knowledge, the first discussion of growth of grains due to condensation by interstellar gas was by Lindblad in about 1935; the refinement which Oort and I tried to make in 1946 takes account of the chemical composition and actual vapor pressures. We found that at these expected temperatures, solid H_2 would boil off immediately. The first discussion of CN as a possible radiometer in space was by Swings in 1939, at a time when nobody knew about black-body cosmological radiation. The first discussion on grain temperatures that I know of was by Eddington in 1923. He made such calculations saying the total radiation field is roughly 10^4 K, and the dilution factor 10^{-14}, so

$$T_{grain} = 10^4 \times (10^{-14})^{1/4} = 3.2\,\text{K}.$$

Greenberg reminded you of the size effect, which can very roughly be taken into account by simply replacing the exponent by $\frac{1}{5}$, because there is a $1/\lambda$ dependence. This makes

$$T_{grain} = 10^4 \times (10^{-14})^{1/5} = 10\,\text{K}.$$

(Oort, J. H. and van de Hulst, H. C.: 1946, *Bull. Astron. Inst. Neth.* **10**, 187.)

25. PROTOSTARS AND OTHER NEUTRAL CONDENSATIONS IN HII REGIONS

Introductory Report

(Thursday, September 18, 1969)

P. G. MEZGER

National Radio Astronomy Observatory*, Green Bank, W.Va., U.S.A.

and

Max-Planck-Institut für Radioastronomie, Bonn, Germany

1. HII Regions and the Formation of Stars

I have been invited to report on globules and OH emission sources. However, I will extend the topic to protostars and other condensations in HII regions, of which globules and OH emission sources are only special, although rather conspicuous, objects.

To comply with the wishes of the program committee, I will not enter into a general discussion of HII regions. However, in Section 2 I will mention some high-angular-resolution radio observations that bear upon the original density distribution of the collapsing protostar or protocluster. Although I make this restriction, I want to stress that it would be very unfortunate if the theoreticians were left with the impression that there is good agreement between theory and observations of HII regions. Let me just point out one example which came up in several discussions at this meeting, i.e., the acceleration of neutral interstellar matter by expanding HII regions which surround O stars. The theory of this process has been meticulously worked out. However, there is little observational evidence to date of uniformly expanding HII regions and no evidence of thin, dense shells of neutral matter surrounding them, and to me the reality of this particular mechanism seems highly doubtful.

With this in mind, let me go on to the problem of star formation where the discrepancy between theory and observations appears to be even larger. The objects which will be discussed in this paper are, in one way or another, related to the process of star formation. How this process works in detail is still quite unclear, but the end result we know: stars with masses ranging from 0.08 $M_\odot$ to about 50 $M_\odot$. More massive stars apparently become unstable; less massive stars never attain a high enough central temperature for nuclear reactions to set in. Averaged over a larger volume of space, the mass spectrum of stars attains a maximum at about 0.5 $M_\odot$. It decreases slowly in the direction of low mass stars but cuts off rapidly towards more massive stars. On the average only about 3 per cent of the total mass of all newly formed stars is contained in O stars. Nevertheless, these O stars play a very important role. Their effective temperatures are so high that they interact with the surrounding interstellar matter and trigger or stop the formation of other stars.

* Operated by Associated Universities, Inc., under contract with the National Science Foundation.

Habing (ed.), Interstellar Gas Dynamics, 336–346. All Rights Reserved. Copyright © 1970 by IAU

In general, star formation may take place as follows. The interstellar gas consists of cool, dense clouds with temperatures of 60 K or less which are embedded in a hot, tenuous intercloud gas with temperatures of several thousand K. These clouds are in pressure equilibrium until some process (e.g., a sudden compression by a density wave or a rapid cooling by an increased dust production) initiates their gravitational contraction. At this time a large fraction of the mass of the cloud is in the form of molecules and dust particles. The subsequent evolution of the cloud is not clear, but in any case it cannot be simple gravitational contraction and fragmentation (as is usually assumed), since all observations indicate that the most massive stars are formed last and that their formation is the most difficult task for nature. For example, there exist T-Tauri associations which do not contain early-type stars, and it is not at all clear that ultimately O stars will form there.* In O associations, star formation appears to take place in subgroups which contain about $2 \times 10^3 \, M_\odot$. Once the O stars of a subgroup are formed they ionize the remnant of the proto-subgroup forming what we now call a 'compact HII region'. The mass ratio of ionized gas to stars in these subgroups is generally less than 10 percent, indicating a very high efficiency of the process of star formation.

As a rule of thumb, stars need about 1 per cent of their main-sequence lifetime for their pre-main-sequence contraction. This means that stars of lower mass, which have been formed together with O stars, should still be on their pre-main-sequence contraction while the O stars have long since reached the main sequence. Compact young HII regions therefore are the obvious places to search for protostars. This, however, does not imply that protostars can only be found in HII regions; the point is that we lack criteria for selecting regions where only stars of lower mass are formed.

In the following sections I will review observations of neutral condensations and protostars associated with HII regions. I will also include observations which may yield information on the original density distribution of the contracting protostar cluster.

2. The Distribution of Ionized Gas Inside HII Regions

The surface brightness of both optical and radio emission from an HII region is proportional to the emission measure which is the square of the electron density integrated along the line of sight. Therefore, these observations should be heavily weighted towards regions of high space density of the plasma. This is certainly correct for radio observations, but in optical observations the extinction by dust plays an intervening role. As a consequence only radio observations can give a correct picture of the surface-brightness distribution (and hence of the emission-measure distribution) of an HII region. However, only very recently did aperture-synthesis observations at radio wavelengths yield an angular resolution comparable to that of optical observations.

---

* See also Section 5 in the following discussion (Chapter 26) on a new mechanism for star formation. (Ed.)

Single-dish radio observations with an angular resolution of a few minutes of arc reveal that most H II regions with high surface brightness consist of compact components with linear dimensions of less than 1 pc, which are embedded in extended low-density H II regions. The average density of the compact components is high, somewhere between 10^2 and 10^4 cm^{-3}, but they contain only a few or a few times 10 $M_\odot$ of ionized hydrogen. Probably these compact components represent the very early evolutionary stages of subgroups in O associations, whose exciting stars in most cases cannot be seen, probably since they are still hidden in a dense circumstellar dust cloud. Figure 1 shows radio contours of the Orion Nebula, observed with an angular resolution of 2′ (or 0.3 pc at 450 pc distance), superimposed on a photograph (Schraml and Mezger, 1969). The Orion Nebula consists of the two compact components M42 and M43. We think we know the exciting stars for these two compact H II regions. The obscuration by dust is relatively low; note, however, the dark bay across M42 which extends close to the Trapezium. I will refer to this dust cloud later in my Report.

Several lines of investigation have shown that in the Orion Nebula (and also in other H II regions) the ionized gas is clumped on an even smaller scale, of angular

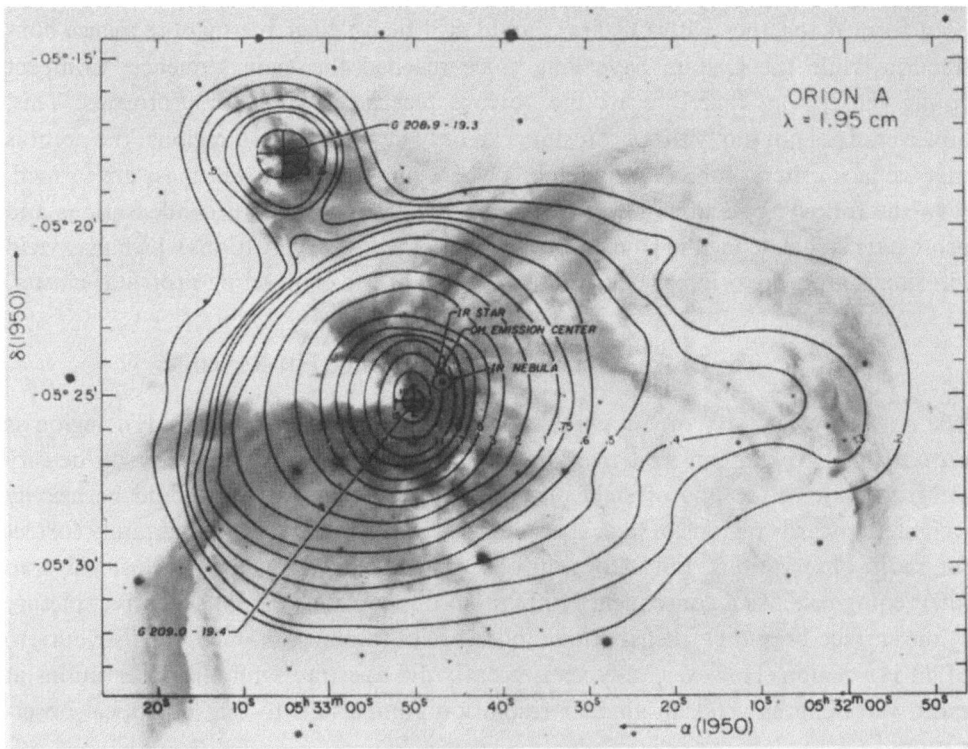

Fig. 1.   Overlay of the radio continuum contours, observed with an angular resolution of 2′ at $\lambda = 2$ cm, on a photograph of the Orion Nebula (Schraml and Mezger, 1969). The source in the upper half of the Figure (G 208.9–19.3) is M43; the source in the lower half (G 209.0–19.4) is M42.

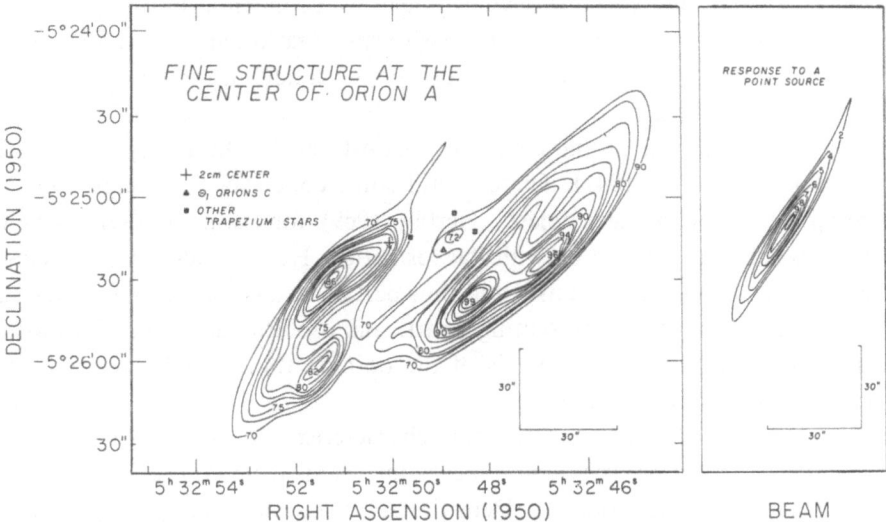

Fig. 2. Aperture synthesis map of the continuum radiation of the compact component M42 in the Orion Nebula, observed at $\lambda = 11$ cm with an angular resolution of about 8″ in right ascension (Webster and Altenhoff, 1970).

dimensions 3″ to 10″, linear dimensions 1.5×10^{-2} pc to 5×10^{-2} pc. Webster and Altenhoff (1970) have observed some such condensations directly, using the National Radio Astronomy Observatory three-element interferometer. Figure 2 shows their result for M42. At least four, perhaps five, superdense condensations are found with angular dimensions of about 4″. Adopting a distance of the Orion Nebula of 450 pc, we obtain $\langle n_e^2 \rangle^{1/2} = 5 \times 10^4$ cm^{-3}, $M_{\text{H II}}/M_\odot = 2 \times 10^{-3}$ and linear dimensions of about 1.6×10^{-2} pc ($= 5 \times 10^{16}$ cm $= 3200$ AU). This is about the diameter of a protostar in its earliest evolutionary stages, where its average density is somewhere between 10^6 and 10^8 cm^{-3}. It has been shown by Hjellming (1969) that such a protostar, exposed to the Lyman continuum flux from an O star, will only be ionized superficially and that it can easily survive during the lifetime of an H II region ($\approx 10^6$ yr). In fact the increased outer pressure may initiate the gravitational contraction of such a star. These superdense condensations contain only about 3 per cent of the total radio-flux density of M42. Further interferometer observations are planned to search for more condensations, which must have smaller angular dimensions and thus could be fragments with masses lower than that of stars.

In the case of M42 the superdense H II condensations do not coincide with very conspicuous optical features. However, in M8 the center of radio emission coincides with a very interesting optical feature, viz., two very bright nebulae of roughly 12″ angular size, together known as the 'Hourglass Nebula'. The exciting O7 star, Herschel 36, is embedded in a dust cloud. The two nebulae apparently contain several stellar components. Recently Smith and Weedman (1969) investigated the kinematics of the two nebulae and found that each of them expanded with velocities of 10 and

7 km sec^{-1}, respectively. Here one obviously deals with condensations of high density, which are being ionized by one or more massive stars that recently have been formed inside the condensation. However, the superdense condensation in M42 may be ionized from the outside.

Globules are observed as dark patches against the bright background of HII regions (Bok and Reilly, 1947). There is still some controversy amongst observers concerning the nature of these objects. Herbig (1969), for example, does not believe that globules have any connection with star formation. He concludes that globules are pinched-off ends of 'elephant trunks' and thus pertain to a certain evolutionary stage of an HII region rather than representing a particular evolutionary stage of protostars. Reddish (1969), on the other hand, is of the opinion that globules represent very early stages of protostars.

Angular size and extinction are the only characteristics of globules which can be estimated directly from observations. Linear dimensions from $(1.5$ to $5 \times 10^{17})$ cm up to 15×10^{17} cm have been obtained. Densities and masses of globules depend on the adopted gas-to-dust ratio, and it is therefore difficult to estimate whether or not these objects are in fact gravitationally unstable.

3. Centers of Anomalous Radio Line Emission*

It has been known for some time that sources of anomalous (i.e., nonthermal) OH emission often have very small linear dimensions ($\ll 1$ pc). This seems to hold also for emission sources of other, recently detected, radio lines (H_2O, NH_3, C). We will call such small-scale sources 'emission centers', although we do not always imply that their dimensions are as small as in the OH case. The OH emission centers are classified as Class I, if the two center lines (at 1665 and 1667 MHz) are predominantly emitted. Class IIa and IIb pertain to the predominant emission of the 1720 MHz and 1612 MHz line, respectively. At present, Class I emission centers are the best investigated; they show a tendency to be located towards the edge of compact HII regions. Using the powerful VLB interferometer technique the MIT group (Moran et al., 1968) succeeded in resolving some of these Class I emission centers. As an example, we consider the OH emission spectrum in IC 1795 (W3). Observed with a single dish, it consists of a number of spikes which are either right- or left-hand circularly polarized. Taking advantage of the frequency shift and different polarization characteristics of the individual spikes the MIT group showed that each individual spike in the spectrum corresponds to an individual emission center. The projected positions of seven emission centers lie in an area of roughly $1.2'' \times 2.3''$. Only one center has a simple disk shape of apparent diameter $0.005''$. The others show structure and might be clusters of two or three emission centers. In one case the observations indicate an elongated structure. At a distance of 2.5 kpc adopted for IC 1795, an angle of $0.005''$ corresponds to a linear size of 2×10^{14} cm $=13$ AU. Mezger and Robinson (1968) investigated the

_____

* See also the Report by Weaver, p. 22. (Ed.)

physical characteristics of these OH emission centers and estimated that densities of the order of 10^{11} atom cm^{-3} and masses of less than $10^{-2}\,M_\odot$ are involved in the OH emission; these objects may have temperatures between 200 and 500K. Such characteristics place these OH emission centers on that part of the evolutionary track of a protostar where it just has become opaque and subsequently undergoes adiabatic contraction. However, the masses involved in the OH emission are too small, and it was therefore concluded that the emission occurs in the outer layers of the protostar only. It is still a puzzle how the over-population is achieved which is necessary for the observed maser effect.

Recently Wilson and Barrett (1968) discovered that Class IIb OH emission centers are associated with a particular type of IR star and exhibit a very strong maser effect. However, accurate positions and sizes of these emission centers have not yet been published.

Anomalous emission of the 22.2 GHz line of H_2O usually comes from the same regions where OH emission centers are observed. In two objects of Class IIb, viz., VY CMa and IRC+10 406, the H_2O emission features are flanked on either side by a range of velocities of OH emission, and there is no OH emission in the velocity range of H_2O emission (Knowles et al., 1969; Turner et al., 1970). VLB observations of the H_2O emission centers have been unsuccessful to date, but the time variation of the H_2O emission occurs with a much shorter time scale than the time variation of OH emission; from this an upper limit of 10^{16} cm can be inferred on the size of the H_2O emission centers (Knowles et al., 1969).

NH_3 emission, to date observed only in two sources close to the galactic center, shows an irregular distribution over an area about 10' wide, with considerable variation in density, velocity, and state of excitation. Kinetic temperatures of the gas are mainly in the range (20 to 100K) (Cheung et al., 1969).

A major step forward in the investigation of HII regions was made through observations of radio recombination lines. In the course of observations of the He109α lines, Palmer et al. (1967) discovered a new and unexpected recombination line which was subsequently identified as that of carbon. Figure 3 shows the 109α recombination spectrum in the Orion Nebula; one recognizes the recombination lines of both He and C. The C109α line is much narrower than both H and He lines, thus indicating that it must be emitted from regions which are considerably cooler than the hydrogen plasma of the HII region. The intensity of that line, on the other hand, is much stronger than would be expected from the normal abundance ratio, and there are further observational indications that the carbon line is also amplified by some sort of maser effect. An investigation of the emission characteristics of the carbon line is very cumbersome and time consuming. It was first done for NGC 2024 by Zuckerman and Palmer (1968) and subsequently for the Orion Nebula (M42) by Churchwell and Mezger (1969). The results were in both cases similar: the line emission appears to come from narrowly confined regions which do not coincide with the center of the associated HII regions. In the case of M42, at least one carbon-line emission center appears to coincide with the upper end of the dark bay, a dust lane which obscures part of the

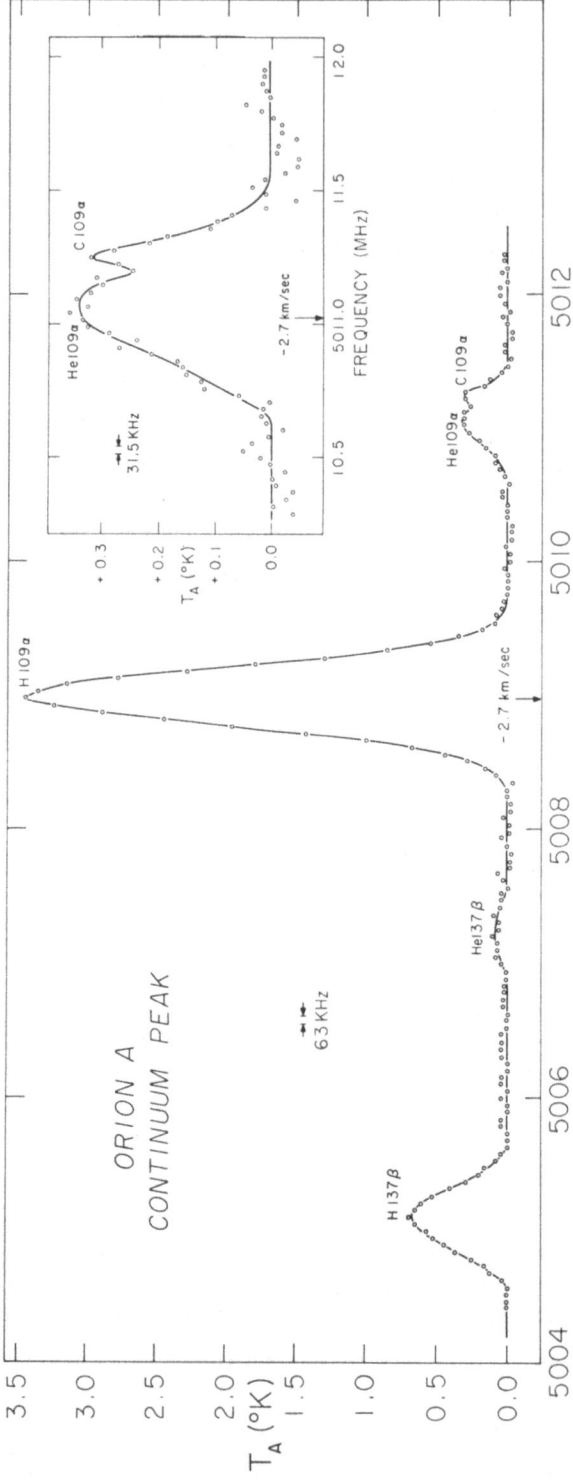

Fig. 3.   Broadband 109α recombination spectrum of the center of the Orion Nebula. The insert has been observed with twice the frequency resolution of the broadband spectrum to obtain a better separation of the C109α line from the He109α line (Churchwell and Mezger, 1969).

optical emission. This result supports the hypothesis put forward by Zuckerman and Palmer (1968) that the carbon-line emission comes from dense dust clouds opaque to the Lyman-continuum UV flux but transparent to the UV radiation between 912 and 1100 Å which is capable of ionizing carbon. [See also in the following discussion, Chapter 26, Section 2 on the C line (Ed.)]

4. IR Objects

I mention IR objects here only for the sake of completeness and since it may help in the following discussion. There are obviously two different types of IR objects found in H II regions. First are the IR stars of the type observed first in the Orion Nebula by Becklin and Neugebauer (1967). The BN star in Orion has a color temperature of about 700 K and linear dimensions of about 4.4×10^{14} cm ($= 30$ AU). This object may be the nucleus of a protostar. Stars of this type are best observed in the wavelength range between 3 and 6 μ. Second are IR nebulae of much larger angular dimensions which are best observed at 10 and 20 μ. The first example of this kind was detected by Kleinmann and Low (1967) in M42, northwest of the Trapezium. At the moment the most probable interpretation of these objects is that they are dust clouds heated by starlight. The existence of circumstellar dust clouds has been inferred from a variety of independent observations (Reddish, 1967; Schraml and Mezger, 1969). Their existence implies that dust can survive for a relatively long time in H II regions, that the dust grains have a high albedo in the UV, and that the dust does not expand as rapidly as the ionized gas. It follows that dust plays an even more important role in the formation of stars than is generally believed. Herbig's (1969) suggestion that dust is a by-product of star formation is of high interest in this context.

5. A Summary of Different Observations Pertaining to the Compact H II Region G209.0–19.4 (M42) in the Orion Nebula

In a paper given at the recent Liège Symposium (Mezger, 1969), I have tried to summarize and to correlate the observations of various neutral condensations and pre-main sequence stellar objects together with the distribution of ionized hydrogen in the compact H II region G209.0–19.4*, which coincides approximately with the H II region M42 in the Orion Nebula. It may be of interest to mention the results here briefly.

Figure 4 shows a schematic representation of the positions and – where possible – of the extension of the various observed objects. The hatched lines indicate the edges of the whole compact H II region which is embedded in a more extended low-density H II region. The diameter of the whole region is 0.65 pc, $\langle n_e^2 \rangle^{1/2} = 2.3 \times 10^3$ cm^{-3}, and the total mass of ionized hydrogen $M_{\text{HII}} = 7 \ M_\odot$. Since the star density in this area is highest, the ratio M_{HII}/M_* must be considerably lower than the value 0.1 derived by Menon (1961) for the whole Orion Nebula. This reiterates that the efficiency of star

---

* This symbol summarizes the position of the radio source in galactic (G) coordinates. (Ed.)

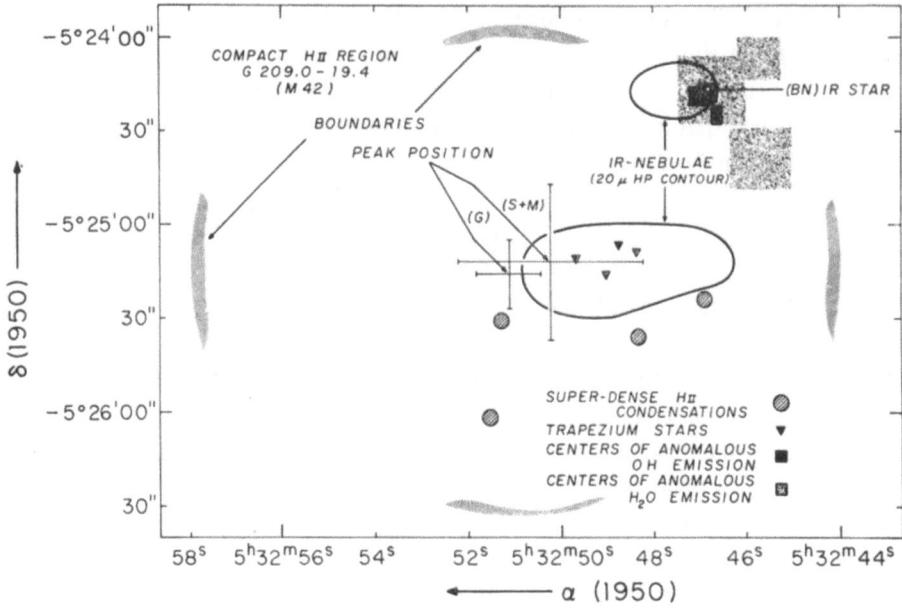

Fig. 4. Schematic representation of various objects observed in the compact H II region M42 in the
Orion Nebula (Mezger, 1969).

formation in subgroups is very high. The center of the radio free-free continuum does
not coincide with the Trapezium but lies slightly to the east of it.

The positions and sizes of four superdense condensations of ionized hydrogen south
of the Trapezium are shown (Webster and Altenhoff, 1970). A fifth condensation may
be located close to the Trapezium. According to recent computations by Larson (1969)
the outer layers of an evolving protostar consist of cool and neutral matter which
undergoes a free-fall contraction until the protostar appears at the lower end of the
Hayashi track. Such a protostar, if exposed to the Lyman continuum flux from a
nearby O star, would be ionized in its outer layers but otherwise should not be
affected in its evolution. It may be that these four superdense condensations are
protostars in such a state; however we must await further observations with higher
angular resolution. It has been pointed out by Osterbrock and Flather (1959) quite
some time ago and recently confirmed independently by Hjellming and Churchwell
(1969) that the clumping in the Orion Nebula must be very high. The ionized gas appears to occupy on the average only $\frac{1}{30}$th of the total volume. We hope that further
interferometer observations will tell us something about the size of these clumps.

Centers of anomalous emission of the OH and H_2O lines have been found in an
area northwest of the Trapezium. Their size is too small to be shown in the diagram.
There are at least three different emission centers observed for each line. Positions of
the OH emission centers have been obtained with an interferometer (Raimond and
Eliasson, 1969); the H_2O positions are obtained with a single-dish radio telescope and
therefore less accurate (Meeks et al., 1969). The shaded areas indicate the positional

uncertainties. In one case OH and H_2O emission centers coincide with the Becklin and Neugebauer IR star within the positional uncertainties of the various objects.

Churchwell and Mezger (1969) have investigated the sources of emission of the carbon recombination line in M42. Although the reduction of these observations is not yet fully completed, it appears that emission comes from one, possibly from two different areas of not very large angular dimensions. One appears to be located SSE and the other one possibly NNE of the Trapezium. The latter could coincide with the upper end of a dust lane. The radial velocities of the carbon-line emitting regions deviate strongly from the mean radial velocity of the ionized hydrogen gas but are rather similar to that of both the surrounding neutral gas and the Trapezium stars.

The two extended nebulae observed in the far IR are shown by solid lines. According to Ney and Allen (1969), the amount of dust needed to account for the observed emission is rather small, $\lesssim 0.06\ M_\odot$. These dust clouds may be the remnants of protostars or protoclusters.

The association of the BN star and the centers of OH and H_2O emission northwest of the Trapezium may indicate a group of very young stellar objects. The fact that these objects are not associated with an increased continuum emission, similar to that of the superdense condensations, is somewhat puzzling. It may well be that these objects are located at quite a distance from the Trapezium stars and their apparent close association with the Trapezium is merely a projection effect.

Acknowledgements

I am indebted to W. Altenhoff, E. Churchwell, and W. Webster for permission to quote their unpublished results.

References

Becklin, E. E. and Neugebauer, G.: 1967, *Astrophys. J.* **147**, 799.

Bok, B. J. and Reilly, E. F.: 1947, *Astrophys. J.* **105**, 255.

Cheung, A. C., Rank, D. M., Townes, C. H., Knowles, S. H., and Sullivan III, W. T.: 1969, *Astrophys. J. Lett.* **157**, L13.

Churchwell, E. and Mezger, P. G.: 1969, paper presented at the 130th American Astronomical Society Meeting, Albany, New York.

Herbig, G. H.: 1969, paper presented at the VIth international astrophysical symposium, Liège.

Hjellming, R. M.: 1969, paper presented at the VIth international astrophysical symposium, Liège.

Hjellming, R. M. and Churchwell, E.: 1969, *Astrophys. Letters* **4**, 165.

Kleinmann, D. E. and Low, F. J.: 1967, *Astrophys. J. Lett.* **149**, L1.

Knowles, S. H., Mayer, C. H., Cheung, A. C., Rank, D. M., and Townes, C. H.: 1969, *Science* **163**, 1055.

Larson, R. B.: 1969, *Monthly Notices Roy. Astron. Soc.* **145**, 405.

Meeks, M. L., Carter, J. C., Barrett, A. H., Schwartz, P. R., Waters, J. W., and Brown III, W. E.: 1969, *Science* **165**, 180.

Menon, T. K.: 1961, *Publ. Nat. Radio Astron. Observ.* **1**, 1.

Mezger, P. G.: 1969, paper presented at the VIth international astrophysical symposium, Liège.

Mezger, P. G. and Robinson, B. J.: 1968, *Nature* **220**, 1107.

Moran, J. M., Burke, B. F., Barrett, A. H., Rogers, A. E. E., Ball, J. A., Carter, J. C., and Cudaback, D. D.: 1968, *Astrophys. J. Lett.* **152**, L97.

Ney, E. D. and Allen, D. A.: 1969, *Astrophys. J. Lett.* **155**, L193.

Osterbrock, D. E. and Flather, E.: 1959, *Astrophys. J.* **129**, 26.

Palmer, P., Zuckerman, B., Penfield, H., Lilley, A. E., and Mezger, P. G.: 1967, *Nature* **215**, 40.

Raimond, E. and Eliasson, B.: 1969, *Astrophys. J.* **155**, 817.

Reddish, V. C.: 1967, *Monthly Notices Roy. Astron. Soc.* **135**, 251.

Reddish, V. C.: 1969, paper presented at the VIth international astrophysical symposium, Liège.

Schraml, J. and Mezger, P. G.: 1969, *Astrophys. J.* **156**, 269.

Smith, M. G. and Weedman, D. W.: 1969, paper given at the 130th American Astronomical Society Meeting, Albany, New York.

Turner, B. E., Buhl, D., Churchwell, E., Mezger, P. G., and Snyder, L. E.: 1970, *Astron. Astrophys.* **4**, 165.

Webster, W. J. and Altenhoff, W.: 1970, *Astrophys. Lett.* **5**, 233.

Wilson, W. J. and Barrett, A. H.: 1968, *Science* **161**, 778.

Zuckerman, B. and Palmer, P.: 1968, *Astrophys. J. Lett.* **153**, L145.

26. DISCUSSION FOLLOWING MEZGER'S REPORT

(Thursday, September 18, 1969)

Chairman: I. S. SHKLOVSKII

Editor's remark: The Discussion has been rearranged into five sections: 1. Density Concentrations, Cocoon Stars, and OH Sources; 2. The Carbon Recombination Line; 3. The He/H Abundance; 4. Various Shorter Comments; 5. A New Mechanism for Star Formation. A part of the Discussion which related to Spiegel's Report has been transferred to Chapter 13. A short exchange on H I shells around H II regions has been deleted and added to Chapter 5. A short part on the dust cloud in Orion at the end of Section 4 has been transferred from Tuesday, September 16.

1. Density Concentrations, Cocoon Stars, and OH-Sources

Menon: Dr. Mezger, I recall that in one of your early papers with Burke and others you refer to a new class of H II regions, called compact H II regions, which you identified as cocoon stars. Now you called the Orion Nebula a compact H II region. Are these two objects the same?

Mezger: More observations with high angular resolution have led us to revise our ideas of the nature of compact H II regions. I now believe that most of the compact H II regions with linear dimensions of less than 0.5 pc, which we observe with an angular resolution of 2′, are the ionized remnants of a protocluster rather than of a protostar. Perhaps in the case of very young objects, like DR 21* or G133.9 + 1.1* in IC 1795, we are seeing the ionized shells of very young individual O stars. In most other cases the compact H II regions are already more expanded and of lower average density than cocoon stars. Often the regions contain a number of condensations, part of which may be protostars of lower mass. If one wants to see these individual condensations in extended compact H II regions, one has to use higher angular resolution, such as the aperture synthesis maps of the Orion Nebula, of M17, of W49, and of IC 1795 recently obtained by Webster and Altenhoff (1970; see also Figure 2 in the preceding chapter). To answer your question concerning Orion: not the Orion Nebula but its two components G209.0–19.4 (M42) and G208.9–19.3 (M43) are considered to be compact H II regions. (Webster, W. J. and Altenhoff, W.: 1970, *Astrophys. Lett.* **5**, 233.)

Verschuur: I do not see how this concept of cocoon stars originated. How do you know that the star is *surrounded* by dust? The only thing you can tell is that, as seen from Earth, the star is behind the dust.

---

* DR 21 is source No. 21 found in the study of the Cygnus X region by D. Downes and R. Rinehart (1966, *Astrophys. J.* **144**, 937). The symbol G133.9 + 1.1 stands for a source with galactic (G) coordinates $l = 133°.9$, $b = +1°.1$. (Ed.)

Mezger: Schraml and I investigated about 20 H II regions with very high surface brightness at $\lambda = 2$ cm (Schraml and Mezger, 1969). About one-half of the sources can be optically identified, but in many cases (e.g., the component G133.7 + 1.2 in IC 1795) we find that the centers of radio emission do not coincide with anything seen in the optical range. In all optically identified H II regions, with the exception of the Orion Nebula, C. B. Stephenson, who is one of the most experienced observers of O stars, searched for the exciting stars but could not find them. Only in two cases did he find highly-reddened OB stars. This shows that the exciting stars are obscured. But also in OB star associations which are associated with young H II regions, e.g., the Cepheus IV OB star association or M16, we do not seem to observe enough exciting stars. This can be shown as follows: the radio continuum flux from an H II region is directly related to the Lyman continuum flux of the exciting stars. In most H II regions, and especially in compact, young H II regions, it turns out that the Lyman continuum flux of the visible stars is not sufficient to explain the radio continuum flux and that there must be some additional sources of ionization, such as for example, stars hidden behind a dust cocoon. (Schraml, J. and Mezger, P. G.: 1969, *Astrophys. J.* **156**, 269).

Verschuur: That is no answer. Why cannot the stars just be behind the dust? Why do you conclude that they are *surrounded* by dust?

Mezger: Well, in the case of the O7 star Herschel 36, in M8, you really see the obscuring dust cloud around the star which is embedded in the H II region with two very bright nebulosities, known as the Hourglass Nebula, at the western rim of the dust cloud.

Verschuur: It is two-dimensional. You are concluding something about three dimensions.

Mezger: It would be rather strange that the dust cloud with a diameter of 18″ just happens to coincide with the star. On a more statistical basis you may consider Reddish's dust-embedded stars in young associations as an indication that most O stars are surrounded by dust clouds.

Let me mention one more piece of information on dust clouds in H II regions. Apart from extinction of light, which gives an indirect method of observing dust clouds, there is a direct method through thermal radiation in the far IR. The mechanism is simple: the dust grains are heated by starlight and then re-radiate in the IR. Typical temperatures of the dust grains appear to be a few hundred K, thus their IR radiation is best observed at 10 or 20 μ. Extended IR nebulae have first been observed northwest of the Trapezium stars in the Orion Nebula by Kleinmann and Low (1967); subsequently a second IR nebula, approximately coincident with the Trapezium stars, has been discovered by Ney and Allen (1969). In the H II region M17, observed at radio wavelengths, two compact radio sources have been found after observations with an angular resolution of 2′, and at the VIth Liège Symposium on Pre-main Sequence Stellar Evolution Low reported that he observed two extended IR nebulae coinciding with the radio components. He further claimed to have observed two stars, which by his IR measurements he can identify as O stars located within the boundaries of the two IR nebulae. This is further strong evidence for the existence of obscuring dust clouds in H II regions which surround the exciting stars. How the stars manage to ionize the gas through these dust cocoons is, however, not yet clear. (Kleinmann, D. E. and Low F. J.: 1967, *Astrophys. J. Lett.* **149**, L1; Ney, E. D. and Allen, D. A.: 1969, *Astrophys. J. Lett.* **155**, L193.)

Pottasch: At present I would not yet draw the conclusion that emission from the dust causes the IR nebulae observed by Low. Nothing is known about the spectra of these objects; other radiation mechanisms are conceivable that would explain the observations. I think that H I gas alone would certainly be enough to produce it. The work of Stein and Gillett (1969a) is perhaps applicable, but is not at all conclusive. (Stein, W. A. and Gillett, F. C.: 1969, *Astrophys. J. Lett.* **155**, L197.)

Mezger: I don't think that is correct, since the spectrum of at least the infrared nebula in Orion has been investigated by Stein and Gillett (1969). They find a bump in the spectrum between 9 and 12 μ which they claim can best be explained by emission of mineral dust grains. That is, in fact, how the silicate grains came into play.

Shklovskii: Does the presence of OH sources in an H II region correlate with other characteristics of H II regions?

Menon: The OH sources in the Orion Nebula could not have been detected in the present surveys, if the Nebula had been twice as distant. So, all H II regions might have similar sources of anomalous OH emission.

Weaver: Strength of source is certainly an important factor in the discovery of anomalous OH emission regions; many faint sources must remain to be found. However, there may be an additional factor.

So far anomalous OH sources have been detected only in very dirty H II regions, that is, in H II regions that are closely associated with large amounts of dark material. It may be that only in H II regions embedded in dark material are conditions appropriate for the production of anomalous OH sources.

Verschuur: Dr. Mezger, do you imply in your discussion on the OH point sources in W3 (= IC 1795), that there are seven protostars, or that these are all aspects of one protostar, because they are after all, within a few thousand AU of one another.

Mezger: That is a good question. In view of their different radial velocities I do not believe that all seven emission centers pertain to one protostar. However, the double and triple sources which form one emission center might probably be part of one protostar.

Menon: Some time ago I used the argument, again in connection with the Orion Nebula OH sources, that the velocity range of the components is very large, of the order of 30 km sec^{-1} or more. If all sources are situated within less than 0.1 pc, as indicated by measurements, their lifetime in terms of separation from each other is less than a few thousand years. So, whatever the detailed nature of the objects might be, the time scale for the phenomenon is only a few thousand years, and the sources could not all be together in such a small space, unless they had been formed a very short time ago. This argument is, I think, still valid.

Mezger: I might add, though, that in discussing the protostar hypothesis Robinson and I (Mezger and Robinson, 1968) did not work from the assumption that OH emission centers are protostars and then tried to find out whether or not the conditions in both types of objects are compatible. It was just the other way around. Making some assumptions, which might be bad or good, we tried to infer the densities, the masses, and the temperatures of OH emission centers. It turned out that the temperatures and the densities agree with those in protostars, but the masses do not. But by assuming that either the maser effect is less effective or that only the outer layers of the protostar are involved in OH emission one gets to stellar masses. If this is true, however, the densities derived are so high that the emission centers must be gravitationally unstable and contract to form a star. (Mezger, P. G. and Robinson, B. J.: 1968, *Nature* **220**, 1107.)

Burke: I hope that someone will tell me what a cosmic maser looks like. There are three models of cosmic masers. Since we have representatives of the protostar school present, we will call one model a protostar model. In this case, the condensed object is the entire maser. In W3 we are seeing, then, a cluster of young stars. The second model says that we are merely looking at filaments, accidentally end-on. In other words, one has a large region corresponding to the entire W3 region; all of this is in an inverted state. Here, there are no protostars at all. The third possible model one could call, I suppose, a condensation model of the maser. This is a model Gold proposed. One has a large region that is a potential maser. Accidental fluctuations cause a few regions to be stronger sources of emission than others. Then, when somewhere radiation is emitted, the outgoing spherical wave grows exponentially and eventually overtakes the inverse square law. Initially, you have many centers competing for a maser, but eventually one center dominates the masering region. Of the three possible models, I think that the observations throw some doubt upon the filamentary model. But the other two are still in the running. I wish to add one problem, which has to do with the tendency of the angular sizes of OH sources to grow with the distance of the sources from us. Take for example W3. It is reasonably close; its distance is 2 kpc. Its angular size is typically of the order of 0.01″ for the individual features. With W49, however, its distance is more like 14 kpc; its angular size is of the order of 0.1″. Thus it is farther away, larger in size. This may refer more to the conditions in the interstellar medium between the source and us, or it may mean that we have more than one kind of cosmic maser.

Mezger: As far as W49 is concerned, Dr. Burke is complaining about the fact that the angular sizes get larger in spite of the larger distance. However, I think one should keep in mind that the size of the OH emission centers in W49 corresponds precisely to the average distance of the individual emission sources found in W3. Could it be that your angular resolution was not sufficient to resolve the individual emission centers?

Burke: From the fringe visibility it is excluded that structure exists of a much smaller angular size.

2. The Carbon Recombination Line

Sorochenko: Dr. Mezger, what are the physical conditions in the regions where you observe the carbon lines?

Mezger: From the line width we conclude that the temperature must be less than 1000 K. Furthermore, the carbon line, when observed at different frequencies, behaves quite differently from the corresponding hydrogen lines. In the center of the Orion Nebula the carbon line has been observed at four wavelengths between 6 cm and 21 cm and its intensity relative to the hydrogen lines increases with increasing wavelength. This behavior suggests that the carbon lines are not emitted in LTE conditions and that some maser action is effective. This effect is by far not as spectacular as the maser amplification observed in the emission of OH and H_2O lines and it may be explained by a weak dependence on n of the b_n factors, as has been first suggested by Goldberg.

Dyson: Would you give us the evidence for the conclusion that C^0 is responsible for the recombination line?

Mezger: The conclusion is partly based on theoretical considerations. First, if the emission comes from neutral clouds, which are only partially ionized, then the ionization potential of the ionized atoms must be lower than that of hydrogen. If the emitter is one of the elements with high cosmic abundance only carbon is left. Second, this so-called carbon line is found in or close to nearly all H II regions whose general radio recombination spectrum is intense enough. (To date the line has been found in all H II regions which have been investigated with sufficient signal-to-noise ratio.) On the assumption that this line is in fact the recombination line of carbon, it never has exactly the same radial velocity as the hydrogen recombination line at the center of the H II region. However, as far as we can say at present the observed deviations in radial velocities are random and not systematic. This is consistent with the idea that this line is in fact the recombination line of carbon and that it is emitted from a small region which has a slightly different radial velocity with respect to the central part of the H II region.

Burke: I seem to remember that Goldberg uses one more argument, namely, that C^+ has a uniquely large dielectronic recombination cross section.

Mezger: I think dielectronic recombination is out, as far as Goldberg is concerned. Recently Mrs. A. K. Dupree, one of Goldberg's associates, has made more quantitative computations (Dupree, 1969). She found that the process of dielectronic recombination, if it is to result in a maser effect, first requires relatively high temperatures, much higher than the electron temperatures derived for normal H II regions. Second, this maser amplification should show a very strong wavelength dependence, with a maximum amplification probably around 5 or 6 cm. The observed increase of the intensity of the carbon line with increasing wavelength rules out an overpopulation by dielectronic recombination, as has been stated by Goldberg himself (1968). [Goldberg, L.: 1968, *Interstellar Ionized Hydrogen* (ed. by Y. Terzian and W. A. Benjamin), New York, p. 373; Dupree, A. K.: 1969, *Astrophys. J.* **158**, 491.]

Sorochenko: Dr. Mezger, I have two questions about the carbon line. First, have calculations been made on the maser effect with realistic parameters and is there agreement between the theoretical calculation and the observed intensities? Second, when this line was observed, it was identified as a carbon line, because of the favorable dielectronic recombination rate. Since this argument seems to be invalid now are other elements becoming possibilities?

Mezger: As far as the first question is concerned, no; there are no computations of the b_n factors for the physical conditions supposedly present in the emitting clouds. *Re* question two: the observers did not base the identification of the line on the dielectronic recombination rate (Palmer *et al.*, 1967); they used the fact that the line widths are so narrow that the kinetic temperatures of the emitting atoms must be less than 1000 K. Thus we concluded that the radiation comes from cool, dense and only partly ionized clouds and consequently the ionization potential of the emitting atom must be lower than that of hydrogen. On the other hand, the emitter must be an abundant element. These were our original arguments, and only a subsequent paper by Goldberg and Dupree (1967) suggested dielectronic recombination as the cause for the maser effect, a theory which has been abandoned in the meantime, as mentioned before. (Goldberg, L. and Dupree, A. K.: 1967, *Nature* **215**, 41; Palmer, P., Zuckerman, B., Penfield, H., Lilley, A. E., and Mezger, P. G.: 1967, Nature **215**, 40).

Van Woerden: Do I understand correctly that the carbon lines come from an H I region?

Mezger: Yes, I think they originate in dense condensations, which are either imbedded in or located at the edge of an H II region. The condensations have densities of the order of 10^5 or 10^6 atom cm^{-3}, and, even if exposed to the Lyman continuum radiation from a very early O star, they will be only superficially ionized. Thus, all the flux below 912 Å will be absorbed in a thin surface layer but the radiation between 912 and 1100 Å, which can ionize C^0, can freely penetrate the condensation.

Dyson: Will your comments regarding carbon remain valid, if you allow the possibility of ionization by cosmic rays?

Mezger: Why should cosmic rays single out one specific element?

3. The He/H Abundance

Shklovskii: A very fundamental problem for cosmogonical studies is the abundance of He relative to H. Dr. Mezger, will you comment on this point?

Mezger: Measurements of He-recombination lines have been started by the Harvard observing group and me (Palmer *et al.*, 1969), and have been pursued in the NRAO-MIT and NRAO-CSIRO-MIT 109α recombination line surveys of the northern and southern sky (in press)* and most recently by Churchwell and me (in preparation).* We have investigated in detail something like 10 H II regions for their He to H abundance. For a much larger number of H II regions we have at least qualitative results. As I mentioned before, the measurement of the He/H abundance, is rather straight-

* (*Note added in proof.*) References: Reifenstein, E. C., Wilson, T. L., Burke, B. F., Mezger, P. G., and Altenhoff, W. J.: 1970, *Astron. Astrophys.* **4**, 357; Wilson, T. L., Mezger, P. G., Gardner, F. F., and Milne, D. K.: 1970, *Astron. Astrophys.* **6**, 364; Churchwell, E. and Mezger, P. G.: 1970, *Astrophys. Lett.* **5**, 227.

forward, since the transition probabilities of the radio recombination lines for He and H are the same. Consequently, a comparison of the intensities immediately yields the number ratio. The present results can be summarized as follows.

(i) Most H II regions are located between 4 kpc and 13 kpc distance from the galactic center. This is the region where star formation occurs and spiral arms are observed. In this region, either no He-lines are observed at all, since the exciting stars are not hot enough (as in NGC 2024), or the observed He abundance is about $N(He^+)/N(H^+) \approx 0.08$. There are, however, two H II regions (DR 21 and NGC 7538) which require early-type O stars for their ionization but nevertheless do not show He109α-emission.

(ii) There are about five 'giant H II regions' (i.e., whose intrinsic flux is equal or larger than four times the radio flux from the Orion Nebula) supposedly located inside the 4-kpc arm. Two out of these five H II regions have been recently investigated by Churchwell and me and no He109α-emission could be detected (the upper limit was $N(He^+)/N(H^+) \lesssim 0.02$). However, we are not yet willing to interpret this as a true deficiency in He for the population II material in the center region of our Galaxy. These H II regions may have different excitation sources and for example the He could be doubly ionized. Therefore we first have to search for the recombination lines of doubly-ionized He before we can interpret the absence of the He109α-line.

(iii) The H II region 30 Doradus in the LMC shows an He abundance of 0.16. Although the error margins on these observations are rather large, our observations (NRAO-CSIRO-MIT) are not compatible with a He abundance of 0.08. There is a paper by Rubin (1968), who investigated the possibility of a non-coincidence of the Strömgren spheres of helium and hydrogen. His conclusion is that there are two possibilities: either the helium is more or less fully ionized within the Strömgren sphere of hydrogen, or it is not ionized at all. The transition range between these two states is very small; it occurs, according to Rubin's computations, at effective star temperatures of about 36 000 K. (Palmer, P., Zuckerman, B., Penfield, H., Lilley, A. E., and Mezger, P. G.: 1969, *Astrophys. J.*, **156**, 887; Rubin, R. H.: 1968, *Astrophys. J.* **153**, 761.)

Ozernoi: The primeval He/H ratio in the Universe is of great importance, as Shklovskii remarked. If the abundance of He should turn out to be low, this does not mean that the big-bang theory is incorrect. It means, perhaps, that the early stages of cosmological expansion are anisotropic or that processes are important which decrease the He abundance. In some quasars the He abundance is ten times smaller than usual, but at present it is unclear whether the lower abundance is due to anisotropic expansion of the Universe or to nonthermal processes in nuclei of quasars. In this connection, I would like to know: Is there any correlation between the regions of low He abundance and the sources of nonthermal radiation such as OH sources, infrared sources, supernovae, and the galactic center? [At this point there were some language problems. Part of Ozernoi's question is answered by the fact that recombination lines and He/H abundances have been detected only in H II regions. (Ed.)]

Mezger: Radio observations of the two H II regions near the galactic center yield the following results. (i) The radio continuum spectrum appears to be thermal. (ii) The H109α line-to-continuum ratio is that observed for typical spiral arm H II regions corresponding to electron temperatures of less than 10 000 K. (iii) The shape of the H109α-lines, however, is rather odd, revealing either the existence of several components or of a very high internal turbulence. Although the two sources look pretty much like normal H II regions, the radio observations cannot rule out a source of ionization different from early-type stars. In this case it would not be unreasonable to expect that most of He is doubly ionized. I feel that only after observations of the He^{++}-recombination line should we continue to speculate on the nature of these thermal sources inside the 4 kpc arm.

Field: The other possibility that you mentioned is that the exciting stars are of spectral type later than O9 and that the He atoms are all He^0. Is that not something we could also verify by computing the number of recombinations in the H II regions and checking that against the spectral type of the star?

Mezger: Yes, it could be done. Of course, you cannot decide from radio observations whether the ionization is caused by a number of B0 stars or by one O5 star. However, I feel it is reasonable to assume that all giant H II regions, i.e., those which have fluxes four times the flux of the Orion Nebula must be ionized by O stars. And these two H II regions located close to the galactic center belong definitely to this class of giants.

Burke: I have only a quick comment on helium abundances. Not only in the galactic center are there H II regions that show very low helium abundances. Several other H II regions (I think NGC 2264 is one of them) have quite low He abundances. There is even one that has an abundance of 3 per cent.

Mezger: To my knowledge in all the H II regions where we have been able to see the He109α-line at all, a He/H ratio of 8 per cent is within the limits set by observational errors.

Field: Finally, there is the question of the excitation of the He lines. Is there not good evidence now from the study of the hydrogen recombination lines that we are dealing with maser effects, and that, therefore, these methods are unreliable for abundance determinations?

Mezger: First, we are doing experiments on the effects of deviations from LTE. I think all observers agree at the moment that deviations from LTE do not amount to more than a few times 10 per cent, if present at all. There are several observational studies possible. Let us consider, as an example, the 109α recombination spectrum. Within 6 MHz one observes the H109α line, the H137β line, the He109α line, and the He 137β line. [See Figure 3 at page 342 (Ed.)]. These four lines are observed simultaneously and a comparison of their intensities should yield considerable information on deviations from LTE and on the He abundances. Some of these observations have already been done, others are under way. Let me summarize our preliminary results. (i) In cases where the H 137β line is too weak (as compared to the LTE case) the He137β line appears to be too weak by the same factor. This would indicate that H and He lines are affected by deviations from LTE in the same way. (ii) In the case of the Orion Nebula we have taken broadband spectrograms in the center as well as in four offset points spaced by one half the HPBW [=antenna beam (Ed.)] of the telescope. Although the ratio H137β/H109α for the two off-set points east and west of the center of the Orion Nebula is pretty close to the LTE-value, the measured He109α/H109α ratio is, within the relatively small observational uncertainties, the same in all five points. (iii) We cannot find significant correlations between deviations from LTE (as measured by the ratio H137β/H109α) and the He abundance (as determined from the ratio He109α/H109α). Further measurements of the He/H ratio at higher and lower frequencies are under way.

Field: I was very interested in Ozernoi's suggestion that mechanisms might exist which destroy helium. Since such mechanisms might operate in nonthermal radio sources where cosmic rays are present, I was reminded of the fact that in the Crab Nebula supernova remnant the He abundance is anomalous. I would like to ask whether or not one can observe the He recombination lines in that Nebula to verify the anomaly.

Pottasch: I thought that in the Crab Nebula the helium abundance is at least twice as high as normal. That was Woltjer's (1958) result. (Woltjer, L.: 1958, *Bull. Astron. Inst. Netherl.* **14**, 39.)

Mezger: As far as the observations are concerned, we have not found recombination lines in any nonthermal source. In general we have found many sources which do not show recombination lines, about 31 per cent for the northern sky, about 13 per cent for the southern sky. It appears that all of the sources which do not emit radio recombination lines have nonthermal continuum spectra. In fact in a few well-known supernova remnants we searched hard for radio recombination lines and did not find any. Either the electron temperatures are too low, and the lines are so wide, owing to high internal velocities, that they become unobservable, or the electron temperatures are too high, and the lines cannot be observed since they are too weak. To be quite specific, we observed the Crab Nebula and found no recombination lines whatsoever.

Field: And would you have expected them from calculations of electron density and temperature?

Mezger: A search for radio recombination lines in supernova remnants was suggested to us by Minkowski, who claims that the electron temperatures are low enough to observe the recombination lines. If the electron temperatures were in fact 10 000 K in the Crab Nebula, it would be very easy for us to observe the lines, unless the internal velocities are of the order of a thousand km sec^{-1}. With such larger velocities it becomes difficult to observe the very wide lines with our 10 MHz bandwidth.

Van Woerden: The Balmer lines are recombination lines, too. Is the fact that you do not find radio recombination lines consistent with the fact that Balmer lines are observed in such sources?

Pottasch: The answer may be that in the optical observations we're observing very thin filaments in the Balmer lines. Perhaps because of their very high relative velocity of a thousand km sec^{-1}, these velocities just smear out in the radio observations.

4. Various Shorter Comments

Menon: You mention a value of about 10 per cent as being the ratio of gas to stars. You don't mean, do you, that this is a constant for all clusters?

Mezger: For all the young compact H II regions this seems an upper limit, although in only two or three cases the limit is based on direct star counts, for example in the Orion Nebula. In all other cases, we can infer the spectral type of the exciting star from the intrinsic radio flux, as discussed before. We then make the assumption that where O stars are formed, stars of lower mass are formed, too, and use, in a wild extrapolation, the mass distribution function. This leads to the estimate that about 30 times the mass of O stars must be present in the form of lower-mass stars. This extrapolated star mass is compared to the directly measured H II mass and in all cases of young compact H II regions we obtain $M_{\rm HII}/M_* \lesssim 0.1$.

Van Woerden: In this 10 per cent figure you included only the gas in the H II region, and not that in the surrounding H I region?

Mezger: That is correct; it only refers to the H II gas and not to the H I gas.

Van Woerden: So, if the H I gas has been driven away by the pressure of the hot H II region, this might change your conclusions?

Mezger: It does not change my conclusion that, at least in the center of an OB star cluster, the efficiency of star formation is very high and that nearly all of the mass goes into the stars. However, I will not exclude the possibility that the H II region is embedded in an H I cloud which made the original cloud gravitationally unstable. But this would not change the conclusion that the efficiency of the formation process is extremely high in the center of the cloud, where most of the star formation is going on.

Van Woerden: You do not believe then that the H I gas can be driven away?

Mezger: I do not object if the gas is driven away, provided that I see it observationally.

Van Woerden: Then you must not look at the surrounding shell which might be too thin for detection, but at the gas that is seen with a different velocity in front of the H II region.

Mezger: Could we leave the question until interferometer measurements of the 21-cm line at Cal Tech or at Groningen really show those shells?

Sorochenko: I want to show some new experimental evidence that the ionized hydrogen in the Orion Nebula does not have a uniform distribution, but is concentrated in clouds with a high electron density n_e. This evidence, which J. J. Berulis and I have obtained (Sorochenko and Berulis, 1969) is based on observations of the hydrogen radio recombination lines. The intensity of these lines depends on the populations of the atomic levels, which in turn depend on the density. For higher densities, the population must be close to LTE. Figure 1 shows how the parameter b_n^* depends on the wavelength (or equivalently on n) for various electron densities n_e; b_n^* is the ratio of the actual intensity of the radio line to the intensity under conditions of LTE. If there is LTE, $b_n^* = 1$, and if there is no LTE, the intensity of the lines is enhanced at $\lambda \gtrsim 5$ cm due to the partial maser effect proposed by Goldberg. At the same time, at

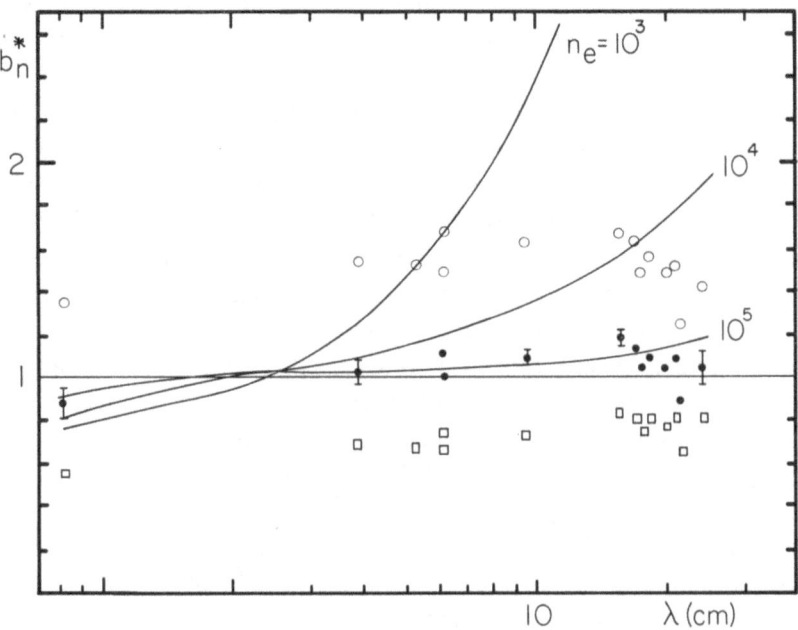

Fig. 1. (See the remark by Sorochenko.) Comparison of the theoretical and observed values of b^*_n, where b^*_n is a parameter characterizing the ratio of the observed intensities of recombination lines and the predicted value for LTE. Solid lines are theoretical values for various electron densities n_e. Open circles are the observed values for $T_e = 10^4$ K, filled circles for $T_e = 7 \times 10^3$ K, and squares for $T_e = 5 \times 10^3$ K. This diagram summarizes results by various observers.

millimeter wavelengths there is underpopulation, and the actual intensity is lower. The three families of points correspond to observed values if we adopt electron temperatures $T_e = 5 \times 10^3$, 7×10^3, and 10^4 K. Good agreement of computed and measured values occurs only with $n_e = 10^5$ cm^{-3} and $T_e = 7 \times 10^3$ K. At the same time, observations of the continuum radiation at $\lambda = 8$ mm give $n_e = 2.7 \times 10^3$ cm^{-3}. Only a small part of this discrepancy may be due to inaccurate values of the cross-sections. Therefore, to reconcile the two results we must conclude that dense concentrations of electrons exist. If they exist we obtain good agreement with the value of $n_e = 5 \times 10^4$ cm^{-3} that has been mentioned by Mezger. This figure also gives the answer to the question asked earlier by Field about deviations from LTE for the radio recombination lines. For $n \gtrsim 100 (\lambda \approx 5$ cm), there is no significant deviation from LTE in the Orion Nebula; and $T_e = 7000$ K seems representative for n_e^2 averaged over the volume of the Nebula. (Sorochenko, R. L. and Berulis, J. J.: 1969, *Astrophys. Lett.* **4**, 173).

Weliachew: I should like to draw attention to temperatures of H II regions determined by Dr. Louise in Marseilles. He used the widths of the Hα and N II lines. On the assumption that both lines originate in the same region, temperatures and mean turbulent velocities follow, since the atoms have different masses. The results will be published soon (Louise, 1970). [At the Symposium a table of results was passed around. (Ed.)] You will note that Louise's results are in fair agreement with the results from the radio recombination lines. (Louise, R.: 1970, *Astron. Astrophys.* **6**, 460. [The following part of the discussion, up to Section 5, actually took place on September 16 (Ed.).]

Menon: I want to comment on the observations of Harwit and Werner (1968) who claim that they observed H_2 infrared radiation from a dark cloud in Orion (see Figure 2 of Mezger's Report, p. 339. The dust cloud is located at $\alpha = 05^h 33^m 00^s$, $\delta = -05° 25'$). I recently observed this same cloud in 21-cm radiation, where it is seen in emission as well as in absorption. Observations with high resolution suggest that the brightness temperature of the 21-cm radiation from that cloud is at least 60 K. It could be that the actual kinetic temperature is higher; the line does not seem saturated. In addition, there are two interesting things. One is that the dark band is approaching the H II region with a velocity of about 14 km sec^{-1}. (The excess emission seen due to the band is $+12$ km sec^{-1}; and the velocity of the H II region is -3 km sec^{-1}, both with respect to the local standard of rest.) Of course this is only a radial motion. One end of the dark band probably has a somewhat lower velocity. This is also the region where the so-called excited line of carbon is being observed. The velocities of the carbon line (if it is carbon) and the cloud are very close to each other. This is important because the velocity observed of the order of $+12$ km sec^{-1} is essentially a peculiar velocity for this neighborhood. The density of neutral hydrogen in this cloud is about 400 or 500 cm^{-3}. The mass, assuming that the cloud extends only over about 4' (and the 21-cm observations show that it does not extend very much farther), is of the order of 10 $M_\odot$, perhaps less. (Werner, M. W. and Harwit, M.: 1968, *Astrophys. J.* **154**, 881).

Burke: Which instrument did you use for these observations?

Menon: They were made with both the 140-foot and the 300-foot telescopes at Green Bank. As a matter of fact, I obtained an upper limit of the brightness temperatures by looking at the excess emission. In the 140-foot it is about 10 K, in the 300-foot it is about 40 K. Correcting from antenna to brightness temperature, I obtain a brightness temperature of about 60 K.

Burke: I asked this question because I see that the angular size of this structure is about 1 or 2' in declination. That is still much less than the resolution of the 300-foot dish. How do you know that *that* 21-cm cloud is associated with *that* dust cloud?

Menon: When you make a complete contour diagram of the 21-cm emission, there is no question about it.

Van Woerden: Do you have an excess brightness temperature over the surroundings? I am asking this because you find 60 K brightness temperature at $+12$ km sec^{-1} everywhere around the Orion Nebula.

Menon: No, you do not find it in this part of the Orion region. When you take a drift curve, you find it only within one beamwidth.

Verschuur: In the very dense cloud in front of the Orion Nebula, I have found a very strong magnetic field of 50 μG. In my Report I suggested that these strong fields were the result of amplification in a contracting medium. The magnetic energy in this cloud, however, is about 10 to 20 times greater than the gravitational energy at present; therefore, the cloud could not have contracted to amplify the field. Thus there must be missing matter, and I hope that this missing matter is molecular hydrogen.

Field: If I remember correctly the density given by Werner and Harwit was about 10^4 cm^{-3}, about ten times greater than you see in H$_I$, so that it is consistent with the remark made by Verschuur.

Menon: I have no comment on the amount of molecular hydrogen. You can use the same argument Heiles (1969) used for his dark clouds (he finds discrepancies of a factor of 5 or 6), simply by saying that the absorption in that cloud is several magnitudes. I have no accurate value, because it is essentially opaque, and following Heiles' calculations, I would expect several times 8×10^{20} cm^{-2} in the line of sight. I find about 7.5×10^{20} cm^{-2}, a discrepancy of about 5 or 6 depending on the total amount of absorption through the cloud. (Heiles, C.: 1969, *Astrophys. J.* **156**, 493.)

5. A New Mechanism for Star Formation

Pikel'ner: May I turn our discussion to the question of star formation? Usually star formation is connected with gravitational instability and with fragmentation of gas due to fast cooling. Indeed, this picture works for old clusters, for instance, globular clusters. But there are problems with this mechanism. For example, stellar associations and clusters have similar stellar-mass distribution functions, although associations have much larger sizes and are less concentrated. From a theoretical point of view one would expect that the gas fragments have much larger masses during the birth of an association than during the birth of a cluster. This difference should affect noticeably the stellar-mass distribution function; but no such effect is observed. One explanation might be that the associations are concentrated during the star birth but expand later on. However, the T Tau associations are younger than O associations, and in the T Tau associations there have never been massive stars or supernovae and it seems unlikely that these associations have much expanded. So we conclude that some other mechanism must lead to star formation, a mechanism which is not similar to that of simple gravitational instability. I would like to suggest such a mechanism. It is connected with the general model of heating of the interstellar gas by X-rays. Let us suppose that somewhere a Rayleigh-Taylor instability develops. The gas flows down the magnetic field lines and the hydrogen density at the bottom of the flux tube is growing (see the Report by Field). When the number of hydrogen atoms (N_H) is about 10^{21} cm^{-2}, the optical depth τ_x for soft X-rays responsible for the heating of the gas will be more than 1. The absorption of X-rays will lead to a decrease in the gas temperature at the bottom of the gas layer. (Ultraviolet radiation of stars is absorbed by dust.) Very schematically we get a two-layer system: a very cold layer at the bottom and a somewhat hotter gas layer with $\tau_x < 1$ on top of the cold layer. This hotter layer of gas will have the usually adopted gas distribution in the z-direction. In the beginning the thickness of the layer of cold gas is small, and the pressure in this cold layer is determined only by the weight of the hotter gas layer. The temperature of the cold gas is several times less than that above. For instance, in the extended hotter layer $T \approx 50\,\mathrm{K}$ and $n_{\mathrm{H}} \approx 500\ \mathrm{cm}^{-3}$, and in the cold layer $T \approx 10\,\mathrm{K}$ and $n_{\mathrm{H}} \approx 2500\ \mathrm{cm}^{-3}$. But the gas continues to flow down. The thickness in the z-direction of the hotter layer remains the same, because the extra gas is added to the cold layer. Ultimately the thickness of the cold layer becomes large enough to produce self-gravitation. Then this cold layer may break into stars, due to horizontal motions. Calculating the mass of these stars one finds about 0.3 to 0.5 $M_{\odot}$. This is the standard value obtained under standard conditions. We know that when star formation sets in, young stars have masses of this kind, perhaps up to 1 $M_{\odot}$. After some time (about 10^{7} yr) the mass of the forming stars becomes higher. This increase of mass can not be explained by the older theory (the gravitational instability mechanism), which predicts that the heaviest stars should be formed first. In this new picture the mass of the forming stars is proportional to T^{2}, where T is the temperature inside the cold layer. If we take into account the heating of the cold layer by the T Tau stars, which can give X-ray and UV emission and perhaps cosmic rays, the temperature will increase several times, perhaps from 5 K in the very cold layer to 10 K or 20 K. The new stars will be increasingly more massive, until they begin to ionize the whole region and to stop the star formation. Another possible mechanism for gradual heating of the cold layer is to have an instability which can develop at the other side of the galactic plane. The gas falling down into the cool layer gives some pressure wave, which crosses the galactic plane and propagates further. At the other side of the plane a hill begins to grow, gas flows from this hill and opens the cold layer from the bottom. Then the layer gets heated and more massive stars can be formed. A more detailed discussion will appear elsewhere (Pikel'ner, S. B.: 1970, *Astrophys. Space Sci.* **7**, 489).

Field: Dr. Pikel'ner, your model is based on the attenuation of X-rays by the interstellar medium. I think you would agree that we are still unclear as to whether the heating in the interstellar medium is due to cosmic rays or to X-radiation. If it is cosmic radiation, then 1 MeV particles will have a free path just about equal to those of the X-rays that you were considering. So we would, perhaps, expect similar effects in that case. Concerning the time-dependence of star formation in a cosmological model, would not your model suggest, if it does in fact depend upon extra-galactic X-rays, that there would be a considerable dependence upon the cosmological time, since as the expansion proceeds the X-radiation is reduced?

Pikel'ner: As for choosing between X-rays and cosmic rays, you are correct. We do not have yet definite arguments to prefer one or the other. But we have more or less reliable X-ray observations, and therefore, X-rays are somewhat preferable. The second question of cosmological time-dependence may be important. However, according to my memory, it seems at present that soft X-rays come mainly from galactic sources, but not from extra-galactic ones, because in the southern hemisphere the intensity of this flux is higher than in the northern hemisphere. In the early states of galaxy formation, the mechanism I propose was hardly important; one expects that at that time gravitational instability was more common.

Silk: What was the X-ray flux that you assumed in getting your critical mass for star formation?

Pikel'ner: I do not assume a certain value for the X-ray flux. The only suggestion is that X-rays heat the gas, and that 10^{21} atom cm^{-2} give a considerable absorption.

Weyman: Dr. Pikel'ner, I did not understand the figure of 0.5 to 0.3 $M_\odot$. Was that the minimum mass, or was that supposed to be the mass-range in which the stars formed? And, if that was the case, why did you not get larger masses?

Pikel'ner: 0.3 to 0.5 $M_\odot$ is the standard mass appearing in the thin cold layer. Star formation starts in the cold layer, when the thickness of this layer is such that its self-gravitation is the same as the external pressure. The average mass of the stars is about two times the mass of a cube with a size like the thickness of the layer (about 0.1 pc); the resulting mass is roughly about 0.3 to 0.5 $M_\odot$. The stellar mass is proportional to T^2 and depends on the molecular weight. An interesting feature is that we get here stars of very low mass without any difficulties. When one tries to explain the formation of low-mass stars from globulae by gravitational instability, one finds that it is very difficult. And Kuhi (1966) quotes evidence that the more massive OB stars appear usually in H II regions containing globulae. But T Tau stars appear only in T associations. Formation of low-mass stars can occur only in a very cold gas; note that Heiles (1970) has observed such very cold regions in the OH lines. (Kuhi, L. V.: 1966, *J. Roy. Astron. Soc.* **60**, 1; Heiles, C.: 1970, *Astrophys. J.* **160**, 51.)

Weymann: You were saying earlier that it was difficult to understand why one has the same mass distribution function, even though initial conditions were different. I do not understand why this is a difficulty when the cloud masses are very much larger than the masses of the stars formed. Take, for example, the case of a turbulent spectrum. At wavelengths comparable to the scale at which energy input takes place, one expects the spectrum to depend strongly upon the details of how the energy was fed in. But at wavelengths which are very much smaller, one does not expect a strong dependence. In fact, it seems to me that we still know almost nothing about the general question of the mass spectrum for stars in various situations. I would suggest that this is one area where the aerodynamicists can, perhaps, make a really significant contribution.

Pikel'ner: I agree with your comment that a turbulence spectrum can give more or less uniform stellar-mass distribution functions, and that this function may be different under different conditions. The main point I wanted to make concerned the average mass of stars which should be larger or smaller in different conditions.

27. FINAL DISCUSSION

(Thursday, September 18, 1969)

Chairman: R. N. Thomas

Editor's remarks: In the present form the Final Discussion is divided into three sections: 1. Discussion on the Overall Energy Flow in the Interstellar Medium; 2. Summaries and Suggestions for Future Research (more or less prepared talks); 3. Summaries and Suggestions for Future Research (more or less free discussion). The discussion in Section 1 actually took place on Saturday, September 13, but it belongs in the Final Discussion. The (invited) summary papers, presented in Section 2, were given by Parker, van Woerden, and Kaplan. Section 3 starts with the remarks by Busemann on the interrelation between interstellar gas dynamics and aerodynamics, as viewed by an aerodynamicist; these remarks were made on Tuesday, September 16. The rest of Section 3 is taken up by the actual discussion held on the final day of the Symposium.

In view of the total length of this Final Discussion I have felt it necessary to shorten several of the longer contributions. I apologize for this, although I am convinced that the text presented here is closer to what was actually said than the texts submitted to me by some of the authors.

1. Discussion on the Overall Energy Flow in the Interstellar Medium

[This section of the discussion actually took place on Saturday, September 13.]

Van de Hulst: In my introduction the first day I mentioned that the energy balance was a major point in earlier Symposia. I have waited until now to see if I could picture the total energy flow from one reservoir into another. Consider the very schematic diagram of Figure 1. The top circle SN represents supernovae but is meant to include also other stellar sources of energy such as OB stars. Below it are the cosmic rays CR (right) and the magnetic fields MF (left). At the bottom I have drawn the energy reservoirs GK containing gravitational (potential) energy and kinetic energy, conveniently combined because there may be a lot of give and take between these two reservoirs by the ordinary laws of motion. The unit rate of energy exchange is 10^{-26} erg cm^{-3} sec^{-1} [$=0.2$ eV cm^{-3} $(10^6$ yr$)^{-1}$] and attention is confined to the solar neighborhood, about which we know most. In earlier symposia the consensus was that energy was supplied from SN directly to GK and then transferred to MF by induction and to CR by a Fermi-type acceleration. It was difficult, however, to match the numbers. During the Symposium I have heard a firm estimate of 5 units as the total loss from CR and a loss of 1 unit from GK, whereas the direct supply from SN to GK would be only 0.1 unit. If these numbers are at all correct, the conclusion must be drawn that a rate of 5 units is supplied by SN directly to CR, of which 4 units are lost by radiation and

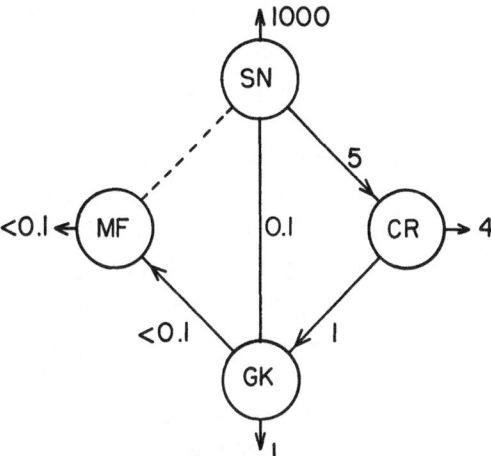

Fig. 1. (See the remark by van de Hulst.) Overall flow of energy in the interstellar medium in the region near the Sun.

escape and 1 unit goes to *GK*. This requires an inverse Fermi mechanism, for which Parker has given us a plausible description. The magnetic fields have estimated losses < 0.1 unit, but whatever the number is, the energy replacement can be supplied to them either from *GK* or directly from *SN*, if the ideas of Piddington and Kardashev are correct. I estimate direct losses of stellar radiation into space at 1000 units. I should like to ask the other speakers if they agree that this is a reasonable synthesis of the numbers they presented.

Pottasch: I am confused about the supply of energy to cosmic rays by supernovae. I thought Woltjer said the other morning (p. 234) that the Crab Nebula supplies less than 10^{48} erg of cosmic rays. Later on people talked about much higher supplies; 10^{51} to 10^{52} erg were mentioned. Does Woltjer agree that supernovae other than the Crab Nebula will supply much higher amounts of energy to cosmic rays?

Woltjer: I agree that there is enough energy present in supernovae to supply all the cosmic rays one needs, possibly in the initial explosion, and certainly in the pulsar that remains. It is possible to convert with very high efficiency the energy of the rotating pulsar into cosmic rays. So there is no energy problem in principle. However, in the Crab Nebula you do not see any evidence that such cosmic rays are being produced. But still I think it is almost certain that supernovae do produce the cosmic rays in one way or another.

Another point with respect to Figure 1 is that probably the energy flow from *SN* (the supernovae plus O and B stars) to *GK* should be multiplied by a factor of five or so.

Shklovskii: In the case of the Crab Nebula, it is quite simple to show that the main part of the cosmic-ray particles is in the form of relativistic electrons and there is some indication that similar situations may exist in other supernova remnants as well. In that situation, it is not possible to connect the present cosmic rays with the relativistic particles produced at the time of the supernova outburst.

Zel'dovich: Should not the nucleus of the Galaxy be included in Figure 1? There is, as we know, an outflow of hydrogen from the nucleus. There is also a large infrared flux. What would happen if we tried to make the energy balance for the nucleus of the Galaxy? Perhaps stellar encounters are important there, although they do not occur in the rest of the Galaxy.

Van de Hulst: Initially I had drawn the galactic nucleus also in the circle containing the supernovae. But I later omitted it because some of the other estimates we know only for the solar neighborhood and it is unfair to mix in one diagram estimates referring to quite different regions of space. But I fully agree there is a lot of energy produced in the galactic nucleus. It may be of the same order or bigger.

Mestel: The question of the galactic non-uniform rotation was raised the other day. A long while ago people wanted to tap this reserve of energy by appealing to shear-flow turbulence. This is no longer popular, largely because, I think, the rotation law satisfied the Rayleigh stability criterion. About ten years ago Hoyle and Ireland (1960) suggested magnetic coupling. Their picture involved twisting of the galactic magnetic field by the non-uniform rotation, buckling of the field-lines into the halo, sliding of gas down the field-lines into the 'galactic flare', etc. Again, one is ultimately drawing on the energy in the gravitational field of the stellar disk population. Is all this now ruled out for some reason, qualitative or quantitative? (Hoyle, F. and Ireland, J. G.: 1960, *Monthly Notices Roy. Astron. Soc.* **120**, 173.)

Field: In answer to the question by Mestel, Hoyle and Ireland proposed that the main energy source of the interstellar gas clouds was the winding up of the interstellar magnetic field. In order for this source to work, the galactic magnetic field must be systematic and have a radial component connecting one spiral arm to another. From the discussion earlier, this does not seem to be the case. Another remark is that van de Hulst has also omitted the infall of gas from intergalactic space, the Oort model of accretion. From Oort's data I estimate that the energy flow is 10^{40} erg sec^{-1}, one solar mass per year coming in with 100 km sec^{-1}.

Woltjer: But again you have to apply an efficiency factor. This gas comes in at 100 km sec^{-1} and if you want to couple that to cloud motions at 10 km sec^{-1}, much of the kinetic energy will be radiated away.

Field: I agree.

Mestel: Van de Hulst glided rather quickly over the problem of the interchange between cloud kinetic energy and their energy in the galactic gravitational field. On the first day he was concerned about the Parker mechanism, and how a gas cloud got back into regions of high potential.

Van de Hulst: Again I tried to summarize what I have understood from the discussions. I understand now that stellar sources, including O and B stars, supernovae, and even the galactic nucleus, will give motions to the gas and push them away. Perhaps there will then be one-sided flow back to the galactic plane. This would differ from the many oscillations back and forth which we envisage for stars but would agree with the raining down and settling in the low pockets of Parker's picture. I'm fairly convinced now that this may be the better picture, compared to the old picture of frequent oscillations.

2. Summaries and Suggestions for Future Research

A. E. N. PARKER, *Department of Physics, University of Chicago, Chicago, Illinois, U.S.A.*

I have jotted down a few ideas that I think this Conference has pointed up. Let me begin with a brief review which will serve as a background for suggesting where future theoretical problems lie.

In the theoretical discussions that you have heard, such things as the density, temperature, scale height, and turbulent velocities of the gas have come in and, as you are aware, have been debated at some length. I hope that in the next few years observations will give more detailed pictures and values for these quantities. The magnetic field, which seems to be a few μG, is being observed in more than one way: by Zeeman broadening, by Faraday rotation, and by polarization. I rather suspect that in the next four or five years our knowledge of the galactic magnetic field may be enormously increased. The cosmic rays are a subject in themselves. There are several things about the cosmic rays that are striking, including the fact that they are very steady, apparently rather stagnant, and not streaming rapidly through the Galaxy. One of the biggest gaps in our knowledge of cosmic rays at the present time is the modulation effect of the solar wind. What really is the cosmic-ray density in interstellar space? There is little uncertainty about the density of the relativistic particles which seem to comprise the major energy in cosmic rays; but in regard to the low-energy cosmic rays, which play other roles besides brute pressure, one really has little more than guesses as to what their intensity is outside the solar system. If space programs continue as presently foreseen, there is a good prospect that within six or eight years suitable instruments will be sent to distances of 10 AU. There is every reason to believe, but no guarantee, that at that distance from the Sun, one will see mainly the interstellar cosmic-ray spectrum without much modulation by the solar wind. That is not to say that the solar wind stops at 10 AU but that most of the modulation of the cosmic-ray intensity takes place between 10 AU and the Sun. Therefore I hope, that, at the next Cosmical Gas Dynamics Symposium, there will be much more definite information on the low-energy cosmic rays. This information will have relevance for the origin of cosmic rays, too. At the present time the explanation for the origin of cosmic rays merely involves objects (supernovae and pulsars) which are very energetic and have suitably anomalous abundances. I will have more to say on that a little later.

Finally, there is the dust in the interstellar space, about which we have had some discussion. I want to emphasize that the gas, field, cosmic rays, and dust all go together to form a coherent system. The common binding agent is the magnetic field. The dust is very tightly bound to the magnetic field. The cyclotron radius of a 10^{-5} cm dust grain, with a velocity of 1 km sec^{-1} in a field of 3 μG, is 2×10^{15} cm. That is a microscopic distance on the galactic scale; the dust is therefore very closely tied to the field. In many cases the dust has a sufficiently small density that its inertial effects are very small and it can be ignored, but in some cases this is not true. It has become increasingly apparent over the past few years that one can consider idealized cases of magnetic field alone, gas alone, cosmic rays alone, or dust alone, or any combination of these. But, in fact, they are all tied together; and the entire system is a composite fluid of field and gas. All the constituents must be treated together if we are to see the overall picture.

Now the overall picture of the interstellar medium can be studied in several parts. One part is the equilibrium of a disk of gas. The question of equilibrium is not simple because one immediately faces the question of the thermal properties of the gas, that is the temperature. Field, van de Hulst, Pikel'ner, and others have elaborated the rather gory details of this thermal equilibrium of the two phases with high and low temperature and low and high density. Then there is the question of mass inventory. If I understand the numbers properly, it is estimated very roughly that 1 $M_\odot$ yr^{-1} is being converted from interstellar gas into stars, give or take a factor of two or three. In the reverse direction, there is ejection from stars of the order of 0.4 $M_\odot$ yr^{-1} again give or take a factor of two or three. The net flow, therefore, is from gas to stars; when you begin to worry about abundances, the flow from stars back into the interstellar gas is of course extremely important. The return of gas from stars back into the interstellar gas can be carried out explosively (as in supernovae) or in less violent but transient phenomena, or in steady winds. While we are on the subject of inventories, I would remind you that the cosmic rays are subject to similar considerations, the present evidence being that cosmic rays in the disk of the Galaxy are replenished about every million years. They are being replaced fairly quickly; in fact, many of the dynamical properties of the disk have characteristic times of 10^7 yr, and the cosmic rays are replaced many times during the growth of, say, a cloud structure. When you consider the dynamical properties of the gas (since, in fact, any equilibrium of the disk is a conspicuous fiction), then the problem becomes rather complicated.

The disk of gas, field, and cosmic rays is unstable, as various speakers including myself have elaborated. It is unstable because it has cosmic rays in it, which give a Rayleigh-Taylor instability; it is unstable because there are magnetic fields in it, which enhance the Rayleigh-Taylor instability; and last, but not least, it is unstable because of thermal instability. All three of these effects combine to give a very complex situation, even under the idealized cases that one considers. The instability tends to clump the gas together, and I think people quite naturally refer to this phenomenon as *clouds*. But the theory also goes on to say that there is no such thing as an equilibrium cloud, and it reminds us once again that a cloud is only a transient shape, something you catch in a snapshot. If one could take snapshots at intervals of 10^5 yr, instead of at intervals of five years, one would in fact see these phantoms, these clouds, changing rather rapidly and probably in a very chaotic manner. Then there is the question of the dynamical balance between the cosmic rays and the magnetic field and, perhaps, *via* plasma turbulence, with the interstellar gas. The magnetic field seems to act as a safety valve. Unknown sources, perhaps supernovae and pulsars, are busy pumping up the magnetic field in the disk of the Galaxy; and the magnetic field is continually inflating and leaking at the surface. We have a pressure balance here. That seems to be as far as we have got at the present time in understanding the interstellar gas.

And now, I would like to say a few words about interesting research problems for the future that have occurred to me and probably to many of you, too, during this discussion. I was most intrigued in the observational discussions by the large negative velocities which seem to occur both at high and at very low galactic latitudes. You have heard the hypotheses for the inflow of gas at high latitudes; I did not hear any explanations for the tendency for negative radial velocities in the disk. Also, I am very interested, and very puzzled, by the present definition of clouds in 21-cm observations. A cloud is defined as any gas in the line-of-sight with a common velocity. It is certainly a very practical way of going about defining clouds, and yet I wonder what it means. The word *cloud* actually means something slightly different; it means 'neighbors in space'. And sometimes, when I see two different lines with different Doppler shifts, I wonder whether or not they might not come from regions of gas which are very close together in space, all lumped together in one cloud; whereas, of course, in this Doppler definition of clouds, they would be treated separately. I do not know the answer to that; I do not know any way of getting around that problem.

So far as theoretical problems are concerned, a lot of work still remains to be done on a variety of effects studied and reported here, such as the dynamics of H II regions and ionization fronts. But there is also the question of dust formation, which, in connection with its formation in lanes along the inside of the spiral arms, is certainly a curious question. And of course there is the classical problem, in which work is going ahead steadily, on star formation: namely, given the chaos that we call the interstellar medium, how do stars form? There again theory and observations are necessarily progressing hand in hand. In a slightly different direction, the structure of shock waves in tenuous gases is one that is still with us; I am speaking of the collisionless shock. I think that great progress in understanding the structure of the collisionless shock has come in the past few years. The collisionless shocks are beginning to be observed in considerable detail in the solar system (the bow shock upstream from the Earth, the shocks associated with blast waves from the Sun, and the collision of streams in the solar wind). Tidman, and a number of other people, have gone a long way in classifying the various kinds of shocks and in exploring the detailed plasma physics which are involved in those shocks. I think that we, as astronomers interested in the Galaxy, should not forget that wealth of information, which may be very useful when applied to the shock waves that we see in the interstellar space.

I can see a number of problems that remain yet to be solved in regard to sources of cosmic rays. A lot of work has been done on supernova dynamics, motivated, at least in part, by a search for the origin of cosmic rays. Recently pulsars have come onto the scene; and, if the neutron-star magnetic rotating model is correct, they may be responsible for much of the cosmic rays, too. The subject is very new, and a great deal of work remains yet to be done.

I commented to you earlier that it seems as though all magnetic fields in nature, and the galactic field in particular, have the stochastic property in which the magnetic lines of force random walk relative to their neighbors. Certainly there is a wealth of observational information that is available on this subject. It is interesting that the dynamical properties of such fields have not really been explored in any great detail, and I think there are quite a number of rather conceptually elementary problems that can and should be pursued in studying the propagation of waves, and the overall dynamical properties, of these stochastic fields.

This brings me to the subject of the general motion of cosmic rays in the disk of the Galaxy, followed by their ultimate escape. This and related problems in plasma turbulence are just beginning to be worked on. Lerche, Kulsrud, Pearce, and Wenzel in the United States and a number of people in the Soviet Union, including Kadomtsev and Tsytovich, are thinking about these problems and have made a number of calculations. Tsytovich has been exploring the effects of plasma turbulence on fast particles, which is so important for understanding the dynamics of the disk of the Galaxy. But he, wisely, has been exploring them in the Sun, the motivation being, if I understand it correctly, that there one can really begin to check one's theories. One knows the temperature and the density of the gas in which the turbulence is taking place. One knows the magnetic field and something about the characteristic scales of the phenomenon. Once one has learned the 'plasma turbulence, fast particle' trade in the solar system in this way, one may then be in a much better position to apply it to some far-off, and obscure, region of the Galaxy.

Finally, I should not fail to mention the question which has gone somewhat unanswered in this Symposium: What is the origin of the galactic magnetic field? It is too soon to answer this question, of course. My own inclination, and this is only an inclination, is that in some way the galactic field is not primordial; it seems too active. We must figure out somehow the motions which produce the field, or whatever process is at work. I think a good deal of very profitable theoretical work can be done in that direction.

I would close with some moralizing, which, of course, is always the temptation of the speaker with the captive audience. I think that, in theoretical work, and to some extent in the direct interpretation of observations, we must have a suitable mix of idealized examples – ridiculously oversimplified examples, if you like – where one effect at a time is studied until it is understood and then the separate effects are put together to form a more composite whole. We may well ask how far one will get with a composite. The interstellar weather we are talking about is even more complicated than the terrestrial weather; and it is well-known that twenty-four hour weather predictions are quantitatively unreliable. So how far can we expect to go in quantitative understanding of interstellar weather? Clearly there are limitations. This brings me to my last point. I hear that some people at Berkeley are beginning to think about the enormous problem of combining the thermal effects in the gas with the magnetic and cosmic-ray effects, to produce a gigantic numerical synthesis of all these things into one or more idealized examples to illustrate all the effects in operation at once. It seems to me that, if we do not have another cosmic-aerodynamical conference for five years, the first tentative results may be reported at that next meeting.

Thank you.

B.   H. VAN WOERDEN, *Sterrekundig Laboratorium Kapteyn, Groningen, The Netherlands*

In these concluding remarks, I shall summarize what I consider the most important recent progress reported at this Symposium and outline a number of problems on which new observations are needed. My viewpoint is that of an observational astronomer, primarily specialized in 21-cm research.

1. *Highlights of Recent Progress*

a. *Observational Information*

As indicated by van de Hulst in his opening review, the amount of *21-cm line* observations has grown enormously in recent years. However, this mass of data has not yet given us a clear understanding of the structure and dynamics of the interstellar medium. The large-scale structure as described by Weaver (p. 22) was a matter of some controversy at the Basel Symposium (Becker and Contopoulos, 1970); and from our discussion here about hydrogen clouds it appeared that a detailed picture cannot be drawn, mainly because the data remain largely undigested.

For the *ionized hydrogen*, too, there is much new information. There is the optical work on Hα by Courtès, and in the radio range many new observations (especially of recombination lines) have been made. Mezger's work on compact H II regions (p. 336) underlined the relationships of the ionized hydrogen to the problem of star formation.

Great progress has been made in the observation of *interstellar molecules*, discussed by Stecher (p. 316). The subject of interstellar molecules was discussed at length during a conference in Cambridge, England, last July (for a summary see Feldman *et al.*, 1969); this conference also devoted much attention to the infrared radiation of the interstellar medium, a subject which was mentioned only in passing during our present Symposium – as were X-rays, gamma rays, and cosmic rays.

Verschuur (p. 150) summarized the impressive amount of data on interstellar *magnetic fields*. I believe this is the subject in which the most striking observational advances have been made since the Noordwijk Symposium three years ago (van Woerden, 1967a), with molecules and recombination lines competing closely.

b. *Theoretical Developments*

It appears that *thermal instability* and *Rayleigh-Taylor instability* play leading roles in the formation of interstellar clouds and of stars. (See the Reports by Field, p. 51, and by Parker, p. 168.) Further development of these theories, together with Lin's density-wave theory of spiral structure, will allow interesting observational tests. A *magnetic-field* strength of 3 to 5 μG in the general solar neighborhood now seems observationally well-established and does no longer present severe theoretical problems. And, while the observational evidence for a *galactic halo* was severely questioned at the Noordwijk Symposium, the theoretical need for it now also appears to have vanished. The *supernovae* may, *via* cosmic rays, supply the major energy source to the interstellar medium, but the details of the energy balance remain poorly understood (see Woltjer's Report, p. 229; Discussion, p. 236).

2. *Problems for Future Research*

Both the reviews and the discussions during this Symposium have indicated a number of problems which call for new observations or for further analysis of existing material. In outlining some of these* I shall discuss 21-cm line research in more detail than other subjects.

a. *Neutral Hydrogen*

The overall distribution of (neutral and ionized) hydrogen in the Galaxy remains an unsolved problem; there is no good agreement on the pattern of *spiral arms and branches*. Kerr, Weaver, and others have followed different methods of defining and placing the structural features; a thorough comparison of assumptions, details of application, and results of these methods would seem valuable. The need to obtain a reliable picture of our Galaxy's spiral structure has become more pressing now that, on the basis of the Lin theory, it appears possible to predict variations of interstellar cloud properties as a function of location in the spiral pattern.

In addition to the structure as projected on the plane of the Galaxy, the *distribution perpendicular to the plane* is of great interest (see the Report by Parker, p. 168). The recent surveys of Kerr and Weaver should soon yield results superseding those from Westerhout's (1957) cross-sections.

The need for more quantitative information about *interstellar clouds* has become obvious from our special discussion about this subject (see Discussion, p. 98). A first, basic requirement is the need of a quantitative way of *defining* clouds. The recognition of clouds in contour diagrams has not been brought on a quantitative basis yet although this could, and should, be done. Least-squares procedures for fitting components into 21-cm profiles represent a first step towards quantitative treatment of the recognition of clouds. Schwarz has overcome some fundamental difficulties associated with existing component-fitting procedures and he succeeded in developing an automatic and fully-quantitative method for the definition of hydrogen clouds. He has made a first application to a region in Camelopardalis (Schwarz, 1970). This method, although laborious, is quite feasible on big computers and deserves further application; in my opinion it represents a major advance in this field, provided that the results of this fully-mathematical process are critically checked for being physically meaningful.

What quantities should we determine for a cloud? One of our aerodynamicists, Willis, listed: mass, size, density, scale of density variations, density contrast across the cloud 'boundary', velocity of the cloud, internal motions, composition. The work in Camelopardalis (Schwarz, 1970), and similar studies in Orion (van Woerden, 1967b, pp. 14–17) and Scorpius (Schwarz and van Woerden, unpublished), do indeed

---

* Field and Habing have kindly allowed me to incorporate here some suggestions which they made during the ensuing general discussion.

supply these quantities, although the distance appears as an unknown parameter affecting the derived masses and densities.

Field has suggested (p. 51) that cloud properties should *vary with location of the clouds* in the spiral arms. Also, he expects systematic motions towards the plane, depending on height; downward motions have indeed been observed (references may be found in Blaauw and Tolbert, 1966), but these have not yet been differentiated with respect to distance.

Now let us discuss the structures smaller than 'clouds'. Heiles (1966, 1967) has, from a high-resolution study, announced the existence of *'cloudlets'*, a new and distinct category of small interstellar clouds. Rots *et al.* (1970) after a critical analysis conclude that many of Heiles' 'cloudlets' must be interpreted as small concentrations or irregularities in larger structures; the number of small, isolated clouds may be only a minor fraction of the number of 'cloudlets'. Obviously, an independent check of these conclusions should be of great value for our knowledge of the spectrum of cloud masses.

The theorists expect *shock fronts* in neutral hydrogen clouds (Field, p. 51) and thin (10^{18} cm) *shells* of neutral hydrogen around H II regions (Discussion, p. 93). The angular resolution of existing 21-cm line equipment is inadequate for the observation of such features in emission. However, aperture-synthesis systems will change the situation. We hope to equip the Synthesis Telescope at Westerbork (The Netherlands) with a line receiver by 1972; its resolution at 21 cm is 22″, or 0.03 pc $= 10^{17}$ cm at a distance of 300 pc. This is only two orders of magnitude above the mean free path (10^{15} cm) of hydrogen atoms at a density of 10 atom cm^{-3}.

The limit for *velocity resolution* is close at hand. For hydrogen atoms a passband of 2 kHz $=0.4$ km sec^{-1} width will resolve the velocity distribution at $T_g = 20$ K. Heiles' survey was done with a 5 kHz bandwidth, and observations with narrower passbands are quite feasible with present low-noise, multi-channel receivers.

Neutral hydrogen shells around H II regions may be observed in *absorption at 21 cm* against the continuum background emitted by the H II region. Clark's (1965) investigation of the absorption spectrum of Orion A may contain some information of this nature, but further similar work is badly needed. Such work will be particularly fruitful if combined with studies of the interstellar sodium and calcium lines.

Combination of emission and absorption measurements is required for a determination of the spin *temperature* of neutral hydrogen. This subject has too long been neglected. Comparison of the internal motions within a cloud with its temperature will show whether these motions are subsonic or supersonic – an item of great importance to the physics and dynamics of the clouds.

b. *Interstellar Sodium and Calcium*

In Adams' (1949) interstellar calcium absorption spectra, the best velocity resolution was about 10 km sec^{-1}. Since then, hardly any data of higher resolution have become available, but just two months ago, Hobbs (1969a) published highly accurate sodium absorption profiles for 77 stars, with a velocity resolution of 0.5 km sec^{-1}. This clearly represents a tremendous advance. One of the important results of this work (Hobbs 1969b – which was mentioned only in passing during the Symposium – is that the velocity dispersions σ of well-resolved individual sodium-line components range between 0.6 and 1.5 km sec^{-1}, while the passband distribution corresponds to $\sigma =$ 0.2 km sec^{-1}. It appears quite feasible (Hobbs, 1969a) to extend such high-resolution measurements to large numbers of stars, including stars of spectral types A and F. This opens great possibilities for detailed studies of the interstellar medium at high velocity *and* angular resolutions.

Of particular interest would be the determination of sodium and calcium absorption in front of early-type stars in H$\textsc{ii}$ regions, for comparison with possible neutral-hydrogen shells or with motions in the H$\textsc{ii}$ regions. [See, e.g., Herbig (1968, *Z. Astrophys.* **68**, 243) on ζ Oph (Ed.).]

c. *Interstellar Molecules*

Hobbs' Fabry-Pérot interferometer would also be quite powerful for the detection of interstellar molecules; the optical information on these (see Stecher's Report, p. 316) is quite meagre so far. Both the velocity resolution and the high accuracy of Hobbs' spectrometer are important in this respect. In addition to measuring molecular lines in the spectra of more stars, observers might also try for molecules so far undetected, for instance OH (see Goss and Spinrad, 1966).

We may, of course, expect much further radio work on the newly discovered poly-atomic molecules and on OH. Also, we look forward to success of the long search for CH. Of far higher priority I would rate the detection of H_2 molecules. Although the anomalous gas/dust ratios in dark clouds are still open for discussion (see p. 331) there is, in my opinion, little doubt that considerable quantities of H_2 must be present.

d. *Supernovae and Supernova Remnants*

In the supernova remnants, further attempts to measure the radio recombination lines should be of interest. The last stage of remnants discussed by Woltjer (p. 229) is that in which they become neutral and merge with the interstellar medium. The 21-cm line observers might try to look for these old remnants, presumably invisible both optically and in the radio continuum. But where? If supernovae of type II (that is, Population I) do occur in OB associations, these old remnants may occur in the surroundings of older associations or – after their dissipation – of galactic clusters. Katgert (1969) and van Woerden and Hack (1970) may have discovered a few such old remnants. Detection of supernova remnants in external galaxies would be valuable. Synthesis telescopes such as those at Cambridge and at Westerbork will here again open a fruitful field of research.

e. *X-Rays*

Field has noted (p. 51) that the hot halo suggested by Spitzer (1956) should be observable through its X-ray radiation. However, I presume it will be difficult to distinguish such a halo from the extragalactic background. Nevertheless, this is a subject that should qualify for future observational attempts. An easier observation may be the X-ray radiation from (stellar and/or interstellar) supernova remnants.

f. *Inter-Arm and Inter-Cloud Conditions*

During this meeting there has been considerable discussion of the properties of regions inside and outside interstellar clouds, or inside and outside spiral arms. I believe we know very little about conditions in inter-arm and inter-cloud regions; in fact, I think they were often confused here, and the difference may not even be physically significant.

Measurements of the non-thermal radio continuum should provide relativistic-electron densities and magnetic-field strengths. The thermal radio continuum, and possibly spectral lines (particularly Hα with Courtès' interferometry), may furnish thermal-electron densities. Further, the 21-cm line can be used to determine atomic-hydrogen densities and temperatures. The work is obviously difficult in our Galaxy, because of the unfavorable location of the Sun. Observation of external galaxies may give more secure answers, although the requirements on sensitivity are quite severe – particularly for synthesis systems.

g. *Extragalactic observations*

I conclude by listing a few problems mentioned above, for which studies of external galaxies appear particularly helpful:

structure and kinematics of spiral galaxies;
scale height of the gas layer in a galaxy;
relative distribution and location of H I complexes;
H II regions and other features;
occurrence and properties of supernovae and their remnants;
haloes of galaxies;
inter-arm and inter-cloud conditions.

These then are the problems I suggest for future research. Successful pursuit of a majority of these would certainly justify another, later symposium in the present series.

References

Adams, W. S.: 1949, *Astrophys. J.* **109**, 354.
Becker, W. and Contopoulos, G. (eds.): 1970, *IAU Symposium No. 38, The Spiral Structure of the Galaxy*, Reidel, Dordrecht, The Netherlands.
Blaauw, A. and Tolbert, C. R.: 1966, *Bull. Astron. Inst. Netherl.* **18**, 405.
Clark, B. G.: 1965, *Astrophys. J.* **142**, 1398.
Feldman, P. A., Rees, M. J., and Werner, M. W.: 1969, *Nature* **224**, 752.
Goss, W. M. and Spinrad, J.: 1966, *Astrophys. J.* **143**, 989.
Heiles, C.: 1966, Observations of the Spatial Structure of Interstellar Hydrogen, Ph.D. Thesis, Princeton University, Princeton, New Jersey.
Heiles, C.: 1967, *Astrophys. J. Suppl. Ser.* **15**, 97.
Hobbs, L. M.: 1969a, *Astrophys. J.* **157**, 135.
Hobbs, L. M.: 1969b, *Astrophys. J.* **157**, 165.

Katgert, P.: 1969, *Astron. Astrophys.* **1**, 54.

Rots, A. H., Schwarz, U. J., and van Woerden, H.: 1970, in preparation.

Schwarz, U. J.: 1970, in preparation.

Spitzer, L.: 1956, *Astrophys. J.* **124**, 10.

Van Woerden, H. (ed.): 1967a, *IAU Symposium No. 31, Radio Astronomy and the Galactic System*, Academic Press, London and New York.

Van Woerden, H.: 1967b, in *IAU Symposium No. 31, Radio Astronomy and the Galactic System* (ed. by H. van Woerden), Academic Press, London and New York, p. 3.

Van Woerden, H. and Hack, M.: 1970, in preparation.

Westerhout, G.: 1957, *Bull. Astron. Inst. Neth.* **13**, 201.

C. S. A. KAPLAN, *Radiophysical Institute of Gorkii University, Gorkii, U.S.S.R.*

Let me begin with some general remarks. First, to many of us the way in which this Symposium worked was completely new, i.e., Introductory Reports to begin and then a long and unprepared discussion to follow. I think that in general this scheme was very satisfying. Many of the U.S.S.R. participants found the Introductory Reports very informative. Second, this Symposium and the previous ones serve the purpose of arranging discussions between astrophysicists and aerodynamicists. It appears to me that such discussions are very valuable. The astrophysicists need help for the interpretation of the observations and for clarification of the physical picture. The aerodynamicists can help in solving the gas-dynamical equations and, in this process, learn something new about their own thoroughly studied equations. I wish that in future symposia in this series the organizing committee will stimulate this discussion more strongly, for example by having an Introductory Report on recent developments in gas dynamics. Third, a new feature of this Symposium was the large role given to plasma physicists (although there were still too few plasma physicists present, I think). This seems a very important development, from which both astrophysicists and plasma physicists can profit.

Now I want to come to direct scientific matters. I will touch only on those problems which are close to my present scientific interests. First of all, we need methods to detect and observe plasma turbulence. Direct measurements can be made only in interplanetary space. Such measurements have been made (see the Report by Lüst, p. 249) but their interpretation is still difficult. Indirectly one can measure plasma turbulence only through the electromagnetic radiation emitted or influenced by the turbulent plasma. One existing mechanism converts plasma wave energy into radiation. In another mechanism plasma waves serve as a catalyst for the conversion of nearly all the energy present in fast and relativistic particles into electromagnetic radiation. Plasma turbulence accelerates charged particles which then generate radiation, including synchrotron radiation. Progress in understanding these processes is at present being made both in theoretical calculations and in the interpretation of the observations. Many radio bursts on the Sun are definitely connected with plasma turbulence phenomena. Plasma physics is able to provide basic mechanisms for cosmic masers. I have mentioned (p. 133) the possibility of the scattering of electromagnetic waves off plasma waves.

The second set of problems I would like to consider concerns changes in the usual hydrodynamic and magnetohydrodynamic equations. These changes are brought about by plasma physics. One of the changes, for example, is the thickness of shock fronts; we have discussed rather intensively the collisionless shock front. Other questions concern the conductivity in turbulent plasmas, which is considerably lower than in non-turbulent plasma. This may have consequences for the dissipation of magnetic fields, for the heating of the plasmas and for an increasing tendency toward instabilities.

Third, in studying plasma turbulence, its excitation in astrophysical objects deserves much attention because the interpretation of the observations will not always be straightforward and unique, and a clear picture is required of all the possible instabilities in plasmas. We require not only theoretical work but also laboratory experiments with cosmical models. Good examples are the latest experiments simulating the solar wind flow around the Earth's magnetic field, which have been discussed by Podgornii (p. 144) and by Karpman (p. 147). I want to stress the importance of the interaction of plasma turbulence and cosmic-ray particles. The latter give high-frequency electromagnetic radiation, are accelerated by the turbulence, and form the main source of excitation of plasma turbulence.

These are some examples of areas where plasma physics and astrophysics interact. It is certain that all the problems of what I would like to call 'plasma astrophysics' may be resolved ultimately by the combined effort of astrophysicists and plasma physicists. A discussion between these two groups is therefore very important. Perhaps we should ask not only the IAU and the IUTAM, but also the International Agency of Atomic Energy to sponsor our next Symposium.

3. Summaries and Suggestions for Future Research

[The first part of the following discussion took place on Tuesday, September 16. At the request of Thomas a short discourse was given by Busemann on the interrelation of interstellar gas dynamics and aerodynamics.]

Busemann: I was asked to speak only a few minutes ago; thus I am completely unprepared, a fact on which I cannot cast doubt by 'accidentally' having a slide with me. The questions which I would like to discuss are: Is there anything the invited aerodynamicists can contribute to help the astrophysicists, or is there something in the more sophisticated problems presented, that can open the eyes of the aerodynamicists, so that they may help the astrophysicists?

I still remember the days when incompressible flow was just established as a young but very exact science of almost mathematical precision while its sister science, the compressible flow or gasdynamics, was rather in its infancy especially at supersonic and transonic speeds. Since gasdynamics was always being compared with its perfect sister, I felt that it was very easy to make disastrous mistakes and I therefore decided to find graphical representations or physical models which could help in the interpretation of the difficult gasdynamical equations. Many such pictures remain still in my mind as valid representations of the facts. When I hear the remarks of some speakers at this Symposium, the pictures instantly come back to me either casting doubt on or reconfirming their statements. However, while working in the field of magneto-hydrodynamics I learned that there exist many domains of very different behavior, not just distinguished by the Reynolds number and the Mach number as in aerodynamics, but also by magnetic field strengths, conductivities, free path lengths compared to Debije lengths, degrees of ionization, etc. Considering those differences, a reasonable criticism needs some thought and time.

Again as an early gasdynamicist, I start with those helpful models to which I already referred. One of these is the shallow water theory. It is a two-dimensional analogy to compressible sub- and supersonic flow (though it corresponds to a ratio of specific heats in perfect gases above the actual values for gases). If I take a pitcher with water and pour the water jet upon a flat bottom of a bowl, the water will rush towards the walls of the bowl and it is stopped there, forming a water jump which is representative of the shock wave in a gas. The rushing flow may be regarded as the supersonic solar wind which is terminated by a shock when joining the interstellar gas. The height of the shallow water indicates pressure, the surface waves simulate Mach waves, and the somewhat wiggling water jump resembles the shock wave of finite amplitude. If I use two pitchers over the same flat bottom, the symmetry line between their points of impact is surrounded by water jumps simulating two solar winds hitting each other head-on. If I let one pitcher pour faster than the other one, the symmetry is spoiled, but the principle effects are still similar. Thus I can instantly demonstrate a large variety of (two-dimensional!) solutions. The water jump near the wall of the bowl produced by using a single pitcher, can be regarded – for a straight wall – as one half of the symmetric arrangement produced by two pitchers. After some time the water accumulation near the wall moves the water jump away from the wall and it simulates, sooner or later, the shock of the solar wind joining the infinite interstellar gas in a coordinate system which contains the interstellar gas at rest.

Such simple and cheap models of our compressible flow problems give a very good introduction for becoming acquainted with combinations of hyperbolic and elliptic non-linear differential equations. However, like this shallow water analogy, the models are only available in two-dimensions while we need solutions in three dimensions. There may even be misleading models. Consider a thin sheet of an electrically con-ducting material and let the electric flux lines correspond to the potential lines of incompressible flow and *vice versa*. Compressibility of the flow is introduced as soon as the conductivity of the material increases with temperature, while the temperature is controlled predominantly by the radiation of heat to the environment. In such an electrical analogy the subsonic or elliptical flow behavior is modelled correctly. But as soon as the electrical potentials are increased to include supersonic flow regimes, the expected hyperbolical flow does not know about upstream and downstream directions. It acts more like lightning, which persists on its original random paths by heating them up more and more.

I come now to my current interest: finding a cure for the 'sonic boom'. To indicate at least the main lines of the sonic boom research, I want to give you some of the results. The problem would completely disappear were we to return to the lighter than air vehicles, the airships which do not transfer any weight to the surrounding air. This, however, seems rather remote as a realistic approach. Removing the airplane weight by centrifugal forces on the great circles around the earth comes automatically at a Mach number of 25, but again this does not appear to be a realistic solution. (On the other hand space vehicles use this speed as earth satellites already.) A more realistic, though not easy solution, is to slow down the pressure rise in the shock wave to a time of about 0.01 sec. Then the noise may be very different for our ears and hopefully not alarming at all. Under this, not quite generally accepted, assumption the appropriate fuselage and wing configurations can be found and the airplane shapes, differing vastly with altitude, can be studied. The concept of finite rise times does not prevent an asymmetrical reception by our ears. The asymmetry is caused by the natural deformation of sound waves along large propagation paths and it corresponds to the deformation of water waves at the beaches. To prevent steepening of the waves there seem to be two approaches possible: by making the airplane nose extremely slender (in which case the noses are probably not stiff enough to carry the plane at supersonic speeds) or to change the surrounding atmosphere in such a way that shock waves do not result. Only a gas, in which the velocity of sound decreases (by van der Waals forces) sufficiently when the isentropical pressure increases never creates shock waves. Such a gas is called 'Chaplygin' or 'Karman-Tsien' gas and has a straight isentropic line in pressure versus volume plots. The mathematical simplifications are obvious, but the existence of the gas is not yet proved and the change of our atmosphere toward such behavior is even less likely. Since it might be difficult to modify the air, changing the nose shape seems a more accessible 'road'. The nose stiffness cannot be increased by deviating from the axially symmetrical shape unless we make the sting hollow. Visualizing the extreme difficulties, it is not surprising that substitutions are proposed for the concrete nose: 'ghost' noses created by electrical charges, laser beams, etc., thrown ahead of the actual nose. But to produce such 'ghost' noses the energy required appears to be prohibitive either in weight or fuel consumption.

When the human brain seems to see no way out, perhaps nature itself can give us a hint. Is there any place in the interstellar gas where the expected shock does not occur? Why don't we hear a boom-boom when a celestial body passes along? If the astrophysicists were fully acquainted with out problems, they might perhaps find examples of electrical effects, radiation effects, special gas properties, etc., which would give us aerodynamicists new hopes and perhaps answers.

In summary I may say that the astrophysicists and the aerodynamicists use the same laws of nature. A striking difference is, however, that the aerodynamicist can control his experiments, but has to find solutions for a given purpose with difficult and sometimes impossible constraints. The astrophysicist observes events far away that have their own natural time scale. Sometimes he may wish he could also put his hands on the phenomena, but all he can hope for are new methods or places of observations. His purpose is to explain the observed phenomena as exactly and precisely as freedom from controversies allows. Apparent contradictions force him to make more and more sophisticated assumptions without always being able to check the results in the laboratory. There is no doubt that looking over each other's shoulders, as in this Symposium, can be very useful for both parties.

Kaplan: Busemann spoke about gas without shockwaves. I would like to mention that in plasmas with magnetic fields, supersonic motions are possible without formation of a shock wave. Such motions are called 'solitons' and their existence has been verified experimentally.

Goldsworthy: I would like to make a comment which I hope will be helpful in trying to bring aerodynamics and astrophysics together. First of all, in reading the proceedings of past symposia and listening to the present one, one cannot help but note that the particular emphasis is placed upon global properties, such as total energy, etc. These are interesting quantities from the astrophysical point of view; but for the aerodynamicist, as he looks at the associated gas dynamics, these quantities act only as parameters in the problem. They do not, in general, say anything precise about the dynamical model. For instance, if we take the problem of an explosion, we may know its energy; but this tells us very little about the nature of the blast damage that results. I therefore appeal that, first, when such global properties are given, the variation about the average be included; and second, where possible, greater emphasis should be placed on velocity and density distribution in the separate objects. It is perhaps the latter that the aerodynamicist has most difficulty in obtaining from the astrophysical literature. [The major part of Goldsworthy's remarks at this point have been transferred to the Discussion following the Reports by Weaver and Field, p. 94 (Ed.).]

Verschuur: Dr. Goldsworthy, is it more useful to study specific clouds in great detail, or to discuss a standard or average cloud? Would you prefer as detailed as possible information on a specific H I region; or should we look at hundreds of these and give you an average?

Goldsworthy: An 'average' really does not mean anything to us. We would prefer more detailed observations on one particular structure, so that we might then be able to analyze associated motions.

Lüst: I would like to make one remark to the aerodynamicists who say that it is difficult to devise experiments in this field. One area where we can do experiments is in the magnetosphere. There we have a magnetic field; we can inject an artificial plasma and observe it from the ground. At higher altitudes the mean free path is quite long, while the gyroradius of the ions and electrons is rather small. We at Munich have carried out several such experiments in recent years. In particular we carried out an experiment at about 70 000 km altitude in the magnetosphere. There we injected about 100 g of barium ions. The density of the surrounding plasma was of the order of about 0.1 proton cm^{-3}, and the magnetic field strength of the order of 5×10^{-4} G. The kinetic energy density of the expanding ions was larger than the magnetic-energy density for about 20 sec, and the barium cloud formed a magnetic cavity. After 20 sec, the initial expansion stopped, and the cloud as a whole moved through the ambient plasma like a comet tail, until the cloud was accelerated to the velocity of the ambient plasma. I feel that experiments of this kind could be of interest to aerodynamicists and plasma physicists, if they want to understand astrophysical phenomena. [A general description of such experiments may be found in Härendel and Lüst, *Scientific American*, November 1968, p. 81 (Ed.).]

[The remaining part of this Final Discussion took place on the last day of the Conference, Thursday, September 18 (Ed.).]

Weymann: As one who dabbled in thermal instability some time ago, the apparent occurrence of clouds at these intermediate temperatures seems most disturbing to me. If I understand correctly, it is not just that the theory predicts that the clouds will be at such-and-such a temperature which is fairly well determined by the excitation potential of the cooling lines; but it also predicts that the clouds cannot be at the other temperature. I think that it would be worth a lot of work to find out observationally whether or not clouds at these intermediate temperatures really exist. If they do, it means at best that one has to have a fairly complicated dynamical theory, and at worst that there is something fundamentally wrong with this picture of heating by cosmic rays or X-rays.

Field: Pikel'ner and I have a running debate on the question of whether or not the intermediate temperatures should occur. The direct formation of a typical cloud requires the gathering in of material by means of instabilities of very long wavelengths. The time scale τ for this is about λ/c, where c is the speed of sound. If $\lambda = 300$ pc and $c = 10$ km sec^{-1}, $\tau = 3 \times 10^{7}$yr. Thus some gas would be in the intermediate phase for observable lengths of time. However, while these instabilities of the order of 300 pc are developing, shorter wavelength instabilities should occur as well, as they have a growth time of only 10^{6} yr; they will tend to wipe out the intermediate phase as soon as it occurs. It is therefore a dynamical problem which should be worked out into the non-linear regime in order to give final results.

Pikel'ner: Field has pointed out correctly that before the gas forms a large dense cloud it should form little droplets of denser phase in the rarefied background. Therefore it is very probable that a cloud with an intermediate *average* density is in fact a system with gas in both of the stable phases. The time of formation of such a system should be about the same as the time of temperature relaxation. Perhaps we can check this theory by observations of different parts of clouds.

Habing: I want to suggest observations that may yield the value of ζ, the ionization rate per hydrogen atom. Actually the method I would like to propose measures ζ in dense, cool, neutral clouds. The method works as follows: Consider a non-thermal radio source. First, determine at a low frequency (say, 20 MHz) the optical depth τ_{ff} for free-free absorption; τ_{ff} is proportional to $\int n_e^2 T_e^{-3/2} \, dl$. Note the heavy weighting of cool electrons. This means that, unless we happen to see an H II region projected on the radio source (a situation that occurs for both Cyg A and Cas A), most of the absorbing electrons are associated with H I regions. Second, determine the 21-cm absorption profile and integrate the optical depth τ over the velocity range. The integral τ_{21} is proportional to $\int n_\mathrm{H} T_e^{-1} \, dl$. If ζ is constant along the line of sight $\zeta n_\mathrm{H} = \alpha n_e^2$, where α is the recombination coefficient. For $T_e < 1000\,\mathrm{K}$ we have $\alpha = \alpha_0 (T/T_0)^{-0.65}$, so that $\zeta \propto n_e^2 n_\mathrm{H}^{-1} \langle T_e^{-0.30} \rangle$. Therefore $\zeta \propto (\tau_{ff}/\tau_{21}) \langle T_e^{-0.30} \rangle^{-1}$. If we can guess T_e within a factor of 5, we can get very accurate values of ζ. This method should give much smaller upper limits than have been known so far.

Pikel'ner: I should like to stress that for many theoretical calculations it is of fundamental importance to know the electron density n_e in the interstellar space; the figure given by Mills ($n_e \approx 0.06\,\mathrm{cm}^{-3}$, see p. 92) is too high from a theoretical point of view (perhaps also from the observational point of view). It is necessary to be very careful, because there are perhaps some H II regions near the Sun which contribute to this number. Also, it is necessary to be careful in comparing high-frequency radio emission and low-frequency radio absorption, because the low-frequency absorption is dependent very strongly upon temperature and will occur mainly in the cold clouds. High frequency radio emission is not so strongly dependent upon temperature, and may originate mainly in the rarefied gas.

Mestel: As I mentioned before (see p. 364) Hoyle and Ireland (1960) suggested that the centrifugal energy in the galactic shear could be tapped by means of the galactic magnetic field. If the field has a large-scale poloidal component, then energy must be fed steadily into the toroidal component until the field becomes too strong for stability, and it buckles into the halo. Some of the energy is thus converted into gravitational potential energy, which in time becomes kinetic energy as the gas streams down into the disk. I am not clear whether or not this model can still be defended as an important source of galactic turbulence. What I want to emphasize is that, if we retain the picture of a large-scale galactic field, then we cannot avoid considering this problem of the winding of the field. The Lindblad-Lin gravitational wave model automatically resolves the classical problem of the winding up of the spiral arms by having them rotate at a fixed angular velocity; but the gas plus magnetic field and the stars still move through the wave pattern with the local angular velocity. As remarked before (p. 364), the shear will produce a dominant toroidal component for the field, which will therefore appear to be nearly parallel to a tightly wound spiral; but as long as there remains a poloidal component, the shear will steadily feed energy into the toroidal component, unless something like the Hoyle-Ireland instability relieves the toroidal field of its excess energy.

Field: In regard to the question raised earlier of the nature of the turbulence and its cause, it seems to me that the outstanding result of 21-cm radio astronomy is that the observed turbulence is highly supersonic. Therefore the energy input must be compressive in nature; $\nabla \cdot V \neq 0$. The relevant energy sources are those which will result in condensation, collapse, and motion along the field lines at a supersonic rate. The three very definite possibilities discussed at this conference are: (i) Rayleigh-Taylor instability; (ii) thermal instability; (iii) expansion of certain objects, notably H II regions and supernova remnants. It seems to me that we have given very little attention here to the essentially different type of motions associated with shear turbulence ($\nabla \times V \neq 0$). In particular, we have not considered the excitation of Alfvén waves in the interstellar medium. On various occasions it was pointed out that large amplitude Alfvén waves can be responsible for certain of the observations. I hope that high resolution work, which can identify scales of 1 pc or less, may yield information on the presence or absence of such motions. The rotation of the Galaxy should be considered as a source of such motions, particularly in connection with the question of the winding of the magnetic field. Another source is the instability caused by streaming of cosmic rays, discussed by Tsytovich (see his Report at p. 108). Although the emphasis here has been on the compressive motions, we should not overlook the fact that the rotational motions are still important.

Lynds: It seems that there are problems in identifying neutral hydrogen clouds. There are no problems whatsoever in identifying dust clouds – at least the densest ones, so that for such clouds direct observational information is available regarding the existence of sharp edges, characteristic patterns, filamentary structures, and so on. Furthermore, in larger dark clouds, there are subcondensations. Semiquantitative estimates of extinction can be made through the clouds, but no information on velocity is available. You therefore have a problem with the dust clouds which is more or less opposite to what you have with the 21-cm clouds. Although there is apparently no correlation of dust clouds with the 21-cm data, studying the characteristics of the dust clouds will perhaps suggest an interpretation for the 21-cm analysis.

Van Woerden: I think that Beverly Lynds' suggestion that boundaries for dust clouds are more easily defined than for hydrogen clouds is due to observational selection; you see the darkest clouds with the sharp edges most easily, and the others you do not see so well. But there are also hydrogen clouds which stick out very easily, and others which are difficult.

Spiegel: We have these 21-cm contour diagrams which were shown at various times. These things are rather similar to spectroheliograms. Leighton made rather sweeping discoveries by treating the spectroheliograms in a statistical way; for example, he would take a negative of one and print it against a positive of another. The two plates he took were only a little bit different in velocity (or wavelength) so that certain small-scale fluctuations of density were suppressed and large-scale structures of kinematic nature emerged (Leighton *et al.*, 1962). In a similar way, he did that with the Zeeman splitting, and saw direct photographs of large-scale magnetic fields. I've been wondering how much of that kind of thing would be useful in 21-cm line work. Of course, the possible range of distance makes the problem rather complicated. But then in the Sun, too, depth variation is very subtle and crucial. How much of this could be done; how easy would it be; and are there such things already? (Leighton, R. B., Noyes, R. W., and Simon, G. W.: 1962, *Astrophys. J.* **135**, 474.)

Field: I think I disagree with everybody on what an interstellar cloud is. In my definition, an interstellar cloud is a region of high density in the three-dimensional space defined by the galactic coordinates, *l* and *b*, and velocity. It is a closed contour in this space. These objects exist. Along the lines of Spiegel's remarks, the possibility exists for computer programming which will do the following. We have a three-dimensional array of data. Ordinarily this is projected on two of the canonical coordinates (say l and v at fixed b) as a contour diagram. But with programs now available one can rotate the coordinate system and view the results on a cathode-ray tube in real time. This added perspective would be helpful to the theorist trying to understand the cause of the flow.

Another point that I want to bring up is the possible existence of what one might call the dust cycle. This cycle is connected with gas flow through a spiral arm and has to do with the freezing out of metal atoms (say, carbon and silicon) on the dust grains. This freezing out affects the chemical composition and the temperature of the (atomic and ionic) gas. If this freezing out occurs during the passage of the dust through the spiral arm (with evaporation of metals after leaving the spiral arm), one would predict that the temperature of H I clouds increases the more a cloud progresses through the spiral arm. In principle this behavior can be checked with 21-cm absorption measurements.

Verschuur: I would like to say something about the possible progress over the next few years in magnetic-field determinations. The efforts, I think, will be spent mainly in looking at the data that is now available, because there is a lot of data available and I do not see that great progress is likely in the collection of data in the near future. To illustrate this in particular, let me say something about the future of the Zeeman effect measurements. It took a long time to measure the first magnetic field, due to the difficulty of the experiments. And now, I already have met with instrumental limitations in measuring the weaker fields. I think that the greatest progress will be made when larger radio telescopes become available, because we need to observe very narrow emission lines at higher latitudes and these narrow emission lines have velocity structure, which broadens the lines when examined with an insufficiently small beamwidth. I have already looked at all the absorption sources available in the Northern Hemisphere. Therefore only the biggest radio telescopes can be useful; we need a fully-steerable dish, greater than the 300 foot.

Van Woerden: We have heard many suggestions for radio observations with large angular resolution. In The Netherlands we are getting a synthesis array which will give such a resolution. I would like to say that we in The Netherlands hope to have frequent visits from foreign colleagues to work with our equipment. Our policy for the coming years is to have two or three positions available on a regular basis for foreign workers.

[At this point the Chairman asked several participants to give their overall impressions (Ed.).]

Ozernoi: I have had very fruitful conversations with Field, Colgate, Mestel, Spiegel, Weymann, Woltjer, and others; and it seems to me that informal conversations are often more useful than the formal sessions. The discussions of the relation between cosmology and the modern state of the Galaxy have been very useful for me. The distinguishing feature of this Symposium was the repeated necessity to speak about phenomena outside the limits of our Galaxy or to look at epochs earlier than the present one. I should like to recall some examples or suggest new ones.

First, concerning the dynamics of gas, the influence of the galactic center on the interstellar gas by ejection of relativistic particles or by outflow of plasma cannot be passed over. These processes, as well as the explosions in the nucleus, are miniature copies of phenomena which are observed in Seyfert galaxies, N-galaxies, and quasars. Thus, the origin of some fraction of stars of the Galaxy is possibly due to explosions in the nucleus, which, as in quasars and radiogalaxies, eject large amounts of gas. In what way could this gas, after cooling and fragmentation, be transformed into stars? An analysis shows that besides star formation by gravitational condensation we have formation of massive rotating magnetoplasma configurations, which eject gas by explosions, transforming some of it into stars. There exist other similarities which connect some unusual phenomena in our Galaxy with the phenomena so distinctly observed in quasars. Second, the fact that QSS's and other strong radiogalaxies probably belong to spherical rather than to flat systems (to which Seyfert galaxies belong) reminds one of the necessity to explore more carefully the properties of stars and gas in the halo of the Galaxy, as well as in elliptical galaxies. This is an important channel of information about the origin and evolution of spherical systems. The comparison of the helium abundance in old halo stars with that in the galactic nucleus and near supernova remnants may help in answering the question of whether the anisotropic character of cosmological expansion or some local nonthermal processes determine the helium abundance. The third and last point refers to the traces of the gas-dynamic past exhibited by the Galaxy in its present state. The whirl model of galaxy formation, which has been discussed earlier (p. 216) shows that dynamical properties of the Galaxy as a whole are the consequences of the turbulent past of the Metagalaxy. Apparently the intriguing problem of the origin of the galactic magnetic field would be resolved in a whirl-turbulent cosmogony, starting from the old idea of Biermann and Schlüter of generating the magnetic field by a velocity field. All the questions mentioned are too difficult to describe briefly. Possibly, a future Symposium could be devoted especially to the gas-dynamic phenomena from the aspect of the origin and evolution of galaxies and quasars.

Toomre: In response to Thomas' inquiry about my personal reactions to this conference, I wish to thank the observers for giving me a good summary of the recent experimental evidence. One of the dynamical aspects that I will be keen on examining involves the two-phase systems emphasized by Field. The dynamical consequences of considering even Burger's one-dimensional model turbulence, while using an equation of state with two phases, may be quite illuminating for this subject. Kadomtsev has further pointed out to many of us that plasma turbulence may provide us with some useful energetic transport mechanisms between quite different length scales. This is an area which we have probably slighted and which deserves some careful study.

Thomas: If you do this, may I just add the suggestion that beyond this simple, two-phase picture, you should look in the Proceedings of the last (Vth) Symposium in this series (Thomas, 1967) – at the introductory-summary-survey paper by Goulard, who discussed the problems of radiative control of aerodynamic flows. There one has this question of the energy equation put in a broader form. If you put it in aerodynamic jargon, the question is one of frozen-in degrees of freedom, where the motions don't have much to do with the equation of state, but where the internal energy is fixed by the radiation field. (Thomas, R. N. (ed.): 1967, *IAU Symposium No. 28, Aerodynamic Phenomena in Stellar Atmospheres*, 5th Symposium on Cosmical Gas Dynamics, Academic Press, New York.)

Now I would ask you to join with me in thanking our hosts, who have done far more than any host for a previous Symposium in this series: they have adopted a language that is not their own. They have made every effort to speak English and to act as a buffer between us and the affairs of the outside world, so that we could concentrate here in this beautiful seaside resort only on science, swimming, and tennis. I cannot say, in English or any other laguage, how grateful we are. I ask you to stand and to thank all of our hosts at this Symposium. (Applause.)

Shklovskii: On behalf of my Soviet colleagues may I express our deep gratitude to Dr. Thomas for his invaluable contribution in the organization of this Symposium. It hardly would be successful without his energy and experience. We are very obliged also to Dr. Gebbie, Mrs. Low, and Miss Thomas for their extremely hard work in transcribing the discussions. We are very obliged to all foreign participants and especially to the invited speakers whose contributions made the Symposium fruitful and enjoyable.

INDEX OF DISCUSSION REMARKS

(Numbers in italics refer to the Invited Reports)

SUBJECT INDEX